ADVANCES IN
MACROFUNGI
Diversity, Ecology and
Biotechnology

Series: Progress in Mycological Research

- Fungi from Different Environments (2009)
- Systematics and Evolution of fungi (2011)
- Fungi from Different Substrates (2015)
- Fungi: Applications and Management Strategies (2016)
- Advances in Macrofungi: Diversity, Ecology and Biotechnology (2019)

ADVANCES IN MACROFUNGI
Diversity, Ecology and Biotechnology

Editors

Kandikere R. Sridhar

Senior Professor, Department of Biosciences
Mangalore University
Mangalore, India

Sunil K. Deshmukh

Nano-Biotechnology Research Centre
The Energy and Resources Institute (TERI)
New Delhi, India

CRC Press
Taylor & Francis Group
Boca Raton London New York

CRC Press is an imprint of the
Taylor & Francis Group, an **informa** business
A SCIENCE PUBLISHERS BOOK

CRC Press
Taylor & Francis Group
6000 Broken Sound Parkway NW, Suite 300
Boca Raton, FL 33487-2742

First issued in paperback 2021

© 2019 by Taylor & Francis Group, LLC
CRC Press is an imprint of Taylor & Francis Group, an Informa business

No claim to original U.S. Government works

Version Date: 20181018

ISBN-13: 978-0-367-78033-3 (pbk)
ISBN-13: 978-1-138-58727-4 (hbk)

Library of Congress Cataloging-in-Publication Data

Names: Sridhar, K. R., editor.
Title: Advances in macrofungi : diversity, ecology and biotechnology /
    editors, Kandikere R. Sridhar, Senior Professor, Department of
    Biosciences, Mangalore University, Mangalore, India, Sunil K. Deshmukh,
    Area Convenor, Nano-Biotechnology Research Centre, The Energy and
    Resources Institute (TERI), New Delhi, India.
Description: Boca Raton : CRC Press, [2018] | "A science publishers book." |
    Includes bibliographical references and index.
Identifiers: LCCN 2018049237 | ISBN 9781138587274 (hardback)
Subjects: LCSH: Macrofungi.
Classification: LCC QK603 .A37 2018 | DDC 579.6--dc23
LC record available at https://lccn.loc.gov/2018049237

Visit the Taylor & Francis Web site at
http://www.taylorandfrancis.com

and the CRC Press Web site at
http://www.crcpress.com

Preface

Mycology is a potentially expanding area involving almost all facets of human health, nutrition and diseases. Diverse and ubiquitous distribution of fungi attained the status of '*Fifth Kingdom*' among the three major evolutionary lines along with plants and animals. Macrofungal diversity (macro- and macro-morphological) is enormous and dependent on a variety of substrates and geographic locations. A rough estimate reveals the macrofungi varieties to number 53,000 to 110,000 globally. They belong to a variety of basidiomycetes and ascomycetes known for their nutrition, medicinal value, novel metabolites, toxins and interaction with higher forms of life (plants and insects). Macrofungi are of special interest to explore their diversity, distribution, ecological functions and ecosystem services.

Interestingly, facts on edibility and medicinal novelty of many wild mushrooms are a product of ethnic knowledge of locals and tribals throughout the world. Mushrooms are a viable alternative non-conventional food source (other than plants and animals) owing to their nutritional, low-fat and low-calorie features (e.g., minerals, amino acids, vitamins and unsaturated fatty acids). However, few mushrooms are cultivated as a source of human nutrition and for therapeutic purposes. Wild and cultivated mushrooms are well known for their bioactive components (e.g., phenolics, tannins, vitamins, flavonoids, carotenoids, phytic acid, pigments and L-DOPA) and antioxidant activities (ferrous ion-chelation, DPPH radical-scavenging, reducing power, antioxidant activity, hydrogen peroxide scavenging, superoxide scavenging, lipid peroxidation, β-carotene-linoleic acid cooxidation and nitric oxide synthase activity). Many wild mushrooms are endowed with therapeutic potential, especially nutraceuticals, and are capable of preventing many lifestyle diseases (e.g., cardiovascular diseases, hypocholesterolaemic properties and anti-cancer agents).

Despite several hundred species of macrofungi having questionable edibility, they are the potential pool of valuable macromolecules and secondary metabolites of pharmacological and industrial interest. Bioactive polysaccharides, proteins, peptides, phenolic compounds and terpenes are known from various macrofungi. Macrofungi are a potential source of compounds responsible for the regulation of blood glucose and can demonstrate hepato-protective, antioxidant, cytotoxic, anti-inflammatory and antimicrobial potential. Some of the bioactive metabolites also serve as specific taxonomic markers. Many novel metabolites of macrofungi pave the way for the synthesis of new compounds. In addition, macrofungi are a valuable tool for green synthesis of nanoparticles, as opposed to the less ideal chemical and physical methods.

Another dimension of the macrofungi is their mutualistic association as ectomycorrhizae. Ectomycorrhizal fungi play an important role in the forestry/ sylviculture/plantations resulting in uptake of water, absorption of minerals, protection against pathogens and below-ground nutrient transfer and prevention of erosion by binding the soil particles. Beyond mutualistic association, ectomycorrhizal fungal association is partly responsible for phytogeographical distribution in several habitats including extreme ecosystems. Besides involvement in biogeochemical cycles, macrofungi are known to degrade environmental pollutants, facilitating bioremediation.

Macrofungal research has become an important global commitment of the 21st century and there is an upsurge in understanding the roles of macrofungi in human, agricultural and environmental processes. In this contribution, we attempted to broadly balance the basic and applied aspects of diversity, ecology and biotechnology of macrofungi. To reflect the recent aspects, many colleagues contributed novel chapters based on their vast experience in consultation with voluminous literature. This book addresses: (1) the diversity and ecology of edible, toxic, medicinal and ectomycorrhizal macrofungi; (2) the impact of ectomycorrhizal fungi in terrestrial ecosystems, forests and plantations; (3) nutritional potential and cultivation of edible wild mushrooms; (4) novel metabolites of macrofungi useful in food, pharmaceutical and cosmecutical industries; (5) eco-friendly synthesis of nanoparticles by mushrooms; (6) proteomics of edible and medicinal mushrooms. In addition, this book also encompasses experimental designs, methodological approaches, biogeochemical cycles, conceptual models, life history strategies and linking mycorrhizal diversity to plant performance.

This contribution hopes to initiate interest among the readers in order to expand their knowledge on macrofungi and to generate new ideas on basic and applied facets with future avenues. It is a valuable resource to graduates, post-graduates and researchers (in botany, microbiology, ecology, biotechnology, forestry, life sciences and environmental sciences) for understanding the diversity, ecology, therapeutic value, mutualistic associations and biotechnological potential of macrofungi. We are indebted to Prof. D.L. Hawksworth for an excellent overview, grateful to all contributors who delivered the chapters on time and for meticulous attention by the publisher to materialize production of this book.

Mangalore, India **Kandikere R. Sridhar**
New Delhi, India **Sunil K. Deshmukh**

Contents

The Macrofungal Resource
Extent, Current Utilization, Future Prospects, and Challenges

David L. Hawksworth

INTRODUCTION

As we progress towards the mid-21st century, concerns increase over the pressure on natural resources, how an exponentially burgeoning population can be fed, kept healthy, wastes disposed of safely, while simultaneously achieving environmental, ecosystem and wildlife protection. As macrofungi become better understood, it is becoming clear that they have the potential for addressing some of these key concerns.

In this overview, I explore aspects of the extent, current utilization, future prospects and challenges relating to the macrofungal resource.

Extent

Seventeen years ago, when 1.5 million was generally accepted as a conservative estimate of the number of species of fungi on Earth, I estimated that this could include around 140 000 mushroom species (Hawksworth, 2001). That figure was based on the number of known macrofungal species, then some 14 000, assuming that they were perhaps around 50% better known than other fungal morphologies, and taking note of the proportion of new species then being discovered in tropical regions, and adding an estimate for the extent of cryptic speciation.

Honorary President, International Mycological Association

Comparative Plant and Fungal Biology, Royal Botanic Gardens, Kew, Surrey TW9 3DS, UK; Department of Life Sciences, The Natural History Museum, Cromwell Road, London SW7 5BD, UK; Jilin Agricultural University, Changchun, Jilin Province, 130118 China; d.hawksworth@kew.org

Since that time, molecular methods have enabled species concepts to be clarified in many genera and the sequencing of environmental samples has led to the discovery of a huge unexpected species diversity; as a result the 1.5 million figure (Hawksworth, 1991) has been revised upwards to between 2.2 and 3.8 million species (Hawksworth and Lücking, 2018). The fungi known only from environmental sequences have conveniently been dubbed "dark taxa" as they are not known from specimens or cultures and remain invisible to the eye (Ryberg and Nilsson, 2018). There are, however, possibilities for visualizing at least some of these new taxa by using sophisticated tailored fluorescent probes (Jones et al., 2011).

If the new estimates of species numbers are accepted, and the same assumptions as in 2001 are adopted as still being reasonable today, that would give an estimate of between 220 000 and 380 000 macrofungal species; that is, that we know perhaps just 3.7–6.4% rather than 10% of the estimated species as suggested in 2001. This is a daunting prospect for field mycologists and taxonomists, but simultaneously a most exciting one for those wishing to explore the ways in which novel mushrooms and other fungi may be exploited to benefit humankind and the environment.

At the regional level, the number of fungal species, growing on all substrates in an area, that can be detected in field surveys, can be expected to be around six times the number of vascular plant species (Hawksworth, 1991; Piepenbring and Yorou, 2017). In the case of India, as there are about 16,000 species of vascular plants recorded, that gives a figure of 96,000 species. In India, the number of known fungal species of all groups (including fungal analogues) currently stands at about 14,500, a figure that suggests there are at least 81,500 species awaiting recognition. Taking the estimate that 18.75% of the known fungi are mushrooms (Hawksworth, 2001), and then applying this percentage to the Indian estimated total fungal biota, there may be as many as 15,300 mushroom species remaining to be recognized in India. This total would include: (1) species new to science; and (2) species that had already been described and named from localities outside India but had not yet been found within India.

Current Utilization

The main current utilization of macrofungi today is for food. While most cultivated mushrooms come from just five genera, *Agaricus, Auricularia, Flammulina, Lentinula,* and *Pleurotus,* a relatively huge number are collected for food from the wild, both for local consumption and trade. A world list of 350 species used for human consumption has been prepared (Boa, 2004), but the actual number is much higher. For example, over 600 species are reported as edible just in Yunnan Province, China, and 60 of these are exploited commercially in that area (Yang, 2002). Clearly, there is a need to extend ethnomycological surveys in remote areas of many parts of the world, and it is pleasing to see that some progress in this direction is being made in India (e.g., Bhaben et al., 2011; Choudhary et al., 2015). A tried and tested model to emulate in carrying out ethnomycological surveys is that developed in Tanzania and Zambia (Härkönen et al., 2015).

In investigations into what local communities eat, edible species that are new to science are discovered quite frequently, as in the case of new *Cantharellus* species in East Africa (Buyck, 1994) and new *Termitomyces* in China, the latter being first discovered on sale in local markets (Wei et al., 2004). While many of the fungi eaten are mycorrhizal, and so may be difficult to exploit on more than a local scale, newly discovered saprobic mushrooms that may be grown on agricultural wastes, may also be found; these will be more suitable for commercial exploitation. For example, native *Pleurotus* species grown on cylinders of paddy straw in Coimbatore, India (seen by me in 1991), and *Lentinus squarroides* cultivated on steam sterilized lignocellulosic substrates, also in India (Atri et al., Chapter 7). As many of these saprobic mushrooms have important ecological roles as decomposers (Dighton, Chapter 5), they have enormous potential to contribute to the sustainable use of resources.

Edible mushrooms form a substantial proportion of the diet in some societies, and there have been many studies of their nutritional value (e.g., Ghate and Sridhar, Chapter 6). They can be of particular value to vegetarians as they contain essential amino acids and a wide range of vitamins, some generally available only through animal products. They also have applications in weight-loss programmes as they are low in carbohydrates and rich in fibre (indigestible chitin), and are commended as dietary supplements or nutriceuticals (Badalyan and Zambonelli, Chapter 9).

Some macrofungal genera have species that are valued for medicinal applications (Ocañas et al., Chapter 8), for example, in enhancing immunological functions, or with anti-tumor properties, as is the case in *Ganoderma* (Paterson, 2015; Papp, Chapter 2), *Phellinus* (Deshmukh et al., Chapter 12), and *Ophiocordyceps* (Baral et al., 2015). Huge numbers of bioactive compounds are formed (Agyare and Agana, Chapter 10), and now their biosynthesis is being investigated by proteonomics (Fung and Razif, Chapter 16). Some are also extremely important toxicologically as poisons or recreational hallucinogens (Benjamin, 1995; Guzmán et al., 2000). In addition, some of the huge array of bioactive compounds may have value as antifeedants against insect pests (Clericuzio and Vizzini, Chapter 11). Particularly exciting has been the discovery of strobilurin fungicides from *Strobilurus tenacellus* that inhibit plant pathogenic fungi, primarily by disrupting their mitochondria (Bartlett et al., 2002).

Ectomycorrhizal macrofungi have a major role in maintaining tree health and so are of vital importance in commercial forestry (Rutz and Luna, Chapter 4), which now extends to the inoculation of containerized trees with spore suspensions prior to planting out (Hall et al., Chapter 13).

Future Prospects

In addition to the established areas utilizing macrofungi and macrofungal products, there are also prospects for novel applications. A particularly exciting new use of mushrooms has been developed during the last decade. Ecovative, a company based in New York State, has started developing a diverse range of materials using mushroom mycelium and a wide range of waste plant materials and gypsum. They have taken out numerous US Patents, including use as an alternative to styrofoam for producing packaging materials, boards, and even insulating brick blocks. This material evidently has the potential largely to replace many uses of plastics globally (Gunther, 2013). The prospects for this new application are phenomenal as it simultaneously addresses the environmental plastics crisis, and recycles waste materials.

The production of the meat substitute mycoprotein Quorn™ from the filamentous fungus *Fusarium venetatum*, cultured in a continuous liquid culture system, has been a remarkable success, and its products provide some 500 000 meals each day in the UK alone. Development of this system started in the mid-1960s, and involved the assessment of around 3000 isolates, strain selection, and complex calculations of fungal growth rate. This strain of *Fusarium* has a doubling time of 3.5–4.1 h so that 300–350 kg biomass is produced each hour (Wiebe, 2004; Finnigan, 2011). If the investment were available, it is conceivable that a similar system could be developed, using mycelium from edible mushrooms which are already familiar to the public. This would avoid the image problem Quorn has experienced with public perception; last year it was ruled in the US that the fungus used should be referred to as a "mould", not a relative of the mushroom. If this vision of using known edible fungi in continuous culture systems could be realized, the contribution to global food security could be enormous—and would also have the environmental advantage of reducing land necessary for animal husbandry, and for disposing of animal waste.

Fungi are an extraordinary source of novel natural products (Cole et al., 2003; Bills and Gloer, 2018), but most of these have been discovered in ascomycetes, and other microfungi rather than macrofungi. In the macrofungi, the focus has been on toxic and neurotropic compounds, and these have been accorded less attention by drug discovery companies because they do not grow as readily in liquid culture systems. The compounds produced can be expected to confer some aspect of fitness to the fungi, or else the biosynthetic pathways that produce them would surely have been lost in the course of evolution. It is likely that most are bioactive, acting against bacteria and fungi that infect basidiomes and cause then to decay more quickly. They might also act as effective poisons or antifeedents against insects, and other invertebrates, that would otherwise feed and lay eggs on them. More intensive screening of mushrooms, in particular those already used for medicinal purposes by local peoples, could lead to the discovery of novel pesticides (as was the case with the strobilurins; see above), or anti-bacterial or anti-fungal antibiotics. The search for new antibiotics is of urgent global concern as pathogenic bacteria and medically important fungi are both developing resistance to the drugs now on the market. On 15 February 2018, the World Health Organization (WHO; http://www.who.int/en/news-room/fact-sheets/detail/antimicrobial-resistance) stressed that antimicrobial resistance was an increasingly serious threat to global public health and that this required action across all government sectors and society. The potential of macrofungi, merits renewed attention in this regard (Suryanarayanan and Hawksworth 2018), though sadly, drug discovery from living organisms has become frustrated by two sets of regulations that have effectively stopped natural product drug discovery operations in many companies: these include the Nagoya Protocol to the Convention on Biological Diversity that came into effect in 2014 (Verkeley, 2015), and misguided national plant health regulations (Hawksworth and Dentinger, 2013). On the positive side, when the genes responsible for the production of compounds of interest have been located in the fungal genome, possibilities for transferring them into other, faster growing, filamentous fungi or yeasts for expression are now conceivable (Keller et al., 2005).

Another emerging promising area for the future is the use of mushrooms in the biosynthesis of nanoparticles of precious metals, discussed here by Tarmizi et al. (Chapter 15).

As increases in greenhouse gasses lead to climate change, and latitudinal movement of vegetation, there may be a major issue with respect to the maintenance of mutualistic ectomycorrhizal relationships. In particular, will tree movements be limited by the absence of appropriate ectomycorrhizal partners in areas being colonized? There is much uncertainty and speculation over the impact upon, and responses of, mycorrhizal associations (Mohan et al., 2014), and this issue is sure to become a major research area in the foreseeable future.

Challenges

As the focus of this book is work by Indian scientists on macrofungi, it is appropriate to draw attention to five challenges which currently constrain realizing their potential in India.

(1) *Cataloguing*: There is a long tradition of cataloguing the fungi recorded from India, of which the first major effort was that of Butler and Bisby (1931). Unfortunately, they omitted the lichen-forming fungi, which was usual at that time, but there have been several supplements and updates published over the years, and the lichen-forming fungi have now been ably dealt with by Awasthi (1991, 2000, 2007). In the 21st century, however, free online databases are needed and, indeed, this is the only way a system can operate in real time and have a chance of catching the vast amount of new data expected to emerge as more and more species are discovered. A start to this end, the "Fungi from India" online database project has now been launched and already holds records on some 6 000 species (Ranadive et al., 2017).

(2) *Barcoding*: DNA barcoding will increasingly facilitate identification (Dutta and Acharya, Chapter 14), but remains limited in that so many already known fungi have never been sequenced; only some 35 000 of the known 120 000 fungal species have sequences deposited in GenBank (Hawksworth and Lücking, 2017). There is a major need to obtain ITS barcode sequences for as many known species as possible and, fortunately, in the case of macromycetes, DNA can often be recovered from dried reference material deposited in fungaria (Brock et al., 2009). This work needs to be expedited, with an emphasis on sequencing type material or designating modern sequenced collections as epitypes where no DNA is recoverable from the original material (Ariyawansa et al., 2014).

(3) *Collections*: Collections of dried reference specimens of fungi (fungaria) and, where possible, permanently preserved living cultures (biological resource collections) underpin all mycological research. These are essential to fix the application of names, facilitate identifications by direct comparisons with reliably named material, and preserve voucher material for the fungi used in all kinds of inventory and experimental investigations. The maintenance of

reference collections is costly and requires considerable space and curatorial staff as collections grow but, regrettably, their scientific value is often not fully appreciated by the institutions where they are housed. In India, collections are currently dispersed through various university and government institutions; in itself, that is not a bad arrangement as it spreads risk, as well as making material accessible in different regions. Now that data on collections can be held in databases (e.g., Ranadive et al., 2017), accessibility is less of an issue than for previous generations. Just how mycology might be better organized in India was an issue close to the heart of the late C.V. Subramanian (1924–2016) who looked forward to a national centre concentrating on tropical mycology (Subramanian 1982, 1986). This issue has been highlighted more recently with respect to genetic resource collections by Suryanarayanan et al. (2015) who stress the importance of fungi to the bioeconomy, but action by pertinent authorities is awaited.

(4) *Conservation*: The penultimate challenge I will highlight is the conservation of fungi of all kinds. Despite the key roles of fungi in ecological processes, and the dependence of plants on them through mycorrhizal associations, it is only in the last 25 years that fungal conservation has started to be taken seriously, especially after the stimulus provided by Moore et al. (2001). Fungi do not, of course, occur in isolation but need to be integrated into conservation programmes (Heilmann-Clausen et al., 2014). Criteria for producing Red Lists of fungi, which assess conservation status species by species at the national or international level, are now available (Dahlberg and Mueller, 2014). Progress has been slow, with just 56 fungi evaluated globally to date by IUCN (International Union for the Conservation of Nature). To make assessments of all the 14 500 species of fungi known from India is a daunting task, yet one that could be overcome with sufficient determination and resources. For example, a Red List assessment of the macrofungi of China, which dealt with almost 10 000 species and involved around 140 mycologists, has just been completed; 97 species were classified as nationally threatened (Fang et al., 2018). National Assessments should be carried out in all countries that are signatories to the Convention on Biological Diversity (United Nations Environment Programme, 1992), of which India is one. Article 4 (b) of the Convention requires nations to monitor the components of biodiversity, paying particular attention to those requiring conservation measures. Assessments of the conservation status of species are a pre-requisite for recognizing those that are of conservation concern. One country that has made enormous progress in this respect in recent years is Chile, which now has fungi embedded in its conservation legislation, and around 30 government mycologists are involved in fungal conservation. This is the result of a project largely driven by one charismatic individual, Giuliana Furci (Anon, 2015).

(5) *Status and training*: The issue of training the next generation of whole organism mycologists is a major problem in many countries, and not one confined to "developing" nations. In many countries, including most European ones, the USA, as well as India, there has been a pattern of distinguished mycologists rising to become heads of botany departments and morphing those into centres

of excellence for mycology. They run post-graduate courses, award PhDs, secure research grants with post-doctoral positions, and generate pertinent research papers and key reference works. A particularly fine example in India was the work of C.V. Subramanian in Chennai (see above) who developed what became the Centre for Advanced Study in Botany of the then University of Madras. He trained or influenced many of the mycologists in senior positions in the country today. When such influential people retire from "Botany" departments and institutions, however, they are rarely replaced by mycologists, so centres of excellence for training new mycologists are lost. Mycologists remain, what I have termed, "orphans" in botany (Hawksworth, 1997). Sadly, perceptions are difficult to change, even if it is pointed out that fungi would, in any case, be better placed in departments of zoology as, genetically, they are closer to animals than plants. Mycologists need to become more active in promoting mycology as an independent discipline. Suggestions as to how that can be done individually, and collectively, have been made (Hawksworth, 2003; Minter, 2011), but these need to be actioned.

Acknowledgment

I am indebted to my wife, Professor Patricia E.J. Wiltshire, for improving my original text.

References

Anon. 2015. Defensores de la Gaia ("Defenders of Gaia") Award. IMA Fungus, 6: 19.

Ariyawansa, H.A., Hawksworth, D.L., Hyde, K.D., Maharachchikumbura, S.S.N., Manamgoda, D.S., Thambugala, K.M., Udayanga, D., Camporesi, E., Daranagama, A., Jayawardena, R., Jones, E.B.G., Liu, J.-K., McKenzie, E.H.C., Phookamsak, R., Senanayake, I.C., Shivas, R.G., Tian, Q. and Xu, J.-C. 2014. Epitypification and neotypification: Guidelines with appropriate and inappropriate examples. Fungal Diversity, 69: 57–91.

Awasthi, D.D. 1991. A key to the microlichens of India, Nepal and Sri Lanka. Bib. Lichenol., 40: 1–337.

Awasthi, D.D. 2000. Lichenology in Indian Subcontinent: A supplement to "A Handbook of Lichens". Bishen Singh Mahendra Pal Singh, Dehra Dun.

Awasthi, D.D. 2007. A Compendium of the Macrolichens from India, Nepal and Sri Lanka. Bishen Singh Mahendra Pal Singh, Dehra Dun.

Baral, B., Shrestha, B. and Teixeira de Silva, J.A. 2015. A review of Chinese *Cordyceps* with special reference to Nepal, focussing on conservation. Environ. Exp. Biol., 13: 61–73.

Bartlett, D.W., Clough, J.M., Godwin, J.R., Hall, A.A., Hamer, H. and Parr-Dobrzanski, B. 2002. The strobilurin fungicides. Pest Manage. Sci., 58: 649–662.

Benjamin, D.R. 1995. Mushrooms: Poisons and Panaceas. W.H. Freeman, New York.

Bhaben, T., Lisha, G. and Chandra, S.G. 2011. Wild edible fugal resources used by ethnic tribes of Nagaland, India. Ind. J. Trad. Know., 10: 512–515.

Bills, G.F. and Gloer, J.B. 2018. Biologically active secondary metabolites from the Fungi. pp. 1087–1119. *In*: Heitman, J., Howlett, B.J., Crous, P.W., Stukenbrock, E.H., James, T.Y. and Gow, N.A.R. (eds.). The Fungal Kingdom, American Society for Microbiology, Washington DC.

Boa, E.R. 2004. Wild Edible Fungi: A Global Overview of their Use and Importance to People. Non-Wood Forest Products Report # 17. Food and Agricultural Organization of the United Nations, Rome.

Brock, P.M., Döring, H. and Bidartondo, M.I. 2009. How to know unknown fungi: The role of a herbarium. New Phytol., 109: 719–724.

Butler, E.J. and Bisby, G.R. 1931. The Fungi of India. Scientific Monograph #1. Imperial Council of Agricultural Research, Calcutta.

Buyck, B. 1994. Ubwoba: Les Champignons Comestibles de l'Ouest du Burundi. Publicacion Agricole # 34. Administration Generale de la Cooperation au Development, Bruxelles.

Choudhary, M., Devi, R., Datta, A., Yadav, A. and Jat, H.S. 2015. Diversity of wild edible mushrooms in Indian subcontinent and its neighbouring countries. Rec. Adv. Biol. Med., 1: 69–76.

Cole, R.J., Schweikert, M.A. and Jarvis, B.B. 2003. Handbook of Secondary Fungal Metabolites. 3 Volumes, Academic Press, Amsterdam.

Dahlberg, A. and Mueller, G.M. 2011. Applying IUCN red-listing criteria for assessing and reporting on the conservation status of fungal species. Fungal Ecol., 4: 147–162.

Fang, R., Kirk, P., Wei, J.-C., Li, Y., Cai, L., Fan, L., Wei, T.-Z., Zhao, R.-L., Wang, K., Yang, Z.-L., Li, T.-H., Li, Y. Phurbu-Dorji and Yao, Y.-J. 2018. Country focus: China. pp. 48–55. *In*: Willis, K.J. (ed.). State of the World's Fungi. Royal Botanic Gardens, Kew.

Finnigan, T.J.A. 2011. Mycoprotein: Origins, production and properties. pp. 335–352. *In*: Phillips, G.O. and Williams, P.A. (eds.). Handbook of Food Proteins. Woodhead Publishing, Cambridge.

Gunther, M. 2013. Can mushrooms replace plastics? The Guardian: https://www.theguardian.com/sustainable-business/mushrooms-new-plastic-ecovative.

Guzmán, G., Allan, J.W. and Gartz, J. 2000. 1998. A worldwide geographical distribution of the neurotropic fungi, an analysis and discussion. Annalidei Museo Civico Rovereto, 14: 189–280.

Härkönen, M., Niemela, T., Mbindo, K., Kotiranta, H. and Pierce, G. 2015. Zambian Mushrooms and Mycology (Norrlinia # 29). Finnish Museum of Natural History, Helsinki.

Hawksworth, D.L. 1991. The fungal dimension of biodiversity: Magnitude, significance, and conservation. Mycol. Res., 95: 641–655.

Hawksworth, D.L. 1997. Orphans in "botanical" diversity. Muelleria, 10: 111–123.

Hawksworth, D.L. 2001. Mushrooms: The extent of the unexplored potential. Int. J. Med. Mushrooms, 3: 333–337.

Hawksworth, D.L. 2003. Monitoring and safeguarding fungal resources worldwide: The need for an international collaborative MycoAction Plan. Fungal Diversity, 13: 29–45.

Hawksworth, D.L. and Dentinger, B.T.M. 2013. Antibiotics: Relax UK import rules on fungi. Nature, 4: 169.

Hawksworth, D.L. and Lücking, R. 2017. Fungal diversity revisited: 2.2 to 3.8 million species. Microbiology Spectrum, 5: FUNK-0052-2016.

Heilmann-Clausen, J., Barron, E.S., Boddy, L., Dahlberg, A., Griffith, G.W., Nordén, J., Ovaskainen, O., Perini, C., Senn-Irlet, B. and Halme, P. 2014. A fungal perspective on conservation biology. Conserv. Biol., 29: 61–68.

Jones, M.D.M., Forn, I., Gadelha, C., Egan, M.J., Bass, D., Massana, R. and Richards, T.A. 2011. Discovery of novel intermediate forms redefines the fungal tree of life. Nature, 474: 200–203.

Keller, N.P., Turner, G. and Bennett, J.W. 2005. Fungal secondary metabolism—From biochemistry to genomics. Nat. Rev. Microbiol., 3: 937–947.

Minter, D.W. 2011. What every botanist and zoologist should know—and what every mycologist should be telling them. IMA Fungus, 2: 14–18.

Mohan. J.E., Cowden, C.C., Baas, P., Dawadi, A., Frankson, P.T., Helmik, K., Hughes, E., Khan, S., Lang, A., Machmuller, M., Taylor, M. and Witt, C.A. 2014. Mycorrhizal fungi mediation of terrestrial ecosystem responses to global change: A mini-review. Fungal Ecol., 10: 3–19.

Moore, D., Nauta, M.M., Evans, S.E. and Rotheroe, M. (eds). 2001. Fungal Conservation: Issues and Solutions. Cambridge University Press, Cambridge.

Paterson, R.R.M. (ed.). 2015. *Ganoderma* phytochemistry. Phytochem., 114: 1–177.

Piepenbring, M. and Yorou, N.S. 2017. Promoting mycology teaching and research on African fungi by field schools on tropical mycology in Benin. IMA Fungus, 8: 74–77.

Ranadive, K., Jagrap, N. and Khare, H. 2017. fungifromindia: The first online initiative to document fungi from India. IMA Fungus, 8: 67–69.

Ryberg, M. and Nilsson, R.H. 2018. New light on names and naming of dark taxa. MycoKeys, 30: 31–39.

Subramanian, C.V. 1982. Tropical mycology: Future needs and development. Current Science, 51: 321–325.

Subramanian, C.V. 1986. The progress and status of mycology in India. Proc. Ind. Acad. Sci. Pl. Sci., 96: 379–392.

Suryanarayanan, T.S., Gopalan, V., Sahal, D. and Sanyal, K. 2015. Establishing a national fungal genetic resource to enhance the bioeconomy. Current Science, 109: 1033–1037.

Suryanarayanan, T.S. and Hawksworth, D.L. 2018. The war against MDR pathogens: move fungi to the front line. Current Science, 115: in press.

United Nations Environment Programme. 1992. Convention on Biological Diversity. United Nations Environment Programme, Nairobi.

Verkeley, G. 2015. How will the Nagoya Protocol affect our daily work? IMA Fungus, 6: 3–5.

Wei, T.-Z., Yao, Y.-J., Wang, B. and Pegler, D.N. 2004. *Termitomyces bulborhizus* sp. nov. from China, with a key to allied species. Mycol. Res., 108: 1458–1462.

Wiebe, M.G. 2004. Quorn™ mycoprotein—overview of a successful fungal product. Mycologist, 18: 17–20.

Yang, Z.-L. 2002. On wild mushroom resources and their utilization in Yunnan Province, southwest China. J. Nat. Resour., 17: 463–469.

Global Diversity of the Genus *Ganoderma*
Taxonomic Uncertainties and Challenges

Viktor Papp

INTRODUCTION

The cosmopolitan polypore genus *Ganoderma* (Polyporales, Basidiomycota) comprises white-rot species, some of them being important pathogens of horticultural plants, such as *G. philippii* of cacao, coffee and tea, or *G. boninense* of oil palm. Greater attention is paid to those *Ganoderma* species which are used for their medicinal properties. The health benefits of the various *Ganoderma* species (e.g., *G. applanatum*, *G. cupreolaccatum*, *G. lingzhi*, *G. lucidum*, *G. resinaceum*, *G. sinense* and *G. tsugae*), and their compounds responsible for beneficial effects are intensly studied wordwide (e.g., Paterson, 2006; Baby et al., 2015; Hapuarachchi et al., 2017). Despite the fact that the genus has an enormous economic value, from the taxonomic point of view, *Ganoderma* is one of the most complex and misunderstood genera among the polypores. The taxonomical and nomenclatural confusion have arisen due to the high morphological variability of *Ganoderma* fruiting bodies. As a result, taxonomists have created many synonyms, or in the other case used wide species concept and merged different species. The difficulties of identification and the different species concepts have resulted in ambiguous species delimitation and identification systems. Names of *Ganoderma* are often misused because of the uncertain identification, making it hard to interpret the results of applied mycological studies; thus it would be important to carry out these studies

Department of Botany, Faculty of Horticultural Science, Szent István University, 1518 Budapest, Hungary
Email: papp.viktor@kertk.szie.hu, agaricum@gmail.com

on a suitable and scientifically correct taxonomical basis. Owing to these disparities, this Chapter discusses: (i) the systematic state of *Ganoderma* genus and of its relative genera; (ii) the possibilities of species separation; (iii) the current taxonomical state and biogeography of the most important *Ganoderma* taxa in the light of most recent research results.

Classification and Nomenclature of *Ganoderma* sensu lato

Ganodermatoid fungi (*Ganoderma* sensu lato) are generally characterized by the unique doublewalled basidiospores with a coloured endosporium ornamented with columns or crests, and a hyaline smooth exosporium. These species were first separated systematically from the other conks-producing polypores in the begining of the 19th century, when the spore morphology became an important taxonomic character. Within Agaricomycetes, the systematic state of species producing typical ganodermatoid spores was unclear due to the various systematic concepts. What follows is a review of the systematic state of ganodermatoid species. The different generic concepts affecting nomenclatural questions will also be discussed.

Systematics of Ganoderma sensu lato in Subgeneric Level

The first step towards circumscribing the ganodermatoid species in subgeneric level was taken by Donk (1933), who proposed the subfamily Ganodermatoideae in the Polyporaceae; and subsequently raised this morphological group to family level (Donk, 1948). Later, Jülich (1981) segregated the family Haddowiaceae from Ganodermataceae sensu Donk and proposed the order Ganodermatales with the two families. However, the results of preliminary phylogenetic studies are not supported by the distinction of Ganodermatales and suggested that *Ganoderma* s. lato does not form a well separated clade in the Polyporales (Moncalvo, 2000). Based on a multi-gene phylogenetic and genomic analysis in the Polyporales, Binder et al. (2013) found that *Ganoderma* and *Amauroderma* species are grouped together with several other poroid taxa in the Core polyporoid clade: for example, *Coriolopsis* spp., *Cryptoporus volvatus* (Peck) Shear, *Dichomitus squalens* (P. Karst.) D.A. Reid, *Donkioporia expansa* (Desm.) Kotl. & Pouzar, *Pachykytospora tuberculosa* (Fr.) Kotl. & Pouzar, and *Perenniporia medulla-panis* (Jacq.) Donk. In the revised phylogenetic overview of the Polyporales, Justo et al. (2017) recomended wide family concept of Polyporaceae and discussed Ganodermataceae as a synonym of the latter. Therefore, the recent phylogenetic studies suggest that despite the unique spore characteristics, ganodermatoid taxa do not form a separated lineage on the family level. The formerly proposed ganodermatoid genera (Table 1) are not consistently accepted in the literature and dividing each group on generic level arises many questions.

What is *Ganoderma* sensu stricto? Generic Concepts and Classification

The generic classification of ganodermatoid species dates back to the second half of the 19th century, when Karsten (1881) established the genus *Ganoderma* based on one single species, *Boletus lucidus* Leyss. (syn. *B. lucidus* Curtis). Some years later,

Table 1. Overview of the described ganodermatoid genera with the number of the proposed names in species rank.

Genus	Year	Type species	Names	Notes
Ganoderma P. Karst.	1881	*Boletus lucidus* Leyss.	345	
Elfvingia P. Karst.	1889	*Boletus applanatus* Pers.	16	Widely accepted as a synonym of *Ganoderma* P. Karst. (see Moncalvo and Buchanan, 2008)
Amauroderma Murrill	1905	*Fomes regulicolor* Berk. ex Cooke	121	Phylogenetically polyphyletic (see Costa-Rezende et al., 2017)
Dendrophagus Murrill	1905	*Polyporus colossus* Fr.	1	Illegitimate under Art. 53.1 (non-*Dendrophagus* Toumey 1900); Synonym of *Tomophagus* Murrill
Tomophagus Murrill	1905	*Polyporus colossus* Fr.	2	
Friesia Lázaro Ibiza	1916	*Boletus applanatus* Pers.	5	Illegitimate under Art. 53.1 (non-*Friesia* Spreng. 1818); synonym of *Ganoderma* P. Karst.
Trachyderma (Imazeki) Imazeki	1952	*Polyporus tsunodae* Yasuda ex Lloyd	2	Illegitimate under Art. 53.1 (non *Trachyderma* Norman 1853)
Haddowia Steyaert	1972	*Polyporus longipes* Lév.	3	
Humphreya Steyaert	1972	*Ganoderma lloydii* Pat. & Har.	4	
Magoderna Steyaert	1972	*Fomes subresinosus* Murrill	3	
Archeterobasidium Koeniguer & Locq.	1979	*A. syrtae* Koeniguer & Locq.	1	Fossil genera; type has ganodermatoid spore (see Koeniguer and Locquin, 1979)
Thermophymatospora Udagawa et al.	1986	*Th. fibuligera* Udagawa et al.	1	Anamorphic synonym of *Tomophagus* Murrill (see Adaskaveg and Gilbertson, 1989)
Ganodermites A. Fleischm et al.	2007	*G. libycus* A. Fleischm. et al.	1	Fossil genera; the type morphologically closely related to the *Elfvingia* group (see Fleischmann et al., 2007)
Foraminispora Robledo et al.	2017	*Porothelium rugosum* Berk.	1	
Furtadoa Costa-Rezende et al.	2017	*F. biseptata* Costa-Rezende et al.	3	Illegitimate under Art. 53.1 (non-*Furtadoa* M. Hotta 1981)

Karsten (1889) proposed the initially monotypic genus *Elfvingia*, typified on *Boletus applanatus* Pers. This morphological group generally comprises of species with perennial basidiocarps and dull pileal surface. Many mycologists doubted that the genus *Elfvingia* should be segregated from *Ganoderma*, and they discussed it under the latter (e.g., Imazeki, 1939; Steyaert, 1980). Although based on morphological traits, it also belongs to this group, a structurally preserved (permineralized) fruiting body fossil from Lower Miocene (Libia, North Africa) was named *G. libycus* A. Fleischm et al. and classified into a new monotypic genus (*Ganodermites* A. Fleischm et al.) (Fleischmann et al., 2007). In contrast to the former opinion of the authors, the spore morphology of another monotypic genus (*Archeterobasidium* Koeniguer & Locq.), described from its fossilized fruiting body (also from Libia), does not suggest relationship with the *Heterobasidion* Bref. genus (Koeniguer and Locquin, 1979) but shows ganodermatoid characteristics. We have very little data regarding ganodermatoid fossils, but in addition to the above mentioned ones, findings are known which are thought to be *G. adspersum*, *G. applanatum* and *G. lucidum* (Fleischmann et al., 2007; Taylor et al., 2015). The taxonomical state of these findings, the possibility that the *Archeterobasidium* and *Ganodermites* genera are identical to each other, and the validity of their segregation from the *Ganoderma* genus should further be investigated.

The genus *Tomophagus*, established by Murrill (1905b), was based on one single species, *Polyporus colossus* Fr. *Tomophagus colossus* (Fr.). Murrill macroscopically differs from other *Ganoderma* s. str. species, by having a soft and light basidiocarp, with thick and pale context (Furtado, 1965). However, based on its spore and microstructural characteristics, several authors have not supported the generic segregation of this species (e.g., Torrend, 1920; Furtado, 1965; Steyaert, 1972, 1980; Corner, 1983; Ryvarden, 1991; Wasser et al., 2006a; Torres-Torres et al., 2015). In contrast, certain phylogenetic studies showed that *Tomophagus* represents an independent lineage (Hong andJung, 2004; Costa-Rezende et al., 2017) composed of two species: *T. colossus* (Fr.) Murrill and *T. cattienensis* Le Xuan Tham & J.M. Moncalvo (Le et al., 2011). It is considered that *Thermophymatospora fibuligera* Udagawa, Awao & Abdullah was described as a thermotolerant hyphomycete from Iraq (Udagawa et al., 1986) and it is identical to the chlamydospore of *Tomophagus colossus* (Adaskaveg and Gilbertson, 1989). In this case, the monotypic genus *Thermophymatospora* Udagawa, Awao & Abdullah is an anamorphic synonym of *Tomophagus*. Imazeki (1939) segregated a new subgenus in *Ganoderma* based on *G. tsunodae* Yasuda, and later it was discussed on a generic level (Imazeki, 1952). Based on the morphological characteristics of the type specimen (lectotype) of *G. tsunodae*, Hattori and Ryvarden (1994) noted that this species resembles *Tomophagus colossus* and the two species are probably congeneric. The close relationship between these species were also suggested by a recent phylogenetic study (Costa-Rezende et al., 2017), but it is still unclear whether these two species belong to the same genus. From the nomenclatural point of view, the generic name *Trachyderma* (Imazeki) Imazeki is illegitimate as it is a homonym of the formerly described lichen genus *Trachyderma* Norm.; therefore, the systematic position and as well as the correct nomenclature of *Ganoderma tsunodae* is unclear. Besides this species, only *Fomes subresinosus* Murrill has been proposed by Imazeki to be placed in *Trachyderma*. This species later was selected by Steyaert (1972) as a type for the new genus *Magoderna* Steyaert. Besides, *M. subresinosum*, two other species were placed to the new genus by Steyaert: *M. infundibuliforme* (Wakef.) Steyaert and *M. vansteenisii* Steyaert. The separation of these species in generic level was questioned by certain authors (e.g., Corner, 1983; Teixeira, 1992; Ryvarden, 1991; Moncalvo and Ryvarden, 1997), although the sequences of *M. subresinosum* (as *Amauroderma subresinosum*) forms a distinct lineage from other ganodermatoid genera (e.g., Gomes-Silva et al., 2015; Costa-Rezende et al., 2017). Based on the ornamentation of the basidiospores, Steyaert (1972) proposed two further stipitate ganodermatoid genera: *Haddowia* and *Humphreya*. The genus *Haddowia* Steyaert was typified on *Polyporus longipes* Lév. and characterized by its unique longitudinally costate spores. Besides the type, one more species was discussed in the genus by Steyaert (1972) described as new from Indonesia: *H. aetii* Steyaert. Later, Teixeira (1992) combined *Ganoderma neurosporum* J.S. Furtado into *Haddowia*, which has been reported throughout tropical America (Moncalvo and Ryvarden, 1997). The species of the genus have not yet been analyzed by molecular methods, so the phylogenetic state of the genus can only be clarified with further investigations. The genus *Humphreya* Steyaert comprises species bearing basidiospores with reticulate, honey-comb or cristulate endosporium (Steyaert, 1972; Costa-Rezende et al., 2017). Formerly four species were accommodated in the genus from Indonesia, South and Central America and tropical Africa: *H. endertii* Steyaert, *H. coffeata* (Berk.) Steyaert, *H. eminii* (Henn.) Ryvarden and *H. lloydii* (Pat. & Har.) Steyaert (Steyaert, 1972; Ryvarden and Johansen, 1980; Decock and Figueroa, 2007). Amongst of these species only *H. coffeata* was studied by phylogenetic perspective, and the systematic position of *Humphreya* at genus level is still uncertain (Costa-Rezende et al., 2017).

Based on the morphological characteristics of the basidiospores, Patouillard (1889) in his monographic study divided *Ganoderma* species into two sections: *Ganoderma* sect. *Ganoderma* and *G.* sect. *Amauroderma*. Torrend (1920) transferred the section *Amauroderma* to generic level, with *Polyporus auriscalpium* Pers. as the type. However, the proposal of Torrend is illegitimate, because Murrill (1905a) used this name earlier and it has priority. Murril introduced a different taxonomic circumscription of the genus when he selected *Fomes regulicolor* Berk. ex Cooke (syn. *Amauroderma schomburgkii* (Mont. & Berk.) Torrend) as the type species, which was not included in the section established by Patouillard. The tropical (or subtropical) genus *Amauroderma* Murrill traditionally circumscribed mainly by the globose to ellipsoid basidiospores, without a truncate apex (Ryvarden, 2004). However, the recent studies using molecular phylogenetic methods have shown that the genus is polyphyletic and will need to be revised (Gomes-Silva et al., 2015; Song et al., 2016; Costa-Rezende et al., 2017). The genus *Polyporopsis* Audet was discussed by Richter et al. (2015) in Ganodermataceae. This genus was described based on one single species, *Albatrellus mexicanus* Laferr. & Gilb. (Laferrière and Gilbertson, 1990). However, in contrast to the original description, the subsequent studies on the holotype revealed that the spores of *A. mexicanus* are not glabrous and have a double wall separated by interwall pillars (Zheng and Liu, 2006), which suggests similarity with the ganodermatoid *Amauroderma* genus (Audet, 2010). Although the phylogenetic state of *Polyporopsis mexicanus* is not clarified, it is possibly related to the albatrelloid lineage, since certain species have similar basidiospores in the *Polyporoletus* clade, which is close to the Albatrellaceae family (Russulales): for instance *Leucophleps spinispora* Fogel, *Mycolevis siccigleba* A.H. Sm., *Polyporoletus sublividus* Snell (Albee-Scott, 2007). In a recent study, Costa-Rezende et al. (2017) segregated two new genera from *Amauroderma* based on morphological observations and molecular evidence. The genus *Foraminispora* Robledo, Costa-Rezende & Drechsler-Santos typified on *Porothelium rugosum* Berk. is morphologically characterised by the unique endosporic ornamentation of basidiospores. The other proposed genus, *Furtadoa* Costa-Rezende, Robledo & Drechsler-Santos is accommodate species with monomitic context and glabrous pilear surface: *F. biseptata* Costa-Rezende, Drechsler-Santos & Reck, *F. brasiliensis* (Singer) Costa-Rezende, Robledo & Drechsler-Santos and *F. corneri* (Gulaid & Ryvarden) Robledo & Costa-Rezende.

The systematics of the ganodermatoid species on genus level is based upon spore morphology, which is fundamentally accepted by the most recent systematic studies dealing with multigene phylogenetic analyses (e.g., Costa-Rezende et al., 2017). However, the phylogenetic reconstructions built from the currently known sequences do not always confirm the suitability of spore morphology for lineage distinction. The classification of the ganodermatoid species on genus level has many questionable points, and currently there is no consensus in the topic. If we follow the narrow genus concept, it will be necessary to separate further genera, although the low level of nucleotide difference in the known barcoding regions and the seemingly monophyletic lineage of the ganodermatoid taxa suggests the possibility that a wide genus concept might be viable. Hereinafter, however, we follow the narrow genus concept, dealing only with the *Ganoderma* s. str. genus.

Species Delimitation in *Ganoderma*

Historically, the species description in the genus *Ganoderma* was mostly based on macro- and micromorphological characteristics of the basidiocarp. Based on a detailed morphological study on laccate *Ganoderma* species, Torres-Torres and Guzmán-Dávalos (2012) considered that the color of the context, resinous deposits, structure of the basidiospores and protuberances of the pileipellis cells are among the most important features for characterization of the species. In addition to morphological examination of the basidiocarp, several authors studied cultural and mating characteristics in *Ganoderma* in order to provide new tools for species-level systematics (e.g., Adaskaveg and Gilbertson, 1986, 1989). The first molecular genetic studies dealing with ribosomal DNA sequence analyses in the *Ganoderma* genus began in the middle of the 90s (e.g., Moncalvo et al., 1995a,b), but until the end of the last decade the new species were still described on the basis of morphological traits (e.g., Wu and Zhang, 1996; Ryvarden, 2000, 2004; Ipulet and Ryvarden, 2005; Torres-Torres et al., 2008). It is a kind of exception, that Smith and Sivasithamparam (2003) described *G. steyaertanum* B.J. Smith & Sivasith., citing one of their previous studies dealing with phylogenetic analyses in which this species is yet discussed under the name "*G.* sp. Grp 6.3" (Smith and Sivasithamparam, 2000), the ITS1 and ITS2 sequences of the type of *G. steyaertanum* only became accessible in the

GeneBank database many years later. In 2009, however, with the description of *G. carocalcareum* Douanla-Meli, barcoding sequences were also published besides the morphological characterisation (Douanla-Meli and Langer, 2009). Following this, further 21 new *Ganoderma* species were described mostly with the addition of ITS sequences, and in some cases, with the addition of further barcoding regions (Table 2). In the molecular era, several markers (e.g., protein-coding genes, rDNA loci, mtSSU rDNA sequence and multilocus marker systems) were used by different authors in order to clarify taxonomic difficulties in the genus *Ganoderma* (e.g., Hong and Jung, 2004; Sun et al., 2006; Zheng et al., 2009; Wang et al., 2012; Thakur et al., 2015; Zhou et al., 2015; Xing et al., 2018), from which, the ITS region became the most popular. In the public databases (GeneBank, UNITE) the ITS sequence of nearly 70 different *Ganoderma* binoms are deposited. Nevertheless, due to the unclear taxonomic interpretation of these *Ganoderma* species and the exclusively morphological based identification, the majority of the *Ganoderma* sequences accessible in the GeneBank are labeled as misidentified or ambiguously (e.g., Jargalmaa et al., 2017; Papp et al., 2017). Therefore, it would be essential to sequence the types of those *Ganoderma* species which were used in the modern scientific literature and previously described based on morphological features. This was carried out only in some cases: for example, *G. microsporum* R.S. Hseu, *G. ahmadii* Steyaert, *G. sichuanense* J.D. Zhao & X.Q. Zhang (Moncalvo et al., 1995a; Cao et al., 2012). Although morphological study combined with DNA sequencing has been the most relevant approach for identification of *Ganoderma* species, an integrative taxonomy combined morphological and phylogenetic methods with secondary metabolite-based chemotaxonomy has presumably attained a more stable taxonomy in the genus (Richter et al., 2015; Welti et al., 2015).

Ganoderma Species Around the World: Taxonomy and Biogeography

The *Ganoderma* genus has a cosmopolitan distribution, but in many cases, due to the taxonomical uncertainties, little is known about the area of the species. Because of the different taxonomical concepts, some of the species were reported from all around the world (e.g., *G. applanatum*, *G. australe* and *G. lucidum*), while many others are only known from type locality (Moncalvo and Ryvarden, 1997). In some recent studies, the biogeography of the *Ganoderma* taxa was investigated through phylogenetic analyses (e.g., Moncalvo and Buchanan, 2008; Zhou et al., 2015). The distribution of the species can also be reevaluated based on the accessible barcoding sequences. In contrast to the earlier theory that certain species have wide distributions and largely unstructured populations, recent genetic and biogeographic studies have indicated that most of the studied *Ganoderma* taxa are geographically restricted. In the majority of cases, we do not have sufficient information about the taxonomical state and so the distribution of most of the described *Ganoderma* species, therefore it has not been possible to comprehensively review the biogeography of the genus yet. Thus, what follows is an attempt to review the most recent results of the taxonomical state and biogeography of the most widely investigated *Ganoderma* species (e.g.,

Table 2. New *Ganoderma* species described in the last twenty years (1998–2017). The formerly described *Ganoderma* species were listed in Moncalvo and Ryvarden (1997). The table includes the proposed name of the new species and the data of the holotype specimens. Herbarium codes: BJFC – Beijing Forestry University (China); DAR – Orange Agricultural Institute (Australia); ENCB – Instituto Politécnico Nacional (Mexico); GACP – Guizhou Agricultural College (China); GDGM – Guangdong Institute of Microbiology, Guangdong Academy of Sciences (China); HKAS – Herbarium of Cryptogams, Kunming Institute of Botany, Chinese Academy of Sciences; IFP – Institute of Applied Ecology, Academia Sinica (China); LIP – Université de Lille (France); MIN – University of Minnesota (USA); MUCL – Université Catholique de Louvain (Belgium); O – Botanical Museum, University of Oslo (Norway); PERTH – Western Australian Herbarium; PREM – National Collection of Fungi in South Africa (South Africa); QCA/QCAM – Pontificia Universidad Catolica del Ecuador (Ecuador); TNM – National Museum of Natural Science (Taiwan); VEN – Universidad Central de Venezuela (Venezuela); (*) isotype; (**) ex-culture.

Species	Year	Geographic origin	Voucher no.	Herbarium code	Published sequence (GenBank no.)	Reference of the original description
G. concinnum	2000	Colombia	Ryvarden 16840	O (NY*)	-	Ryvarden (2000)
G. longistipitatum	2000	Venezuela	Ryvarden 40558	VEN (O*) (NY*)	-	Ryvarden (2000)
G. multicornum	2000	Venezuela	G. J. Samuels	NY (O*)	-	Ryvarden (2000)
G. steyaertanum	2003	Indonesia	DAR73780 PERTH 05509114	DAR** PERTH	EU239395/EU239396 (ITS)	Smith and Sivasithamparam (2003)
G. citriporum	2004	Venezuela	Ryvarden 40466	VEN (O*)	-	Ryvarden (2004)
G. elegantum	2004	Ecuador	Ryvarden 44573	O (QCA*)	-	Ryvarden (2004)
G. guianense	2004	French Guiana	MUCL 43922	MUCL	-	Ryvarden (2004)
G. turbinatum	2005	Uganda	Ipulet 477	O	-	Ipulet and Ryvarden (2005)
G. vivianmercedianum	2008	Mexico	E. Bastidas-Varela	ENCB	-	Torres-Torres et al. (2008)
G. carocalcareum	2009	Cameroon	DMC 322	HUYI (O*)	EU089969 (ITS) EU089968 (mtSSU)	Douanla-Meli & Langer (2009)
G. martinicense	2010	Martinique	SW 55	LIP	KF963256 (ITS)	Welti and Courtecuisse (2010)
G. neogibbosum	2010	Martinique	SW 37	LIP	Not published	Welti and Courtecuisse (2010)
G. parvigibbosum	2010	Martinique	SW 22	LIP	Not published	Welti and Courtecuisse (2010)
G. ryvardenii	2010	Cameroon	HKAS 58053	HKAS	HM138671 (ITS)	Kinge and Mih (2011)

Table 2 contd....

...Table 2 contd.

Species	Year	Geographic origin	Voucher no.	Herbarium code	Published sequence (GenBank no.)	Reference of the original description
G. lingzhi	2012	China, Hubei	Wu 1006-38	TNM (BJFC*) (IFP*)	JQ781858 (ITS) JX029989 (mtSSU) JX029984 (RPB1) JX029980 (RPB2) JX029976 (TEF1-α)	Cao et al. (2012)
G. mutabile	2012	China, Yunnan	Yuan 2289	IFP	JN383977 (ITS)	Cao and Yuan (2012)
G. austroafricanum	2014	South Africa	PREM 61074	PREM	KM507324 (ITS) KM507325 (LSU)	Crous et al. (2014)
G. destructans	2015	South Africa	PREM 61265	PREM	KR183856 (ITS) KR183860 (LSU)	Coetzee et al. (2015)
G. enigmaticum	2015	South Africa	PREM 61264	PREM	KR183855 (ITS) KR183859 (LSU)	Coetzee et al. (2015)
G. leucocontextum	2015	China, Tibet	GDGM 40200	GDGM	KF011548 (ITS)	Li et al. (2015)
G. wiiroense	2015	Ghana	MIN 938704	MIN	KT952363 (ITS) KT952364 (LSU)	Crous et al. (2015)
G. aridicola	2016	South Africa	Dai12588	BJFC (IFP*)	KU572491 (ITS) KU572502 (TEF1-α)	Xing et al. (2016)
G. ecuadoriense	2016	Ecuador	QCAM3430	QCAM	KU128524 (ITS) KX228350 (LSU)	Crous et al. (2016)
G. mbrekobenum	2016	Ghana	MIN 850481	MIN	KX000896 (ITS) KX000897 (LSU)	Crous et al. (2016)
G. wuzhishanensis	2016	China, Hainan	GACP14081689	GACP	KU994772 (ITS)	Li et al. (2016)
G. mizoramense	2017	India, Mizoram	MIN 948145	MIN	KY643750 (ITS) KY747490 (LSU)	Crous et al. (2017a)

G. podocarpense	2017	Ecuador	QCAM6422	QCAM	MF796661 (ITS) MF796660 (LSU)	Crous et al. (2017b)
G. angustisporum	2018	China, Fujian	Cui 13817	BJFC	MG279170 (ITS) MG367563 (*TEF1-α*) MG367507 (*RPB2*)	Xing et al. (2018)
G. casuarinicola	2018	China, Guangdong	Dai 16336	BJFC	MG279173 (ITS) MG367565 (*TEF1-α*) MG367508 (*RPB2*)	Xing et al. (2018)
G. ellipsoideum	2018	China, Hainan	GACP14080966	GACP	MH106867 (ITS)	Hapuarachchi et al. (2018)
G. dunense	2018	South Africa	CMW42157	PREM	MG020255 (ITS), MG020150 (β-tubulin), MG020227 (TEF1-α)	Tchotet Tchoumi et al. (2018)

G. applanatum, *G. boninense*, *G. lingzi*, *G. lucidum* and *G. sinense*) and morpho-groups in the applied research.

The Ganoderma lucidum Aggregate

In the second half of the 18th century, the species scientifically named *G. lucidum* (Curtis) P. Karst. (Fig. 2d,e) was described many times under several names. Among the 14 different binomials given to this species by pre-Friesian European mycologists, the epithet *lucidus* proposed by Curtis (1781) and accepted by Fries (1821) proved to be the most popular name, thereafter spreading worldwide. Currently *G. lucidum* seems to be the most often incorrectly used name within the genus and taxonomically represents a difficult complex with uncertain species boundaries (Papp et al., 2017). The type of *G. lucidum* was described from Europe, where currently three morphologically similar species have been accepted (Ryvarden and Melo, 2014). Among these, *G. carnosum* Pat., was described from France (Patouillard, 1889) and considered to be identical to *G. atkinsonii* H. Jahn, Kotl. & Pouzar which is typified on a specimen collected in the Czech Republic (Jahn et al., 1980, 1986). In the subsequent literature, *G. carnosum* had often been misidentified as *G. valesiacum* Boud., which was described by Boudier (1895) from a collection found on *Larix decidua* in Switzerland. The microscopic structures of *G. valesiacum* were similar to those of *G. carnosum* and *G. lucidum* and only macroscopical features separate the species in this complex (Wasser et al., 2006a). Preliminary molecular genetic studies indicated that *G. carnosum* and *G. valesiacum* are not separated from *G. lucidum* on species level, however without type studies and detailed phylogenetic analysis, the species delimitation of the European *G. lucidum* complex is not fully understood.

In the early 20th century, four *Ganoderma* species were described from various coniferous trees in North America by Murrill (1902, 1908), which are considered as *G. tsugae* complex. The type specimen of *Ganoderma tsugae* Murrill was found on *Tsuga canadensis* and in the description it is characterized as a species that is closely related to *G. lucidum* (as *G. pseudoboletus*) (Murrill, 1902). The close relationship beetwen *G. tsugae* and the European *G. lucidum* aggregate was confirmed by morphological examinations (Steyaert, 1977; Wasser et al., 2006a), cultural studies (Stalpers, 1978; Adaskaveg and Gilbertson, 1986) and phylogenetic analysis based on ribosomal DNA markers (Moncalvo et al., 1995a,b; Hong and Jung, 2004). The most recent molecular study based on multilocus phylogenetic analysis confirmed that *G. tsugae* is an independent species, and grouped together with *G. oregonense* Murril and *G. lucidum* s. str. (Zhou et al., 2015). Murrill (1908) differentiated *G. oregonense* from *G. tsugae* mainly by its host species preference (found on *Picea sitchensis*) and geographic distribution. The two species morphologically are rather similar, but the large size of the basidiocarp, the generally duplex context, the larger spores and the long-shafted pileocystidia without apical projections seem to be the main morphological features that distinguish *G. oregonense* from *G. tsugae* (Adaskaveg and Gilbertson, 1988; Torres-Torres et al., 2015). The other two conifer-inhabiting species (*G. nevadense* Murrill and *G. sequoiae* Murrill) described by Murrill (1908) are characterized by sessile basidiocarps and separated based on host

Fig. 1. Basidiocarps of *Ganoderma* species in natural habitat: a–b, *G. cupreolaccatum*: c, *G. applanatum*; d–e, *G. lucidum* (Photocredit: V. Papp).

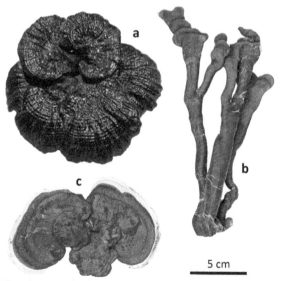

Fig. 2. Cultivated *Ganoderma* species in East Asia: a, basidiocarp of *G. sinense*; b–c, basidiocarps of *G. lingzhi* (Photo credit: V. Papp).

preference and the rimose or unbroken crust of the pileus. However, the taxonomic concept described by Murrill, which was based on geographic distribution and host specificity, was not supported in further studies, and based on morphological characteristics *G. nevadense* and *G. sequoiae* were considered to be synonyms for *G. oregonense* (Steyaert, 1980; Ryvarden, 1985).

Besides North America, *G. tsugae* was also reported from Asia (e.g., Zhao, 1989), and according to morphological observations, a new variety of this species (var. *jannieae* S. Wasser et al.) was isolated based on a collection found in Northeast China (Wasser et al., 2006b). From temperate Asia three further *Ganoderma* species have been described, which are closely related to *G. lucidum*. Based on a specimen originating from Pakistan, Steyaert (1972) described *G. ahmadii* Steyaert, which was also reported from India and China (Steyaert, 1972; Zhao, 1989). The ITS and LSU-D2 regions of type specimen were sequenced by Moncalvo et al. (1995a) and in his phylogeny *G. ahmadii* is nested together with the European and North American coniferously related species, although it was collected on a legume hardwood species *Dalbergia sissoo*. The other species, *G. mongolicum* Pilát described from Hebei Province in China (Zhao, 1989), is synonymised with *G. tsugae* by Steyaert (1980) based on uniform basidiospore morphology. By contrast, other mycologists accepted *G. mongolicum* as a separate species (Zhao, 1989; Wu and Dai, 2005); however, *G. mongolicum* is only known from the type locality (Moncalvo and Ryvarden, 1997) and no DNA sequence data has been published to date for this species. The morphological features and phylogenetic position based on ITS analysis of the recently described *G. leucocontextum* T.H. Li et al., placed this species in the *G. lucidum* complex. This new taxa known from the Tibet Autonomous Region and Sichuan Province of China is characterised by its white context and slightly smaller spores (Li et al., 2015).

According to the literature, *G. lucidum* has been reported to have a worldwide distribution, nevertheless, phylogenetic studies have shown that the examined strains labeled as "*G. lucidum*" were polyphyletic according to geographic origin. Moncalvo et al. (1995) explained that the *G. lucidum* aggregate (incl. *G. tsugae* complex) might be too young to have spread worldwide, so this taxa is restricted to the temperate region. The results of recent phylogenetic studies also indicate that *G. lucidum* s. str. (excluding the other members of the *G. lucidum* complex) is a Eurasian species, with widespread distribution in Europe toward temperate Eurasia (Europe, northwestern and northeastern China) to Sichuan and Yunnan Provinces (southwestern China) (e.g., Yang and Feng, 2013; Zhou et al., 2015; Papp et al., 2017). According to Zhou et al. (2015), the representatives of the *G. lucidum* aggregate in North America are *G. tsugae* and *G. oregonense*. Besides of these two species, surprisingly, the occurrences of *G. lucidum* sensu stricto was also confirmed by Loyd et al. (2018) in the United States. However, they noted that *G. lucidum* s. str. was found only in disturbed habitats in geographically restricted areas, which suggests that this species is not native in North America and the collections possibly derived from mushroom growers producing *G. lucidum* outdoors. Further studies should be carried out in order to prove the European and Asian distribution of these species, since the taxonomy of the group is unclear.

Formerly, Adaskaveg and Gilbertson (1986) have found that the North American taxon identified as "*G. lucidum*" are interfertile with *G. resinaceum* Boud. Basidiocarps of *G. resinaceum* is somewhat similar to *G. lucidum* s. str., but it differs in terms of the color of the context and the characteristics of the basidiospores (Steyaert, 1980). Molecular phylogentic studies showed that the European specimens of *G. resinaceum* grouped in a well separated lineage from the *G. lucidum* aggregate (incl. *G. ahmadii*, *G. carnosum*, *G. leucocontextum*, *G. lucidum* s. str., *G. oregonense*, *G. tsugae* and *G. valesiacum*) (e.g., Zhou et al., 2015; Papp et al., 2017). However, the literature discusses at least 13 *Ganoderma* species described from North America (California, Florida, New York, Ohio), Middle America (Honduras, Mexico), the Carribean region (Cuba, Jamaica, Grenada), South America (Argentina, Venezuella) and Australia, which are considered to be the synonyms of *G. resinaceum* based on previous morphological studies (e.g., Steyaert, 1972; Bazzalo and Wright, 1982; Ryvarden, 1985; Buchanan and Ryvarden, 1993). The taxonomical status of this group was discussed by Steyaert (1980), who reported specimens under *G. resinaceum* from several countries in Europe, temperate and tropical Asia, Northern and Southern America (incl. South America and the Caribian region) and Africa. Amongst the synonymised *Ganoderma* species, *G. sessile* Murrill specimens from the type locality and other USA states were studied by phylogenetic perspective and were found to be different from the European *G. resinaceum* (Zhou et al., 2015). The majority of the other species are only known from their type locality, and further investigations are needed to clarify the taxonomical state of the *G. resinaceum* complex and the distribution of *G. resinaceum* s. str.

The Prized East Asian Medicinal Mushroom

The correct taxonomical status of the highly prized medicinal *Ganoderma* species distributed in East Asia and that it has been described in traditional Asian medicine under several popular names (such as "Ling-zhi" in China, "Mannentake" or "Reishi" in Japan, and "Yeongji" in Korea) are intensively studied in the molecular era. Based on similarities in morphology, in therapeutic practice and in the literature, the scientific binomial "*G. lucidum*" has widely been accepted for the commercially cultivated East Asian medicinal mushroom. However, early molecular studies in *Ganoderma*, based on analyses of ITS1, ITS2, and the D2 domain of LSU (Moncalvo et al., 1995a), and mtSSU rDNA sequences (Hong and Jung, 2004) have indicated that the European collections of *G. lucidum* are clearly different from the East Asian ones. However, in these works the Asian "*G. lucidum*" samples were not identified on species level, and without a proposed scientific name, the "Ling-zhi" medicinal mushroom has continuously been referred to as *G. lucidum* in the scientific literature. Based on comprehensive morphological observations and multilocus phylogenetic analyses, Wang et al. (2012) concluded that the species representing "Ling-zhi" is identical to *G. sichuanense* J.D. Zhao & X.Q. Zhang, originally described from the Sichuan Province in Southwest China. By contrast, in a similar study published at almost the same time, Cao et al. (2012) found that the ITS sequence from the holotype of *G. sichuanense* did not belong to the lingzhi clade and grouped together with specimens identified as *G. weberianum* (Bres. & Henn. ex Sacc.) Steyaert. The authors claimed that lingzhi was hitherto an undescribed species, therefore the binomial *G. lingzhi* Sheng H. Wu et al. (Fig. 2b,c) was proposed, reflecting the traditional Chinese name. Yao et al. (2013) queried the published ITS sequence of the *G. sichuanense* holotype and supposed that DNA from the holotype and paratype of *G. sichuanense* was degraded irreparably and, thus, impossible to sequence. In favor of fixing the application of the name *G. sichuanense*, they designated an epitype and its ITS sequence; however, this is identical to the holotype of *G. lingzhi*. By contrast, Paterson and Lima (2015) emphasized that Yao et al. did not provide supporting data, that the DNAs of the type and authentic materials of *G. sichuanense* were damaged, thereby undermining their interpretations. Zhou et al. (2015) rejected the epitypification of *G. sichuanense*, and suggested that both species are independent and taxonomically valid. This statement was also confirmed by Dai et al. (2017) in a recent study, in which the taxonomy and nomenclature of the "Lingzhi" mushroom has been thoroughly discussed. They noted, that the first possible name applied to "Lingzhi" in East Asia was *Boletus dimidiatus* Thunb., which was described by Thunberg (1784) based on specimens collected from Japan. *Ganoderma dimidiatum* (Thunb.) V. Papp is definitely the first validly published laccate *Ganoderma* species from East Asia and the valid name of *G. japonicum* (Fr.) Sawada as pointed out by Papp (2016). However, the correct taxonomic status of this species remains unclear, due to the fact that the lectotype of *G. dimidiatum* was not available for loan. Therefore, Dai et al. (2017) concluded that the rejection of the name *Boletus dimidiatus* is better in order to stabilize the scientific binomial for "Lingzhi".

The type specimen of *G. lingzhi* was designated from Hubei Province located in central China, and it was also reported from other Chinese Provinces, viz. Anhui, Henan, Hunan, Jiangsu, Liaoning, Shandong, Sichuan, Tianjin, Yunan and Zhejiang (Cao et al., 2012; Wang et al., 2012). Outside China, *G. lingzhi* was reported from Bangladesh, Japan and Korea (Cao et al., 2012; Wang et al., 2012; Kwon et al., 2016) and recently it is published under the name *G. sichuanense* from Thailand (Thawthong et al., 2017). Moreover, the ITS sequences of 4 specimens collected from legume trees (*Cassia* spp., *Delonixregia*) in southern India differ only in 2 nucleotide positions from the holotype of *G. lingzhi*. These are probably the westernmost known localities of *G. lingzhi* so far (Papp et al., 2017). The natural range of *G. lingzhi* seems to be restricted to East and Southeast Asia (i.e., central to eastern China, Japan, Korea and Thailand), and it also occurs in Bangladesh and South India.

The Phaeonema Group: Medicinal Mushrooms, and Plant Pathogens

The section *Phaeonema* Zhao et al. was typified on *Ganoderma sinense* J.D. Zhao et al. (Fig. 2a) and morphologically characterised by laccate pilear surface, and uniformly brown to deep red context (Zhao et al., 1979). *Ganoderma sinense* was recorded as one of the "Ling-zhi" sources in Chinese pharmacopoeia (besides *G. lingzhi*), and long has been used in traditional Chinese medicine (TCM). This species was described from Hainan (South China) and characterised by annual and stipitate basidiocarp with purplish-black to black laccate pileus and mostly uniform brown context. *Ganoderma sinense* is frequently reffered in the scientific literature to *G. japonicum* Fr. The nomenclatural status of this familiar name was discussed by Papp (2016), who concluded that *G. japonicum* is a superflous name of the formerly described species from Japan, named *G. dimidiatum* (Thunb.) V. Papp. Chang and Chen (1984) described *G. formosanum* T.T. Chang & T. Chen as a new species from Taiwan based on a specimen growing on formosan sweetgum (*Liquidambar formosana*). This species morphologically is similar to *G. sinense*, but it is formerly accepted by certain authors as a separate species based on its duplex context and ovoid basidiospores with persistent apex (Zhao and Zhang, 2000; Wu and Dai, 2005). The type specimens of both species was studied by Wang et al. (2005), who concluded that the discrimination between these two species resulted from variable characters and incomplete description, thus, they stated that *G. formosanum* is a later synonym of *G. sinense*. This proposal confirmed by preliminary molecular results, but it should be noted that only one material was sequenced, which was labeled "*G. fornicatum*" by Moncalvo et al. (1995b). Without type sequencing, the taxonomical position of the popular medicinal mushroom which was generally identified as *G. sinense* remains unclear. To clarify the taxonomy and nomenclature of numerous East-Asian stipitate *Ganoderma* species, particularly of those which have blackish and laccate pileus (e.g., *G. atrum* J.D. Zhao et al., *G. austrofujianense* J.D. Zhao et al., *G. formosanum* T.T. Chang & T. Chen, *G. neojaponicum* Imazeki and *G. sinense*), further morphological and more serious molecular studies and the inclusion of more specimens and type materials is required.

The morphological group, *Phaeonema* sensu Zhao, comprises several important *Ganoderma* species besides *G. sinense*, such as *G. boninense* Pat., *G. capense* (Lloyd) Teng, *G. cupreolaccatum* (Kalchbr.) Z. Igmándy, *G. tropicum* (Jungh.) Bres. or *G. zonatum* Murrill. Among these, the species named as "*G. boninense*" has an enormous economic value, because it is the causal agent of basal stem rot, one of the most devastating diseases of the oil palm (Mercière et al., 2015). *Ganoderma boninense* was described by Patouillard (1889), based on a collection from Bonin Island. The correct taxonomical state of this species is rather complicated; as a result, several similar, but taxonomically poorly known species were described by taxonomists. Moncalvo and Ryvarden (1997) discussed *G. boninense* in the "*G. chalceum–boninense* complex", in which 27 different *Ganoderma* species were listed. The synonymy of *G. boninense* with the formerly described *G. chalceum* (Cooke) Steyaert was suggested by Corner (1983) in agreement with Ryvarden (1983). Later, Ryvarden (2000) synonymized *G. boninense* with *G. orbiforme* (Fr.) Ryvarden, a species described from Guinea (West Africa). The taxonomical state of *G. orbiforme* and allies was recently discussed by Wang et al. (2014) based on morphological observations and phylogenetic evidence. They concluded, that *G. boninense* is presumably not identical with *G. orbiforme*, although five other *Ganoderma* species, namely *G. cupreum* (Cooke) Bres., *G. densizonatum* J.D. Zhao & X.Q. Zhang, *G. limushanense* J.D. Zhao & X.Q. Zhang, *G. mastoporum* (Lév.) Pat. and *G. subtornatum* Murrill are synonyms of the latter. They emphasized that the recently described *G. ryvardenii* Tonjock & Mih (as *G. ryvardense*), which is a pathogenic species of oil palm in Cameroon (Central Africa) shows a close relationship to *G. boninense* and *G. orbiforme*, thus the relationships among these species require further clarification.

The peculiar *Ganoderma* species with perennial basidiocarp covered by laccate surface and usually named as *G. pfeifferi* Bres. is most likely identical to the formerly described *G. cupreolaccatum* (Fig. 1a,b), which is based on a specimen collected from Austria (Central Europe) (Steyaert, 1980; Papp and Szabó, 2012). This perspective medicinal species (Lindequist et al., 2015) is mostly reported from natural beech forests and its distribution is presumably restricted to Europe (Ryvarden and Melo, 2014). However, Corner (1983) described *G. pfeifferi* var. *bornense* from Mt Kinabalu (Borneo), what should be confirmed, or taxonomically revised. A specimen found on *Quercus* sp. in Belgium and described as *G. soniense* Steyaert is later considered to be identical to *G. cupreolaccatum* (Steyaert, 1967). Another species described from Italy under the name *G. puglisii* Steyaert has similar cutis to *G. cupreolaccatum*, but according to Steyaert (1972) the basidiospores are larger. *Ganoderma puglisii* is known only from the type locality and the taxonomical state of this species is still unclear. Phylogenetically, the sequenced specimens of *G. cupreolaccatum* (as *G. pfeifferi*) are not grouped with other *Phaeonema* (e.g., *G. sinense* and *G. boninense*) and are closely related to *G. mutabile* Y. Cao & H.S. Yuan and other species belonging to the *Elfvingia* group (Papp et al., 2017).

The Elfvingia Group: Ganoderma applanatum and Related Species

The morphological group traditionally labeled as *Elfvingia* is mostly comprised of the species with sessile or substipitate perennial basidiocarps with dull pileal surface. Moncalvo and Ryvarden (1997) listed 51 species in the *Elfvingia* group, of which they considered 21 to be synonyms. The taxonomical uncertainty is increased by the fact that the majority of the listed species are represented by one or few collections, all restricted to the type locality and adjacent regions: For example, *G. chilense* (Fr.) Pat. (from Chile), *G. dubiocochlear* (Lloyd) Sacc. & Trotter (from Madagascar), *G. wuhuense* X.F. Ren (from Anhui province, Eastern China), *G. luteicinctum* Corner (from Singapore), *G. dejongii* Steyaert and *G. hoehnelianum* Bres. (from Java, Indonesia). Within the *Elfvingia* group, Steyaert (1980) separated three sub-genera based on the microscopic structure of the pileipellis: (i) *Ganoderma* subgen. *Elfvingia* (*G. applanatum*), (ii) subgen. *Anamixoderma* (*G. adspersum*), and (iii) subgen. *Plecoderma* (*G. philippii*). Wasser et al. (2006a) proposed to discuss *Anamixoderma* and *Plecoderma* as sections benath the *Elfvingia* subgenus. They noted, that besides the differences in the pileipellis structure all of these species are rather uniform in their basic fomitoid structure and basidiospores with smooth type of the surface. Because the phylogenetic state of the elfvingioid *Ganoderma* species is not yet sufficiently clear, in the majority of cases it is discussed under the name "*Elfvingia* group" or "*Ganoderma applanatum-australe* species complex" in the literature (Moncalvo and Buchanan, 2008).

The central species of the *Elfvingia* group is *G. applanatum* (Pers.) Pat. (Fig. 1c), which is considered to be identical to *G. lipsiense* (Batsch) G.F. Atk. Although the latter name was previously described, eventually the Nomenclature Committee for

Fungi sanctioned the basionym *Boletus applanatus* Pers. against *B. lipsiensis* Batsch (Redhead et al., 2006; Demoulin, 2010; Norvell, 2011). The neotype of *B. applanatus* s selected by Redhead et al. (2006) was examined by Niemelä and Miettinen (2008). They proved that it conforms morphologically with the current European concept of *G. applanatum* and is distinct from the superficially similar species, *G. adspersum* (Schulzer) Donk. It is also suggested by the phylogenetic studies that only two non-laccate, and perennial *Ganoderma* species are known from Europe, therefore the other species with similar fruiting body described from the continent might be the synonym of *G. adspersum* or *G. applanatum*: *G. europaeum* Steyaert, *G. gelsicola* (Berl.) Sacc., *G. kosteri* Steyaert, *G. linhartii* (Kalchbr.) Z. Igmándy, *G. lipsiense* and *G. vegetum* (Fr.) Bres. Species considered to be synonyms of *G. applanatum* even originated from other continents than Europe. The type specimen of *G. incrassatum* (Berk.) Bres., a species originally described from Australia was studied by Ryvarden (1984), who concluded that it is identical to *G. applanatum*. Three other species described from the United States, namely *G. brownii* (Murrill) Gilb., *G. leucophaeum* (Mont.) Pat. and *G. megaloma* (Lév.) Bres. are also considered as a synonyms of *G. applanatum* in the literature (Gottlieb and Wright, 1999; Lowe and Gilbertson, 1961; Ryvarden, 1982). However, amongst these, the correct taxonomical state of *G. brownii* is the most controversial (see Moncalvo and Ryvarden, 1997); and certain authors accepted it as a distinct species apparently restricted to California (e.g., Gilbertson and Ryvarden, 1986; Zhou et al., 2016). In the literature, *G. applanatum* is reported as a cosmopolitan species, although Seo and Kirk (2000) emphasised that "*G. applanatum*" (besides *G. lucidum*) is probably the most frequently misapplied name in the genus. Therefore, to circumscribe the real distribution of *G. applanatum* s. str., the taxonomic revision of morphologically similar species and specimens identified as *G. applanatum* from different geographical regions (especially from the tropics) would be necessary.

Future Challenges

The following statement of Moncalvo (2005) should continue to be emphasized: "incorrect taxonomic identification of *Ganoderma* strains hampers comprehensive strategies for drug discovery as well as for monitoring and managing diseases caused by *Ganoderma* in woody crops and forest ecosystems". However, a lot of new taxonomical information on *Ganoderma* was published recently, due to the rapid spread of molecular genetic methods; the extremely diverse nomenclature, caused by the difficulties in the identification and by the different taxonomical concepts, makes it very hard to understand the new results of the applied mycological research carried out on *Ganoderma* species. In the upcoming scientific papers, it would be important to report barcoding sequences (i.e., ITS or *Tef1-α*) from the examined *Ganoderma* samples as many times as it is possible, because it could help greatly to understand the published results correctly. Besides describing new species from those *Ganoderma* samples which can not be identified by unambiguous morphological characteristics nor by molecular genetic analyses, greater efforts should be made to analyse the types of the previously described species. In order to clarify the nomenclatural and taxonomical problems, first the valid *Ganoderma* names should be typified, and these representative collections should be sequenced and studied using molecular genetic methods.

Acknowledgements

The work of the author was supported by the ÚNKP-17-4 New National Excellence Program of the Ministry of Human Capacities.

References

Adaskaveg, J.E. and Gilbertson, R.L. 1986. Cultural studies and genetics of sexuality of *Ganoderma lucidum* and *G. tsugae* in relation to the taxonomy of the *G. lucidum* complex. Mycologia, 78: 694–705.

Adaskaveg, J.E. and Gilbertson, R.L. 1988. Basidiospores, pilocystidia and other basidiocarp characters in several species of the *Ganoderma lucidum* complex. Mycologia, 80: 493–507.

Adaskaveg, J.E. and Gilbertson, R.L. 1989. Cultural studies of four North American species in the *Ganoderma lucidum* complex with comparisons to *G. lucidum* and *G. tsugae*. Mycol. Res., 92: 182–191.

Albee-Scott, S. 2007. The phylogenetic placement of the Leucogastrales, including *Mycolevis siccigleba* (Cribbeaceae), in the Albatrellaceae using morphological and molecular data. Mycol. Res., 111: 653–662.

Atkinson, G.F. 1908. On the identity of *Polyporus "applanatus"* of Europe and North America. Annals Mycol., 6: 179–191.

Audet, S.A. 2010. Essai de découpage systématique du genre *Scutiger* (Basidiomycota): *Albatrellopsis*, *Albatrellus*, *Polyporoletus*, *Scutiger* et description de six nouveaux genres. Mycotaxon, 111: 431–464.

Baby, S., Johnson, A.J. and Govindan, B. 2015. Secondary metabolites from *Ganoderma*. Phytochemistry, 114: 66–101.

Bazzalo, M.E. and Wright, J.E. 1982. Survey of the Argentine species of the *Ganoderma lucidum* complex. Mycotaxon, 16: 293–325.

Binder, M., Justo, A., Riley, R., Salamov, A., Lopez-Giraldez, F., Sjökvist, E., Copeland, A., Foster, B., Sun, H., Larsson, E., Larsson, K.H., Townsend, J., Grigoriev, I.V. and Hibbett, D.S. 2013. Phylogenetic and phylogenomic overview of the Polyporales. Mycologia, 105: 1350–1373.

Boudier, E. 1895. Description de quelques espèces récoltées en août 1894 dans les régions élevées des Alpes du Valais. Bull Trimest Soc. Mycol Fr., 11: 27–30.

Buchanan, P.K. and Ryvarden, L. 1993. Type studies in the Polyporaceae. 24. Species described by Cleland, Rodway and Cheel. Aust. Syst. Bot., 6: 215–235.

Cao, Y., Wu, S.H. and Dai, Y.C. 2012. Species clarification of the prized medicinal *Ganoderma* mushroom "Lingzhi." Fungal Divers, 56: 49–62.

Cao, Y. and Yuan, H.S. 2013. *Ganoderma mutabile* sp. nov. from southwestern China based on morphological and molecular data. Mycol. Prog., 12: 121–126.

Chang, T. and Chen, T. 1984. *Ganoderma formosanum* sp. nov. on Formosan sweet gum in Taiwan. Trans. Br. Mycol. Soc., 82: 731–733.

Coetzee, M.P.A., Marincowitz, S., Muthelo, V.G. and Wingfield, M.J. 2015. *Ganoderma* species, including new taxa associated with root rot of the iconic *Jacaranda mimosifolia* in Pretoria, South Africa. IMA Fungus, 6: 249–256.

Corner, E.J.H. 1983. Ad Polyporaceas I. *Amauroderma* and *Ganoderma*. Beiheft. Nova Hedw., 75: 1–182.

Costa-Rezende, D.H., Robledo, G.L., Góes-Neto, A., Reck, M.A., Crespo, E. and Drechsler-Santos, E.R. 2017. Morphological reassessment and molecular phylogenetic analyses of *Amauroderma* s. lat. raised new perspectives in the generic classification of the Ganodermataceae family. Persoonia, 39: 254–269.

Crous, P.W., Wingfield, M.J. and Schumacher, R.K. et al. 2014. Fungal Planet description sheets: 281–319. Persoonia, 33: 212–289.

Crous, P.W., Wingfield, M.J., Le Roux, J.J. et al. 2015. Fungal Planet description sheets: 371–399. Persoonia, 35: 264–327.

Crous, P.W., Wingfield, M.J., Richardson, D.M. et al. 2016. Fungal Planet description sheets: 400–468. Persoonia, 36: 316–458.

Crous, P.W., Wingfield, M.J., Burgess, T.I. et al. 2017a. Fungal Planet description sheets: 558–624. Persoonia, 38: 240–384.

Crous, P.W., Wingfield, M.J., Burgess, T.I. et al. 2017b. Fungal Planet description sheets: 625–715. Persoonia, 39: 270–467.

Curtis, W. 1781. Flora Londinensis: or plates and descriptionsof such plants as grow wild in the environs of London. London: n.p.

Dai, Y.C., Zhou, L.W., Hattori, T., Cao, Y., Stalpers, J.A., Ryvarden, L., Buchanan, P., Oberwinkler, F., Hallenberg, N., Liu, P.G. and Wu, S.H. 2017. *Ganoderma lingzhi* (Polyporales, Basidiomycota): the scientific binomial for the widely cultivated medicinal fungus Lingzhi. Mycol Progress, 16: 1051–1055.

Decock, C. and Figueroa, S.H. 2007. Studies in Ganodermataceae (Basidiomycota): The concept of *Ganoderma coffeatum* in the Neotropics and East Asia. Cryptogam., Mycol., 28: 77–89.

Demoulin, V. 2010. Why conservation of the name *Boletus applanatus* should be rejected. Taxon, 59: 283–286.

Donk, M.A. 1933. Revision der Niederlandischen Homobasidiomycetes. Aphyllophoraceae 2. Mededelingen van het botanisch Museum en Herbarium van de Rijksuniversiteit Utrecht, 9: 1–278.

Donk, M.A. 1948. Notes on Malesian fungi. I. Bulletin du Jardin Botanique de Buitenzorg, 17: 473–482.

Douanla-Meli, C. and Langer, E. 2009. *Ganoderma carocalcareus* sp. nov., with crumbly-friable context parasite to saprobe on *Anthocleista nobilis* and its phylogenetic relationship in *G. resinaceum* group. Mycol Prog., 8: 145–155.

Fleischmann, A., Krings, M., Mayr, H. and Agerer, R. 2007. Structurally preserved polypores from the Neogene of North Africa: *Ganodermites lybicus* gen. et sp. nov. (Polyporales, Ganodermataceae). Rev. Palaeobot. Palynol., 145: 159–172.

Fries, E.M. 1821. Systema mycologicum. Vol. 1. Lundæ: Ex Officina Berlingiana.

Furtado, J.S. 1965. *Ganoderma colossum* and the status of *Tomophagus*. Mycologia, 57: 979–984.

Gilbertson, R.L. and Ryvarden, L. 1986. North American Polypores.Vol. 1. *Abortiporus–Lindtneria*. Fungiflora, Oslo. pp. 1–433.

Gomes-Silva, A.C., Lima-Júnior, N., Malosso, E., Ryvarden, L. and Gibertoni, T. 2015. Delimitation of taxa in Amauroderma (Ganodermataceae, Polyporales) based in morphology and molecular phylogeny of Brazilian specimens. Phytotaxa, 227: 201–228.

Gottlieb, A. and Wright, J. 1999. Taxonomy of *Ganoderma* from southern South America: sub-genus *Elfvingia*. Mycol Res., 103: 1289–1298.

Hapuarachchi, K.K., Cheng, C.R., Wen, T.C., Jeewon, R. and Kakumyan, P. 2017. Mycosphere Essays 20: Therapeutic potential of *Ganoderma* species: Insights into its use as traditional medicine. Mycosphere, 8: 1653–1694.

Hapuarachchi, K.K., Karunarathna, S.C., Raspé, O., De Silva, K.H.W.L., Thawthong, A., Wu, X.L., Kakumyan, P., Hyde, K.D. and Wen, T.C. 2018. High diversity of *Ganoderma* and *Amauroderma* (Ganodermataceae, Polyporales) in Hainan Island, China. Mycosphere 9(5): 931–982.

Hattori, T. and Ryvarden, L. 1994. Type studies in the Polyporaceae. 25. Species described from Japan by R. Imazeki & A. Yasuda. Mycotaxon, 50: 27–46.

Hong, S.G. and Jung, H.S. 2004. Phylogenetic analysis of *Ganoderma* based on nearly complete mitochondrial small-subunit ribosomal DNA sequences. Mycologia, 96: 742–755.

Imazeki, R. 1939. Studies on *Ganoderma* of Nippon. Bull. Nat. Sci. Mus. Tokyo, 1: 29–52.

Imazeki, R. 1952. A contribution to the fungus flora of Dutch New Guinea. Bull. Govt. Forest. Exp. St. Tokyo, 57: 87–128.

Ipulet, P. and Ryvarden, L. 2005. New and interesting polypores from Uganda. Synop Fung., 20: 87–99.

Jahn, H., Kotlaba, F. and Pouzar, Z. 1980. *Ganoderma atkinsonii* Jahn, Kotl. & Pouz., spec. nova, a parellel species to *Ganoderma lucidum*. Westf Pilzbriefe, 10–11: 97–121.

Jahn, H., Kotlaba, F. and Pouzar, Z. 1986. Notes on *Ganoderma carnosum* Pat. (*G. atkinsonii* Jahn, Kotl. & Pouz.). Westf Pilzbriefe, 10–11: 378–382.

Jargalmaa, S., Eimes, J.A., Park, M.S., Park, J.Y., Oh, S.Y. and Lim, Y.W. 2017. Taxonomic evaluation of selected *Ganoderma* species and database sequence validation. PeerJ., 5: e3596.

Jülich, W. 1981. Higher taxa of basidiomycetes. Bibliotheca Mycologica 85, J. Cramer, Vaduz, 485 pp.

Justo, A., Miettinen, O., Floudas, D., Ortiz-Santana, B., Sjökvist, E., Lindner, D., Nakasone, K., Niemelä, T., Larsson, K.H., Ryvarden, L. and Hibbett, D.S. 2017. A revised family-level classification of the Polyporales (Basidiomycota). Fungal Biol., 121: 798–824.

Karsten, P.A. 1881. Enumeratio Boletinearum et Polyporearum Fennicarum, systemate novo dispositarum. Rev Mycol., Toulouse. 3: 16–19.

Karsten, P.A. 1889. Kritisk öfversigt af Finlands Basidsvampar (Basidiomycetes; Gastero- & Hymenomycetes). Bidr till Kännedom av Finlands Natur och Folk, 48: 1–470.

Kinge, T.R. and Mih, A.M. 2011. *Ganoderma ryvardense* sp. nov. associated with basal stem rot (BSR) disease of oil palm in Cameroon. Mycosphere, 2: 179–188.

Koeniguer, J.C. and Locquin, M.V. 1979. Un polypore fossile à spores porées du Miocène de Libye: *Archeterobasidium syrtae*, gen. et sp. nov. pp. 323–329. *In*: Ministère des Universités Comité des (Travaux Historiques et Scientifiques (eds.). Comptes Rendus du 104e Congrès National des Sociétés Savantes, Bordeaux, fasc. I (Paléobotanique). Bibliothèque Nationale: Paris (France).

Kwon, O.C., Park, Y.J., Kim, H.I., Kong, W.S., Cho, J.H. and Lee, C.S. 2016. Taxonomic position and species identity of the cultivated Yeongji "*Ganoderma lucidum*" in Korea. Mycobiology, 44: 1–6.

Laferrière, J.E. and Gilbertson, R.L. 1990. A new species of *Albatrellus* (Aphyllophorales: Albatrellaceae) from Mexico. Mycotaxon, 37: 183–186.

Le, X.T., Nguyen, L.Q.H., Pham, N.D., Duong, V.H., Dentinger, B.T.M. and Moncalvo, J.M. 2011. *Tomophagus cattienensis* sp. nov., a new Ganodermataceae species from Vietnam: Evidence from morphology and ITS DNA barcodes. Mycol Progress, 11: 775–780.

Li, T.H., Hu, H.P., Deng, W.Q., Wu, S.H., Wang, D.M. and Tsering, T. 2015. *Ganoderma leucocontextum*, a new member of the *G. lucidum* complex from southwestern China. Mycoscience, 56: 81–85.

Li, G.J., Hyde, K.D., Zhao, R.L. et al. Fungal diversity notes 253–366: Taxonomic and phylogenetic contributions tofungal taxa. Fungal Divers., 78: 1–237.

Lindequist, U., Jülich, W.D. and Witt, S. 2015. *Ganoderma pfeifferi*—A European relative of *Ganoderma lucidum*. Phytochemistry, 114: 102–108.

Lowe, J.L. and Gilbertson, R.L. 1961. Synopsis of the Polyporaceae of the Western United States and Canada. Mycologia, 53: 474–511.

Loyd, A.L., Barnes, C.W., Held, B.W., Schink, M.J., Smith, M.E., Smith, J.A. and Blanchette, R.A. 2018. Elucidating "lucidum": Distinguishing the diverse laccate *Ganoderma* species of the United States. PLoS ONE 13(7): e0199738.

Mercière, M., Laybats, A., Carasco-Lacombe, C., Tan, J.S., Klopp, C., Durand-Gasselin, T., Alwee, S.S.R.S., Camus-Kulandaivelu, L. and Breton, F. 2015. Identification and development of new polymorphic microsatellite markers using genome assembly for *Ganoderma boninense*, causal agent of oil palm basal stem rot disease. Mycol Progress, 14: 103.

Moncalvo, J.M. 2000. Systematics of ganoderma. pp. 23–46. *In*: Flood, J., Bridge, P.D. and Holderness, M. (eds.). *Ganoderma* Diseases of Perennial Crops. CABI Publishing. Wallingford, UK.

Moncalvo, J.M. 2005. Molecular systematics of *Ganoderma*: What is Reishi? Int. J. Med. Mushrooms, 7: 452.

Moncalvo, J.M., Wang, H.F. and Hseu, R.S. 1995a. Gene phylogeny of the *Ganoderma lucidum* complex based on ribosomal DNA sequences: Comparison with traditional taxonomic characters. Mycol. Res., 99: 1489–1499.

Moncalvo, J.M., Wang, H.H. and Hseu, R.S. 1995b. Phylogenetic relationships in *Ganoderma* inferred from the Internal Transcribed Spacers and 25S ribosomal DNA sequences. Mycologia, 87: 223–238.

Moncalvo, J.M. and Ryvarden, L. 1997. A nomenclatural study of the Ganodermataceae Donk. Synop Fung., 11: 1–114.

Moncalvo, J.M. and Buchanan, P.K. 2008. Molecular evidence for long distancedispersal across the Southern Hemisphere in the *Ganoderma applanatum-australe* species complex (Basidiomycota). Mycol. Res., 112: 425–436.

Murrill, W.A. 1902. The Polyporaceae of North America. I. Thegenus *Ganoderma*. Bull. Torrey Bot. Club., 29: 599–608.

Murrill, W.A. 1905a. The Polyporaceae of North America: XI. A synopsis of the brown pileate species. Bull. Torrey Bot. Club., 32: 353–371.

Murrill, W.A. 1905b. *Tomophagus* for *Dendrophagus*. Torreya, 5: 197.

Murrill, W.A. 1908. Polyporaceae, part 2. North Am Flora., 9: 73–131.

Niemelä, T. and Miettinen, O. 2008. The identity of *Ganoderma applanatum* (Basidiomycota). Taxon, 57: 963–966.

Norvell, L.L. 2011. Report of the Nomenclature Committee for Fungi: 18. Taxon, 60: 1199–1201.

Papp, V. and Szabó, I. 2013. Distribution and host preference of poroid basidiomycetes in Hungary I.— *Ganoderma*. Acta Silv. Lign. Hung., 9: 71–83.

Papp, V. 2016. The first validly published laccate *Ganoderma* species from East Asia: *G. dimidiatum* comb. nov., the correct name for *G. japonicum* (Ganodermataceae, Basidiomycota). Studia Bot. Hung., 47: 263–268.

Papp, V., Dima, B. and Wasser, S.P. 2017. What is *Ganoderma lucidum* in the molecular era? Int. J. Med. Mushrooms, 19: 575–593.

Paterson, R.R.M. 2006. *Ganoderma*—a therapeutic fungal biofactory. Phytochemistry, 67: 1985–2001.

Paterson, R.P.M. and Lima, N. 2015. Failed PCR of *Ganoderma*type specimens affects nomenclature. Phytochemistry, 114: 16–17.

Patouillard, N. 1889. Le genre *Ganoderma*. Bull. Soc. Mycol. Fr., 5: 64–80.

Redhead, S.A., Ginns, J. and Moncalvo, J.M. 2006. Proposal to conserve the name *Boletus applanatus* against *B. lipsiensis* (Basidiomycota). Taxon, 55: 1029–1030.

Richter, C., Wittstein, K., Kirk, M.P. and Stadler, M. 2015. An assessment of the taxonomy and chemotaxonomy of *Ganoderma*. Fungal Divers., 71: 1–15.

Ryvarden, L. 1982. Type studies in the Polyporaceae. 11. Species described by J.F.C. Montagne, either alone or with other authors. Nord. J. Bot., 2: 75–84.

Ryvarden, L. 1983. Type studies in the Polyporaceae. 14. Species described by N. Patouillard, either alone or with other mycologists. Occas. pap. Farlow Herb., 18: 1–39.

Ryvarden, L. 1984. Type studies in the Polyporaceae. 16. Species described by J.M. Berkeley, either alone or with other mycologists from 1856 to 1886. Mycotaxon, 20: 329–363.

Ryvarden, L. 1985. Type studies in the Polyporaceae. 17. Species described by W.A. Murrill. Mycotaxon, 23: 169–198.

Ryvarden, L. 1991. Genera of Polypores. Nomenclature and taxonomy. Synopsis Fungorum 5, Fungiflora, Oslo, Norway, 363 pp.

Ryvarden, L. 2000. Studies in neotropical polypores 2: a preliminary key to neotropical species of *Ganoderma* with a laccate pileus. Mycologia, 92: 180–191.

Ryvarden, L. 2004. Neotropical polypores, Part 1. Introduction, Ganodermataceae and Hymenochaetaceae. Synop Fung., 19: 1–228.

Ryvarden, L. and Johansen, I. 1980. A preliminary polypore flora of East Africa. Fungiflora, Oslo, Norway, 636 pp.

Ryvarden, L. and Melo, I. 2014. Poroid fungi of Europe. Synop Fung., 31: 1–455.

Seo, G.S. and Kirk, P.M. 2000. Ganodermataceae: nomenclature and classification. pp. 3–22. *In*: Flood, J., Bridge, P.D. and Holderness, M. (eds.). *Ganoderma* Diseases of Perennial Crops. CABI: Wallingford, UK.

Smith, B.J. and Sivasithamparam, K. 2000. Internal transcribed spacer ribosomal DNA sequence analysis of five species of *Ganoderma* from Australia. Mycol. Res., 104: 943–951.

Smith, B.J. and Sivasithamparam, K. 2003. Morphological studies of *Ganoderma* (Ganodermataceae) from the Australasian and Pacific regions. Aust. Syst. Bot., 16: 487–503.

Song, J., Xing, J.H., Decock, C., He, X.L. and Cui, B.K. 2016. Molecular phylogeny and morphology reveal a new species of *Amauroderma* (Basidiomycota) from China. Phytotaxa, 260: 47–56.

Stalpers, J.A. 1978. Identification of wood-inhabiting fungi inpure culture. Stud Mycol., 16: 1–248.

Steyaert, R.L. 1967. Considération générale sur le genre *Ganoderma* et plus spécialement sur les espèces Européenes. Bull. Soc. Roy. Bot. Belg., 100: 189–211.

Steyaert, R.L. 1972. Species of *Ganoderma* and related generamainly of the Bogor and Leiden Herbaria. Persoonia, 7: 55–118.

Steyaert, R.L. 1977. Basidiospores of two *Ganoderma* species andothers of two related genera under the scanning electron microscope. Kew Bull., 31: 437–42.

Steyaert, R.L. 1980. Study of some *Ganoderma* species. Bull. Jard Bot. Nat. Belg., 50: 135–86.

Sun, S.J., Gao, W., Lin, S.Q., Zhu, J., Xie, B.G. and Lin, Z.B. 2006. Analysis of genetic diversity in *Ganoderma* populations with a novel molecular marker SRAP. Appl. Microbiol. Biotechnol., 72: 537–543.

Taylor, T.N., Krings, M. and Taylor, E.L. 2015. Fossil fungi. London: Elsevier/Academic Press Inc, 398 pp.

Tchotet Tchoumi, J.M., Coetzee, M.P.A., Rajchenberg, M., Wingfield, M.J. and Roux, J. 2018. Three Ganoderma species, including *Ganoderma dunense* sp. nov., associated with dying Acacia cyclops trees in South Africa. Australasian Plant Pathology 47: 431–447.

Teixeira, A.R. 1992. New combinations and new names in the Polyporaceae. Rev. Bras Bot., 15: 125–127.

Thakur, R., Kapoor, P., Sharma, P.N. and Sharma, B.M. 2015. Assessment of genetic diversity in *Ganoderma lucidum* using RAPD and ISSR markers. Indian Phytopath., 68: 316–320.

Thawthong, A., Hapuarachchi, K.K., Wen, T.C., Raspé, O., Thongklang, N., Kang, J.C. and Hyde, K.D.2017. *Ganoderma sichuanense* (Ganodermataceae, Polyporales) new to Thailand. MycoKeys, 22: 27–43.

Torrend, C. 1920. Les Polyporacées du Brésil. I. Polyporacées stipités. Brotéria Série Botânica, 18:121–143.

Torres-Torres, M.G., Guzmán-Dávalos, L. and Gugliotta, A.M. 2008. *Ganoderma vivianimercedianum* sp. nov. and the related species, *G. perzonatum*. Mycotaxon, 105: 447–454.

Torres-Torres, M.G. and Guzmán-Dávalos, L. 2012. The morphology of *Ganoderma* species with a laccate surface. Mycotaxon, 119: 201–216.

Torres-Torres, M.G., Ryvarden, L. and Guzmán-Dávalos, L. 2015. *Ganoderma* sub-genus *Ganoderma* in Mexico. Rev Mex. Micol., 41: 27–45.

Udagawa, S., Awao, T. and Abdullah, S.K. 1986. *Thermophymatospora*, a new thermotolerant genus of basidiomycetous hyphomycetes. Mycotaxon, 27: 99–106.

Wang, D.M., Zhang, X.Q. and Yao, Y.J. 2005. Type studies of some *Ganoderma* species from China. Mycotaxon, 93: 61–70.

Wang, D.M., Wu, S.H. and Yao, Y.J. 2014. Clarification of the concept of *Ganoderma orbiforme* with high morphological plasticity. PLoS ONE, 9: e98733.

Wang, X.C., Xi, R.J., Li, Y., Wang, D.M. and Yao, Y.J. 2012. The species identity of the widely cultivated *Ganoderma*, "*G. lucidum*" (Ling-zhi), in China. PLoS One, 7: e40857.

Wasser, S.P., Zmitrovich, I.V., Didukh, M.Y., Spirin, W.A. and Malysheva, V.F. 2006a. Morphological traits of *Ganoderma lucidum* complex highlighting *G. tsugae* var. *jannieae*: the current generalization. Ruggel (Germany): A.R.A. Gantner Verlag K.-G., 187 pp.

Wasser, S.P., Zmitrovich, I., Didukh, M. and Malysheva, V. 2006b. New medicinal *Ganoderma* mushroom from China: *G. tsugae* Murrill var. *jannieae* var. nov. (Aphyllophoromycetidae). Int. J. Med. Mushrooms, 8: 161–172.

Welti, S. and Courtecuisse, R. 2010. The Ganodermataceae in the French West Indies (Guadeloupe and Martinique). Fungal Divers., 43: 103–126.

Welti, S., Moreau, P.A., Decock, C., Danel, C., Duhal, N., Favel, A. and Courtecuisse, R. 2015. Oxygenated lanostane-type triterpenes profiling in laccate *Ganoderma* chemotaxonomy. Mycol Prog., 14: 45.

Wu, X. and Zhang, X. 1996.*Ganoderma* spp. and *G. cupulatiprocerum* sp. nov. from Guizhou Province. Acta Mycologica Sinica, 15: 4–8.

Wu, X.L. and Dai, Y.C. 2005. Coloured illustrations of Ganodermataceae of China. Beijing (China): Science Press.

Xing, J.H., Song, J., Decock, C. and Cui, B.K. 2016. Morphological characters and phylogenetic analysis reveal a new species within the *Ganoderma lucidum* complex from South Africa. Phytotaxa, 266: 115–124.

Xing, J.H., Sun, Y.F., Han, Y.L., Cui, B.K. and Dai Y.C. 2018. Morphological and molecular identification of two new *Ganoderma* species on *Casuarina equisetifolia* from China. MycoKeys, 34: 93–108.

Yang, Z.L. and Feng, B. 2013. What is the Chinese "Lingzhi"? A taxonomic mini-review. Mycology, 4: 1–4.

Yao, Y.J., Wang, X.C. and Wang, B. 2013. Epitypification of *Ganoderma sichuanense* J.D. Zhao & X.Q. Zhang (Ganodermataceae). Taxon, 62: 1025–1031.

Zhao, J.D., Xu, L.W. and Zhang, X.Q. 1979. Taxonomic studies of the subfamily Ganodermatoideae of China. Acta Microbiol Sin., 19: 265–279.

Zhao, J.D. 1989. The Ganodermataceae in China. Bibl. Mycol., 132: 1–176.

Zhao, J.D. and Zhang, X.Q. 2000. Flora Fungorum Sinicorum 18: Ganodermataceae. Beijing: Science Press.

Zheng, H.D. and Liu, P.G. 2006. *Albatrellus yunnanensis*, a new species from China. Mycotaxon, 97: 145–151.

Zheng, L., Jia, D., Fei, X., Luo, X. and Yang, Z. 2009. An assessment of the genetic diversity within *Ganoderma* strains with AFLP and ITS PCR-RFLP. Microbiol Res., 164: 312–21.

Zhou, L.W., Cao, Y., Wu, S.H., Vlasák, J., Li, D.W., Li, M.J. and Dai, Y.C. 2015. Global diversity of the *Ganoderma lucidum* complex (Ganodermataceae, Polyporales) inferred from morphology and multilocus phylogeny. Phytochemistry, 114: 7–15.

3

Auriculoscypha
A Fascinating Instance of Fungus-Insect-Plant Interactions*

Patinjareveettil Manimohan

INTRODUCTION

Fungus-insect symbioses provide challenging and fascinating systems for studying biotic interactions between fungi, insects and other associated organisms (Klepzig et al., 2001). The types of interactions among these organisms are often difficult to pigeonhole as mutualism, antagonism, parasitism or commensalism. A remarkable but little-known interaction involving a fungus, a coccid and some tree species in southwest India is highlighted here. The sap of the plant is fed by the coccid that in turn is parasitized by the fungus, thus involving interactions between three trophic levels.

The Fungus

Taxonomic History

Auriculoscypha D.A. Reid & Manim. (Basidiomycota, Pucciniomycetes, Septobasidiales) is a monotypic genus that seems to be endemic to southwest India. This fungus is usually restricted to the bark of a few trees (mostly Anacardiaceae) and is invariably associated with a coccid. *Auriculoscypha* was erected as a new genus by Reid and Manimohan (1985). They erected it as a monotypic genus with *A. anacardiicola* as the only species. Although they assigned it to the order

Department of Botany, University of Calicut, Kerala, 673 635, India.
Email: pmanimohan@gmail.com
* This paper was presented as a poster at the 10th International Mycological Congress, Bangkok in August 2014.

Auriculariales, Reid and Manimohan (1985) acknowledged its affinities to the order Septobasidiales. But at that time, an insect-association was not known for *Auriculoscypha* and, hence, their reluctance to place it in that order. The evidence of an obligate relationship with a coccid was provided by Lalitha and Leelavathy (1990) and they suggested a transfer of *Auriculoscypha* from Auriculariales to Septobasidiales. This was followed by the observation by Lalitha et al. (1994) of a yeast phase in the life-cycle of *Auriculoscypha,* which supported its redisposition in Septobasidiales. Finally, based on molecular and ultrastructural characters, Kumar et al. (2007) unequivocally placed *Auriculoscypha* in the Septobasidiales. The only family of the order, the Septobasidiaceae, contains five genera, all of which are phytoparasitic insect symbionts (Swann et al., 2001). All species are associated with scale insects on living plants. *Auriculoscypha* differs from all other genera of the family in having woody, stipitate-cupulate basidiomata. The basidiomata of all the other genera are simple mycelial mats that are resupinate on the host plant (Kumar et al., 2007).

Geographical Distribution

Lalitha (1992) provided a detailed account of the distribution of *Auriculoscypha* in southwest India. According to her, the fungus occurs in all Districts of Kerala and also in some parts of Karnataka and Goa. It's occurrence in other parts of India and the world has not yet been reported. At present, it seems to be endemic to southwest India.

Morphology

Fruit bodies of *Auriculoscypha* (Fig. 1A) are rather woody or leathery and are stipitate-cupulate with a cylindrical stipe (2–16 × 1.5–3.5 mm) and a cup-shaped or saucer-shaped pileus (3–25 mm). Stipe of *Auriculoscypha* always arises from a basal tubercle containing mostly a single or rarely 2–3 crawlers (juveniles) of a coccid (Fig. 1B). This tubercle is either partially or fully embedded in the bark of the host tree and, on removal, it leaves a cavity in the underlying woody tissue (Fig. 1C). The tubercles are ovo-ellipsoid to sub-globose, 2–5 mm in diameter, with a smooth surface. Except at the top, the walls of the tubercle are woody, made up of tree tissues in combination with fungal mycelia. At the top, each tubercle has a mycelial roof that formed a mat or cushion on the surface of the tree, from which one or more fruit bodies arise. The cavity of the tubercle is internally lined with a waxy material secreted by the coccid. Fruit bodies of *Auriculoscypha* are seen more on the under surface and lateral surfaces of horizontal and inclined branches of the host trees. Mature fruit bodies nearly always show a tendency to be pendent with the hymenial surface facing the ground (Reid and Manimohan, 1985; Lalitha, 1992). The hymenium shows transversely septate basidia giving rise to allantoid basidiospores (Fig. 1H). These basidiospores germinate by producing bud-cells (Fig. 1I) that, in culture, continue to grow as yeasts for some time before becoming mycelial.

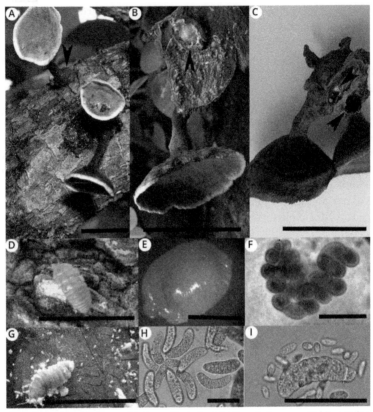

Fig. 1. *Auriculoscypha* and the associated coccid. A. Fruit bodies of *A. anacardiicola* growing on the bark of cashew tree showing the basal tubercle partially immersed in the bark (arrow). B. Fruit body of *A. anacardiicola* with the basal tubercle cut open to reveal the enslaved coccid (arrow). C. Fruit bodies of *A. anacardiicola* with the basal tubercles cut open and coccids removed (arrows). D. Adult female of the coccid on the bark of cashew tree. E. Juvenile female taken out from the tubercle of a fruit body of *A. anacardiicola*. F. Haustorial coils of *A. anacardiicola* seen inside the body of the coccid. G. Adult female of the coccid held in captivity in the lab laying eggs. H. Basidiospores of *A. anacardiicola*. I. Basidiospores of *A. anacardiicola* producing bud-cells (Scale bars: A, B, C, D & G: 10 mm; E: 2.5 mm; F: 10 µm; H and I: 25 µm).

The Insect

The insect symbiont of *Auriculoscypha* is a little-known coccid (soft scale) *Neogreenia zeylanica* (Green) McGillivray (syn. *Kuwania zeylanica* Green), belonging to the scale-insect family Margarodidae (syn. Kuwaniidae) (Lalitha, 1992). This coccid, originally reported from Sri Lanka, is a minor pest of cashew and mango trees. Adult females of this coccid occur in cracks on the trunk or underneath the bark of anacardiaceous trees and are rarely found on exposed surfaces (Fig. 1D). They have an oblong-ellipsoid body ($3–7 \times 2–3$ mm) with marked segmentation and are bright reddish-orange coloured. They have a pair of antennae and three pairs of legs. They tend to secrete a white, waxy coating around them (Fig. 1G). Although males have been reported for this species, they have not been observed during studies on *Auriculoscypha*. Adult females lay numerous eggs parthenogenetically when incubated in isolation (Fig. 1G). Eggs hatch out the juveniles (crawlers) that feed on the plant sap but adults do not feed.

The Plants

Auriculoscypha is seen mostly on *Anacardium occidentale* L. (cashew) and occasionally on *Mangifera indica* L. (mango). These are large, long-lived and evergreen tree species belonging to the angiosperm family Anacardiaceae. Very rarely, the fungus has been spotted on some other trees species of both Anacardiaceae and Euphorbiaceae (Reid and Manimohan, 1985; Lalitha, 1992).

The Interactions

Fungus-Insect Interactions

Auriculoscypha enslaves 1–3 juvenile females (Fig. 1E) within the tubercle found at the base of each fruit body and develops extensive coiled haustoria within the coccid's body (Fig. 1F). The enslaved juvenile coccids are devoid of both antennae and legs but have a long (3–6 mm) and very slender (< 1 um) sucking tube on their ventral side that pierces both the wall of the tubercle at its bottom and the woody tissue beneath the bark of the tree. Slender, hyaline septate hyphae of the fungus enter the body of the juvenile coccids held captive inside the tubercles through their dermal pores and form coiled haustoria inside. These haustoria are irregularly-shaped entities formed of densely coiled, 2–5 μm wide hyphae. Fungal infections prevent the juveniles from attaining the adult stage and renders them immobile; however, the infected juveniles continue to live in this arrested condition. Members of the juvenile population of the coccids that escape fungal infection eventually develop into mature adult females. Fungal haustoria are never seen inside the bodies of the free-living adult females. Although basidiospores of *Auriculoscypha* germinate and produce rudimentary mycelia in solid culture media, these mycelia fail to grow in a purely synthetic medium devoid of any growth factors (Lalitha, 1992; Kumar et al., 2011). Addition of yeast extract favours the growth of the cultured mycelium to a certain extent. However, all mycelial cultures of *Auriculoscypha* lose viability within a few months and they never produce fruit bodies (Lalitha, 1992; Kumar et al., 2011). It is obvious from all these observations that *Auriculoscypha* is nutritionally dependant on the coccid.

Insect-Plant Interactions

Neogreenia zeylanica juveniles are parasites of plants, feeding on sap drawn directly from the plant's vascular system with the help of their sucking tube. Although *N. zeylanica* is seen usually on anacardiaceous plants, it is not a case of family-level monophagy as, very rarely, the coccid and the associated fungus are seen on trees of Euphorbiaceae as well (Lalitha, 1992). Observations reveal that the coccids are seen on those trees where the bark has a particular texture with lots of cracks and crevices that provide protection for the coccids from their enemies. Thus, bark texture seems to be a major factor involved in host selection of this coccid. Apparently, the plant is a loser in this association, while the coccid gets both nutrition and a safe dwelling place. However, the plants do not seem to be negatively affected to any great extent by being directly or indirectly parasitized by the coccid and the fungus. There are no visible symptoms that indicate an appreciable drain of the plant sap. This may be because the plants are large trees which are parasitized at any time by only a small number of minute coccids. The associated plants seem to be no way benefited by their associations with the coccid.

Fungus-Plant Interactions

Unlike the association between the fungus and the coccid, the association between the fungus and the host tree is not very specific. *Auriculoscypha* does not directly parasitize the host trees and its hyphae never invade the plant tissues. The fungus draws its nourishment from the coccid, which in turn feeds on trees, showing an indirect form of phytoparasitism. It seems *Auriculoscypha* is seen on those trees where the coccids are common. The fungus is obligatorily and specifically associated with the coccid, it seems to have an incidental association with the tree species. There is a possibility that *Auriculoscypha* is indirectly helping the plant by controlling the population size of the coccid by preventing at least some juveniles from reaching adulthood.

Grey Areas

The impact of *Auriculoscypha* on populations of *Neogreenia* in nature is not clearly known. We need to understand at what stage in the life-cycle of both the coccid and the fungus that the former gets infected by the latter. Another unresolved aspect is how the fungus is capturing the coccid inside the tubercle? It is likely that the fungus attacks the juveniles which have established a feeding relationship with the trees after inserting their sucking tube into the tissues of the plants.

References

Klepzig, K.D., Moser, J.C., Lombardero, M.J., Ayres, M.P., Hofstetter, R.W. and Walkinshaw C.J. 2001. Mutualism and antagonism: Ecological interactions among bark beetles, mites and fungi. pp. 237–267. *In*: Jeger, M.J. and Spence, N.J. (eds.). Biotic Interactions in Plant-Pathogen Associations. CABI Publishing, New York.

Kumar, T.K.A., Celio, G.J., Matheny, P.B., McLaughlin, D.J., Hibbett, D.S. and Manimohan, P. 2007. Phylogenetic relationships of *Auriculoscypha* based on ultrastructural and molecular studies. Mycol. Res., 111: 268–74.

Kumar, T.K.A., Jisha, K.C., Jisha, E.S. and Manimohan, P. 2011. Isolation and *in vitro* cultivation of *Auriculoscypha anacardiicola* D.A. Reid et Manim., An insect-associated and potentially medicinal fungus from India. Int. J. Med. Mushr., 13: 273–280.

Lalitha, C.R. and Leelavathy, K.M. 1990. A coccid-association in *Auriculoscypha* and its taxonomic significance. Mycol. Res., 94: 571–72.

Lalitha, C.R. 1992. Studies on the biology and taxonomy of the genus *Auriculoscypha* (PhD thesis). Calicut: University of Calicut, India.

Lalitha, C.R., Leelavathy, K.M. and Manimohan, P. 1994. Patterns of basidiospore germination in *Auriculoscypha anacardiicola*. Mycol. Res., 98: 64–66.

Reid, D.A. and Manimohan, P. 1985. *Auriculoscypha*, a new genus of Auriculariales (Basidiomycetes) from India. Trans. Brit. Mycol. Soc., 85: 532–35.

Swann, E.C., Frieders, E.M. and McLaughlin, D.J. 2001. Urediniomycetes. pp. 37–56. *In*: McLaughlin, D.J., McLaughlin, E.G. and Lemke, P.A. (eds.). The Mycota. Volume II Part B. Systematics and Evolution. Springer-Verlag, Berlin.

Biodiversity and Ecology of Boreal Pine Woodlands Ectomycorrhizal Fungi in the Anthropocene

Luis Villarreal-Ruiz [1,]* and *Cecilia Neri-Luna* [2]

INTRODUCTION

Climate change has been an evolutionary driving force (Walochnik et al., 2010) since early microbial activity began on Earth ~3,700 to 4.1 million years ago (Bell et al., 2015; Nutman et al., 2016). Biocomplex interactions with physiochemical exterior factors as well as with geodynamic planetary processes lead to the evolution from microbial prokaryotic (Archaebacteria, Eubacteria) to eukaryotic lineages (Animals, Fungi and Plants) that produce emergent functional complex systems (Hazen et al., 2007; Koonin, 2015) which we here call the "hologenomicrobiome" (Lederberg, 2001; Schlaeppi and Bulgarelli, 2015; van der Heijden and Hartmann, 2016; Koskella et al., 2017). Universal symbiogenesis (Gontier, 2007; Ku et al., 2015) along with symbiosis ("any long-term, intimate association between two organisms" *sensu* van der Heijden et al., 2015) has possibly been an important evolutionary and ecological phenomenon (Moran, 2007; Neri-Luna and Villarreal-Ruiz, 2012) complementary to the Darwinian "Universal" selectionist theory (Dawkins, 1983) that gradually

---

[1] Laboratorio de Recursos Genéticos Microbianos & Biotecnología (LARGEMBIO), PREGEP-Genética, Colegio de Postgraduados, Campus Montecillo, Texcoco, C.P. 56230, Edo. México, México.
[2] Laboratorio de Ecofisiología Vegetal, Departamento de Ecología, CUCBA, Universidad de Guadalajara, Camino Ing. Ramón Padilla Sánchez No. 2100, C.P. 45110, Zapopan, Jalisco, México.
* Corresponding author: luisvirl@colpos.mx

adapted or provided the resilience to the ~1% of living organisms that survived five global major extinctions (Barnosky et al., 2011). As a result, life gradually shapes the biosphere with the addition of oxygen (~2.45 billion years ago) that allows more complex life-forms to evolve and diversify, interacting and functioning as a single unit (Villarreal-Ruiz, 1996; Farquhar et al., 2000).

Terrestrial biomes are a core biosphere component that evolved from: (1) bare land colonized predominantly by early microbial groups in which "...the charophycean green algae, the ancestors of land plants, were terrestrial long before the emergence of land plants" (Harholt et al., 2016). (2) "...Terrestrialization probably began more than one billion years ago and irreversibly altered biogeochemical processes at planetary scale..." (Gerrienne et al., 2016). Whether the evolution of land plants was due to a single event of colonization or combined events is still a matter of debate, although, it was thought that most early terrestrial fungi might have lived as symbiotic endophyte (Krings et al., 2007; Vries and Archibald, 2018). Feasibly endophytism mode may evolve into balanced mycorrhizal symbiosis where fungi played a central role in plant root evolution, plant adaptation and possibly in terrestrial biome diversification (Redecker et al., 2000; Brundrett, 2002; Bidartondo et al., 2011; Desirò et al., 2017). This major event of land colonization by plant, fungi (mycorrhizal), bacteria and arthropods turned our planet from blue to green and, since then, terrestrial life has become the major influence on the geochemical carbon cycle (Kenrick et al., 2012; Martin et al., 2017; Morris et al., 2018; Strullu-Derrien et al., 2018).

However, since the spread of agriculture and deforestation of terrestrial biomes, a distinctive human signature has recently been reported in sediments and ice, which are stratigraphically and functionally different to the Holocene; leading to the conclusion that a new Anthropocene ('Age of Man' *sensu* Crutzen, 2002) epoch has begun (Lewis and Maslin, 2015; Waters et al., 2016). This new insight claims that because human technology is altering geological planetary processes (Rocha et al., 2015) by reducing global biological diversity, such human impact has triggered the sixth major extinction of planetary life history (Chapin et al., 2000; Ceballos et al., 2015, 2017). This concern leads to the conclusion that humans need to understand that the actual biodiversity and planetary conditions are the heritage of evolution and environmental history, the "hologenomicrobiome" (is the microbial gene pool that we inherited has a valuable source of genetics resources) (Fig. 1) (Léveque and Mounolou, 2003). The future of planetary sustainability and restoration will depend on the adoption of global bioethics measures in order to minimize the anthropogenic climate change in which microbial communities (such as mycorrhizal fungi) may play an important role (Hens and Lefevere, 2007; Macpherson, 2013; Gillings and Paulsen, 2014). This Chapter assessed the relevance of ectomycorrhizal fungi from a biocomplex perspective in Boreal pine woodlands in the Anthropocene.

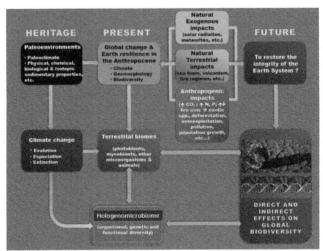

Fig. 1. Climate change as an evolutionary force in the Anthropocene from a biocomplex planetary perspective, underlining the relevance of the "hologenomicrobiome" as a heritage resilient gene pool aiding (at present) to mitigate anthropogenic impacts and restore the integrity of Earth System in the future.

The Significance of Boreal Pine Forests in the Anthropocene

In spite of the long-term human impact on biodiversity due to deforestation and agricultural activities (Brown, 1997), the Boreal forest or taiga ecosystems are today distributed between 47 to 70° northern latitude representing a circumpolar belt with 1,200 million ha that constitute between 11 to < 17% of terrestrial biomes (Bourgeau-Chavez et al., 2000; Kasischke, 2000). Goldammer and Stocks (2000) estimate that 920 million ha of the boreal zone are closed forests of which 87% contain an exploitable growing stock of 95 billion m³. This is significant because these forests constitute 29% of Earth's forest surface, 73% of its coniferous woodlands and generate 47% of the world total forest products. In addition, the fact that boreal forests contain between 30–37% of the total terrestrial carbon pool suggests that they are a potential sink or source of atmospheric carbon and may play a significant role in the global climate system (Kasischke, 2000; Goldammer and Stocks, 2000) in the Anthropocene. The biodiversity contained above- and below-ground represents another important value of the boreal biome, particularly in the case of ectomycorrhizal (EcM) fungi with a species richness higher than host trees species richness such as the widespread *Pinus sylvestris* L. populations (Allen et al., 1995; Jonsson et al., 1999a; Tedersoo et al., 2014).

Scots pine (*Pinus sylvestris* L.) is the second most important tree of boreal forests, after *Picea abies*, which is the climax species (Richardson and Rundel, 1998). In addition, *P. sylvestris* has the widest distribution of any pine in the world covering a surface of 108 million ha (Nikolov and Helmisaari, 1992). The natural range of *Pinus sylvestris* is 37° N and 50° N east of the Caspian Sea and 10° W from Scotland to 150° W in northeastern Asia (Agee, 1998). In Britain, Scots pine native

Fig. 2. Native Caledonian pine woodland (Scots pine *Pinus sylvestris-Hylocomium splendens*), with the dominant pine overstorey *old growth phase* (260-yr-old stand) and the ericaceous understorey vegetation, at Glen Tanar Nature Reserve, Scotland, UK.

populations are confined to the Scottish Highlands from Shieldaig in the west to Glen Tanar in the east (Mason et al., 2004). The native Caledonian pinewoods of Scotland represent an outlier in the extreme north-west of the species range, 500 km to the west of the main distribution in Continental Europe and Scandinavia (Vidaković, 1991; Sinclair et al., 1998). However, the present *Pinus sylvestris* population has declined in some countries, such as Scotland, where the total area of *Pinus sylvestris* has been dramatically reduced from 1.5 million ha to 16,000 ha due to centuries of human impact and changes in climatic conditions producing a fragmented forest with a scattered distribution (Vickers and Palmer, 2000).

Chapin (2003) proposed that vegetation productive capability and the rates of decomposition and nitrogen mineralization are determined by plant traits related to size and growth rate, affecting most of the key ecosystem processes such as carbon and nutrient cycling with effects on water and energy exchange and disturbance regime at landscape and regional scales. Consequently, the assemblage, survival and coexistence of Scots pine "heathy acid pinewoods" (*P. sylvestris-Hylocomium splendens* [W18], Rodwell, 1991) with an overstorey dominated by *P. sylvestris* and an understorey composed of dominant calcifugous species such as *Calluna vulgaris*, *Vaccinium myrtillus*, *V. vitis-idaea*, *Deschampsia flexuosa* and a ground floor covered by *Hylocomium splendens* and other bryophytes might be determined by adaptive plants traits in close association with mycorrhizal fungi (Fig. 2; Table 1).

Scots Pine Stand Dynamics and Above- and Below-ground Processes

Native pine woodlands are characterized by poor quality substrate leaf litter produced by overstorey and understorey ericaceous plants with high C:N ratio, high content of polyphenols and a low pH (Helmisaari, 1995; Lindahl et al., 2002). As a result, plant community species composition is strongly related to the limited nitrogen supply, the high degree of nutrient immobilization in the organic soil layers and the activities of typical ErM fungi (Read, 1991; Tamm, 1991). Although the function of ericaceous plants as understorey vegetation in forest ecosystems is unclear (Smith

Table 1. Adaptive plant traits that may determine coexistence and survival in *Pinus sylvestris*-*Hylocomium splendens* woodlands in the Anthropocene.

Plant characteristics and ecophysiological parameters	Overstorey Species: *Pinus sylvestris*	*Calluna vulgaris*	Understorey Species: *Vaccinium vitis-idaea*	*Vaccinium myrtillus*	*Deschampsia flexuosa*	Ground floor Species: *Hylocomium splendens*
Morphology						
Life form	Tree[12]	Shrub[8]	Shrub[17]	Shrub[18]	Rhizomatous grass[9]	Ectohydric moss[21]
Stem structure	Woody[3]	Woody[8]	Woody[17]	Woody[18]	Herbaceous[5]	Chain of segments[13]
Leaf form	Needle clusters[1]	Xeromorphic-inear[9]	Ovovate-emarginate[17]	Acute-ovate[18]	Hair-like[16]	Pleurocarpic[20]
Root habit	Heterorrhizic[2]	Hair-Roots[8]	Hair-roots	Hair-roots[18]	Shallow[9]	Rhizoids[10]
Root deep	1–2 m[24]	50–84 cm[8]	10–20 cm[17]	15–20 cm[18]	> 50 cm[9]	> 1 cm[23]
Life-history						
Strategies	Stress-tolerant competitor[9]	Stress-tolerant competitor[9]	Intermediate Stress-tolerant/Stress-tolerant competitor[9]	Stress-tolerant competitor[9]	Intermediate Stress-tolerant/Stress-tolerant competitor[9]	Stress-tolerant[11]
Life span	Long (300 yr)[4]	Short (< 30 yr)[9]	Long? (almost undeterminable)[17]	Long? (difficult to determine[18])	Perennial[16] (No information available)	Short (8–20 yr)[10,20]
Fire regime	Moderate-severity[3]	Moderate[8]	Moderate[17]	Moderate[18]	Moderate[16]	Intolerant[10]
Regenerative strategies[a]	S,W [4,15]	Bs[9]	V, ?Bs[9]	V, Bs[9]	V,S[9]	W,V[10]
Leaf phenology	Evergreen[3]	Evergreen[9]	Evergreen[9]	Deciduous[9]	Winter green[9]	Evergreen[10]
Flowering phenology	May–June[12]	August–September[9]	Early summer[9]	Early summer[9]	June–July[9]	-
Seed weight	5.59 mg[5]	0.03 mg[9]	0.26 mg[9]	0.26 mg[9]	0.43 mg[9]	-
Seed longevity	16 months[6,7]	150 years[19]	-	-	14–15 months[16]	-
Mycorrhizal status	EcM[2]	ErM[9]	ErM[9]	ErM[9]	AM[9]	-

Physiology						
Photosynthesis	C3[22]	C3[22]	C3[22]	C3[22]	C3[22]	C3[21]
Seedling RGR	< 0.5[9]	< 0.5[9]	< 0.5[9]	< 0.5–0.9[9]	< 0.5–0.9[9]	-
A_{max}	12.9 ± 1.7[14,a]	-	-	-	-	-
P_{max}	-	-	-	-	-	1.6–2.4[21]
Plant uptake						
N	68; 3 ;29[25]	-	73; 10; 17[25]	38; 17; 45[25]	-	-
aa	0.65 ± 0.16[26]	1.28 ± 0.28[26]	0.45 ± 0.04[26]	3.76 ± 0.21[26]	4.32 ± 0.61[26,b]	-

V, vegetative; S, seasonal regeneration in vegetation gaps; W, numerous small wind-dispersed seed or spores; Bs, persistent seed bank; Bsd, persistent seedling bank (Adapted from Grime, 2001); ?, need further confirmation, -, information not available at the moment of the review. [a]Mean ± SD ($n = 9$). [b]data from *D. caespitosa*. Seedling RGR, seedling relative growth-rate (week[-1]). A_{max}, Maximum net photosynthesis (μmol m[-2] s[-1]). P_{max}, mg CO_2 g dw[-1] h[-1]. aa, Aminoacid uptake rates in μmol g[-1] root DW h[-1] (Mean ± SD). EcM, ectomycorrhizal; ErM, ericoid mycorrhizal; AM, arbuscular mycorrhiza. [1]Richardson and Rundel (1998); [2]Read (1998); [3]Agee (1998); [4]Nikolov and Helmisaari (1992); [5]Grime (2001); [6]Zasada et al. (1992); [7]Granström (1987); [8]Gimingham (1960); [9]Grime et al. (1988); [10]Rook (1998); [11]Eckstein (2000); [12]Vidaković (1991); [13]Ross et al. (1998); [14]Rundel and Yoder (1998); [15]Lanner (1998); [16]Scurfield (1954); [17]Ritchie (1956); [18]Cumming and Legg (1995); [20]Rydgren et al. (2001); [21]Valamne (1984); [22]Larcher (2003); [23]Richards (1984); [24]Persson (1984); [25]Nordin et al. (2001), values are based on total plant N uptake (100%) distributed on NH_4^+, NO_3^- and glycine respectively; [26]Persson and Näsholm (2001).

et al., 1992), Grime (1994) pointed out that slow growing plants in unproductive habitats depend on above- and below-ground resources capture by long-lived tissues able to maintain viability under extreme conditions. This is particularly important in the resource supply when canopies of *P. sylvestris* may cause deep shade and provide major sinks for mineral nutrients and potential mycorrhizal networks might firmly fix the mineral nutrient cycle, generating a nutrient stress environment for the understorey vegetation (Mencuccini and Grace, 1996). In this layer, evergreen plants (*C. vulgaris, V. vitis-idaea* and *D. flexuosa*) and deciduous plants (*V. myrtillus*) are under limited low irradiance and sunflecks may be of great importance as well as their internal N cycling and the symbiotic community of organisms associated with their root systems for efficient nutrient acquisition (Cornelissen et al., 2001). In the understorey reinitiation stage, the C:N ratio increases as a result of the slow turnover of large amounts of organic debris accumulated after tree canopy opening (Waring and Running, 1998). These authors suggest that in a high C:N ratio and a less demanding old-growth stage the understorey vegetation may play a central functional role, capturing and immobilizing nutrients that otherwise will be lost by leaching. In the same line, Villarreal-Ruiz et al. (2004; 2012) demonstrated for the first time the notion that a strain from the *Meliniomyces bicolor* (isolate LVR4069) has the potential to produce both EcM and what appear to be ErM with *Vaccinium myrtillus* and that a typical basidiomycete (*Laccaria bicolor*) EcM fungi was capable to form which appears to be ErM with *V. macrocarpon.* "This raises the intriguing possibility that understorey and overstorey plants are linked by a common mycorrhizal fungal mycelium. Identifying the functional significance, if any, of such linkages will require careful experimentation in the field" (Villarreal-Ruiz et al., 2004) (Fig. 3).

Shugart (1997) found that Scots pine stand dynamics corresponds with Strategy 3 in which recruitment is preceded by previous cohort senescence and a biomass reduction. This assumption is supported by Helmisaari's (1995) nutrient cycling chronosequence study on naturally regenerated Scots pine woodlands:

(1)   The number of soil nutrient pools is related to the organic matter accumulation as the stand ages with more available and total N in the thickest humus layers.

(2)   Nutrient distribution in tree structures across the stands is a function of tree age.

(3)   Above- to below-ground biomass ratio decreases with stand age.

(4)   Three-fold more N is needed in mature stands than in 35-yr-old stands for biomass production.

(5)   Root/shoot ratio is proportionally inverse to nutrient uptake per fine root unit mass.

(6)   The relative size of the nutrient pool in the overstorey biomass increases and that in the understorey decreases, with stand age.

(7)   Understorey vegetation is more efficient in its use of N than overstorey trees, especially in mature stands.

Consequently, the structural and functional changes on native and planted pine woodlands have an impact on stand dynamics of above- and below-ground EcM fungi. Current views on how communities of EcM fungi change as forest stands age are based primarily on surveys of sporome occurrence (Egger, 1995). Following this

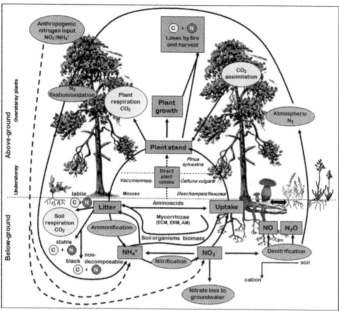

Fig. 3. Nitrogen cycle processes above- and below-ground in native pinewoods, considering the potential involvement of overstorey, understorey plants, and plant direct aminoacids uptake and their micorrhizal habits (Redrawn and adapted from Näsholm et al., 1998; Schulze, 2000; Persson and Näsholm, 2001). Solid arrows refer C fluxes, broken arrows refer the paths of added anthropogenic N and dotted broken arrow refers to direct aminoacids uptake by plants. Thick double arrow refers to the tripartite fungal interactions with overstorey and understorey (Ericaceous) plants reported by Villarreal-Ruiz et al. (2004; 2012).

line of research, the "early- and late-stage" model was proposed in order to describe spatial and temporal patterns of EcM fungal succession on *Betula* spp. planted in agricultural soils (Mason et al., 1982, 1983). The model was supported by below-ground sampling of ectomycorrhizas on *Betula* plants, and beneath the sporomes growing around trees (Deacon et al., 1983).

Dighton and Mason (1985) speculated that changes in vegetation occur during the development of natural and planted forests, and that this alters soil conditions and resource availability to photo- and myco-bionts. The perceived increase in EcM species diversity up until canopy closure, and the subsequent decline, was explained as a function of stand age and of the amount and quality of litter driving soil organisms to adopt "*k*" instead of "*r*" strategies. In addition, they hypothesized that tree size/age governed photoassimilate production, which tended to be greater in older trees; consequently, changes in community of mycobionts on root systems might be expected (Fig. 4a).

This "early- and late-stage" model was supported to some extent by studies in first-rotation forest plantations (Ricek, 1981; Termorshuizen and Schaffers, 1989), although other researchers found no correlations between sporome fungal diversity and stand age (Blasius and Oberwinkler, 1989; Keizer and Arnolds, 1994). In addition, below-ground surveys in natural forests show that the distribution and abundance of fungi on root systems may bear little resemblance to the production

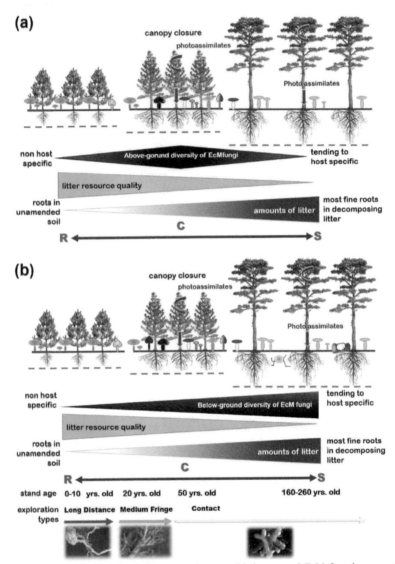

Fig. 4. Pinewood stand dynamics influence on above- and below-ground EcM fungal communities. (a) Last et al. (1987) above ground hypothetical model. (b) Villarreal-Ruiz (2006) above- and below-ground EcM fungal communities combined methodological approaches. Life strategies; R, ruderal; C, competitive; S, stress tolerant (Grime, 1977). Exploration types of ectomycorrhizae (Agerer, 2001).

of obvious sporomes (Gardes and Bruns, 1996; Jonsson et al., 1999b). Few studies directly address the question of possible successional processes in EcM communities, such as Visser (1995), who reported an increase of species diversity of sporomes and ectomycorrhizas after a fire in a *Pinus banksiana* chronosequence. In the same line of research, Rao et al. (1997) found that the diversity index of EcM fungi was related to stand age in a *Pinus kesiya* chronosequence. In both reports, the chronosequence was unreplicated and the results were important but inconclusive insights into successional patterns of EcM fungi in natural ecosystems. Later, Villarreal-Ruiz's (2006) unpublished study compared above- and below-ground EcM fungal richness across a replicated chronosequence (0–10; 10–20; 50; 160 and 260-yr-old stands) in Scots pine, and found that above-(sporome) richness supported previous findings. However, when the below-ground EcM fungal communities were studied, the species richness increased with stand age and a shift on EcM exploration types was demonstrated, which contradicted the Last et al. (1987) succession model (Fig. 4b).

Biodiversity and EcM Fungal Diversity in Boreal Pine Forests

Global biodiversity is the variety of life on Earth represented by around 12.5 million living organisms, their interactions within biotic process interrelated with the abiotic environment, and its products in all its manifestations (World Conservation Monitoring Centre, 1992). However, some researchers recently predicted that globally ~ 8.7 million (± 1.3 million SE) eukaryotic species probably do exist, of which ~ 2.2 million (± 0.18 million SE) are marine (Mora et al., 2011).

Three-fold definitions of biodiversity have been proposed in terms of genes, species, and ecosystems (Solbrig, 1991; Zak et al., 1994; Dodson et al., 1998; Gaston and Spicer, 1998). However, Swingland (2001) pointed out that because different living kingdoms are involved, including fungi, no universal agreements are possible, and that biodiversity is in practice defined according to the context and the author's particular purposes.

As a result, a pyramidal three-fold conceptual model for EcM fungal biodiversity is presented in Fig. 5, by deconstructing the potential three main components:

(1) *organismal diversity*, the species richness, relative abundance and composition (taxonomic diversity) of all EcM species coexisting in a given area and their range of adaptive traits (adapted from Swingland, 2001; Tilman, 2001; Tilman and Lehman, 2001).

(2) *Functional diversity*, the value and variety of EcM organismal adaptive functional traits that influence ecosystem processes (adapted from Tilman, 2001; Allen et al., 2003).

(3) *Genetic diversity*, the unique homo- vs. heterokaryotic mechanisms that allow molecular hereditary differences within or between EcM fungal populations to adapt and evolve in response to environmental changes and selective pressures (adapted from Nevo, 2001; Swingland, 2001; Morris and Robertson, 2005).

Organismal Diversity of EcM Fungi

The ectomycorrhizal symbiosis is a balanced mutualistic association, horizontally (or "pseudo-vertically") transmitted between the lateral fine roots of woody perennial plants enclosed by a hydrophilic or hydrophobic multilayered mantle of densely packed or loosely interwoven fungal hyphae (Wilkinson, 1998; Agerer, 2001; Brundrett, 2004). The hyphae "outgrow" as specialized structures (i.e., cystidia, extramatrical mycelium, rhizomorphs, sporomes), and "ingrow" forming a

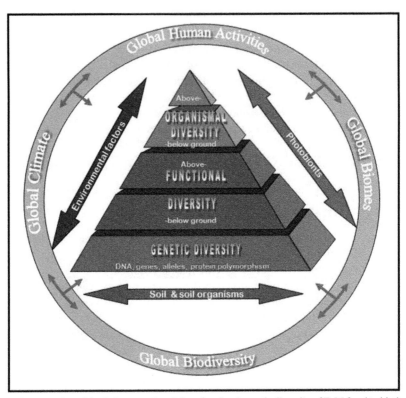

Fig. 5. Conceptual model relating organismal, functional and genetic diversity of EcM fungi to biotic and abiotic environments (thick double arrows) in an interconnected and reciprocal impacted (triple arrowed lines) global biodiversity, climate, biome and human activities context (Allen, 1991; Chapin, 2003; Swift, 2005).

labyrinthine Hartig net between cortical and/or epidermal cells towards a synchronic developed plant-fungus interface where bidirectional transfer of nutrients (i.e., C, N, P) takes place (Allen, 1991; Agerer, 2001; Brundrett, 2004; Smith and Read, 2008).

The organismal diversity of EcM fungi represented by what we can directly perceive and measure above-ground, i.e., epigeous sporome (Fig. 6a,e) production is only the "tip of the iceberg". Below-ground there are hypogeous sporomes (Fig. 6c,g), mycorrhizal tips on tree roots (Fig. 6b,d,f,h,i,j) and fungal mycelium in the forest soils, plus their interactions with other forest organisms (Shaw, 1992; Vogt et al., 1992; Landeweert et al., 2003). The use of combined approaches with the aid of classical (culture, morpho-anatomy) and modern (first- and second-generation technology, genomics) methods of research has led to the accurate profiling, identification and quantification of EcM biodiversity in order to understand the life history, adaptive traits and role of these fungi in global ecosystem functioning (Villarreal-Ruiz et al., 2004; Bidartondo and Gardes, 2005; Schmit and Lodge, 2005, Villarreal-Ruiz et al., 2012; Tedersoo et al., 2014; Villarreal-Ruiz and Neri-Luna, 2018; Villarreal-Ruiz et al., 2018).

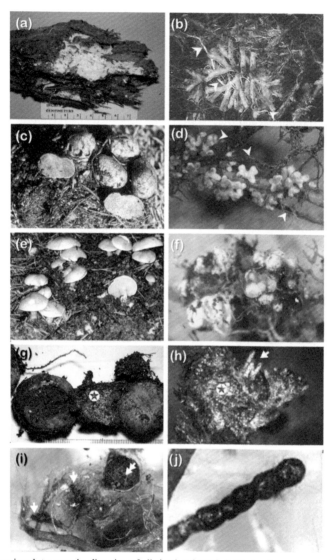

Fig. 6. Organismal taxonomic diversity of distinctive boreal EcM fungi sporome and their EcM morphotypes and exploration types (ET). (a) *Tomentellopsis submollis* resupinate basidiome. (b) *T. submollis* pink mycorrhizal system with rhizomorph (arrowhead). (c) *Rhizophogon luteolus* hypogeous basidiome. (d) *R. luteolus* EcM long distance ET with rhizomorphs (arrowhead). (e) *Suillus bovinus* epigeous basidiome. (f) *S. bovinus* long distance ET. (g) *Elaphomyces granulatus* hypogeous ascome with basket (✪) mycorrhizal system. (h) *E. granulatus* basket (✪) mycorrhizal system and EcM root tips (arrow). (i) *Cenococcum geophilum* short distance ET (arrow) and sclerotia (double arrow). (j) *Meliniomyces bicolor* with short distance ET.

Life-History Strategies in EcM Fungi

The following Begon et al. (1996) organismal life-history concept adapted to EcM fungi is the lifetime pattern of mycelial growth on plant roots and their differentiation into exploration organs for colonization, nutrient uptake, storage and reproduction. A sequence involving 10 stages in mycorrhizal formation was characterized by Brundrett (1991), in which at least four stages represent a complementary and interconnected functional phenomenon in ectomycorrhizal fungal life-history (Fig. 7).

(1) *Propagule-bank* is composed of potential "diaspores" dispersal units, i.e., sexual and asexual spores, sclerotia, old mycorrhizas, mycelial and rhizomorph fragments (Malloch and Blackwell, 1992), deposited in: (a) airborne spore-bank involved on primary succession establishment (Fig. 7, 1a) (Hodkinson et al., 2002) and (b) forest soils as a "dormant spore bank" (Fig. 7, 1b) involve in secondary succession establishment (Glassman et al., 2016). This propagule-

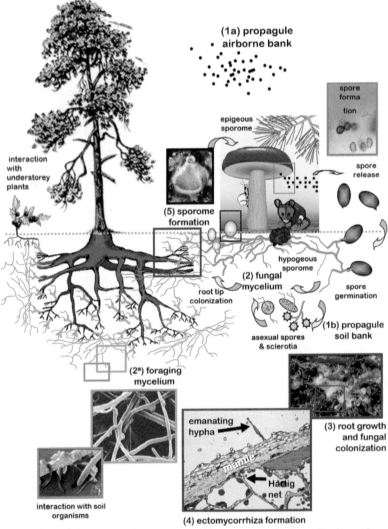

Fig. 7. Ectomycorrhizal fungal life-history with overstorey an understorey plant interactions.

bank may constitute a source of genetic diversity or a back-up colonization pool for major disturbances, i.e., fire (Fries, 1987; Baar et al., 1999; Taylor and Bruns, 1999; Jumpponen, 2003).

(2) *Colonization and foraging* is performed by the (dikaryotic) extraradical mycelial network, produced by (monokaryotic) hyphal anastomosis of compatible germinated sexual spores that grow and branch just behind the tip and colonize the plant fine roots, giving rise to large "fan-like" structures that explore the soil at rates of 2–4 mm/day with a hyphal front density of 250 individual filaments per linear millimeter (Fig. 7, 2a) (Read, 1992a; Deacon, 2006). In most EcM fungi, these individual hyphae are composed of elongated tubes with rigid thin walls and perforated cross walls (termed dolipore septa) at regular intervals that allow the protoplasm to pass through, retaining the nucleus in each cell compartment. However, in Ascomycetes and mitosporic fungi (Fig. 6i,j) the nuclei can pass through the septal pores and produce multinucleate hyphae, with significant implications for genetic variation (Deacon, 2006).

The capability of individual hypha to fuse and differentiate into contact organs (called rhizomorphs) capable of water and nutrient uptake and transport was documented by Agerer (1992). Ogawa (1985) recognized three mycelial "life types" in EcM fungi growing in forest soils:

(i) fairy ring type, observed in *Cantharellus*, *Collybia*, *Lyophyllum*, *Ramaria* and *Tricholoma*.

(ii) Irregular mycelial mat type, present in *Calodon*, *Cortinarius*, *Hydnellum*, *Rozites* and *Suillus*.

(iii) Dispersed colony type in *Amanita*, *Boletus*, *Clavulina*, *Gomphidius*, *Lactarius*, *Paxillus* and *Russula*.

Recent studies (Agerer, 1987–2002; Agerer et al., 1998–2004; Warwick et al., 2000; Tedersoo et al., 2003) confirmed that some genera (i.e., *Clavulina*, *Collybia* (*Rhodocollybia butyracea*), *Lyophyllum*, *Ramaria*, *Tricholoma matsutake*) previously considered by Read (1992b) as "litter decomposers" or "parasitic" are indeed mycorrhizal. Leake et al. (2004) stressed the ecological importance of the extraradical mycelia through this influence on biogeochemical cycling, plant community composition and ecosystem functioning.

(3) *Storage and nutrient exchange* are performed in the ectomycorrhizal tip (Fig. 7, 3) which displays considerable structural diversity, development and function because of plant-fungus interactions (Smith and Read, 2008). The presence of a Hartig net (Fig. 7, 4) is the primary characteristic to designate the ectomycorrhizal symbiosis because it constitutes the main pathway for nutritional exchange between both partners because of a synchronized development (Bonfante-Fasolo and Scannerini, 1992; Brundrett, 2004).

As soon as the host fine root is in contact with the mycobiont a loose hyphal weft is formed with a subsequent intercellular penetration between cortical or epidermal cells leading to morphogenetic, biochemical and biophysical events as follows (Fig. 7, 4):

(i)   increase of exchangeable surface in both partners (i.e., hyphal branching between cells forming a labyrinthine Hartig net).

(ii) Changes in the enzymatic and electrophysiological properties of host-mycobiont plasma membranes.

(iii) Continuous modification in the permeability of both partners' exchanges zones (Blasius et al., 1986; Bonfante-Fasolo and Scannerini, 1992; Kottke, 1992).

Successful fungal colonization produces a multilayered mantle that is a highly differentiated fungal-dependent structure producing two main patterns of multilayered hyphal arrangements, i.e., mantle of densely packed or loosely interwoven fungal hyphae. Agerer (1995) used these structural attributes for mycorrhiza classification and Harley and Smith (1983) documented their functional nutrient storage importance. In addition, the whole EcM may play an important functional role as nutrient storage organs in tuberculate ectomycorrhizas (Fig. 6f) (Smith and Read, 2008).

(4) *Reproductive and propagule dispersal strategies.* Reproduction in fungi is a genetically regulated process of differentiation and development of monokaryotic or dikaryotic hyphae, producing asexual or sexual dispersal structures (Fig. 7, 5) and propagules, triggered by environmental and nutritional conditions (Delmas, 1987). Most EcM fungi produce sexual reproductive structures above- or below-ground, called epigeous (Fig. 6a,e) (82% of species) or hypogeous (18%) sporomes (Fig. 6c,g) (the term sporome will be used instead of "fruiting bodies", following Holden et al. (2001), because fungi do not "fruit" technically), displaying a diverse morphology and producing massive quantities of spores (Molina et al., 1992). A few species, mainly mitotic Ascomycota, i.e., *Cenococcum geophilum* (Fig. 6i), and *Meliniomyces bicolor* (Fig. 6j), are unable to form sporomes and produce asexual spores or sclerotia (Fig. 6i, with double arrow), although some Basidiomycota forming sporomes, i.e., *Sarcodon imbricatus* and *Paxillus involutus*, are also reported to produce chlamydospores and sclerotia (Agerer, 1987–2002).

Three-propagule dispersal strategies can be recognized in EcM fungi:

(i)   Intramatrical dispersal: Colonization and intense proliferation on a recently available substrate.

(ii) Intermatrical dispersal: Colonization of a new substrate after nutrient depletion of a previous substrate; the dispersal agent differs from the one involved in the intramatrical dispersal.

(iii) Dormancy: Fungal resting stage (i.e., sclerotia as in Fig. 6i with double arrow) brought on by the plant host life cycle, seasonality or disturbance (Malloch and Blackwell, 1992).

The physical dispersal agents reported so far are air, splash dispersal or water dispersal (Ingold, 1971; Allen et al., 1993). However, dispersal through mycophagy involving invertebrates (i.e., slugs, snails, earthworms, mites), insects and vertebrates (i.e., birds and mammals) is well documented (Fig. 7) (Claridge and Trappe, 2005; Trappe and Claridge, 2005).

Patterns in Species Richness, Composition and Relative Abundance

At a global scale fungal species richness is estimated at 0.6–1.5 million species, of which 74,000 to 120,000 are currently described and around 55–60% are still unknown from Antarctica and Europe respectively, and > 95% from each of the American, Asian, African and Australian continents (Hawksworth and Mueller, 2005). Recently, Mora et al. (2011) estimated that at global scale ~ 611,000 (± SE 297,000) are fungi. More recently, Tedersoo et al. (2014) analyzed a total of 1,019,514 quality-filtered sequences of which 94.5% sequences and 85.4% OTUs were classified as Fungi. The study group was reduced to 44,563 OTUs after removing singletons of which ~ 30,000 OTUs could be new species from the 365 sites sampled worldwide. Because of that, Hibbett (2016) considered as "…the largest study of fungal diversity to date…". In the same line Tedersoo et al. (2014) conclude that global fungal richness is overestimated 1.5- to 2.5-fold as speculated before (Hawksworth and Mueller, 2005; Mora et al., 2011).

Mycorrhizal fungi are important functional components of global biodiversity, with 5,000 to 6,000 species estimated by Molina et al. (1992). In addition, Taylor and Alexander (2005) considered this figure conservative and speculated that between 7,000–10,000 EcM fungi may exist worldwide, associated with around 8,000 plant species. However, Tedersoo et al. (2014) documented 10,334 OTUs EcM fungi that contributed 34.1% of all taxa in the northern temperate deciduous forests.

Patterns of species richness in EcM fungal communities seem to increase along a latitudinal gradient from the tropics to the boreal zone, inversely to the general trend observed in plants (Tedersoo et al., 2014). In boreal forests, the proportion appears to be 10–100 times higher, if compared with the low species richness of host plants, i.e., > 500 macro-fungal EcM species and 4 dominant host trees in Sweden (Jonsson et al., 1999a). Allen et al. (1995) suggested that this pattern of high taxonomic diversity observed in northern EcM fungal communities is due to their relative selectivity to host plant species. General trends of hierarchical species richness (α-, β-, γ-diversity) in EcM fungi suggest that 8 to 20 "types" colonize individual root systems (α-diversity), with a turnover of 4 to 6 species between communities (β-diversity) and γ-diversity values of 40 to 50 species at a landscape scale (Allen et al., 2003). Recent estimations in native Scots pine forests suggest that between 15 to 19 putative EcM taxa colonized individual mature pine trees (Saari et al., 2005), in line with the above α-diversity values.

Differences in species composition of sporome-forming EcM fungi are observed between European and North American forests with a low percentage (27%, using Sorensen's index) of species in common, although the genera overlap by 100% between both continents (Allen et al., 1995). In addition, the relative abundance in below-ground EcM fungal communities is typically represented elsewhere by a few common species colonizing 50–70% of fine roots and a large tail of rare species (Erland and Taylor, 2002; Villarreal-Ruiz and Neri-Luna, 2018). Horton and Bruns (2001) observed this pattern of inverse relationship between abundance and rarity after analyzing several studies of EcM fungi on roots from European and North American forests. In addition, Villarreal-Ruiz and Neri-Luna (2018) demonstrate that methods of scoring ectomycorrhizas by number of root tips versus DW provide a different picture of community structure and species composition; however, a relationship between the two measures was found and they conclude that the number of EcM root tips is more suitable and less time consuming.

Functional Diversity

One of the most striking features of EcM fungi is the morphological simplicity of their extraradical hyphae and hyphal apex compared with the functional diversity displayed in growth (theoretically unlimited), foraging, nutrient uptake and immobilization, sensory recognition of host plants, response to environmental changes and somatic and reproductive differentiation (Read, 1992a; Burnett, 2003). The wide range of molecules (i.e., antibiotics, enzymes, hydrophobins, organic acids, polysaccharides and siderophores) produced by the hyphal tips potentially displayed through a mycelial network in order to mobilize nutrients from complex organic or even inorganic substrates that benefit host plants nutrient acquisition and survival are likely to be influential in ecosystem processes (Landeweert et al., 2001; Lindahl et al., 2005; Sinsabaugh, 2005). Cairney (1999) presented a comprehensive discussion on functional diversity in EcM fungi.

Isolated attempts to classify EcM fungi into functional guilds, according to their ecological, physiological, biochemical and structural attributes are well documented (Mason et al., 1982; Danielson, 1984; Ogawa, 1985; Abuzinadah and Read, 1986; Newton, 1992; Agerer, 2001). However, the narrow range of species tested and the lack of information on physiological adaptive traits and life-history of the vast majority of species and the high intraspecific variation found in fungal populations made it difficult to integrate a synthetic functional classification linked with their potential impact on ecosystem functioning (Allen et al., 2003; Lindahl et al., 2005). Nevertheless, Allen et al. (2003) proposed a set of fungal traits for a functional classification of EcM fungi based on their potential effects on ecosystem functioning. In Table 2, an adaptation is presented considering that adaptive traits related to host specificity and stand developmental stage, structure and biomass, demography and epidemiology and physiology-biochemistry are the most relevant for EcM fungi.

Relating Diversity-Functioning in EcM Fungi

Ectomycorrhizas are "balanced" mutualistic associations between plants and fungi, involving the exchange of commodities that prompts their growth, fitness and survival (Brundrett, 2004). However, ectomycorrhizal fungal species richness itself is less relevant for an understanding of ecosystem functioning compared to an understanding of the role that each fungus plays when performing certain "tasks" and the rate at which these "tasks" are accomplished (Allen et al., 2003; Morris and Robertson, 2005). Consequently, functional attributes (i.e., physiological and biochemical) of groups of EcM fungal species with similar responses to the environment or similar effects on major ecosystem processes need to be considered

Table 2. Adaptive functional traits in ectomycorrhizal fungi and their potential influence in ecosystem processes.

Adaptive traits in EcM fungi and Ecosystem processes	A	B	C	D	E	F	G	H	I
I. Host specificity and stand developmental stage	•	•	•	•	•	•	•	•	•
II. Structure and biomass									
Exploration type, mantle thickness and cystidia	•				•	•			•
Extraradical mycelium architecture, hyphal branching and density	•	•	•	•					
Presence of mycelial mats, fans or rhizomorphs	•	•	•	•					
Total biomass per soil volume	•	•	•	•				•	
Sporome production/abundance or propagules	•							•	
III. Demography and epidemiology									
Life span of genets, longevity of EcM and extraradical mycelium	•	•		•				•	
Fungal genets' size and spatial distribution	•	•	•	•				•	
Mycelium relative growth rate	•	•						•	
Ability of mycelia to proliferate in resources rich spots	•	•							
Main strategy of root colonization (spores vs. mycelium)	•	•							
Percent of root colonization	•	•	•		•	•	•		
Mycelium distance away from roots	•	•	•	•					
IV. Physiology and biochemistry									
Mycelial hydrophobicity properties			•	•					
Enzymatic capabilities (i.e., organic source use, nutrient mobilization from minerals)	•	•							
Ability to excrete organic acids and modify soil pH	•	•		•					
Membrane transporters present in hyphae, differential capability and efficiency to absorb nutrient in different form		•							
Stable isotope values of fungal tissue (i.e., $\delta^{13}C$, $\delta^{14}C$, $\delta^{15}N$)	•	•							
Metabolic rate and nutrient uptake efficiency	•	•							
Nutrient immobilization in fungal tissues	•	•		•				•	
Saprotrophic capabilities	•								
Production of secondary metabolites (i.e., antibiotics, phenolics)	•					•	•	•	
Production of phytohormones	•								
Production of heavy metal chelating compounds									•
Enzymatic capabilities to degrade toxic organic compounds (biotic or xenobiotic)	•								•
Exudation of organic compounds to the hyphosphere	•			•				•	

A, Carbon sink; B, soil nutrient uptake and transfer to plants; C, soil water uptake and transfer to plants; D, effects on soil physical-chemical properties and carbon sequestration; E, plant protection against pathogens; F, plant protection against herbivores; G, phytohormones production; H, food sources for other organisms; I, plant protection against toxic compounds. (After Allen et al., 2003 with modifications).

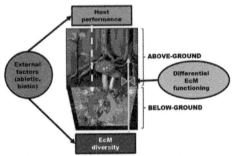

Fig. 8. Conceptual model linking ectomycorrhizal diversity and host plant performance by the differential functioning of individual ectomycorrhizal species. External factors such as pollution, disturbance and drought may alter host performance by direct influence of EcM diversity and EcM fungi may be affected by external factors via host performance. Dashed arrow refers to the feedback pathway of carbon allocation from plant host to EcM diversity (After Baxter and Dighton, 2005).

in order to classify functional groups (Hooper et al., 2002). Ecosystem functional types "are aggregated components of ecosystems whose interactions with one another and with the environment produce differences in patterns of ecosystem structure and dynamics" (Shugart, 1997).

These individual adaptive traits can be used to classify species or group of species with similar responses to a set of environmental conditions into functional types (Gitay and Noble, 1997). Therefore, Baxter and Dighton (2005) proposed a conceptual model linking ectomycorrhizal diversity to plant host performance, in line with the differential physiological functioning of EcM species under disturbances and anthropogenic influences (Fig. 8).

The model is based on the high degree of fungal diversity on individual host plants and how their physiological traits may influence the host performance. They assume that any change in EcM species composition may lead to changes in function, following the conceptual framework proposed by Allen et al. (2003) and invoking Tilman's (1994) niche complementarity. The diversity-functioning model predicts positive, negative or neutral responses of the host due to changes in EcM diversity with a positive relationship between diversity-functioning.

Genetic Diversity

The genetic diversity is placed at the bottom of the model, representing the "pyramid base" where the genetic information is concentrated and gives the fungus the ability to form, control and inherit the ectomycorrhizal habit in forest ecosystems (Tagu et al., 2002).

Biological Attributes of Genetic Relevance

The main biological features of fungal mycelium of genetic importance are clonal habit, phenotypic plasticity and pleomorphism (Burnett, 2003):

(1) The clonal habit observed in fungi follows the rule of modular organisms in which the genetic individual (genet, i.e., extraradical mycelium) is sessile with an indeterminately branched growth by the iteration of a unit of construction

(modules, i.e., individual hyphal tips) which may remain attached or become detached, physiologically independent individuals (ramets) (Andrews, 1992).

(2) Phenotypic plasticity is the variance in genotypic expression caused by environmental constraints that allow structural and physiological changes in the fungal mycelium to evolve and adapt to different micro-environments, generating many phenotypes from one genotype (Bradshaw, 1965; Stearns, 1982).

(3) Pleomorphism is the separation in time and space of asexual (anamorph) and sexual (teleomorph) phases in one fungal species in response to different environmental conditions (Burnett, 2003). All these biological attributes in fungi provide the adaptive traits to follow the next rules of growth:

   (i)   To maintain a high phenotypic plasticity in body shape, size, growth as a population and reproductive potential.

   (ii) The same genet is exposed to different environments and selective pressures.

   (iii) Somatic mutation may play a potentially important role.

   (iv) Developmental processes are produced in resource depletion zones (Andrews, 1992).

The Fungal Genome

Following Anderson et al. (1992) the overall physical attributes of the fungal genome are:

 (i)   Small nuclear fungal genome (2.6–4.3×10^4 kilobase pairs kb) if compared with other eukaryotes but 6 to 10-fold larger than in prokaryotes. The size of the genome in the ectomycorrhizal fungus *Paxillus involutus* is reported as approximately 4×10^4 kb (Deacon, 2006).

 (ii) Repeated DNA is mostly clustered and represents $< 20\%$ of the nuclear genome.

 (iii) The bulk DNA is concentrated in typical eukaryotic chromosomes.

 (iv) Additional DNA can be found in circular mitochondrial plasmids (20–200 kb) and "B" chromosomes.

 (v)   Coding information can be found in some dsRNAs. Van der Heijden et al. (2015) presented an updated list of genomic studies in EcM fungal species.

Genetic Mechanisms

Anderson et al. (1992) recognize two events during the fungal life cycle and the genetic mechanisms involved. Prezygotic events associated with the haploid homokaryotic or heterokaryotic stages before karyogamy, and postzygotic events produced after meiosis and their immediate survival. The main genetic mechanisms are vegetative incompatibility, sexual incompatibility (heterothallism and homothallism), meiotic mutants and spore killers, intersterility, prezygotic and postzygotic barriers. Somatic incompatibility is reported in EcM fungi, i.e., *Laccaria bicolor, Pisolihtus arhizus, Suillus bovinus, S. granulatus, S. luteus, S. variegatus* (Fries, 1987; Sen, 1990;

Dahlberg and Stenlid, 1994; Dahlberg, 1995). Smith and Read (2008) discussed the intraspecific and interspecific variability in EcM fungi, i.e., *Pisolitus tinctorius, Laccaria bicolor* and *Hebeloma cylindrosporum* and the genetic control of mating systems involved during their life cycle. Tagu et al. (2001) demonstrated that ectomycorrhiza formation in *Laccaria bicolor* is genetically determined by a dominant and male inherited trait. Tagu et al. (2002) reported that the ability to form ectomycorrhiza is a heritable trait and probably polygenically controlled in fungi. The molecular and genetic approaches to understanding the genetic variation on development and functioning in ectomycorrhizas have been discussed by Kropp and Anderson (1994) and Tagu et al. (2000).

Boreal Biomes EcM Fungi and Ecosystem Services

Ecosystem services, on which humans depend for their livelyhoods and for their ability to cope with and adapt to global change, have recently emerged as a strong element of environmental policy and ecosystem management (Perrings et al., 2011). Moreover, soil biodiversity research is growing among scientists and policy-makers as a result of the importance of soil organisms for the supply of ecosystem goods and services to human society (Bardgett and van der Putten, 2014). Given the beneficial functions of mycorrhizal fungi on plant fitness, resilience against environmental stresses, nutrient cycling, soil quality and plant community dynamics, mycorhizal symbiosis is now known to play a fundamental role as a provider of ecosystem services (Table 3). In addition, van der Heijden et al. (2015) presented a list of ecosystem services provided by mycorrhizal fungi.

Table 3. Main roles and ecosystem goods and services delivered by EcM fungi.

Mycorrhizal function	Ecosystem service
Root morphology modification and development of a complex, ramifying mycelial network in soil [1,2,3]	Increase plant/soil adherence, soil stability (binding action and improvement of soil structure) and water retention
Organic carbon deposition in the rhizosphere [2,3]	Changes in distribution, chemical composition and amount of organic inputs into soil
Increasing mineral nutrient and water uptake by plants [1,2,3,4]	Promote plant growth while reducing fertilizer requirement
Buffering effect against abiotic stresses [2,3]	Increased plant resistance to drought, salinity, heavy metal pollution and mineral nutrient depletion
Alleviating plant stress caused by biotic factors [1,2,3]	Increased plant resistance or tolerance against biotic stresses (i.e., pest and diseases) while reducing phytochemical input
Modification of plant metabolism and physiology [2,5,6]	Bioregulation of plant development and increase in plant quality for human health, industrial purposes and food production
Influencing plant fitness and sustainability of the whole plant-soil system [5,7]	Conservation of biodiversity Protection of landscape
Mushrooms' habitat and source of food [8,9,10]	Animal shelter and human consumption

Adapted from: [1]Barrios (2007); [2]Gianinazzi et al. (2010); [3]Smith et al. (2010); [4]Smith and Smith (2011); [5]Vosátka et al. (2012); [6]Pellegrino and Bedini (2014); [7]Bardgett and van der Putten (2014); [8]Claridge and Trappe (2005); [9]Trappe and Claridge (2005); [10]Villarreal and Gomez (1997).

Finally, linking functional traits of plants and soil microbes, such as EcM fungi, with the delivery of multiple ecosystem services is currently considered a rational mean for assessing the functioning of a given ecosystem (Rapparini and Peñuelas, 2014), as well as in determining the ecological and evolutionary responses of terrestrial ecosystems to current and future environmental change in this new climate change Anthropocene epoch (Bardgett and van der Putten, 2014; Gillings and Paulsen, 2014).

References

Abuzinadah, R.A. and Read, D.J. 1986. The role of proteins in the nitrogen nutrition of ectomycorrhizal plants: I. Utilization of peptides and proteins by ectomycorrhizal fungi. New Phytol., 103: 481–494.

Agee, J.K. 1998. Fire and pine ecosystems. pp. 193–218. *In*: Richardson, D.M. (ed.). Ecology and Biogeography of *Pinus*. Cambridge University Press, Cambridge.

Agerer, R. 1987–2002. Colour atlas of ectomycorrhizae: 1st-12th delivery. Einhorn-Verlag. Schwäbisch Gmünd, Germany.

Agerer, R. 1992. Ectomycorrhizal rhizomorphs: Organs of contact. pp. 84–90. *In*: Read, D.J., Lewis, D.H., Fitter, A.H. and Alexander, I.J. (eds.). Mycorrhizas in Ecosystems. CAB International, Wallingford.

Agerer, R. 1995. Characteristics of identified ectomycorrhizas: An attempt towards a natural classification. pp. 685–734. *In*: Varma, A.K. and Hock, B. (eds.). Mycorrhiza: Structure, Function, Molecular Biology and Biotechnology. Springer, Berlin, Heidelberg.

Agerer, R., Danielson, R.M., Egli, S., Ingleby, K., Luoma, D. and Treu, R. 1998–2004. Descriptions of Ectomycorrhizae. 1/8. Einhorn-Verlag, Schwäbisch Gmünd.

Agerer, R. 2001. Exploration types of ectomycorrhizae: A proposal to classify ectomycorrhizal systems according to their patterns of differentiation and putative ecological importance. Mycorrhiza, 11: 107–114.

Allen, E.B., Allen, M.F., Helm, D.J., Trappe, J.M., Molina, R. and Rincon, E.M. 1995. Patterns and regulation of mycorrhizal plant and fungal diversity. Plant Soil, 170: 47–62.

Allen, M.F. 1991. The ecology of mycorrhizae. Cambridge University Press, Cambridge.

Allen, M.F., Allen, E.B., Dahm, C.N. and Edwards, F.S. 1993. Preservation of biological diversity in mycorrhizal fungi: Importance and human impacts. pp. 81–108. *In*: Sundnes, G. (ed.). Human Impacts on Self-Recruiting Populations. 3rd International Knogsvoll Symposium, Tapir Press, Trondheim.

Allen, M.F., Swenson, W., Querejeta, L.M., Egerton-Warburton, L.M. and Treseder, K.K. 2003. Ecology of mycorrhizae: A conceptual framework for complex interactions among plants and fungi. Ann. Rev. Phytopathol., 41: 271–303.

Anderson, J.B., Kohn, L.M. and Leslie, J.F. 1992. Genetic mechanisms in fungal adaptation. pp. 73–98. *In*: Carroll, G. and Wicklow, D.T. (eds.). The Fungal Community: Its Organization and Role in the Ecosystem, 2nd Edition. Marcel Dekker, New York.

Andrews, J.H. 1992. Fungal life-history strategies. pp. 119–145. *In*: Carroll, G. and Wicklow, D.T. (eds.). The Fungal Community: Its Organization and Role in the Ecosystem, 2nd Edition. Marcel Dekker, New York.

Baar, J., Horton, T.R., Kretzer, A.M. and Bruns, T.D. 1999. Mycorrhizal colonization of *Pinus muricata* from resistant propagules after a stand-replacing wildfire. New Phytol., 143: 409–418.

Bardgett, R.D. and van der Putten, W.H. 2014. Below-ground biodiversity and ecosystem functioning. Nature, 515: 505–511.

Barnosky, A.D., Matzke, N., Tomiya, S., Guinevere, O.U. et al. 2011. Has the Earth's sixth mass extinction already arrived? Nature, 471: 51–57. doi:10.1038/nature09678.

Barrios, E. 2007. Soil biota, ecosystem services and land productivity. Ecol. Econ., 64: 269–285.

Baxter, J.W. and Dighton, J. 2005. Diversity-functioning relationships in ectomycorrhizal fungal communities. pp. 383–398. *In*: Dighton, J., White, J.F. and Oudemans, P. (eds.). The Fungal Community: Its Organization and Role in the Ecosystem, 3rd Edition. Taylor and Francis, Boca Raton.

Begon, M., Harper, J.L. and Townsend, C.R. 1996. Ecology: Individuals, Populations and Communities, 3rd Edition. Blackwell Science, Oxford.

Bell, E.A., Boehnke, P., Harrison, T.M. and Mao, W.L. 2015. Potentially biogenic carbon preserved in a 4.1 billion-year-old zircon. Proc. Nat. Acad. Sci., 112: 14518–14521.

Bidartondo, M.I. and Gardes, M. 2005. Fungal diversity in molecular terms: Profiling, identification, and quantification in the environment. pp. 215–239. *In*: Dighton, J., White, J.F. and Oudemans, P. (eds.). The Fungal Community: Its Organization and Role in the Eecosystem, 3rd Edition. Taylor and Francis, Boca Raton.

Bidartondo, M.I., Read, D.J., Trappe, J.M., Merckx, V., Ligrone, R. and Duckett, J.G. 2011. The dawn of symbiosis between plants and fungi. Biol. Lett., 7: 574–577. doi:10.1098/rsbl.2010.1203.

Blasius, D., Feil, W., Kottke, I. and Oberwinkler, F. 1986. Hartig net structure and formation in fully ensheathed ectomycorrhizas. pp. 251–258. *In*: Gianinazzi-Pearson, V. and Gianinazzi, V. (eds.). Physiological and Genetical Aspects of Mycorrhizae. Proc. 1st SEM.-INRA, Paris.

Blasius, D. and Oberwinkler, F. 1989. Succession of mycorrhizae: A matter of tree age or stand age? Annales des Sciences Forestieres, 46: 758s–761s.

Bonfante-Fasolo and Scannerini, S. 1992. The cellular basis of plant-fungus interchanges in mycorrhizal associations. pp. 65–101. *In*: Allen, M.F. (ed.). Mycorrhizal functioning: An Integrative Plant-Fungal Process. Chapman and Hall, London, UK.

Bourgeau-Chavez, L.L., Alexander, M.E., Stocks, B.J. and Kasischke, E.S. 2000. Distribution of forest ecosystems and the role of fire in the North American boreal region. pp. 111–131. *In*: Kasischke, E.S. and Stocks, B.J. (eds.). Fire, Climate Change, and Carbon Cycling in the Boreal Forest. Springer, New York.

Bradshaw, A.D. 1965. Evolutionary significance of phenotypic plasticity in plants. Advances in Genetics, 13: 115–155.

Brown, T. 1997. Clearances and clearings: Deforestation in Mesolithic/Neolithic Britain. Oxford J. Archaeol., 16: 133–146.

Brundrett, M.C. 1991. Mycorrhizas in natural ecosystems. Adv. Ecol. Res., 21: 171–313.

Brundrett, M.C. 2002. Coevolution of roots and mycorrhizas of land plants. New Phytol., 154: 275–304.

Brundrett, M.C. 2004. Diversity and classification of mycorrhizal associations. Biological Review 79: 473–495.

Burnett, J. 2003. Fungal population and species. Oxford University Press, Oxford.

Cairney, J.W.G. 1999. Intraspecific physiological variation: Implications for understanding functional diversity in ectomycorrhizal fungi. Mycorrhiza, 9: 125–135.

Ceballos, G., Ehrlich, P.R., Barnosky, A.D, García, A., Pringle, R.M. and Palmer, T.M. 2015. Accelerated modern human-induced species losses: Entering the sixth mass extinction. Sci. Adv., 1:e1400253, dOI: 10.1126/sciadv.1400253.

Ceballos, G., Ehrlich, P.R. and Dirzo, R. 2017. Biological annihilation via the ongoing sixth mass extinction signaled by vertebrate population losses and declines. Proc. Nat. Acad. Sci., 114: 1–8.

Chapin III, S.F., Zavaleta, E.S., Eviner, V.T. et al. 2000. Consequences of changing biodiversity. Nature, 405: 234–242. doi:10.1038/35012241.

Chapin III, S.F. 2003. Effects of plant traits on ecosystems and regional processes: A conceptual framework for predicting the consequences of global change. Ann. Bot., 91: 455–463.

Claridge, A.W. and Trappe, J.M. 2005. Sporocarp mycophagy: nutritional, behavioral, evolutionary, and physiological aspects. pp. 599–611. *In*: Dighton, J., White, J.F. and Oudemans, P. (eds.). The Fungal Community: Its Organization and Role in the Rcosystem, 3rd Edition. Taylor and Francis, Boca Raton.

Cornelissen, J.H.C., Aerts, R., Cerabolini. B., Werger, M.J.A. and van der Heijden, M.G.A. 2001. Carbon cycling traits of plant species are linked with mycorrhizal strategy. Oecologia, 129: 611–619.

Crutzen, P.J. 2002. Geology of mankind. Nature, 415: 23.

Cumming, G. and Legg, C. 1995. Longevity of the *Calluna vulgaris* seed bank determined from a history of lead smelting at Leadhills and Wanlockhead, Scotland. pp. 135–139. *In*: Thompson, D.B.A., Hester, A.J. and Usher, M.B. (eds.). Heats and Moorlands: Cultural Landscapes. HMSO, Edinburgh.

Dahlberg, A. and Stenlid, J. 1994. Size, distribution and biomass of genets in populations of *Suillus bovinus* (L.: Fries) Roussel revealed by somatic incompatibility. New Phytol., 128: 225–234.

Dahlberg, A. 1995. Somatic incompatibility in ectomycorrhizae. pp. 115–136. *In*: Varma, A. and Hock, B. (eds.). Mycorrhiza: Function, Molecular Biology and Biotechnology. Springer-Verlag, Heidelberg.

Danielson, R.M. 1984. Ectomycorrhizal associations in jack pine stands in north- eastern Alberta. Can. J. Bot., 62: 932–939.

Dawkins, R. 1983. Universal Darwinism. pp. 15–35. *In*: Hull, D.L. and Ruse, M. (eds.). The Philosophy of Biology. Oxford University Press, New York.

Deacon, J. 2006. Fungal biology, 4th Edition. Blackwell Publishing, Malden, MA, USA.

Deacon, J.M., Donaldson, S.J. and Last, F.T. 1983. Sequences and interactions of mycorrhizal fungi on birch. Plant Soil, 71: 257–262.

Delmas, J. 1987. Fruiting requirements of fungi under natural and artificial conditions. Indian Mushroom Sci., 2: 219–229.

Desirò, A., Rimington, W.R., Jacob, A., Pol, N.V., Smith, M.E., Trappe, J.M., Bidartondo, M.I. and Bonito, G. 2017. Multigene phylogeny of Endogonales, an early diverging lineage of fungi associated with plants. IMA Fungus, 8: 245–257.

Dighton, J. and Mason, P.A. 1985. Mycorrhizal dynamics during forest tree development. pp. 117–139. *In*: Moore, D., Casselton, L.A., Wood, D.A. and Frankland, J.C. (eds.). Developmental Biology of Higher Fungi. Cambridge University Press, Cambridge.

Dodson, S.I., Allen, T.F.H., Carpenter, S.R., Ives, A.R., Jeanne, R.L., Kitchell, J.F., Langston, N.E. and Turner, M.G. 1998. Ecology. Oxford University Press, New York.

Eckstein, R.L. 2000. Nitrogen retention by *Hylocomium splendens* in a subarctic birch woodland. J. Ecol., 88: 506–515.

Egger, K.N. 1995. Molecular analysis of ectomycorrhizal fungal communities. Can. J. Bot., 73: S1415–S1422.

Erland, S. and Taylor, A.F.S. 2002. Diversity of ecto-mycorrhizal fungal communities in relation to the abiotic environment. pp. 163–200. *In*: van der Heiden, M.G.A. and Sanders, I.R. (eds.). Mycorrhizal Ecology, Ecological Studies. Springer, Berlin.

Farquhar, J., Bao, H. and Thiemens, M. 2000. Atmospheric influence of Earth's earliest Sulfur Cycle. Science, 289: 756–758.

Fries, N. 1987. Somatic incompatibility and field distribution in *Suillus luteus* (Boletaceae). New Phytol., 107: 735–739.

Gardes, M. and Bruns, T.D. 1996. Community structure of ectomycorrhizal fungi in a *Pinus muricata* forest: Above-and below-ground views. Can. J. Bot., 74z: 1572–1583.

Gaston, K.J. and Spicer, J.I. 1998. Biodiversity: An Introduction. Blackwell Science, Oxford.

Gerrienne, P., Servais, T. and Vecoli, M. 2016. Plant evolution and terrestrialization during Palaeozoic times: The phylogenetic context. Rev. Palaeobot. Palynol., 227: 4–18.

Gianinazzi, S., Gollotte, A., Noëlle Binet, M., van Tuinen, D., Redecker, D. and Wipf, D. 2010. Agroecology: The key role of arbuscular mycorrhizas in ecosystem services. Mycorrhiza, 20: 519–530.

Gillings, M.R. and Paulsen, I.T. 2014. Microbiology of the Anthropocene. Anthropocene, 5: 1–8.

Gimingham, C.H. 1960. Biological flora of the British Isles. *Calluna vulgaris* L. Hull. J. Ecol., 48: 455–483.

Gitay, H. and Noble, I.R. 1997. What are functional types and how should we seek them? pp. 3–19. *In*: Smith, T.M., Shugart, H.H. and Woodward, F.I. (eds.). Plant Functional Types. Cambridge University Press, Cambridge.

Glassman, S.I., Levine, C.R., DiRocco, A.M., Battles, J.J. and Bruns, T.D. 2016. Ectomycorrhizal fungal spore bank recovery after a severe forest fire: Some like it hot. The ISME Journal, 10: 1228–1239.

Goldammer, J.G. and Stocks, B.J. 2000. Eurasian perspective of fire: dimension, management, policies, and scientific requirements. pp. 49–65. *In*: Kasischke, E.S. and Stocks, B.J. (eds). Fire, climate change, and carbon cycling in the boreal forest. Springer, New York.

Gontier, N. 2007. Universal symbiogenesis: An alternative to universal selectionist accounts of evolution. Symbiosis, 44: 167–181.

Granström, A. 1987. Seed banks in five boreal forest stands originating between 1810 and 1963. J. Ecol., 75: 321–331.

Grime, J.P., Hodgson, J.G. and Hunt, R. 1988. Comparative Plant Ecology: A Functional Approach to Common British Species. Unwin Hyman, London.

Grime, J.P. 1977. Evidence for the existence of three primary strategies in plants and its relevance to ecological and evolutionary theory. The Am. Nat., 111: 1169–1194.

Grime, J.P. 1994. The role of plasticity in exploiting environmental heterogeneity. pp. 1–19. *In*: Calwell, M.M. and Pearcy, R.W. (eds.). Exploitation of Environmental Heterogeneity by Plants: Ecophysiological Processes Above- and Belowground. Academic Press, San Diego.

Grime, J.P. 2001. Plant Strategies, Vegetation Processes and Ecosystem Properties, 2nd Edition. John Willey & Sons, Chichester.

Harholt, J., Moestrup, Ø. and Ulvskov, P. 2016. Why plants were terrestrial from the beginning. Tr. Pl. Sci., 21: 96–101.

Harley, J.L. and Smith, S.E. 1983. Mycorrhizal Symbiosis. Academic Press, London.

Hawksworth, D.L. and Mueller, G.M. 2005. Fungal communities: Their diversity and distribution. pp. 27–37. *In*: Dighton, J., White, J.F. and Oudemans, P. (eds.). The Fungal Community: Its Organization and Role in the Ecosystem, 3rd Edition. Taylor and Francis, Boca Raton.

Hazen, R.M., Griffin, P.L., Carothers, J.M. and Szostak, J.W. 2007. Functional information and the emergence of biocomplexity. Proc. Nat. Acad. Sci., 104: 8574–8581.

Helmisaari, H.-S. 1995. Nutrient cycling in *Pinus sylvestris* stands in eastern Finland. Plant Soil, 168-169: 327–336.

Hens, L. and Lefevere, R. 2007. Global bioethics and human ecology. *In*: Susanne, C., Cambron, A., Casado, M., Cascais, F., Rebato, E., Salona, M., Sanchez, A., Simitopoulou, K., Szente, M., Toth, J. and Xirotiris, N. (eds.). Bioethics Global and Societal Aspects. European Association of Global Bioethics, Laboratory of Anthropology D.U.TH., Komotini, Greece.

Hibbett, D. 2016. The invisible dimension of fungal diversity. Science, 351: 1150–1151.

Hodkinson, I.D., Webb, N.R. and Coulson, S.J. 2002. Primary community assembly on land – the missing stages: why are the heterotrophic organisms always there first. J. Ecol., 90: 569–577.

Holden, E., Newton, A., Davy, L. and Watling, R. 2001. Survey of Woodland Fungi, Mar Lodge Estate. Final Report to the National Trust for Scotland.

Hooper, D.U., Solan, M., Symstad, A., Diaz, S., Gessner, M.O., Buchmann, N., Degrange, V., Grime, P., Hulot, F., Mermillod-Blondin, F., Roy, J., Spehn, E. and van Perr, L. 2002. Species diversity, functional diversity and ecosystem functioning. pp. 195–281. *In*: Loreau, M., Naeem, S. and Inchausti, P. (eds.). Biodiversity and Ecosystem Functioning: Synthesis and Perspectives. Oxford University Press, Oxford.

Horton, T.R. and Bruns, T.D. 2001. The molecular revolution in ectomycorrhizal ecology: Peeking into the black-box. Mol. Ecol., 10: 1855–1871.

Ingold, C.T. 1971. Fungal Spores: Their Liberation and Dispersal. Clarendon Press, Oxford.

Jonsson, L., Dahlberg, A., Nilsson, M.-C., Karen, O. and Zackrisson, O. 1999a. Continuity of ectomycorrhizal fungi in self-regenerating boreal *Pinus sylvestris* forests studied by comparing mycobiont diversity on seedlings and mature trees. New Phytol., 142: 151–162.

Jonsson, L., Dahlberg, A., Nilsson, M.-C., Zackrisson, O. and Kårén, O. 1999b. Ectomycorrhizal fungal communities in late successional Swedish boreal forest, and their composition following wildfire. Mol. Ecol., 8: 205–215.

Jumpponen, A. 2003. Soil fungal community assembly in a primary successional glacier forefront ecosystem as inferred from rDNA sequence analyses. New Phytol., 158: 569–578.

Kasischke, E.S. 2000. Boreal ecosystems in the global carbon cycle. pp. 19–30. *In*: Kasischke, E.S. and Stocks, B.J. (eds.). Fire, Climate Change, and Carbon Cycling in the Boreal Forest. Springer, New York.

Keizer, P.J. and Arnolds, E. 1994. Succession of ectomycorrhizal fungi in roadside verges planted with common oak (*Quercus robur* L.) in Drenthe, The Netherlands. Mycorrhiza, 4: 147–159.

Kenrick, P., Wellman, C.H., Schneider, H. and Edgecombe, G.D. 2012. A timeline for terrestrialization: Consequences for the carbon cycle in the Palaeozoic. Phil. Trans. Royal Soc. B., 367: 519–536.

Koonin, E.V. 2015. Origin of eukaryotes from within archaea, archaeal eukaryome and bursts of gene gain: Eukaryogenesis just made easier?. Phil. Trans. Royal Soc. B., 370: 20140333.

Koskella, B., Hall, L.J. and Metcalf, C.J.E. 2017. The microbiome beyond the horizon of ecological and evolutionary theory. Nat. Ecol. Evol., 1: 1606–1615.

Kottke I. 1992. Ectomycorrhizas—organs for uptake and filtering of cations. pp. 316–322. *In*: Read, D.J., Lewis, D.H., Fitter, A.H. and Alexander, I.J. (eds.). Mycorrhizas in ecosystems. CAB International, Wallingford.

Krings, M., Taylor, T.N., Hass, H., Kerp, H., Dotzle, N. and Hermsen, E.J. 2007. An alternative mode of early land plant colonization by putative endomycorrhizal fungi. Pl. Signal. Behavior, 2: 125–126.

Kropp, R.B. and Anderson, A.J. 1994. Molecular and genetic approaches to understanding variability in mycorrhizal formation and functioning. pp. 309–336. *In*: Pfleger, F.L. and Lingerman, R.G. (eds.). Mycorrhizae and Plant Health. APS Press, St Paul.

Ku, C., Nelson-Sathi, S., Roettger, M., Sousa, F.L., Lockhart, P.J., Bryant, D., Hazkani-Covo, E., McInerney, J.O., Landan, G. and Martin, W.F. 2015. Endosymbiotic origin and differential loss of eukaryotic genes. Nature, 524: 427–437.

Landeweert, R., Hoffland, T.W., Finlay, R.D., Kuyper, P. and van Breemen, N. 2001. Linking plants to rocks: Ectomycorrhizal fungi mobilize nutrients from minerals. Tr. Ecol. Evol., 16: 248–254.

Landeweert, R., Leeflang, P., Kuyper, P., Hoffland, T.W., Rosling, A., Wernar, K. and Smit, E. 2003. Molecular identification of ectomycorrhizal mycelium in soil horizons. Appl. Environ. Microbiol. 69: 327–333.

Lanner, R.M. 1998. Seed dispersal in *Pinus*. pp. 281–295. *In*: Richardson, D.M. (ed.). Ecology and Biogeography of *Pinus*. Cambridge University Press, Cambridge.

Larcher, W. 2003. Physiological Plant Ecology: Ecophysiology and Stress Physiology of Functional Groups. Springer, Berlin.

Last, F.T., Dighton, J. and Mason, P.A. 1987. Succession of sheathing mycorrhizal fungi. Tr. Ecol. Evol., 2: 157–161.

Leake, J., Johnson, D., Donnelly, D., Muckle, G., Boddy, L. and Read, D. 2004. Networks of power and influence: The role of mycorrhizal mycelium in controlling plant communities and agroecosystem functioning. Can. J. Bot., 82: 1016–1045.

Lederberg, J. 2001. 'Ome Sweet 'Omics – a genealogical treasury of words. The Scientist, 15: 8.

Léveque, C. and Mounolou, J.C. 2003. Biodiversity. John Wiley & Sons, West Sussex.

Lewis, S.L. and Maslin, M.A. 2015. Defining the Anthropocene. Nature, 519: 171–180.

Lindahl, B.D., Finlay, R.D. and Cairney, J.W.G. 2005. Enzymatic activities of mycelia in mycorrhizal fungal communities. pp. 331–348. *In*: Dighton, J., White, J.F. and Oudemans, P. (eds.). The Fungal Community: Its Organization and Role in the Ecosystem, 3rd Edition. Taylor and Francis, Boca Raton.

Lindahl, B.O., Taylor, A.F.S. and Finlay, R.D. 2002. Defining nutritional constraints on carbon cycling in boreal forests-towards a less "phytocentric" perspective. Plant Soil, 242: 123–135.

Macpherson, C.C. 2013. Climate change is a bioethics problem. Bioethics, 271: 305–308.

Malloch, D. and Blackwell, M. 1992. Dispersal of fungal diaspores. pp. 147–171. *In*: Carroll, G.C. and Wicklow, D.T. (eds.). The Fungal Community: Its Organization and Role in the Ecosystem, 2nd Edition Marcel Dekker, New York.

Martin, F.M., Uroz, S. and Barker, D.G. 2017. Ancestral alliances: Plant mutualistic symbioses with fungi and bacteria. Science, 356, eaad4501. DOI: 10.1126/science.aad4501.

Mason, P.A., Last, F.T., Pelham, J. and Ingleby. K. 1982. Ecology of some fungi associated with ageing stand of birches (*Betula pendula* and *Betula pubescens*). For. Ecol. Manag., 4: 19–39.

Mason, P.A., Wilson, J., Last, F.T. and Walker, C. 1983. The concept of succession in relation to the spread of sheathing mycorrhizal fungi on inoculated tree seedlings growing in unsterile soils. Plant Soil, 71: 247–256.

Mason, W.L., Hampson, A. and Edwards, C. 2004. Managing the Pinewoods of Scotland. Forestry Commission, Edinburgh.

Mencuccini, N. and Grace, J. 1996. Hydraulic conductance, light interception, and needle nutrient concentration in Scots pine stands (Thetford, U.). Tree Physiol., 16: 459–469.

Molina, R., Massicote, H. and Trappe, J.M. 1992. Specificity phenomena in mycorrhizal symbiosis: Community-ecological consequences and practical implications. pp. 357–423. *In*: Allen, M.F. (ed.). Mycorrhizal Functioning: An Integrative Plant-Gungal Process. Chapman and Hall, London.

Mora, C., Tittensor, D.P., Adl, S., Simpson, A.G.B. and Worm, B. 2011. How many species are there on earth and in the ocean? PLoS Biology, 9: e1001127.

Moran, N.A. 2007. Symbiosis as an adaptive process and source of phenotypic complexity. Proc. Nat. Acad. Sci., 104: 8627– 8633.

Morris, J.L., Puttick, M.N., Clark, J.W., Edwards, D., Kenrick, P. et al. 2018. The timescale of early land plant evolution. Proc. Nat. Acad. Sci. DOI: 10.1073/pnas.1719588115.

Morris, S.J. and Robertson, G.P. 2005. Linking function between scales of resolution. pp. 13–26. *In*: Dighton, J., White, J.F. and Oudemans, P. (eds.). The Fungal Community: Its Organization and Role in the Ecosystem, 3rd Edition Taylor and Francis, Boca Raton.

Näsholm, T., Akblad, A., Nordin, A., Geisler, R., Högberg, M. and Högberg, P. 1998. Boreal forest plants take up organic nitrogen. Nature, 392: 914–916.

Neri-Luna, C. and Villarreal-Ruiz, L. 2012. Simbiosis micorrícica: Un análisis de su relevante función ecosistémica y en la provisión de servicios ambientales. pp. 37–61. *In*: Huerta-Martínez, F.M. and Castro-Félix, L.P. (eds.). Interacciones Ecológicas. Universidad de Guadalajara. Jalisco. México.

Nevo, E. 2001. Genetic diversity. *In*: Levin, S.A. (ed.). Encyclopedia of Biodiversity, 3: 195–213.

Newton, A.C. 1992. Towards a functional classification of ectomycorrhizal fungi. Mycorrhiza, 2: 75–79.

Nikolov, N. and Helmisaari, H. 1992. Silvics of the circumpolar boreal forest tree species. pp. 13–84. *In*: Shugarte, H.H., Leemans, R. and Bonan, G.B. (eds.). A Systems Analysis of the Global Boreal Forest. Cambridge University Press, Cambridge.

Nordin, A., Högberg, P. and Näsholm, T. 2001. Soil nitrogen form and plant nitrogen uptake along a boreal forest productivity gradient. Oecologia, 129: 125–132.

Nutman, A.P., Bennett, V.C., Friend, C.R.L., Van Kranendonk, M.J. and Chivas. A.R. 2016. Rapid emergence of life shown by discovery of 3,700-million-year-old microbial structures. Nature, 537.

Ogawa, M. 1985. Ecological characters of ectomycorrhizal fungi and their mycorrhizas: An introduction to the ecology of ectomycorrhizal fungi. JARQ 18: 305-314.

Pellegrino, E. and Bedini, S. 2014. Enhancing ecosystem services in sustainable agriculture: Biofertilization and biofortification of chickpea (*Cicer arietinum* L.) by arbuscular mycorrhizal fungi. Soil Biol. Biochem., 68: 429–439.

Perrings, Ch., Duraiappah, A., Larigauderie, A. and Mooney, H. 2011. The Biodiversity and Ecosystem Services Science-Policy Interface. Science, 331: 1139–1140.

Persson, A. 1984. The dynamic fine roots of forest tree. *In*: Ågren, G.I. (ed.). State and Change of Forest Ecosystems-indicators in Current Research. Swedish University of Agricultural Sciences, Department of Ecology and environmental Research, report # 13: 193–204.

Persson, J. and Näsholm, T. 2001. Amino acid uptake: A widespread ability among boreal forest plants. Ecol. Lett., 4: 434–438.

Rao, C.S., Sharma, G.D. and Shukla, A.K. 1997. Distribution of ectomycorrhizal fungi in pure stands of different age groups of *Pinus kesiya*. Can. J. Bot., 85: 85–91.

Rapparini, F. and Peñuelas, J. 2014. Mycorrhizal fungi to alleviate drought stress on plant growth. pp. 21–42. *In*: Mohammad Miransari (ed.). Use of Microbes for the Alleviation of Soil Stress, Volume 1. Springer-New York.

Read, D.J. 1991. Mycorrhiza in Ecosystems. Experientia, 47: 376–391.

Read, D.J. 1992a. The mycorrhizal mycelium. pp. 102–133. *In*: Allen, M.F. (ed.). Mycorrhizal Functioning: An Integrative Plant-Fungal Process. Chapman and Hall, London, UK.

Read, D.J. 1992b. The mycorrhizal fungal community with special reference to nutrient mobilization. pp. 631–668. *In*: Carroll, G. and Wicklow, D.T. (eds.). The Fungal Community: Its Organization and Role in the Ecosystem, 2nd Edition. Marcel Dekker, New York.

Read, D.J. 1998. The mycorrhizal status of *Pinus*. pp. 324–340. *In*: Richardson, D.M. (ed.). Ecology and Biogeography of *Pinus*. Cambridge University Press, Cambridge.

Redecker, D., Koder, R. and Graham, L.E. 2000. Glomalean Fungi from the Ordovician. Science, 289: 1920–1921.

Ricek, E.W. 1981. Die pilzgesellschaften heranwaschsender fichtenbestande auf ehemaligen wiesenflachen. Zeska. Mykologia, 47: 123–148.

Richards, P.W. 1984. Introduction. pp. 1–8. *In*: Dyer, A.F. and Duckett, J.G. (eds.). The Experimental Biology of Bryophytes. Academic Press, London.

Richardson, D.M. and Rundel, P.W. 1998. Ecology and biogeography of *Pinus*: And introduction. pp. 3–46. *In*: Richardson, D.M. (ed.). Ecology and Biogeography of *Pinus*. Cambridge University Press, Cambridge.

Ritchie, J.C. 1955. Biological flora of the British Isles: *Vaccinium vitis-idaea* L. J. Ecol., 44: 701–708.

Ritchie, J.C. 1956. Biological flora of the British Isles: *Vaccinium myrtillus* L. J. Ecol., 44: 291–299.

Rocha, J.C., Peterson, G.D. and Biggs, R. 2015. Regime Shifts in the Anthropocene: Drivers, Risks, and Resilience. PLoS ONE, 10: e0134639.

Rodwell, J.S. (ed). 1991. British plant communities: volume 1. Woodlands and scrub. Cambridge University Press, Cambridge, pp. 300–315.

Rook, E.J.S. 1998. *Hylocomium splendens*: Mountain Fern Moss.

Ross, S.E., Callaghan, T.V., Ennos, A.R. and Sheffield, E. 1998. Mechanics and growth form of the moss *Hylocomium splendens*. Ann. Bot., 82: 787–793.

Rundel, P.W. and Yoder, B.J. 1998. Ecophysiology of *Pinus*. pp. 296–323. *In*: Richardson, D.M. (ed.). Ecology and Biogeography of *Pinus*. Cambridge University Press, Cambridge.

Rydgren, K., Kroon, D.H., Økland, R.H. and Groenendael, J.V. 2001. Effects of fine-scale disturbances on the demography and population dynamics of the clonal moss *Hylocomium splendens*. J. Ecol., 89: 395–405.

Saari, S.K., Campbell, C.D., Russell, J., Alexander, I.J. and Anderson, I.C. 2005. Pine microsatellite markers allow roots and ectomycorrhizas to be linked to individual trees. New Phytol., 165: 295–304.

Schlaeppi, K. and Bulgarelli, D. 2015. The plant microbiome at work. MPMI, 28: 212–217.

Schmit, J.P. and Lodge, D.J. 2005. Classical methods and modern analysis for studying fungal diversity. pp. 193–214. *In*: Dighton, J., White, J.F. and Oudemans, P. (eds.). The Fungal Community: Its Organization and Role in the Ecosystem, 3rd Edition. Taylor and Francis, Boca Raton.

Schulze, E.-D. 2000. The carbon and nitrogen cycle of forest ecosystem. pp. 3–13. *In*: Schulze, E.-D. (ed.). Carbon and Nitrogen Cycling in European Forest Ecosystems. Ecological Studies, Volume 142. Springer-Verlag, Berlin Heidelberg.

Scurfield, G. 1954. Biological flora of the British Isles: *Deschampsia flexuosa* (L.) Trin. J. Ecol., 42: 225–233.

Sen, R. 1990. Intraspecific variation in two species of *Suillus* from Scots pine (*Pinus sylvestris* L.) forests based on somatic incompatibility and isoenzyme analyses. New Phytol., 114: 603–612.

Shaw, P.J.A. 1992. Fungi, fungivores, and fungal food webs. pp. 295–310. *In*: Carroll, G. and Wicklow, D.T. (eds.). The Fungal Community: Its Organization and Role in the Ecosystem, 2nd Edition. Marcel Dekker, New York.

Shugart, H.H. 1997. Plant and ecosystem functional types. pp. 20–43. *In*: Smith, T.M., Shugart, H.H. and Woodward, F.I. (eds.). Plant functional types. Cambridge University Press, Cambridge.

Sinclair, W.T., Morman, J.D. and Ennos, R.A. 1998. Multiple origins for Scots pine (*Pinus sylvestris* L.) in Scotland: Evidence from mitochondrial DNA variation. Heredity, 80: 233–240.

Sinsabaugh, R.L. 2005. Fungal enzymes at the community scale. pp. 349–360. *In*: Dighton, J., White, J.F. and Oudemans, P. (eds.). The Fungal Community: Its Organization and Role in the Ecosystem, 3rd Edition. Taylor and Francis, Boca Raton.

Smith, F.A. and Smith, S.E. 2011. What is the significance of the arbuscular mycorrhizal colonization of many economically important crop plants? Plant Soil, 348: 63–79.

Smith, J.E., Perry, D.A. and Molina, R. 1992. Occurrence of ecto- and ericoid mycorrhizas on *Gaultheria shallon* and *Rhododendron macrophyllum* seedlings grown in soils from the Oregon Coast Range. pp. 401–402. *In*: Read, D.J., Lewis, D.H., Fitter, A.H. and Alexander, I.J. (eds.). Mycorrhizas in Ecosystems. CAB International, Wallingford.

Smith, S.E. and Read, D.J. 2008. Mycorrhizal Symbiosis, 3rd Edition. Academic Press, London.

Smith, S.E., Facelli, E., Pope, S. and Smith, F.A. 2010. Plant performance in stressful environments: interpreting new and established knowledge of the roles of arbuscular mycorrhizas. Plant Soil, 326: 3–20.

Solbrig, O.T. 1991. From genes to ecosystems: A research agenda for biodiversity. Report of a IUBS-SCOPE-UNESCO workshop. The International Union of Biological Science, Paris.

Stearns, S.C. 1982. The role of development in the evolution of life-histories. pp. 237–258. *In*: Bonner, J.T. (ed.). Evolution and Development. Springer-Verlag, New York.

Strullu-Derrien, C., Selosse, M.A., Kenrick, P. and Martin, F.M. 2018. The origin and evolution of mycorrhizal symbioses: From palaeomycology to phylogenomics. New Phytol., https://doi.org/10.1111/nph.15076.

Swift, M.J. 2005. Human impacts on biodiversity and ecosystem services: An overview. pp. 627–641. *In*: Dighton, J., White, J.F. and Oudemans, P. (eds.). The Fungal Community: Its Organization and Role in the Ecosystem, 3rd Edition. Taylor and Francis, Boca Raton.

Swingland, I.R. 2001. Biodiversity, definition of. pp. 377–391. *In*: Levin, S.A. (ed.). Encyclopedia of Biodiversity, Volume 1, University of Minnesota Press, Minneapolis.

Tagu, D., Lapeyrie, F., Ditengou, F., Lagrange, H., Laurent, P., Missoum, N., Nehls, U. and Martin, F. 2000. Molecular aspects of ectomycorrhiza development. pp. 69–89. *In*: Podila, G.K. and Douds, D.D. (eds.). Current Advances in Mycorrhizae Research. APS Press, St Paul.

Tagu, D., Faivre Rampant, P., Lapeyrie, F., Frey-Klett, P., Vion, P. and Villar, M. 2001. Variation in the ability to form ectomycorrhizas in the Fi progeny of an interspecific poplar (*Populus* spp.) cross. Mycorrhiza, 10: 237–240.

Tagu, D., Lapeyrie, F. and Martin, F. 2002. The ectomycorrhizal symbiosis: Genetics and development. Plant Soil, 10: 97–105.

Tamm, C.O. 1991. Nitrogen in terrestrial ecosystems: Questions of productivity, vegetational changes, and ecosystem stability. Ecological studies 81. Springer-Verlag, Berlin, Heidelberg.

Taylor, A.F.S. and Alexander, I. 2005. The ectomycorrhizal symbiosis: Life in the real world. Mycologist, 19: 102–112.

Taylor, D.L. and Bruns, T.D. 1999. Community structure of ectomycorrhizal fungi in a *Pinus muricata* forest: Minimal overlap between the mature forest and resistant propagules communities. Mol. Ecol., 8: 1837–1850.

Tedersoo, L., Koljalg, U., Hallenberg, N. and Larsson, K.-H. 2003. Fine scale distribution of ectomycorrhizal fungi and roots across substrate layers including coarse woody debris in a mixed forest. New Phytol., 159: 153–165.

Tedersoo, L., Bahram, M., Põlme, S., Kõljalg, U., Yorou, N.S., Wijesundera, R., Villarreal-Ruiz, L. et al. 2014. Global diversity and geography of soil fungi. Science, 346: 1256688.

Termorshuizen, A.J. and Schaffers, A.P. 1989. Succession of mycorrhizal fungi in stands of *Pinus sylvestris* in the Netherlands. Agric. Ecosys. Environ., 28: 503–507.

Tilman, D. 1994. Competition and biodiversity in spatially structured habitats. Ecology, 75: 2–16.

Tilman, D. and Lehman, C. 2001. Biodiversity, composition and ecosystem processes: Theory and concepts. pp. 9–41. *In*: Kinzig, A.P., Pacala, S.W. and Tilman, D. (eds.). The Functional Consequence of Biodiversity: Empirical Progress and Theoretical Extensions. Princeton University Press, Princeton and Oxford.

Tilman, D. 2001. Functional diversity. *In*: Levin, S.A. (ed.). Encyclopedia of Biodiversity, Volume 3: 109–120.

Trappe, J,M, and Claridge, A.W. 2005. Hypogeous fungi: Evolution of reproductive and dispersal strategies through interactions with animals and mycorrhizal plants. pp. 613–623. *In*: Dighton, J., White, J.F. and Oudemans, P. (eds.). The Fungal Community: Its Organization and Role in the Ecosystem, 3rd Edition. Taylor and Francis, Boca Raton,

Valanne, N. 1984. Photosynthesis and photosynthetic products in mosses. pp. 257–273. *In*: Dyer, A.F. and Duckett, J.G. (eds.). The Experimental Biology of Bryophytes. Academic Press, London.

van der Heijden, M.G.A., Martin, F.M., Selosse, M.-A. and Sanders, I.R. 2015. Mycorrhizal ecology and evolution: the past, the present, and the future. New Phytol., 205: 1406–1423.

van der Heijden, M.G.A. and Hartmann, M. 2016. Networking in the Plant Microbiome. PLoS Biol., 14: e1002378.

Vickers, A.D. and Palmer, S.C.F. 2000. The influence of canopy and other factors upon the regeneration of Scots pine and its associated ground flora within Glen Tanar National Nature reserve. Forestry, 73: 37–49.

Vidaković, M. 1991. Conifers: Morphology and Variation. Grafički Zavod Hrvatske, Zagreb.

Villarreal, L. and Gomez, A. 1997. Inventory and monitoring wild edible mushrooms in Mexico: Challenge and opportunity for sustainable development. pp. 99–109. *In*: Palm, M.E. and Chapela, I.H. (eds.). Mycology in Sustainable Development: Expanding Concepts, Vanishing Borders. Parkway Publishers, Boone, North Carolina.

Villarreal-Ruiz, L. 1996. Gaia la madre tierra: Una nueva vision del mundo vivo? IREGEP Informa (Boletín informativo del Instituto de Recursos Genéticos y Productividad del Colegio de Postgraduados en Ciencias Agrícolas). Año 2, Noviembre-Diciembre (8): 1–3.

Villarreal-Ruiz, L., Anderson, I.C. and Alexander, I.J. 2004. Interaction between an isolate from the *Hymenoscyphus ericae* aggregate and roots of *Pinus* and *Vaccinium*. New Phytol., 164: 183–192.

Villarreal-Ruiz, L. 2006. Biodiversity and ecology of native pinewood ectomycorrhizal fungi across a chronosequence and their in vitro interactions with ericaceous plants. Ph.D. Dissertation, University of Aberdeen, Scotland, UK.

Villarreal-Ruiz, L, Neri-Luna, C., Anderson, I.C. and Alexander, I.J. 2012. *In vitro* interactions between ectomycorrhizal fungi and ericaceous plants. Symbiosis, 56: 67–75.

Villarreal-Ruiz, L. and Neri-Luna, C. 2018. Testing sampling effort and relative abundance descriptors of below-ground ectomycorrhizal fungi in a UK planted scots pine woodland, Mycology, 9 (2): 106–115.

Villarreal-Ruiz, L., Neri-Luna, C., Tedersoo, L. and Koljalg, U. 2018. Global and Mexican local diversity of mycorrhizal fungi in forest soils revealed by Next-Generation Sequencing: A preliminary aproach. pp. 2–5. *In*: Méndez-Vilas, A. (ed.). Exploring Microorganisms: Recent Advances in Applied Microbiology. Brown Walker Press, Irvine & Boca Raton.

Visser, S. 1995. Ectomycorrhizal fungal succession in jack pine stands following wildfire. New Phytol., 129: 389–401.

Vogt, K.A., Bloomfield, J., Ammirati, J.F. and Ammirati, S.R. 1992. Sporocarp production by Basidiomycetes, with emphasis on forest ecosystems. pp. 563–599. *In*: Carroll, G. and Wicklow, D.T. (eds.). The Fungal Community: Its Organization and Role in the Ecosystem, 2nd Edition. Marcel Dekker, New York.

Vosátka, M., Látr, A., Gianinazzi, S. and Albrechtová, J. 2012. Development of arbuscular mycorrhizal biotechnology and industry: Current achievements and bottlenecks. Symbiosis, 58: 29–37.

Vries, J. and Archibald, J.M. 2018. Plant evolution: Landmarks on the path to terrestrial life. New Phytol., 217: 12–14.

Walochnik, J., Harzhauser, M. and Aspöck, H. 2010. Climate change as a driving force for evolution. Nova Acta Leopoldina NF 111, Nr., 381: 21–32.

Waring, R.H. and Running, S.W. 1998. Forest ecosystems: Analysis at multiple scales, 2nd Edition. Academic Press, San Diego.

Warwick, M.G., Guerin-Laguette, A., Lapeyrie, F. and Suzuki, K. 2000. Matsutake-morphological evidence of ectomycorrhiza formation between *Tricholoma matsutake* and roots in a pure *Pinus densiflora* stand. New Phytol., 147: 381–388.

Waters, C.N., Zalasiewicz, J., Summerhayes, C., Barnosky, A.D., Poirier, C. et al. 2016. The Anthropocene is functionally and stratigraphically distinct from the Holocene. Science, 351: aad2622-1. DOI: 10.1126/science.aad2622.

Wilkinson, D.M. 1998. Mycorrhizal fungi and Quaternary plant migrations. Glob. Ecol. Biogeogr. Lett., 7: 137–140.

World Conservation Monitoring Centre. 1992. Global biodiversity: status of the Earth's living resources. Chapman & Hall, London.

Zak, J.C., Willig, M.R., Moorhead, D.L. and Wildman, H.G. 1994. Functional diversity of microbial communities: a quantitative approach. Soil Biol. Biochem., 26: 1101–1108.

Zasada, J.C., Sharik, T.L. and Nygren, M. 1992. The reproductive process in boreal forest trees. pp. 85–125. *In*: Shugarte, H.H., Leemans, R. and Bonan, G.B. (eds.). A Systems Analysis of the Global Boreal Forest. Cambridge University Press, Cambridge.

5

The Roles of Macro Fungi (Basidomycotina) in Terrestrial Ecosystems

John Dighton

INTRODUCTION

It would appear that the evolution of the Basidiomycota is not monophyletic. This fungal group split from Ascomycota some 550 Ma ago, with a subsequent split between the plant pathogenic and macrofungal groups around 445 Ma ago. A later split from the Sebacei, around 200 Ma ago, gave rise to the Homobadiomyces (Oberwinkler, 2012). In the Basidiomycota, the dikaryotic stage of the life cycle is dominant and, although characteristic, the development of clamp connections during cell division is not essential for hyphal extension. Clamps in Basidiomycota and croziers in Ascomycota are homologous (Anderson and Kohn, 2007). A good outline of basidiomycete taxonomy may be found in Watling (1982).

In terms of material processing, Basidiomycota may be classified into functional groups, as discussed by Oberwinkler (1993) and Cairney (2005). Basidiomycota are well known for saprotrophic and mycorrhizal roles, as macrofungi, and plant pathogenicity in its micro-fungal forms. Additionally, the ability of these fungi to form extensive hyphal networks and rhizomorphs make them important contributors to the translocation of materials through ecosystems and connecting components of ecosystems together (Jennings, 1991; Cairney, 2005). Basidiomycetes are often non-resource unit restricted and possess a variety of hyphal exploration strategies, such as diffusal or dense hyphal growth and rhizomorphs, are adapted in order to colonize

Rutgers University Pinelands Field Station, Department of Ecology, Evolution and Natural Resources (SEBS) and Biology (Camden), PO Box 206, 501 Four Mile Road, New Lisbon, NJ 08064, 74 USA. Email: dighton@camden.rutgers.edu new and diverse resources and to translocate carbon and nutrients between these resources (Fricker et al., 2008). Ecosystem services of Basidomycota, and other fungal groups, are discussed by Frankland et al. (1982) and Dighton (2016).

Basidiomycota Supporting Primary Production

Creating Soil Fertility

The mineral component of soil, derived from underlying rock, results from both fragmentation and dissolution of rock. In addition to physical processes, fungi also play a role in this ecosystem service. Although a number of lichen species are important for rock breakdown and dissolution (Banfield et al., 1999), most basidio-lichens have an epiphytic life style and are more important as primary producers (Oberwinkler, 2012).

Burford et al. (2003), Finlay et al. (2009) and Rosling et al. (2009) provide comprehensive reviews of rock weathering by fungi. These reviews show that a number of fungal taxa are associated with the breakdown of rock and subsequent release on mineral ions. Micro-fungi are highlighted by Burford et al. (2003) and saprotrophic and ectomycorrhizae Basidiomycota by Finlay et al. (2009). Increased water holding capacity of fungal colonized rock fissures (Finlay et al 2009) may increase freeze-thaw expansion and contraction to effect rock fracturing (Fomina et al., 2010). Small holes (3–10 μm diameter) in feldspars and hornblende containing fungal hyphae of saprotrophic and ectomycorrhizal fungi of the overlying pine forest are created by micromolar concentrations of organic acids (succinic, citric, oxalic, formate and malate) produced by the fungi (Jongmans et al., 1997; Hoffland et al., 2001). The effect of this organic acid production is enhanced by hyphal turgor pressure in order to allow fungal hyphae to penetrate tiny fractures in rock (Martino and Perotto, 2010).

Rock dissolution by ectomycorrhizal fungi, and the release of plant essential K, Mg and Ca ions, is reviewed by Landeweert et al. (2001) and Finlay et al. (2009). This ability varies by fungal species. The ectomycorrhizal fungi *Hysterangium setchellii*, *Rhizopogon vinicolor* and *Suillus bovinus* were found to solubilize limestone, marble and calcium phosphate, whereas *Cenococcum geophilum*, *Hebeloma crustuliniforme*, *Laccaria laccata* and *Piloderma croceum* could not (Chang and Li, 1998). Solubilization of strontianite sand by the white rot wood decomposing fungus, *Resinicium bicolor,* released strontium mineral ions Connolly et al. (1998). Thus, in association with rhizospheric bacterial communities, mycorrhizae may access a greater number of nutrient elements from rock than previously thought (Hoffland et al., 2001; Thompson et al., 2001). This activity is not restricted to bedrock, but also in association with stones embedded in the soil profile, termed the 'stonosphere' (Koele, 2012). Fungal attack of stones not only provides mineral nutrients, but the physical shape of stones in soil can influence porosity, water percolation rates and provide habitats of refuge for fungi and bacteria during stressful conditions, such as drought (Koele, 2012).

Additional nutrients made available for plant growth come from the decomposition of dead organic matter, through the process of mineralization (see Box 1). The decomposition process is carried out by both bacteria and saprotrophic fungi, in conjunction with the soil fauna. Fungi secrete extracellular enzymes (Sinsabaugh and Liptak, 1997; Sinsabaugh, 2005; Baldrian, 2008) which degrade the complex organic molecules contained in organic resources and increase more labile elements for plant uptake. The rate at which decomposition can occur depends largely on the ratio of complex carbohydrates to nitrogen and phosphorus, usually expressed as carbon:nitrogen or lignin C:N ratios (Melillo et al., 1982; Thomas and Asakawa, 1993; Silveira et al., 2011). Where the C:N or lignin:N ratios are high, rates of decomposition are reduced, compared to resources containing lower ratios. Basidiomycota are most influential in the decomposition of high C:N ratio material, such as wood. Basidiomycota are the main wood rotting fungal group, which can be of economic importance when that wood is part of buildings (Levy, 1982; Dickinson, 1982; Held, 2017).

During the decomposition of plant residues in soil there are defined successions of fungi in relation to the resources available and the enzymatic competence of the fungal community (Frankland, 1998; Ponge, 1990, 1991, 2005; Lindahl and Boberg, 2008; Voříšková and Baldrian, 2013). In general, the labile materials of low C:N ratio are utilized first and recalcitrant high C:N ratio materials are utilized later by fungi with the specific enzymes capable of degrading them. These later fungal colonizers are dominated by Basidiomycota. In a culture-based study, basidiomycetes caused significantly greater lignin loss from litter than the ascomycetes ($41.6 \pm 3.0\%$ and $50.2 \pm 0.4\%$ lignin remaining, respectively), though this did not result in a significant difference in nitrogen loss from litter between basidiomycetes and ascomycetes (Osono et al., 2003).

In initial phases of decomposition of high lignin-C:N material, such as coniferous litter, there is net accumulation of nitrogen to support fungal biomass to breakdown complex resources, such as lignin (Jensen, 1929). Translocation of exogenous nutrients resources into woody residues can have important practical consequences for forestry. In fast growing tree plantations, there can be competition for essential nutrients between wood decomposing fungi in logging residues and the nutrient demand for the growth of the next rotation tree crop. Such nutrient lock-up may result in a check of growth of the second rotation crop, resulting in economic losses, particularly if one year's growth is lost in a 5 to 7-year rotation forest (Parfitt et al., 2001). Post-harvest residue management should be such as to avoid nutrient loss by burning (Dighton, 1995) to a process of chopping and incorporating woody remains into the soil in order to enhance wood decomposition and N mineralization (Jones et al., 1999; Shammas et al., 2003).

The fungal communities of live trees are usually different from that of dead tree material and the communities in standing dead stems differ from those of standing dead twigs (Boddy et al., 2017). Fungal communities in the canopy of oak wood consist of a number of basidiomycetes, with pioneer communities dominated by *Phellinus ferreus*, *Sterium gausapatum* and *Vuilleminia comendens* on partially living branches, secondary colonizers of *Phlebia adiata* and *Coriolus versicolor*

Schema of the roles of macrofungi in forest ecosystems (after Dighton, 2016).

On the left, dead plant parts are contributed to the soil system where basidiomycete fungi are of greatest importance in the decomposition of low resource quality litter (high lignin C:N) and in effecting nutrient mineralization. Together with their role in solubilizing nutrients from rock, ectomycorrhizae are essential for the uptake and accumulation of mineral nutrients into plants via their root systems. These mycorrhizal fungi also provide defenses against microbial and faunal pathogens and provide drought tolerance.

Production of macro-fungal fruit bodies and the hyphae of these fungi provide food to both soil invertebrates and above- and below-ground vertebrates. In some cases, fungi form a large proportion of the diet of vertebrates. Through consumption of fruit bodies, spores can be disseminated via mammals.

Through the mycelial and rhizomorph structures of both mycorrhizal and saprotrophic basidiomycetes, there are facilities allowing translocation of material over short or large distances. For example, trees of the same or different species can exchange carbon and nutrients allowing individual trees to benefit from the whole tree community. In the decomposition of low resource quality woody debris, saprotrophic fungi can immobilize nutrients (particularly N) from soil into the decomposing wood, thus competing with adjacent plants for a limited nutrient resource.

Basidiomycetes can translocate heavy metals and radionuclides through their hyphae and immobilize these pollutants in their cell walls. This can lower the availability of the metals to the host plant and detoxify the environment in ectomycorrhizae. These fungi can accumulate metals and radionuclides in fruit bodies, sometimes to higher concentrations than the surrounding soil and vegetation (hyper-accumulation). This can have importance for transfer of metals through the food chain of mycophagous animals and the potential mycoremediation of the ecosystem. With their enzyme competence, wood rotting basidiomycetes can play an important role in the decomposition of organic pollutants, since the same enzyme suite is effective in both systems.

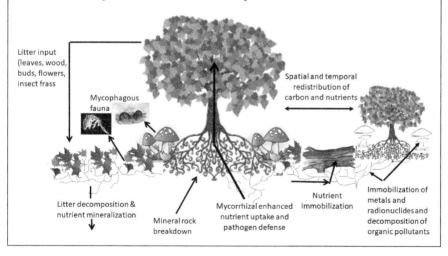

and *Hyphoderma setigerum* and *Sterium hirsutum* related to insect activity (Boddy and Rayner, 1983). Interactions between fungi can be clearly observed in decaying wood where zones of interaction between adjacent, competing fungi are delineated by the deposition of secondary metabolites. Fomajorins and increased oxidative activity stain the wood, allowing the progress of colonization to be mapped in three dimensions (Rayner, 1978; Rayner and Boddy, 1988; Jensen et al., 2017; Boddy et al., 2017). Process rates in resource utilization are influenced by competition between basidiomycete fungal species, where respiratory loss of carbon and release of phosphate by fungi in competition is greater than fungal species alone (Woodward and Boddy, 2008).

Some of the wood rotting basidiomycete fungi produce rhizomorphs or cords, which provide long-lived connections between islands of woody residues and allow reallocation of resources, such as water, nitrogen and phosphorus by translocation (Jennings, 1982; Cairney, 1992). Mycelial networks in general, hyphal and rhizomorphic, play a major role in acting as nutrient reservoirs and translocation of materials through heterogenous soil environments (Tlalka et al., 2008).

Diffusion accounted for the translocation of ^{14}C and ^{32}P through hyphal systems of *Rhizopus*, *Trichoderma* and *Stemphylium* (Olsson and Jennings, 1991). The rate of translocation and the directional flow of carbon to the building phases of the hyphae reacts in real time to source-sink strengths within the mycelial network (Olsson, 1995). With high demand for nutrients and carbon at advancing hyphal fronts, nutrients and carbon are translocated acropetally through cytoplasmic flow and diffusion in the cytoplasm and apoplasm. In contrast, translocation of ^{137}Cs through hyphae of *Schizophyllum commune* was slower than diffusion rates, suggesting incorporation into cytoplasm or hyphal wall, reducing translocation rate (Gray et al., 1995). This explains the potential for basidiomycete fungi to readily accumulate radiocesium and heavy metals (Dighton and Horrill, 1988; Yoshida and Muramatsu, 1994) where sequestion is in cell walls (Denny and Wilkins, 1987; Wilkins, 1991; González-Guerrero et al., 2008).

Some 75% (*Phanerochaete velutina*) and 13% (*Phallus impudicus*) of exogenous phosphorus added to a decomposed wood resource is translocated to newly colonized wood via mycelial cord systems at a rate of 7 μ mol/cm^2/d (Wells and Boddy, 1990). As cord systems only develop in unsterile soil conditions, field experiments showed translocation over distances of up to 75 cm between decomposing resources on the forest floor (Wells and Boddy, 1995a). Phosphorus translocation through mycelial cords of *Phanerochaete velutina* is greater at higher temperatures (Wells and Boddy, 1995b) and reduced by drying (Wells et al., 2001).

Prolific growth of fungi in litter of tropical forests can result in the development of basidiomycete fungal mats. These mycelial mats develop upwards on steeply sloping ground, sequentially colonizing newly recruited leaf litter and, in doing so, reduce downslope soil and litter erosion, thus retaining nutrients in the system (Lodge et al., 2008).

As saprotrophs, basidiomycetes are more commonly associated with decomposition of forest tree litter and woody residues (see Box 1). However, they also play an important role in grasslands, where they contribute to lignocellulose decomposition and dung decomposition (Griffith and Roderick, 2008). Specific fungal genera are found in association with dung decomposition (e.g., *Coprinus, Panaeolus, Psathyrella, Stropharia* and *Psilocybe*) and lignicolous decomposers with forming fairy rings (Marasmius, Lepista, Agaricus, etc.) where the rings are associated with macronutrient deficiencies and competition for these between the fungi and grasses (Griffith and Roderick, 2008). Due to changes in grassland management, some basidiomycete groups, such as the Hygropheraceae (wax caps), have been reduced in abundance and have been identified as a group of concern for conservation efforts (Watling, 2005).

Basidiomycete fungi are important for the development of the mineral component of soil and the release of essential plant nutrients from rock and decomposing organic residues to provide soil fertility. These fungi work in concert with other fungal groups and bacteria to affect these ecosystem services. Due to their specific enzyme competency, the basidiomycetes are functionally dominant in later stages of plant litter decomposition and wood decomposition. Their propensity to form long-distance translocatory mycelial cords provides them with the ability to spatially regulate flows of carbon, nutrients and water at a scale of meters to tens of meters.

Mycorrhizae and Primary Productivity

Mycorrhizae are symbiotic associations between fungi and plant roots (Harley, 1969; Harley and Smith, 1983; Smith and Read, 1997; Peterson et al., 2004; Brundrett, 2009; Bonfante and Genre, 2010; Dighton, 2011). Approximately 95% of all vascular plants have a mycorrhizal association (Brundrett, 1991). The ecology and role of mycorrhizae in ecosystems is discussed by Allen (1991); Read et al. (1992); Varma and Hock (1995); Mukerji (1996); Itoo and Reshi (2013). The main function of mycorrhizae is to support their host plants by allowing access to nutrients and water in exchange for energy supplied by the host plant to the fungus. In these ways, fungi help to underpin primary production. In addition to these functions, mycorrhizae also protect their host from pathogens and root herbivory and can induce halotolerance of their host plants.

Of the different types of mycorrhizae, members of the Basidiomycotina and Ascomycotina form ectomycorrhizae with trees. It is estimated that some eight thousand plant species associated with between seven to ten thousand fungal species forming ectomycorrhizae (Taylor and Alexander, 2005). The fungus penetrates between the root cortical cells to form a Hartig Net. The surface sheath has varying degrees of complexity and changes in root branching, and color as well as surface structures allow morphological identification of these mycorrhizae (Agerer, 1987–1999; Ingleby et al., 1990; Goodman et al., 1996–2000). As with wood rotting fungi, some ectomycorrhizal fungi develop mycelial strands or rhizomorphs that are important in long distance transport of nutrients and water (Duddridge et al., 1980).

Edaphic controls have been implicated in the distribution of mycorrhizae in relation to climatic zones (latitude) or altitude Read (1991a,b). In general, arbuscular mycorrhizae dominate in the temperate and tropical grasslands, tropical forests and desert communities (or at low altitudes); ectomycorrhizae dominate in temperate and arctic zones (or mid-high altitudes) in forested ecosystems and ericoid mycorrhizae are most common in the boreal heathland ecosystems (or very high altitudes).

Mycorrhizal distributions are related to the resource quality of litter input to the decomposer community and climatic limitations to decomposition rate in relation to the ability of the mycorrhizae to access nutrients. At equatorial and temperate latitudes and low altitudes, plant litters have high resource quality and decompose readily in rarely limiting climatic conditions. Here soil has a high availability of labile nutrients for plant uptake, where arbuscular mycorrhizae are most efficient and dominate. In contrast, at high latitudes and altitudes litter decomposition is slow due to climatic conditions and plant responses to stress with high levels of complex organic compounds to restrict herbivory. Thus, soils have a high nutrient capital in recalcitrant forms but low turnover and availability of labile nutrients. Here, ericoid and ectomycorrhizal mycorrhizae dominate as they have the enzymatic capabilities to access nutrients directly from the more recalcitrant resources (Went and Stark, 1968).

In nutrient-stressed ecosystems of high nutrient capital but low turnover, eicoid mycorrhizae possess protease enzymes and can access some 95% of their N needs from organic sources (Stribley and Read, 1980; Bajwa and Read, 1985; Leake and Read, 1989, 1990a,b; Read and Kerley, 1995; Kerley and Read, 1995). They can also derive phosphate from inositol hexaphosphate with acid phosphatases (Leake and Miles, 1996; Mitchell and Read, 1981; Myers and Leake, 1996).

In forested ecosystems, much of the phosphorus in soil is associated with organic forms. This can be as much as 50% of the total phosphorus (Häussling and Marschner, 1989). In the organic horizons of these soils, the colonization of root tips by ectomycorrhizae is increased (Repáč, 1996a,b), where there is a 2- to 2.5-fold increase in acid phosphatase activity in the rhizosphere compared to the bulk soil (Häussling and Marschner, 1989). Access to organic forms of P, such as inositol hexaphosphate, by ectomycorrhizae have often been reported (Dighton, 1983; Mousin and Salsac, 1986; Antibus et al., 1992, 1997) and the regulation of the acid phosphatase enzyme expression is regulated by a negative feedback from the external orthophosphate concentration in soil (MacFall et al., 1991).

Ectomycorrhizae of *Paxillus involutus* can access phosphorus from complex inorganic forms of phosphate, such as calcium phosphate, but only where available ammonium or nitrate nitrogen is adequate (Lapeyrie et al., 1991; Chang and Li, 1998). Exploitation of phosphorus from deep soil layers is enhanced by the ectomycorrhizal fungal communities on deep roots, which increases the retention of phosphorus in some tropical legume tree species and buffers the seasonal changes in P demand by the host tree (Newberry et al., 1997). There is a strong interaction between environmental controls of phosphate release, seasonal demands for P by the trees and the mycorrhizal phosphate acquisition capacity. This is of special importance during mast years which are highly P demanding. This complex interaction is called a phenological and climatic ectomycorrhizal response (PACER), which optimizes phosphate utilization and minimizes phosphate leaching loss (Newberry et al., 1997).

In forested ecosystems, with litters of diverse resource quality, ectomycorrhizae produce protease enzymes in order to access organic forms of nitrogen (Abuzinada and Read, 1986a,b, 1989; Read et al., 1989; Leake and Read, 1990a,b; Zhu et al., 1994; Tibbett et al., 1999; Anderson et al., 2001). Acid phosphatase enzymes and other enzymes (Bartlett and Lewis, 1973; Dighton, 1983; Antibus et al., 1992, 1997;

Leake and Miles, 1996; Giltrap, 1982; Durall et al., 1994; Joner and Johansen, 2000) allow ectomycorrhizae to utilize forest floor carbon.

The ectomycorrhizal fungus *Hebeloma crustuliniforme* can incorporate up to 53% of the total N contained in proteins or peptides (Abuzinada and Read, 1986a,b). In arctic regions, Tibbett et al. (1998a) suggest that *Hebeloma* species is pre-adapted to utilize nitrogen from proteins and glutamic acid, that are released from organic matter during freezing (Tibbett et al., 1998b,c). Also, *Hebeloma* only produces cold active phosphomonoesterase enzyme when grown at 6°C. *Laccaria bicolor* can access nitrogen from ammonium, nitrate and urea sources of nitrogen and *Hebeloma* spp. from bovine serum albumin. However, neither fungi could utilize nitrogen from ethylenediamine or putrescine, suggesting they could not compete with saprotrophic fungi for access to nitrogen from decaying animal carcasses (Yamanaka, 1999).

Naturally, there is variability among fungal species in their enzyme production (Dighton, 1983; 1991; Lapeyrie et al., 1991). Read (1991) suggests that *Laccaria laccata* and *Pisolithus tinctorius* are poor enzyme producers, relying on labile nutrients in soil. Other species (*Paxillus involutus, Lactarius* spp., *Amanita* spp. and *Suillus* spp.) have a greater degree of enzyme competency. *Suillus bovinus* accumulated more nitrogen from bovine serum albumin than *Lactarius controversus, Paxillius involutus, Piloderma croceum* and *Pisolithus tinctorus* mycelia (Bending and Read, 1996). These differences in enzyme production are likely to explain the observed succession of ectomycorrhizal fungal communities during forest growth, where in early seral succession labile nutrients are more readily available than in later successional stages where more complex, tree-derived litter is more recalcitrant (Last et al., 1987).

The efficiency of nutrient acquisition by ectomycorrhizae may depend on the searching strategy of extraradical mycelium and rhizomorphs; exploration types (Agerer, 2001). Although ectomycorrhizae of higher exploration type show more degradative enzyme activity than those with fewer hyphae, phylogeny of the mycorrhizal fungus proved a better predictor of nutrient acquisition in an African lowland forest (Tedersoo et al., 2012). The morphological and physiological plasticity of extraradical hyphae makes them ideally suited for the exploitation of patchily distributed nutrient resources (Bending and Read, 1995a,b; Cairney and Burke, 1996; Tibbett, 2000). These exploration types appear to be related to root density with long distance exploration being more common in low root density, such that short exploration would be more common in more mature forests with high root density (Peay et al., 2011).

Where soils are poorly developed, and nutrient availability is limited, mycorrhizae become more important for plant growth. In tropical and sub-tropical trees with high epiphyte loading, such as *Nothofagus menziesii* in New Zealand, the 'soil' created in the canopy by dead epiphytes has a very high organic content (86%) when compared to soil on the ground (10%). Adventitious roots of *Nothofagus* in this 'canopy soil' are colonized by a variety of ectomycorrhizal fungi (*Clavulina, Cortinarius, Cenococcum, Leotia, Inocybe, Laccaria, Lactarius, Russula* and *Thelephora*) that are likely using organic nutrient sources to supplement nutrients gained from soil on the ground (Orlovich et al., 2013).

Much of the basic understanding of ectomycorrhizal function has been derived from laboratory and greenhouse studies and demonstrations of the effects of different ectomycorrhizae in the field are more difficult to obtain (Miller, 1995). From field studies, *Laccaria proxima* induced a higher level of tissue phosphorus content in willow (*Salix viminalis*) than did *Thelephora terrestris*. Jones et al. (1990), using injections of radioactive phosphorus into defined root zones of field grown birch, found that incorporation of P into leaves was higher when influenced by mycorrhizal communities dominated by *Hebeloma* spp. than by communities dominated by either *Laccaria* spp. or *Lactarius* spp. (Dighton et al., 1990).

Given that individual forest trees can maintain a community of many ectomycorrhizal fungal species at the same time (Zak and Marx, 1964; Gibson and Deacon, 1988; Palmer et al., 1994; Allen et al., 1995; Shaw et al., 1995), it is likely that the ectomycorrhizal community is functionally plastic. Changes in the physiological function of the community can be altered at a spatial and temporal scale in order to optimize resource utilization, as the local environmental conditions change (Peay et al., 2011). Tibbett (2000) indicates that in both ericoid and ectomycorrhizal symbioses, the extraradical hyphae exhibit significant morphological and physiological plasticity (Bending and Read, 1995a,b; Cairney and Burke, 1996) that makes them ideally suited for the exploitation of patchily distributed nutrient resources.

However, function of the ectomycorrhizal community may be a result of the number of species rather than the composition. Increased plant growth was seen with higher numbers of mycorrhizal fungal species in Radiata pine seedlings (Chu-Chou and Grace, 1985) and Douglas fir seedlings (Perry et al., 1989). Two ectomycorrhizal species were found to produce larger host plants than one (Parladé and Alvarez, 1993; Reddy and Natarajan, 1997) and under low fertility, birch growth was higher in association with eight ectomycorrhizal with single fungal species (Jonsson et al., 2001).

Changes in birch seedling biomass (root and shoot) and plant phosphate content were significantly correlated with ectomycorrhizal diversity, rather than the level of root colonization. Furthermore, phosphorus uptake from organic P sources was enhanced by increased mycorrhizal diversity (Baxter and Dighton, 2005a,b). However, Jonsson et al. (2001) suggests that the effects of mycorrhizal fungal diversity may depend on the context in which they exist, and the community structure may be beneficial, detrimental or neutral, depending on the nutritional conditions of the soil, plant age or other factors. Hence, Fitter (1985, 1991), Cairney (1999) and Leake (2001) say that it is not always obvious what the function of mycorrhizae is in a natural ecosystem, and difficult to replicate in a laboratory situation where the fungal species may be difficult or impossible to grow (Cairney and Burke, 1996). However, new techniques may allow us to explore more of the mycorrhizal function in a field setting (Courty et al., 2010), where heterogeneity of function drives the ability of ectomycorrhizae to exploit resources, enzyme expression, nutrient uptake and translocation within the mycorrhizal system (Cairney, 1992).

In forest succession and forest growth the function of mycorrhizae is related to changes in edaphic factors (Last et al., 1987) and may have a more dramatic effect on plant growth in oligotrophic systems than in fertile systems. In localized areas of nutrient poor soils, such as volcanic fields and glacial outwash (Gehring and Whitham, 1994; Jumpponen et al., 1998, respectively), growth of pinyon pine on cinder soils was doubled by the addition of ectomycorrhizae, compared to the effect of mycorrhizae on adjacent loam soil. The development of the ectomycorrhizal community depends on the establishment of an inoculum source and biotic filters imposed by the compatibility of mycorrhizae and host, host filters of the physiology of the host plant and environmental filters of nutrient availability, etc. (Jumpponen and Egerton-Warburton, 2005). These interactions between the ectomycorrhizal community, edaphic heterogeneity and exploration characteristics of the extraradical hyphae of fungal species are explored by Peay et al. (2011).

Mycelial Networks

The potential importance of a below-ground network of mycorrhizal hyphae and inter-plant translocation of information (carbon and nutrients) stems from studies of arbuscular mycorrhizae. Phosphorus transfer between decapitated and intact plants of *Lolium* and *Plantago* was observed between adjacent plants of the same or different species (Heap and Newman 1980a,b). This flow was assumed to be through arbuscular mycorrhizal connections as no translocation occurred between mycorrhizal plants and the non-mycorrhizal cabbage (*Brassica oleracea*) (Newman and Eason, 1989), or between arbuscular- and ecto-mycorrhizal plants (Eason et al., 1991). In the same way, ^{14}C label applied to pine tree seedlings was preferentially transferred to neighboring, unlabelled pine tree seedlings, rather than to neighboring plant species associated with arbuscular mycorrhizae (Eason et al., 1991). Transfer of phosphorus and carbon between pine trees and feather moss (*Pleurozium schreberi*) communities in the understory, over distances of several centimeters, was attributed to the interconnecting ectomycorrhizal fungi associated with the roots of both plant species (Carleton and Read, 1990).

Demonstration of the transfer of carbon between plants in the field was achieved by Simard et al. (1997a–c), where transfer of carbon was transferred from paper birch (*Betula papyrifera*) to Douglas-fir (*Pseudotsuga menziesii*) seedling growing in partial or deep shade. The amount of carbon transferred between plants was a significant proportion of the carbon contained in the shoots (13% for *P. menziesii* and 45% for *B. papyrifera*), suggesting carbon supplementation in limited photosynthesis in the shaded plant. In trenching experiments, Douglas fir seedlings planted into trenched areas in a birch dominated community had approximately half the ectomycorrhizal diversity of seedlings planted into un-trenched plots. The increased mycorrhizal diversity significantly increased the photosynthetic capacity of the Douglas fir seedlings (Simard et al., 1997a), showing the importance of a common mycorrhizal network.

Indeed, Wu et al. (2002) have shown that some 24% of ^{14}C label occurring in the underground parts of pine seedlings is allocated to the extraradical hyphal component of ectomycorrhizal associations. Where plants are under stress, access to a common mycorrhizal network may be of importance. For example, an experimental 50% defoliation of lodgepole pine in a mixed lodgepole pine (*Pinus contorta*) Engelmann spruce (*Picea engelmannii*) forest in Yellowstone National Park resulted in no change in mycorrhizal colonization (142 mycorrhizal tips/core) or species richness (~ 5.0 species/core), compared to control plots (Cullins et al., 2001). However, the ecosystem dominant ectomycorrhizal species, *Inocybe* sp. became rare in defoliation plots, and Agaricoid and Suilloid species, associated with both trees, became dominant in both the defoliated and control plots. A similar response was found in conspecifics where the mycorrhizal benefit to defoliation was greatest closest to adjacent intact plants and decreased with distance from the intact plant (Pietkäinen and Kytöviita, 2007). Competitive dominance of dwarf birch (*Betula nana*) in the arctic may be enhanced by warming climate trends as conspecific exchange of carbon by mycorrhizal networks and root grafting increase with increased soil warming (Desleppe and Simard, 2011).

Hyphal growth in the ectomycorrhizal network may be rapid and allow mycelia to aggregate into cords or rhizomorphs (Read, 1991). These rhizomorphs allow translocation of nutrients from distal parts of the extraradical mycelial network to the root (Fitter et al., 1986a,b,c) in an analogous way to mycelial cord systems of wood rotting fungi (Rayner et al., 1985; Wells and Boddy, 1990, 1995; Cairney, 1992; Boddy, 1999). The mycorrhizal mycelium 'mats' (Griffiths et al., 1990), in these humic soil horizons can be so dense that they form 10 to 20% of the top 10 cm of soil in a temperate forest ecosystem (Cromack, 1988) and account for 45 to 55% of the total soil biomass (Cromack et al., 1979). These mat forming ectomycorrhizal communities in Douglas fir forests are important in increasingly removing organic nitrogen from the soil pool and immobilizing it into high C:N ratio fungal tissue as forest growth progresses (Aguilera et al., 1993). Although the forest soil becomes enriched with organic nitrogen as the forest matures, this N becomes increasingly less available to plant growth. The patchy existence of nutrients or accessible resources for mycorrhizal utilization in soil would indicate that these fungi would be adaptable to being able to exploit a variety of resources as and when they become available (Peay et al., 2011).

Stress

Severe fire damage in forests can push the ecosystem back to a previous point in successional history and, thus, change the ectomycorrhizal community composition after the fire (Visser, 1995; Jonsson et al., 1999a) and during re-establishment of a mature forest (Frankland, 1992, 1998; Boerner et al., 1996). If reestablishment of the forest is rapid, dying roots of the previous forest can act as a source of mycorrhizal inoculum to maintain a similar species diversity on the new forest (Baar et al., 1999; Jonsson et al., 1999b). In lower intensity burns, designed to manage fuel load, ectomycorrhizal diversity and richness can be reduced, especially where burns re-occur frequently (Tuininga and Dighton, 2004). These effects are linked to changes in nutrient availability but are likely to be short-lived and the forest returns to a status quo within a year or so (Tuininga and Dighton, 2004; Tuininga et al., 2002).

In ectomycorrhizae, drought tolerance is increased more by *Cenococcum geophilum* than by *Lactarius* spp. (Jany et al., 2003). This is possibly associated with melanin deposition in fungal cell walls of *Cenococcum*, since by comparing a melanin inhibited with a control strain of *Cenococcum geophilum*, Fernandez and

Koide (2013) showed reduced fungal growth in the melanin inhibited strain under water stress and related this to the significantly reduced hyphal diameter of melanin-inhibited mycelium. However, the relationship of these finding to the tolerance of mycorrhizal plants was not tested.

Access to a common mycorrhizal network may be of importance where plants are under stress. A 50% defoliation of lodgepole pine (*Pinus contorta*) mixed with Engelmann spruce (*Picea engelmannii*) in Yellowstone National Park resulted in no change in the degree of mycorrhizal colonization of roots, or fungal species richness compared to intact forest (Cullins et al., 2001). Ectomycorrhizal fungal species associating with both tree species were affected more by defoliation than more host specific fungi, suggesting possible nutrient and carbon exchange between trees through a common mycorrhizal network. A similar response was found where the mycorrhizal benefit to defoliation was greatest close to an intact conspecific plant, but the effect decreased with distance from the intact plant (Pietkäinen and Kytöviita, 2007). Conspecific exchange of carbon by mycorrhizal networks may explain dominance of dwarf birch (*Betula nana*) in the arctic with climate-induced increased soil warming (Disleppe and Simard, 2011).

Much of the work on the roles of ectomycorrhizal communities have originated from field observations of mycorrhizal species distributions and laboratory/greenhouse studies of function. New techniques that are becoming available may allow us to obtain much more information on the function of ectomycorrhizal communities *in situ* (Courty et al., 2010).

Basidiomycetes Supporting Secondary Production

Fungi are comprised of 70–94% water, some 8–40% proteins, 40% lipids, 28–85% carbohydrates (mainly complex), high concentrations of P, K and Se and high concentrations of vitamins A, B complex, C, D and K (Fogel and Trappe, 1978; Kinnear et al., 1979; Claridge and Trappe, 2005). Thus, it is not surprising that a number of vertebrate and invertebrate animals consume mushrooms as part of their diet (Cave, 1997). The nutrient content of fungal fruitbodies (mushrooms of both mycorrhizal and saprotrophic basidiomycetes) of a *Nothofagus* forest floor and showed that all elements other than calcium are more concentrated in fungal tissue than the forest floor material (Clinton et al., 1999). However, much of the nitrogen they contain is in complex forms, such as indigestible cell walls (Cork and Kenagy, 1989a). For animals to effectively utilize the nutrients in fungi, they are required to have a complex community of gut symbionts that assist in the breakdown of these compounds. Indeed, experiments conducted by Cork and Kenagy (1989b) showed that the weight of ground squirrels declined when fed entirely upon fruit bodies of the hypogeous ectomycorrhizal fungus, *Elaphomyces granulatus*, as more than 80% of the nitrogen was locked up in complex forms and could not be made available in the digestive tract of these animals.

The brush-tailed potoroo (*Potorus longipes*) has about 25% of its diet as fungi throughout the year. In contrast, the smoky mouse (*Pseudomus femeus*) and bush rats (*Rattus fuscipes*) relied heavily on fungi during the winter when other foods are scarce (Claridge and May, 1994; Claridge and Trappe, 2005). Potoroos consume the most varied fungal diet of any animal (36 fungal taxa), of which most are hypogeal.

The large range of a family group of Goeldi's monkeys (*Callimico goeldii*) in Bolivia, some 115–150 ha, is attributable to the low density and patchy distribution of the ephemoral fungal fruiting bodies (*Auricularia* and bamboo fungi, *Ascopolyporus*) which prove to be about 35% of their diet (Porter and Garber, 2010). This is important to the behavior of this monkey which devotes up to 63% of its feeding time consuming fungi. This is in contrast to most other mycophagous primates, who spend less than 5% of their feeding time on fungi (Hanson et al., 2003).

Through mycophagy, small rodents are important vectors of fungal spores (Kotter and Farentinos, 1984), where above-ground fruiting mycorrhizal fungi such as *Laccaria* and *Suillus* can be an important part of the diet of mice (*Peromyscus*) (Pérez et al., 2012). Passage of spores through an animal's gut can have different effects on spore survival. Gut processing of spores of *Suillus tomentosus* increases their germination and ability to form mycorrhizae, whereas spores of *Laccaria trichodermophora* are reduced in their ability to form mycorrhizae.

Availability of mushrooms, for faunal grazing, can be influenced by forest management. Mulching significantly reduces the production of saffron milk cap mushrooms (*Lactarius deliciosus*) in radiata pine plantations (Guerin-Laguette et al., 2014). Control burning to reduce fuel load in *Eucalyptus* forests in Tasmania reduces fungal fruitbody abundance to 2 and 5% that of the unburned forest, reducing food availability for a variety of small mammals dependent on fungi for food, (Trappe et al., 2006).

Numerous invertebrates inhabit basidiomycete mushrooms (Courtney et al., 1990). Consumption of large, fleshy mushrooms by dipteran larvae bears little correlation between the human toxicity of the fungus and those consumed by invertebrates. Many species of the fly *Drosophila* are tolerant of the toxic compound, α-amanitn, of *Amanatia* spp. (Jaenike et al., 1983). High densities of collembola graze on the surface and spores of *Laccaria* spp. Hanski (1989) suggests that the spatial distribution of fungal fruiting bodies can influence the feeding activities of above-ground fungivorous insects and the seasonal appearance of particularly basidiomycete fruit bodies can influence the growth and development of insect larvae and the fecundity of the adult insect. Variable spatial and temporal availability of any one fungal species is a likely determining factor as to why most fungal feeding insects are polyphagous, rather than monophagous. Invertebrate grazing below-ground can disrupt the mycelial network and translocation of nutrients and carbon (Johnson et al., 2005) and, by selective grazing (Hiol Hiol et al., 1994) alter basidiomycete fungal competition and spatial relationships (Newell, 1984a,b).

In wood feeding beetles (bark beetles and ambrosia beetles), the insect larvae feed on wood colonized by fungi that are usually imported by the adult beetle. Fungal colonization of the wood increases the nitrogen content of wood, making it more nutritious (Ayers et al., 2000) and, although many beetles carry a community of fungi, ambrosia beetles (*Ambrosiodmus rubicollis*, *A. minor Ambrosiophilus atratus* and *A. subnepoltus*) all depend on the basidiomycete fungus *Flavodon ambrosius* (Li et al., 2017). In a similar way, all species of the termite genus *Macrotermes* are dependent on their cultivation of a basidiomycete fungus (*Termitomyces* sp.), which rarely produces fruiting bodies (Rikkinen and Vesala, 2017).

The value of mushrooms as both food and medicine has led to the development of mushroom farming around the world, with increasing biomass production of an increasing diversity of species. Discussion of this is outside the remit of this chapter, but it worth noting that individual collection and local industries based on harvesting wild mushrooms has led to over harvesting and social interactions between harvesters, land owners and scientists to understand the conservation of suitable mushroom habitat (Barron, 2017; Barron and Emery, 2012).

Basidiomycetes and Pollution

Acid Rain

In contrast to analyses of concentrations of toxic elements in ecosystems, 'critical loads' of pollutants can be inferred from bioindicators, such as changes in the abundance of macrofungal fruitbodies, particularly ectomycorrhizal Basidiomycotina (Fellner, 1989; Colpaert and Van Tichelen, 1996). Part of the reason for the forest die-back in Bavarian forests (the 'Waldsturben' effect) was due to damage of roots and their ectomycorrhizae by acidifying pollutants (Sobotka, 1964; Ulrich et al., 1979; Hüttermann, 1982, 1985; Blaschke et al., 1985; Stroo and Alexander, 1985). Trees and their ectomycorrhizae were affected by both reduced carbohydrate supply (leaf cuticle thinning and reduced photosynthesis), nutrient levels in soil (especially increased Al lability) and hormone levels in roots (Nylund, 1989; Dighton and Jansen, 1991). Belowground, the acidifying pollutants reduced soil pH to make toxic metals (aluminum, manganese and magnesium) more soluble and plant available (Van Breemen and Van Dijk, 1988; Skeffington and Brown, 1986; Tyler et al., 1987; Ruark et al., 1991). This increased toxicity leads to reduced root growth, root dieback and reduced mycorrhizal fungal growth and root colonization. Much of the effect was observed in changes in abundance and community composition of forest mushrooms. The overall results of the work on the effects of acid rain on mycorrhizae are reviewed in Jansen et al. (1988), Jansen and Dighton (1990) and Dighton and Jansen (1991).

Acid rain caused a decline in mycorrhizal formation on roots of trees and reduced root vigor in forests (Sobotka, 1964; Liss et al., 1984; Meyer, 1987; Blaschke, 1988), along with reduction in ectomycorrhizal fruitbody production (Arnolds, 1985, 1988; Jansen and Van Dobben, 1987; Felner, 1988). Although the effect of the acidifying pollutants was different between ectomycorrhizal fungal species, there was a general trend of greater effect of reduced abundance in mycorrhizal than saprotrophic fungal species. Some 45 to 50% of all fruitbodies in healthy forest ecosystems were ectomycorrhizal, but this was reduced to about 10% in acid rain affected forests (Arnolds, 1988). This change in mycorrhizal to saprotrophic fungal ratio has been used as an index of damage (Fellner and Pešková, 1995). However, addition of sulfur at 50 kg/ha to a Sitka spruce forest in Scotland caused no change in mycorrhizal fruitbody production or colonization of roots (Carfrae et al., 2006).

Acid rain induced changes in soil chemistry which affected the ectomycorrhizal community structure of tree roots (Dighton and Skeffington, 1987; Markkola and

Ohtonen, 1988). This is attributable to increased solubility of Al and changes in Al:Ca, Al:Mg and Al:PO$_4$ (Schier, 1985; Thompson and Medve, 1984; Jongbloed and Borst-Pauwels, 1988, 1989; Kottke and Oberwinkler, 1990). Increased Al availability induced increased superoxidase dismutase, peroxidase and glutathione production in *Pisolithus tincitorius* mycorrhizal roots of *Pinus massoniana* and consequent oxidative damage (Kong et al., 2000).

In a field fumigation of Scots pine, Sitka spruce and Norway spruce with SO$_2$ and O$_3$ (McLeod et al., 1992) resulted in reduced occurrence of *Paxillus involutus* mycorrhizae (Shaw et al., 1992, 1993). However, *Paxillus involutus* fruited more abundantly in Scots pine plots that received high SO$_2$ loading; the converse of the root colonization. Ozone tolerant provenances of Loblolly pine (*Pinus taeda*) showed less effect of reduced ectomycorrhizal root colonization than the ozone intolerant provenance (Qui et al., 1993). Most of these effects occurred from aboveground damage to host trees, in contrast to the more direct effect of acid rain through changes in soil chemistry.

Nitrogen Deposition

Fossil fuel combustion releases nitrogen into the atmosphere, which is deposited as both wet and dry deposition into terrestrial ecosystems. Additional nitrogen deposition acts as a fertilizer to increase plant growth (McNulty and Aber, 1993). At some point, N will reach saturating conditions and N in excess of plant demand will leach down the soil profile or through lateral flow into watercourses, leading to eutrophication of waterways and groundwater (Harrison et al., 1995). Eutrophication of aquatic systems has adverse effects on organisms and, at extreme levels, results in production of the greenhouse gasses N$_2$O and CH$_4$ (Aber, 1992; Tietma et al., 1993).

The decline in appearance of ectomycorrhizal fruitbodies and increase in saprotrophic and pathogenic fungal fruitbodies in the Netherlands was attributed to a combination of acidifying pollutants and nitrogen deposition (Arnolds, 1989a,b; 1991, 1997; Termorshuizen and Schaffers, 1987, 1991; Kårén and Nylund, 1997). These significant changes in fruiting of basidiomycetes, especially the loss of some species, has led to the adoption of 'red data' lists for the conservation of fungal species (Arnolds, 1989b, 1997). The call for inclusion of fungi and non-vascular plants in the lists of species for conservation has been adopted in the UK (Watling, 1999, 2005; Dahlberg and Mueller, 2011; Barron, 2017).

However, the appearance or non-appearance of fruitbodies of ectomycorrhizal fungi may bear little relation to the abundance of that mycorrhizal morphotype on the root system of the tree (Termorshuizen and Schaffers, 1989; Egli et al., 1993; Jonsson, 1999a). Addition of ammonium sulfate decreased tree fine root biomass, but not ectomycorrhizal colonization (Termorshuizen and Schaffers, 1991; Kårén and Nylund, 1997). However, in other studies, reduction in ectomycorrhizal colonization of tree roots and changes in mycorrhizal community composition have been observed (Arnebrant and Söderström, 1992; Brandrud, 1995). Complete cessation of mycorrhizal fungal fruiting occurred with the addition of three times ambient N deposition (65 and 198 kg N/ha/y) as ammonium nitrate in beech woodlands (Rühling and Tyler, 1991). However, many leaf litter-inhabiting, saprotrophic fungal species increased fruiting, including species of the genera *Mycena, Clitocybe, Lepista, Agaricus* and *Lycoperdon*. Differing responses between ecosystems could be due to soil fertility, as in an oligotrophic pine dominated ecosystem in the New Jersey pine barrens, very low N deposition velocities of between 4 and 8 kg N/ha/y resulted in a decrease of both abundance of ectomycorrhizal root colonization and mycorrhizal species richness (Dighton et al., 2004). However, in a comparison between oligotrophic pine ecosystems in New Jersey and Florida, acute N application of 35 and 70 N/ha/y equivalent caused little effects on ectomycorrhizal community composition (Adams-Krumins et al., 2009). Thus, there may be differing responses to chronic and acute N application.

Under acidifying conditions, organic nitrogen accumulates in the soil due to slowing decomposition rates (Oulehle et al., 2006). Remediation by the addition of lime has an influence on the fruiting of lignicolous fungi. For example, occurrence of *Amphinema byssoides*, *Hyphodontia breviseta*, *Hypochnicium geogenium* and *Sitotrema octosporum* increased and *Trechyspora farinacea* decreased in a Scots pine forest following liming on acidic soils (Veerkamp et al., 1997).

Heavy Metals and Radionuclides

Metal pollutants cannot be eradicated by metabolic processes, although their availability to organisms, and occasionally their valency, can be altered. Thus, they remain a toxic contaminant of ecosystems. Due to this toxicity, abundance and diversity of fungi decreased with increasing pollutant loading (As, Cu, Cd, Pb and Zn) in the humic soil horizon where ectomycorrhizal species were more tolerant of high metal concentrations than saprotrophs (Rühling and Söderström, 1990). For example, the ectomycorrhizal fungus *Laccaria laccata* was the most tolerant macro-fungus to heavy metal pollutants, but growth of a number of soil microfungi isolates on agar did not decrease with increasing metal loading (Rühling et al., 1984). Ectomycorrhizal colonization (total number of mycorrhizal root tips), but not plant biomass, of Loblolly pine, was reduced by increasing concentrations of Pb in soil (Chappelka et al., 1991). The same effect on ectomycorrhizae of Scots pine, but not growth suppression, was found in Cd and Zn contaminated soils (Hartley-Whittaker et al., 2000a). Metals in interaction can affect toxicity. For example, both Pb and Sb ameliorate toxicity of Cd to *Suillis granualtus*, and the combination of Cd + Pb + Zn was less toxic to *Lactarius deliciosus* than the individual or paired metals (Hartley et al., 1997a). Differences in tolerance of metals varies by both fungal species and metal. *Pisolithus tinctorius*, *Suillus luteus* and *Suillus variegatus* were more tolerant of the heavy metals, Cu, Cd and Zn than *Paxillus involutus*, however, *Paxillus* was more resistant to Ni (Blaudez et al., 2000).

Effective cadmium concentration for reducing root colonization by 50% (EC_{50}) was 3.7 µg/g for *Paxillus involutus* and 2.3 µg/g for *Suillus variegatus* (Hartley-Whitaker et al., 2000b). This phenomenon had been identified in studies using X-ray diffraction (EDAX) (Denny and Wilkins, 1987b). Heavy metals (Zn) accumulated in fungal hyphae in the extraradical hyphal network, the fungal sheath and Hartig net of ectomycorrhizae, preventing translocation of the metal into the host cortex. This is due to the high zinc binding on extracellular slime formed by the hyphae of the ectomycorrhizal fungus *Pisolithus tinctorius* (Denny and Ridge, 1995). A variety of heavy metal binding sites are reported in the ectomycorrhizal fungal mycelia of *Hymenogaster* sp., *Scleroderma* sp. and *Pisolithus tinctorius*, which were tolerant of high concentrations of Al, Fe, Cu and Zn (Tam, 1995). Fungal immobilization of the metals Cu and Zn is linked to polyphosphate granules (Turneau et al., 1993; Kottke et al., 1998) and the cystine-rich proteins in the outer pigmented layer of the cell wall of *Pisolithus tinctorius* (Turneau et al., 1994). When colonized by mycorrhizae, the fungi reduced the toxicity of heavy metals by preventing translocation of plant toxic levels into the host plant tissue (Hartley-Whitaker et al., 2000b). Rizzo et al. (1992) showed that, despite the fact that melanized rhizomorphs of *Armillaria* spp. are able to take up heavy metals from the environment, some elements were 50–100 times more concentrated in fungus than surrounding soil, with Al, Zn. Fe, Cu, Pb in rhizomorphs reaching up to 3440, 1930, 1890, 15 and 680 µg/g, respectively. X-ray dispersal electron microscopy (EDAX) showed that metals accumulate in the outer portion of the rhizomorph rather than the interior.

There is little information about the interactions between fungi and mercury. Ectomycorrhizal colonization of pitch pine seedlings showed a threshold response to mercury, showing no reduction in root colonization until after 37 µg/g Hg (Crane et al., 2012). However, growth of pure cultures of ectomycorrhizal species on agar showed changes in growth trajectories with increasing concentrations of Hg which could change outcomes of fungal-fungal competition, resulting in different mycorrhizal communities developing on roots at different mercury concentrations (Crane et al., 2010).

Leyval et al. (1997) postulated two possible evolutionary routes that mycorrhizal fungi have taken to cope with heavy metals. Fungal siderophores, such as ferricrocin or fusigen, are produced at low metal concentrations in order to sequester specific metal species. At higher concentrations of heavy metals, siderophore production is suppressed but the host plant is still protected against the heavy metal. These fungi benefit the host plant by accumulating metals in cell walls or vacuoles or by sequestering metals in fungal siderophores, complex metals to metallothioneins and phytochelatins (González-Guerrero et al., 2008), and by assisting in metal extrusion by transportins associated with fungal membranes.

The potential application of mycorrhizae for heavy metal resistance and contaminated site restoration has attracted interest. Initial studies of the importance of ectomycorrhizae in protecting host plants from heavy metal comes from studies by Marx (1975, 1980). Survival of pine trees in mine spoil soils showed that inoculating tree seedlings with ectomycorrhizal fungi improved both tree survival and growth. The ectomycorrhizal fungus, *Pisolithus tinctorius*, appeared more frequently in polluted sites than in other habitats and tree seedling inoculation with *P. tinctorius* resulted in tree volumes 250% greater than those trees assuming natural inoculum from the site or inoculation with *Thelephora terrestris*. Trees also had higher foliar phosphate levels, but reduced levels of Ca, S, Fe, Mn, Zn, Cu and Al, suggesting that the effect of this mycorrhizal fungus may reduce the uptake of heavy metals into the host tree (Denny and Wilkins, 1987a,b). Thus, ectomycorrhizal trees may have a role of mycorrhizal fungi in restoration and reclamation (Senior et al., 1993; Donnelley and Fletcher, 1994).

The explosion of the Chernobyl Atomic Electric Station in the Ukraine in 1986 and, subsequently, the accident at Fukushima Daiichi, Japan in 2011, has renewed focus on the accumulation of radioactive fallout in a variety of biotic components and the role of organisms in influencing radionuclide retention in the ecosystem. In the International Commission on Radiological Protection (Coughtree, 1983), the importance of organic soil horizons and their microbial communities as potential accumulators of nutrient elements and radionuclides in terrestrial systems was highlighted by Heal and Horrill (1983). This is particularly true for forest ecosystems, where the important role of fungi in radionuclide cycling in forest ecosystems was highlighted by Steiner et al. (2002). Haselwandter (1978); Eckl et al. (1986); Haselwandter et al. (1988) and Byrne (1988) also showed that lichens and mushroom-forming fungi took up and accumulated radionuclides in their fruiting structures. Byrne (1988) paid especial attention to members of the Cortinaraiacea, which are known to be Cs accumulators. The European Community set a limit of radioactivity in foodstuffs at 600 Bq/kg. Byrne found that the levels of ^{134,137}Cs radioactivity in fungi in Slovenia ranged between 0.5 kBq/kg dry weight (*Cortinarius praestans*) to 43 and 44 kBq/kg (*Laccaria amethystina* and *Cortinarius armillatus* respectively), up to 80 times the limit considered safe to consume. In addition to radiocesium, fungi have been shown to take up ^{7}Be, ^{60}Co, ^{90}Sr, ^{95}Zr, ^{95}Nb, ^{100}Ag, ^{125}Sb, ^{144}Ce, ^{226}Ra and ^{238}U (Haselwandter and Berreck, 1994), which contributes more to the radionuclide content than the natural radioisotope of potassium (40K), where potassium may be between 0.15 and 11.7% of the dry weight of fungal tissue. They cite values of ^{137}Cs concentration from a variety of basidiomycete fungal species at between 266 and 25160 Bq/kg dry weight before and between 95 and 947400 Bq/kg after the Chernobyl explosion. Although a variety of radionuclides were released from Chernobyl, most research has focused on radiocesium.

There is considerable variation in the accumulation of radiaonuclides both between and within fungal species. In Poland, Mietelski et al. (1994) reported a difference between 300 Bq/kg dry weight of ^{137}Cs in *Macrolepiota procera* to 20000 Bq/kg in *Xercomius badius* and a range of 300–1800 Bq/kg within Boletus edulis in the same location. The variation in activity of the α-emitting isotopes 90Sr and $^{239+240}$Pu ranged from 0.6 to 4 Bq/kg in *Leccinum* sp. for Sr, and from undetectable to 90 MBq/kg for *Boletus edulis*. Wide ranges of accumulation of radiocesium (< 3 to 1520 Bq/kg) were also found in mushrooms collected in Japan (Muramatsu et al., 1991), with the lowest levels of activity (< 50 Bq/kg) found in the edible species *Lentinus edodes*, *Flammulina velutipes*, *Pleurotus ostreatus* and *Pholiota nameko*. Cesium was taken up by *Suillius granulatus* and *Lactarius hatsudake* in preference to potassium, which is in contradiction to Olsen et al. (1990) who reports the affinity of fungi to cations decreases in the order K > Rb > Cs > Na > Li (Muramatsu et al., 1991). Accumulation of radionuclides into fungi is dependent on the external concentration of a number of cationic elements. By varying the availability of stable potassium, cesium, strontium and calcium in growth medium, uptake of ^{134}Cs, ^{85}Sr, ^{60}Co and ^{210}Pb, ^{134}Cs content in the fungus *Pleurotus eryngii* increased with raised concentration of stable Cs. However ^{85}Sr content decreased with increased concentration of stable Cs, and transfer of ^{60}Co and ^{210}Pb was unaffected by changes in cation availability (Guillén et al., 2012).

A large proportion of [137]Cs (25 to 92%) from sources occurring prior to the accident at Chernobyl accumulated in a variety of ectomycorrhizal fruitbodies (Dighton and Horrill, 1988). Similar figures (13 to 69%) of pre-Chernobyl accumulation of radiocesium were calculated from the data presented by Byrne (1988) and Giovani et al. (1990). This information suggests that fungi could be long-term accumulators and retainers of radionuclides in the environment. Together with an ecological half-life of 8–13 years (Vinichuk et al., 2013c) it is anticipated that fungi will continue to be important in the retention of radiocesium in the ecosystem.

Much effort has been invested in the use of plants to accumulate pollutants (mainly heavy metals) in the process of phytoremediation (Raskin and Ensley, 2000). The accumulation of radionuclides by fungal mycelia and, particularly in fruitbodies, could be a means to attempt both heavy metal and radionuclide clean-up from contamination environments. The formation of large and harvestable fruiting structures (mushrooms) provides a potential means of removal of metals and radionuclides that have accumulated in the mushrooms. Mycorrhizal fungi are a major component of the radionuclide accumulation (radiocesium concentration) in a boreal coniferous ecosystem in Sweden (Guillitte et al., 1990) although its relative contribution to the total standing crop biomass is probably not large (Vogt et al., 1982; Fogel and Hunt, 1983). The mass of fungal mycelia in upper soil horizons correlates with the accumulation of radiocesium in the upper soil and reduced leaching of radionuclides due toimmobilization in fungal tissue (Rommelt et al., 1990; Guillette et al., 1990). This attribute, together with evidence of translocation of radionuclides from soil to basidiomycete fruitbodies (Gray, 1995, 1996) points to the potential utility of fungal mycelia in soil to prevent leaching loss of radionuclides and remediation by removal mushrooms that have accumulated metals or radionuclides (Gray, 1998; Rozpadek et al., 2017; Dighton, 2018).

Organic Pollutants

A list of fungal taxa of importance in the degradation of specific organic compounds and the mechanisms whereby they metabolize organic pollutants, using exoenzymes and intracellular biochemical reactions, is detailed in Harms et al. (2011). They also suggests that fungal hyphae may stimulate the activity of bacteria, by the production of exudates as an energy source, thus synergizing the process of pollutant degradation.

2,4–D and atrazine is incorporated into the mycelium of the ectomycorrhizal fungus *Rhizopogon vinicolor* and the ericoid mycorrhiza *Hymenoscyphus ericae*. Of 21 mycorrhizal fungi, 14 species could metabolize some of the PCBs by at least 20% (Donnelley and Fletcher, 1994) and *Radiigera atrogleba* and *Hysterangium gardneri* were able to degrade 80% of 2,2'-dichlorobiphenyl. In contrast Scots pine seedlings associated with the ectomycorrhizal fungus *Suillus bovinus* impedes degradation of poly-aromatic hydrocarbons in soil (Joner et al., 2006).

Numerous fungi utilize toluene as a carbon source, although the rate of decomposition of toluene is concentration dependent, where fungal activity decreased by 50% when toluene concentration exceeds 2.5 mM (Prenafeta-Boldu et al., 2001). Dead mycelium of the basidiomycete *Pleurotus sajor caju* can be used as a biofilter for phenols. Adsorption rates of 0.9 m mol/g for phenol, 1.2 for *o*-chlorophenol, 1.4

for *p*-chlorophenol and 1.86 for 2,4,6-trichlorophenol have been achieved in aquatic systems (Denizli et al., 2005).

Research into the role of higher fungi in the degradation of organic pollutants is an emerging field. It seems likely that interaction between ectomycorrhizal and saprotrophic fungi and bacteria may synergistically enhance decomposition and warrants further study (Cairney and Meharg, 2002). The roles of higher fungi are important here due to their more complex enzymes systems adapted, for example, for wood decay, allowing them to mineralize a wide range of xenobiotics (Harms et al., 2017).

Conclusions

Macrofungi, and Basidiomycota in particular, have a more persistent presence in the environment. With longer-lived mycelial structures (hyphae and rhizomorphs) allowing them to develop long lived mycelial connections between different components of, especially, forest floors. With their ability to form these stable hyphal structures, translocatory pathways can be established, allowing the movement of carbon and nutrients between plants (mycorrhizae) and between patches of litter of contrasting resource quality (saprotrophs). This allows for both reducing forest floor resource heterogeneity and communication channels between plants (Bahram et al., 2014). This communication has been established for carbon and nutrient channels, but recent interesting data from arbuscular mycorrhizae suggest that other mutual benefits may be conferred that have yet to be explored in ectomycorrhizal associations (Babikova et al., 2013).

The role of macrofungi in rock dissolution and the formation of the mineral component of soil is another recent area of research. Although current data suggest a role, we do not have enough information to compare the fungal impact in relation to physical rock erosion rates. However, combined with the saprotrophic role of fungi, both processes are vital in providing soil fertility and underpinning primary production. Also, mycorrhizal assemblages created by macrofungi are important in supporting primary production. Here too, the ability of both saprotrophic and mycorrhizal macrofungi to form long-lived mycelial networks is important in stabilizing the mobility of metal and radionuclide pollutants by adsorption and absorption. With absorption, the subsequent translocation to fruiting bodies may result in hyper-accumulation of these elements. Long-term stabilizing pollutants in mycelial structures and harvesting metal-accumulating fruiting structures are both potential ways to affect contaminated site remediation. However, there are few documented cases of this function being implemented.

Basidiomycete saprotrophs have a wide range of enzyme competence. Wood rotting fungi particularly have laccase and polyphenol oxiadase activity that not only breaks down complex plant material but is also effective in decomposing a number of organic pollutants. Understanding the potential of the fungi in organic pollutant breakdown is still in its infancy and could be important in helping the health of our ecosystems.

References

Aber, J. 1992. Nitrogen cycling and nitrogen saturation in temperate forest ecosystems. Trends Ecol. Evol., 7: 220–236.

Abuzinadah, R.A. and Read, D.J. 1986a. The role of proteins in the nitrogen nutrition of ectomycorrhizal plants. I. Utilization of peptides and proteins by ectomycorrhizal fungi. New Phytol., 103: 481–493.

Abuzinadah, R.A. and Read, D.J. 1986b. The role of proteins in the nitrogen nutrition of ectomycorrhizal plants. III. Protein utilization by *Betula*, *Picea* and *Pinus* in mycorrhizal association with *Hebeloma crustuliniforme*. New Phytol., 103: 507–514.

Abuzinadah, R.A. and Read, D.J. 1989. The role of proteins in the nitrogen nutrition of ectomycorrhizal plants. V. The utilization of peptides by birch (*Betula pendula* L.) infected with different mycorrhizal fungi. New Phytol., 112: 55–60.

Adams-Krumins, J., Dighton, J., Gray, D., Franklin, R.B., Morin, P. and Roberts, M.S. 2009. Soil microbial community response to nitrogen enrichment in two scrub oak forests. For. Ecol. Manage., 258: 1383–1390.

Agerer, R. 1987. Colour Atlas of Ectomycorrhizae. Einhorn-Verlag, Munich.

Agerer, R. 2001. Exploration types of ectomycorrhizae. Mycorrhiza, 11: 107–114.

Aguilera, L.M., Griffiths, R.P. and Caldwell, B.A. 1993. Nitrogen in ectomycorrhizal mat and non-mat soils of different-age Douglas-fir forests. Soil Biol. Biochem., 25: 1015–1019.

Allen, E.B., Allen, M.F., Helm, D.J., Trappe, J.M., Molina, R. and Rincon, E. 1995. Patterns and regulation of mycorrhizal plant and fungal diversity. Plant Soil, 170: 47–62.

Allen, M.F. 1991. The Ecology of Mycorrhizae. Cambridge University Press, Cambridge.

Anderson, I.C., Chambers S.M. and Cairney, J.W.G. 2001. Variation in nitrogen source utilization by *Pisolithus* isolates maintained in axenic culture. Mycorrhiza, 11: 53–56.

Anderson, J.B. and Kohn, L.M. 2007. Dikaryons, diploids and evolution. *In*: Heitman, J., Kronstad, J., Taylor, J. and Caselton, L. (eds.). Sex in Fungi: Molecular Determination and Evolutionary Implications. American Society for Microbiology Press, USA.

Antibus, R.K, Sinsabaugh, R.L. and Linkins, A.E. 1992. Phosphatase activities and phosphorus uptake from inositol phosphate by ectomycorrhizal fungi. Can. J. Bot., 70: 794–801.

Antibus, R.K., Bower, D. and Dighton, J. 1997. Root surface phosphatase activities and uptake of 32P-labelled inositol phosphate in field-collected gray birch and red maple roots. Mycorrhiza, 7: 39–46.

Arnebrant, K. and Söderström, B. 1992. Effects of different fertilizer treatments on the ectomycorrhizal colonization potential in two Scots pine forests in Sweden. For. Ecol. Manag., 53: 77–89.

Arnolds, E. 1985. Veranderingen in de paddestoelenflora (mycoflora). Wet. Meded., 167.

Arnolds, E. 1988. The changing macromycete flora in The Netherlands. Trans. Br. Mycol. Soc., 90: 391–406.

Arnolds, E. 1989a. Former and present distribution of stipitate hydnaceous fungi (Basidiomycetes) in the Netherlands. Nova Hedwigia, 1–2: 107–142.

Arnolds, E. 1989b. The influence of increased fertilization on the macrofungi of a sheep meadow in Drenthe, the Netherlands. Opera Bot., 100: 7–21.

Arnolds, E. 1991. Decline of ectomycorrhizal fungi in Europe. Agric. Ecosys. Environ., 35: 209–244.

Arnolds, E.J.M. 1997. Biogeography and conservation. pp. 47–63. *In*: Wicklow, D.T. and Soderstrom, B. (eds.). The Mycota IV: Environmental and Microbial Relationships. Springer-Verlag, Berlin Heidelberg.

Baar, J., Horton, T.R., Kretzer, A.M. and Bruns, T.D. 1999. Mycorrhizal colonization of *Pinus muricata* from resistant propagules after a stand-replacing wildfire. New Phytol., 143: 409–418.

Babikova, Z., Gilbert, L., Bruce, T.J.A., Birkett, M., Caulfield, J.C. et al. 2013. Underground signals carried through common mycelial networks warn neighbouring plants of aphid attack. Ecol. Lett., 16: 835–843.

Bahram, M., Harend, H. and Tedersoo, L. 2014. Network perspectives in ectomycorrhizal associations. Fung. Ecol., 7: 70–77.

Bajwa, R. and Read, D.J. 1985. The biology of mycorrhiza in the Ericaceae. IX. Peptides as nitrogen sources for the ericoid endophyte and for mycorrhizal and non-mycorrhizal plants. New Phytol., 101: 459–467.

Baldrian, P. 2008. Enzymes of saprotrophic basidiomycetes. pp. 19–41. *In*: Boddy, L., Frankland, J. C. and van West, P. (eds.). Ecology of Saprotrophic Basidiomycetes, Elsevier Academic Press, Amsterdam.

Banfield, J.F., Barker, W.W., Welch, S.A. and Taunton, A. 1999. Biological impact of mineral dissolution: Application of the lichen model to understanding mineral weathering in the rhizosphere. Proc. Natl. Acad. Sci. USA, 96: 3404–3411.

Barron, E.S. 2017. Who cares? The human perspective on fungal conservation. pp. 321–329. *In*: Dighton, J. and White, J.F. (eds.). The Fungal Community: Its Organization and Role in the Ecosystem, 4th Edition, CRC Press, Boca Raton.

Barron, E.S. and Emery, M. 2012. Implications of variation in social-ecological systems for the development of United States fungal management policy. Social Natural Res., 25: 996–1011.

Bartlett, E.M. and Lewis, D.H. 1973. Surface phosphatase activity of mycorrhizal roots of beech. Soil Biol. Biochem., 5: 249–257.

Baxter, J.W. and Dighton, J. 2001. Ectomycorrhizal diversity alters growth and nutrient acquisition of gray birch (*Betula populifolia* Marshall) seedlings in host-symbiont culture conditions. New Phytol., 152: 139–149.

Baxter, J.W. and Dighton, J. 2005a. Phosphorus source alters host plant response to ectomycorrhizal diversity. Mycorrhiza, 15: 513–523.

Baxter, J.W. and Dighton, J. 2005b. Diversity-functioning relationships in ectomycorrhizal fungal communities. pp. 383–398. *In*: Dighton, J., White, J.F. and Oudemans, P. (eds.). The Fungal Community: Its Organization and Role in the Ecosystem, 3rd Edition. CRC Taylor and Francis, Boca Raton.

Bending, G.D. and D.J. Read. 1996. Nitrogen mobilization from protein-polyphenol complex by ericoid and ectomycorrhizal fungi. Soil Biol. Biochem., 28: 1603–1612.

Bending, G.D. and Read, D.J. 1995a. The structure and function of the vegetative mycelium of ectomycorrhizal plants. V. Foraging behaviour and translocation of nutrients from exploited litter. New Phytol., 130: 401–409.

Bending, G.D. and Read, D.J. 1995b. The structure and function of the vegetative mycelium of ectomycorrhizal plants. VI. Activities of nutrient mobilizing enzymes in birch litter colonized by *Paxillus involutus* (Fr.) Fr. New Phytol., 130: 411–417.

Blaschke, H. 1988. Mycorrhizal infection and changes in fine root development of Norway spruce influenced by acid rain in the field. pp. 112–115. *In*: Jansen, A.E., Dighton, J. and Bresser, A.H.M. (eds.). Ectomycorhiza and Acid Rain, Bilthoven, The Netherlands.

Blaschke, H., Brehmer, U. and Schwartz, H. 1985. Wurzelschäden und Waldsterben: Zur Bestimmung morphometrischer Kenngrössen von Feinwurtzelsytemen mit dem IBAS - erset Ergebnisse. Forstw. Cbl., 104: 199–205.

Blaudez, D., Jacob, C., Turnau, K., Colpaert, J.V., Ahonen-Jonnarth, U., Finlay, R., Botton, B. and Chalot, M. 2000. Differential responses of ectomycorrhizal fungi to heavy metals *in vitro*. Mycol. Res., 104: 1366–1371.

Boddy, L. 1999. Saprotrophic cord-forming fungi: Meeting the challenge of heterogenous environments. Mycologia., 91: 13–32.

Boddy, L. and Rayner, A.D.M. 1983. Ecological roles of basidiomycetes forming decay communities in attached oak branches. New Phytol., 93: 177–188.

Boddy, L., Hiscox, J., Gilmartin, E.C., Johnston S.R. and Heilmann-Clausen, J. 2017. Wood decay communities in angiosperm wood. pp. 169–189. *In*: Dighton, J. and White, J.F. (eds.). The Fungal Community: Its Organization and Role in the Ecosystem, 4th Edition, CRC Press, Boca Raton.

Boerner, R.E., DeMars, B.G. and Leicht, P.N. 1996. Spatial patterns of mycorrhizal infectiveness of soils long a successional chronosequence. Mycorrhiza, 6: 79–90.

Bokhorst, S. and Wardle, D.A. 2014. Snow fungi as a food source for micro-arthropods. Eur. J. Soil Biol., 60: 77–80.

Bonfante, P. and A. Genre. 2010. Mechanisms underlying beneficial plant-fungus interactions in mycorrhizal symbiosis. Nature Comm., 1: 48 doi: 10.1038/ncomms1046.

Brandrud, T.E. 1995. The effects of experimental nitrogen addition on the ectomycorrhizal fungal flora in an oligotrophic spruce forest at Gardsjon, Sweden. For. Ecol. Manag. 71: 111–122.

Brundrett, M. 1991. Mycorrhizas in natural ecosystems. Adv. Ecol. Res., 21: 171–313.

Brundrett, M. 2009. Mycorrhizal associations and other means of nutrition of vascular plants: Understanding the global diversity of host plants by resolving conflicting information and developing reliable means of diagnosis. Plant Soil, 320: 37–77.

Burford, E.P., Kierans, M. and Gadd, G.M. 2003. Geomycology: Fungi in mineral substrata. Mycologist, 17: 98–107.

Byrne, A.R. 1988. Radioactivity in fungi in Slovenia, Yugoslavia, following the Chernobyl accident. J. Environ. Radioact., 6: 177–183.

Cairney, J.W.G. 1992. Translocation of solutes in ectomycorrhizal and saprotrophic rhizomorphs. Mycol. Res., 96: 135–141.

Cairney, J.W.G. and Burke, R.M. 1996. Physiological heterogeneity within fungal mycelia: An important concept for a functional understanding of the ectomycorrhizal symbiosis. New Phytol., 134: 685–695.

Cairney, J.W.G. 1999. Intraspecific physiological variation: implications for understanding functional diversity in ectomycorrhizal fungi. Mycorrhiza, 9: 125–135.

Cairney, J.W.G and A.A. Meharg. 2002. Interactions between ectomycorrhizal fungi and soil saprotrophs: Implications for decomposition of organic matter in soils and degradation of organic pollutants in the rhizosphere. Can. J. Bot., 80: 803–809.

Cairney, J.W.G. 2005. Basidiomycete mycelia in forest soils: Dimensions, dynamics and roles in nutrient distribution. Mycol. Res., 109: 7–20.

Carfrae, J.A., Skene, K.R., Sheppard, L.J., Ingleby, K. and Crossley, A. 2006. Effects of nitrogen with and without acidified sulphur on an ectomycorrhizal community in a Sitka spruce (*Picea sitchensis* Bong. Carr) forest. Environ. Poll., 141: 131–138.

Carleton, T.J. and Read, D.J. 1990. Ectomycorrhizas and nutrient transfer in conifer-feathermoss ecosystems. Can. J. Bot., 69: 778–784.

Cave, B. 1997. Toadstools and springtails. The Mycologist, 11: 154.

Chang, T.T. and Li, C.Y. 1998. Weathering of limestone, marble, and calcium phosphate by ectomycorrhizal fungi and associated microorganisms. Taiwan J. For. Sci., 13: 85–90.

Chappelka, A.H., Kush, J.S., Runion, G.B., Meir, S. and Kelley, W.D. 1991. Effects of soil-applied lead on seedling growth and ectomycorrhizal colonization of loblolly pine. Environ. Pollution 72: 307–316.

Cheal, D.C. 1987. The diets and dietary preferences of *Rattus fuscipes* and *Rattus lutreolus* at Walkerville in Victoria. Aust. Wildl. Res., 14: 35–44.

Chu-Chou, M. and Grace, L.J. 1985. Comparative efficiency of the mycorrhizal fungi *Laccaria laccata*, *Hebeloma crustuliniforme* and *Rhizopogon* spp. on the growth of radiata pine seedlings. N. Z. J. Bot., 23: 417–424.

Claridge, A.W. and T.W. May. 1994. Mycophagy among Australian mammals. Aut. J. Ecol., 19: 251–275.

Claridge, A.W. and Trappe, J.M. 2005. Sporocarp mycophagy: nutritional, behavioral, evolutionary, and physiological aspects. pp. 599–611. *In*: Dighton, J., White, J.F. and Oudemans, P. (eds.). The Fungal Community; Its Organization and Role in the Ecosystem, CRC Taylor and Francis, Boca Raton.

Clinton, P.W., Buchanan, P.K. and Allen, R.B. 1999. Nutrient composition of epigeous fungal sporocarps growing on different substrates in a New Zealand mountain beech forest. N. Z. J. Bot., 37: 149–153.

Colpaert, J.V. and Van Tichelen, K.K. 1996. Mycorrhizas and environmental stress. pp. 201–216. *In*: Frankland, J.C., Magan, N. and Gadd, G.M. (eds.). Fungi and Environmental Change. Cambridge University Press, Cambridge.

Connolly, J.H., Shortle, W.C. and Jellison, J. 1998. Translocation and incorporation of strontium carbonate derived strontium into calcium oxalate crystals by the wood decay fungus *Resinicium bicolor*. Can. J. Bot., 77: 179–187.

Cork, S.J. and Kenagy, G.J. 1989a. Nutritional value of a hypogeous fungus for a forest-dwelling ground squirrel. Ecology, 70: 577–586.

Cork, S.J. and Kenagy, G.J. 1989b. Rates of gut passage and retention of hypogeous fungal spores in two forest-dwelling rodents. J. Mammal., 70: 512–519.

Coughtree, P.J. Ed. 1983. Ecological Aspects of Radionuclide Release. Blackwell Scientific Publication, Oxford.

Courtney, S.P., Kibota, T.T. and Singleton, T.S. 1990. Ecology of mushroom-feeding Drosphilidae. Adv. Ecol. Res., 20: 225–275.

Courty, P-E., Buée, M., Diedhiou, A.G., Frey-Klett, P., Le Tacon, F. et al. 2010. The role of ectomycorrhizal communities in forest ecosystem processes: New perspectives and emerging concepts. Soil Biol. Biochem., 42: 679–698.

Crane, S., Dighton, J. and Barkay, T. 2010. Growth responses to and accumulation of mercury by ectomycorrhizal fungi. Fungal Biol., 114: 873–880.

Crane, S., Barkay, T. and Dighton, J. 2012. The effect of mercury on the establishment of *Pinus rigida* seedlings and the development of their ectomycorrhizal communities. Fungal Ecol., 5: 245–251.

Cromack, K., Fichter, B.L., Moldenke, A.M. and Ingham, E.I. 1988. Interactions between soil animals and ectomycorrhizal fungal mats. Agric. Ecosyst. Environ., 24: 161–168.

Cromack, K., Sollins, P., Granstein, W.C., Speidel, T., Todd, A.W. et al. 1979. Calcium oxalate accumulation and soil weathering in mats of the hypogeous fungus *Hysterangium crassum*. Soil Biol. Biochem., 11: 463–487.

Cullings, K.W., Vogler, D.R., Parker, V.T. and Makhija, S. 2001. Defoliation effects on the ectomycorrhizal community of a mixed *Pinus contorta/Picea engelmannii* stand in Yellowstone Park. Oecologia, 127: 533–539.

Dahlberg, A. and G.M. Mueller. 2011. Applying IUCN red-listing criteria for assessing and reporting on the conservation status of fungal species. Fung. Ecol., 4: 147–162.

Denizli, A., Cihangir, N., Tüzmen, N. and Alsancak, G. 2005. Removal of chlorophenols from aquatic systems using the dried and dead fungus *Pleurotus sajor caju*. Bioresource Technol., 96: 59–62.

Denny, H.J. and Ridge, I. 1995. Fungal slime and its role in the mycorrhizal amelioration of zinc toxicity to higher plants. New Phytol., 130: 251–257.

Denny, H.J. and Wilkins, D.A. 1987a. Zinc tolerance in *Betula* spp. I. Effects of external concentration of zinc on growth and uptake. New Phytol., 106: 517–524.

Denny, H.J. and Wilkins, D.A. 1987b. Zinc tolerance in *Betula* spp. IV. The mechanism of ectomycorrhizal amelioration of zinc toxicity. New Phytol., 106: 545–553.

Deslippe, J.R. and Simard, S.W. 2011 Below-ground carbon transfer among *Betula nana* may increase with warming in Arctic tundra. New Phytol., 192: 689–698.

Dickinson, D.J. 1982. The decay of commercial timbers. pp. 179–190. *In*: Frankland, J.C., Hedger, J.N. and Swift, M.J. (eds.). Decomposer Basidiomycetes: Their Biology and Ecology. Cambridge University Press.

Dighton, J. 1983. Phosphatase production by mycorrhizal fungi. Plant Soil, 71: 455–462.

Dighton, J. and Skeffington, R.A. 1987. Effects of artificial acid precipitation on the mycorrhizas of Scots pine seedlings. New Phytol., 107: 191–202.

Dighton, J. and Horrill, A.D. 1988. Radiocaesium accumulation in the mycorrhizal fungi *Lactarius rufus* and *Inocybe longicystis*, in upland Britain. Trans. Br. Mycol. Soc., 91: 335–337.

Dighton, J., Mason, P.A. and Poskitt, J.M. 1990. Field use of 32P tracer to measure phosphate uptake by birch mycorrhizas. New Phytol., 116: 655–661.

Dighton, J. 1991. Acquisition of nutrients from organic resources by mycorrhizal autotrophic plants. Experientia, 47: 362–369.

Dighton, J. and Jansen, A.E. 1991. Atmospheric pollutants and ectomycorhizas: more questions than answers? Environ. Poll., 73: 179–204.

Dighton, J. 1995. Nutrient cycling in different terrestrial ecosystems in relation to fungi. Can. J. Bot., 73: S1349–S1360.

Dighton, J., Tuininga, A.R., Gray, D.M., Huskins, R.E. and Belton, T. 2004. Impacts of atmospheric deposition on New Jersey pine barrens forest soils and communities of ectomycorrhizae. For. Ecol. Manage., 201: 131–144.

Dighton, J. 2016. Fungi in Ecosystem Processes, 2nd Edition, CRC Taylor and Francis, Boca Raton.

Dighton, J. 2018. Fungi and remediation of radionuclide pollution. *In*: Tomasini, A. and Leon-Santiesteban, H. (eds.). Fungal Bioremediation: Fundamentals and Applications. RC Press (in press).

Donnelley, P.K. and Fletcher, J.S. 1994. Potential use of mycorrhizal fungi as bioremediation agents. Am. Chem. Soc. Symp. Ser., 563: 93–99.

Duddridge, J.A., Malibari, A. and Read, D.J. 1980. Structure and function of mycorrhizal rhizomorphs with special reference to their role in water transport. Nature, 287: 834–836.

Durall, D.M., Todd, AW. and Trappe, J.M. 1994. Decomposition of ^{14}C-labelled substrates by ectomycorrhizal fungi in association with Douglas fir. New Phytol., 127: 725–729.

Eason, W.R., Newman, E.I. and Chuba, P.N. 1991. Specificity of interplant cycling of phosphorus: The role of mycorrhizas. Plant Soil, 137: 267–274.

Eckl, P., Hoffman, W. and Turk, R. 1986. Uptake of natural and man-made radionuclides by lichens and mushrooms. Rad. Environ. Biophy., 25: 43–54.

Egli, S., Amiet, R., Zollinger, M. and Schneider, B. 1993. Characterization of *Picea abies* (L) Karst. ectomycorrhizas: Discrepancy between classification according to macroscopic versus microscopic features. TREE, 7: 123–129.

Fellner, R. 1988. Effects of acid deposition on the ectotrophic stability of mountain forest ecosystems in central Europe (Czechoslovakia). *In*: Jansen, A.E., Dighton, J. and Bresser, A.H.M. (eds.). Ectomycorhiza and Acid Rain. Bilthoven, The Netherlands.

Fellner, R. and Pesková, V. 1995. Effects of industrial pollutants on ectomycorrhizal relationships in temperate forests. Can. J. Bot., 73: S1310–S1315.

Fernandez, C.W. and Koide, R.T. 2013. The function of melanin in the ectomycorrhizal fungus *Cenococcum geophilum* under water stress. Fung. Biol. 6: 479–486.

Finlay, R., Wallander, H., Smits, M. et al. 2009. The role of fungi in biogenic weathering in boreal forest soils. Fung. Biol. Rev., 23: 101–106.

Finlay, R.D. and Read, D.J. 1986a. The structure and function of the vegetative mycelium of ectomycorrhizal plants I. Translocation of ^{14}C-labelled carbon between plants interconnected by a common mycelium. New Phytol., 103: 143–156.

Finlay, R.D. and Read, D.J. 1986b. The structure and function of the vegetative mycelium of ectomycorrhizal plants II. The uptake and distribution of phosphorus by mycelial strands interconnecting host plants. New Phytol., 103: 157–165.

Finlay, R.D. and Read, D.J. 1986c. The structure and function of the vegetative mycelium of ectomycorrhizal plants. III The uptake and distribution of phosphorus by mycelial strands interconnecting host plants. New Phytol., 103: 157–165.

Fitter, A. 1985. Functioning of vesicular-arbuscular mycorrhizas under field conditions. New Phytol. 99: 257–265.

Fitter, A.H. 1991. Cost benefits of mycorrhizas: Implications for functioning under natural conditions. Experientia, 47: 350–355.

Fogel, R. 1976. Ecological studies of hypogeous fungi. II: Sporocarp phenology in a western Oregon Douglas-fir stand. Can. J. Bot., 54: 1152–1162.

Fogel, R. and Trappe, J.M. 1978. Fungus consumption (mycophagy) by small animals. Northwest Sci. 52: 1–31.

Fomina, M.E.P. Burford, Hillier, S., Klerans, M. and Gadd, G.M. 2010. Rock-building fungi. Geomicrobiol. J., 27: 624–629.

Frankland, J.C., Hedger, J.N. and Swift, M.J. 1982. Decomposer Basidiomycetes: Their Biology and Ecology, Cambridge University Press, Cambridge.

Frankland, J.C. 1992. Mechanisms in fungal succession. pp. 403–426. *In*: Carroll, G. and Wicklow, D.T. (eds.). The Fungal Community: Its Organization and Role in the Ecosystem, 2md Edition. Marcel Dekker Press, New Your.

Frankland, J.C. 1998. Fungal succession - unravelling the unpredictable. Mycol. Res. 102: 1–15.

Fricker, M.D., Bebber, D. and Boddy, L. 2008. Mycelial networks: structure and dynamics. pp. 3–18. *In*: Boddy, L., Frankland, J.C. and van West, P. (eds.). Ecology of Saprotrophic Basidiomycetes, Elsevier Academic Press, Amsterdam.

Gehring, C.A. and Whitham, T.G. 1994. Comparisons of ectomycorrhizae on Pinyon pines (*Pinus edulis*; Pinaceae) across extremes of soil type and herbivory. Am. J. Bot., 81: 1509–1516.

Gibson, F. and Deacon, J.W. 1988. Experimental study of establishment of ectomycorrhizas in different regions of birch root systems. Trans. Br. Mycol. Soc., 91: 239–251.

Giltrap, N.J. 1982. Production of polyphenol oxidases by ectomycorrhizal fungi with special reference to *Lactarius* spp. Trans. Br. Mycol. Soc., 78: 75–81.

Giovani, C., Nimis, P.L. and Padovani, R. 1990. Investigation of the performance of macromycetes as bioindicators of radioactive contamination. pp. 485–491. *In*: Desmet, G., Nassimbeni, P. and Belli, M. (eds.). Transfer of Radionuclides in Natural and Semi-natural Environments. Elsevier Applied Science, London.

González-Guerrero, M., Melville, L.H., Ferrol, N., Lott, J.N.A., Azcón-Aguiar, C. and Peterson, R.L. 2008. Ultrastructural localization of heavy metals in the extraradical mycelium and spores of the arbuscular mycorrhizal fungus *Glomus intraradicis*. Can J. Microbiol., 54: 103–110.

Goodman, D., Durall, D.M., Trofymow, J.A. and Berch, S.M. 1996–2000. A Manual of Concise Descriptions of North American Ectomycorrhizae. Mycologue Publications, Victoria, BC.

Gray, S. 1998. Fungi as potential bioremediaton agents in soil contaminated with heavy or radioactive metals. Biochem. Soc. Trans., 26: 666–670.

Gray, S.N., Dighton, J., Olsson, S. and Jennings, D.H. 1995. Real-time measurement of uptake and translocation of 137Cs within mycelium of *Schizophyllum commune* Fr. by autoradiography followed by quantitative image analysis. New Phytol., 129: 449–465.

Gray, S.N., Dighton, J. and Jennings, D.H. 1996. The physiology of basidiomycete linear organs III. Uptake and translocation of radiocaesium within differentiated mycelia of *Armillaria* spp. growing in microcosms and in the field. New Phytol., 132: 471–482.

Griffith, G.W. and Roderick, K. 2008. Saprotrophic basidiomycetes in grasslands: Distribution and function. pp. 277–299. *In*: Boddy, L., Frankland, J.C. and van West, P. (eds.). Ecology of Saprotrophic Basidiomycetes, Elsevier Academic Press, Amsterdam.

Griffiths, R.P., Caldwell, B.A., Cromack, K. and Morita, R.Y. 1990. Microbial dynamics and chemistry in Douglas fir forest soils colonised by ectomycorrhizal mats: I. Seasonal variation in nitrogen chemistry and nitrogen cycle transformation rates. Can. J. For. Res., 20: 211–218.

Grönwall, O. and Pehrson, Å. 1984. Nutrient content in fungi as primary food of the red-squirrel, *Sciurua vulgaris*. Oecologia, 64: 230–231.

Guerin-Laguette, A., Cummings, N, Butler, R.C., Willows, A., Hesom-Williams, N., Li, S. and Wang, Y. 2014. *Lactarius deliciosus* and *Pinus radiata* in New Zealand: Towards the development of innovative gourmet mushroom orchards. Mycorrhiza, 24: 511–523.

Guillén, J., Baez, A. and Salas, A. 2012. Influence of alkali and alkaline earth elements on the uptake of radionuclides by *Pleurotus eryngii* fruit bodies. Appl. Rad. Isot., 70: 650–655.

Guillette, O., Fraiture, A. and Lambinon, J. 1990. Soil-fungi radiocaesium transfers in forest ecosystems. pp. 468–478. *In*: Desmet, G., Nassimbeni, P. and Belli, M. (eds.). Transfer of Radionuclides in Natural and Semi-natural Environments. Elsevier Applied Science, London.

Hanski, I. 1989. Fungivory: fungi, insects and ecology. pp. 275–324. *In*: Wilding, M., Collins, M.M., Hammond, P.M. and Webber, J.F. (eds.). Insect-Fungus Interactions. Academic Press, London.

Hanson, A.M., Hodge, K.T. and Porter, L.M. 2003. Mycophagy among primates. Mycologist 17: 6–10.

Harley, J.L. 1969. The Biology of Mycorrhiza. Leonard Hill, London.

Harley, J.L. and Smith, S.E. 1983. Mycorrhizal Symbiosis. Academic Press, London.

Harms, H. 2011. Untapped potential: exploiting fungi in bioremediation of hazardous chemicals. Nature Rev. Microbiol., 9: 177–192.

Harms, H., Wick, L.Y. and Schlosser, D. 2017. The fungal community in organically polluted systems. pp. 459–469. *In*: Dighton, J. and White, J.F. (eds.). The Fungal Community: Its Organization and Role in the Ecosystem, 4th Edition. CRC Press, Boca Raton.

Hartley, J., Cairney, J.W.G. and Meharg, A.A. 1997a. Do ectomycorrhizal fungi exhibit adaptive tolerance to potentially toxic metals in the environment? Plant Soil, 189: 303–319.

Hartley, J., Cairney, J.W.G., Sanders, F.E. and Meharg, A.A. 1997b. Toxic interactions of metal ions (Cd^{2+}, Pb^{2+}, Zn^{2+} and Sb^{3-}) on *in vitro* biomass production of ectomycorrhizal fungi. New Phytol., 137: 551–562.

Hartley-Whitaker, J., Cairney, J.W.G. and Meharg, A.A. 2000a. Sensitivity to Cd or Zn of host and symbiont of ectomycorrhizal *Pinus sylvestris* L. (Scots pine) seedlings. Plant Soil, 218: 31–42.

Hartley-Whitaker, J., Cairney, J.W.G. and Meharg, A.A. 2000b. Toxic effects of cadmium and zinc on ectomycorrhizal colonization of scots pine (*Pinus sylvestris* L.) from soil inoculum. *Environ. Toxicol. Chem.*, 19: 694–699.

Haselwandter, K. and Berreck, M. 1994. Accumulation of radionuclides in fungi. pp. 259–277. *In*: Winkelmann, G. and Winge, D.R. (eds.). Metal Ions in Fungi. Marcel Dekker, New York.

Haselwandter, K. 1978. Accumulation of the radioactive nuclide 137Cs in fruitbodies of basidiomycetes. Health Phy., 34: 713–715.

Haselwandter, K., Bereck, M. and Brunner, P. 1988. Fungi as bioindicators of radiocaesium contamination. Pre- and post-Chernobyl activities. Trans. Br. Mycol. Soc., 90: 171–176.

Häussling, M. and Marschner, H. 1989. Organic and inorganic soil phosphates and acid phosphatase activity in the rhizosphere of 80-year old Norway spruce [*Picea abies* (L.) Karst.] trees. Biol. Fertil. Soils, 8: 128–133.

Heal, O.W. and Horrill, A.D. 1983. Terrestrial ecosystems: An ecological context for radionuclide research. pp. 31–46. *In*: Coughtree, P.J. (ed.). Ecological Aspects of Radionuclide Release. Oxford: Blackwell Scientific Publications.

Heap, A.J. and Newman, E.I. 1980. The influence of vesicular arbuscular mycorrhizas on phosphorus transfer between plants. New Phytol., 85: 173–179.

Heap, A.J. and Newman, E.I. 1980. Links between roots by hyphae of vesicular arbuscular mycorrhizas. New Phytol., 85: 169–171.

Held, B.W. 2017. Decomposition of wooden structures by fungi. pp. 491–499. *In*: Dighton, J. and White, J.F. (eds.). The Fungal Community: Its Organization and Role in the Ecosystem, 4th Edition. CRC Press, Boca Raton.

Hiol Hiol, F., Dixon, R.K and Curl, E.A. 1994. The feeding preference of mycophagous Collembola varies with ectomycorrhizal symbiont. Mycorrhiza, 5: 99–103.

Hoffland, E., Landeweert, R., Kuyper, T.W. and van Breemen, N. 2001. (Further) links from rocks to plants. Tr. Ecol. Evol., 16: 544.

Hütterman, A. 1982. Fruhdiagnose von Immissionsschaden im Wurzelbereich von Waldbaumen. Landesanst. f. Okologie, Landschaftsentw. u. Forstpl. Nordhein-Westfalen: 26–31.

Hüttermann, A. 1985. The effects of acid deposition on the physiology of the forest ecosystem. Experientia, 41: 585–590.

Ingleby, K., Mason, P.A., Last, F.T. and Fleming, L.V. 1990. Identification of Ectomycorrhizas. Institute of Terrestrial Ecology Research Pub. # 5, London.

Itoo, Z.A. and Reshi, Z.A. 2013. The multifunctional role of ectomycorrhizal associations in forest ecosystem processes. Bot. Rev., 79: 371–400.

Jaenike, J., Grimaldi, D., Shuder, A.E. and Greenleaf, A.L. 1983. Alpha-amanitin tolerance in mycophagous *Drosophila*. Science, 221: 165–167.

Jansen, A.E. and Dighton, J. 1990. Effects of air pollutants on ectomycorrhizas. CEC Air Pollution Research Report # 30.

Jansen, A.E, Dighton, J. and Bresser, A.H.M. 1988. Ectomycorrhiza and Acid Rain. Brussels: CEC Air Pollution Research Report # 12.

Jansen, A.E. and Van Dobben, H.F. 1987. Is the decline of *Cantharellus cibarius* in the Netherlands due to air pollution? Ambio, 16: 211–213.

Jany, J.-L., Martin, F. and Garbaye, J. 2003. Respiration activity of ectomycorrhizas from *Cenococcum geophilum* and *Lactarius* sp. in relation to soil water potential in five beech forests. Plant Soil, 255: 487–494.

Jennings, D.H. 1982. The movement of *Serpula lacrimans* from substrate to substrate over nutritionally inert surfaces. pp. 91–108. *In*: Frankland, J.C., Hedger, J.N. and Swift, M.J. (eds.). Decomposer Basidiomycetes: Their Biology and Ecology. Cambridge University Press, Cambridge.

Jennings, D.H. 1991. Techniques for studying the functional aspects of rhizomorphs of wood-rotting fungi: Some possible applications to ectomycorrhizae. pp. 309–329. *In*: Norris, J.R. et al. (eds.). Techniques for the Study of Mycorrhiza, Methods in Microbiology, Volume 23. Academic Press, Londo.

Jensen, D.F., Karlsson, M. and Lindahl, B.D. 2017. Fungal-fungal interactions: from natural ecosystems to managed plant production, with emphasis on biological control of plant diseases. pp. 549–562.

In: Dighton, J. and White, J.F. (eds). The Fungal Community: Its Organization and Role in the Ecosystem, 4th Edition. CRC Press, Boca Raton.

Jensen, H.L. 1929. On the influence of the carbon: Nitrogen ratios of organic material on the mineralization of nitrogen. J. Agric. Sci., 71–82.

Johnson, D., Krsek, M., Wellington, E.M.H. et al. 2005. Soil invertebrates disrupt carbon flow through fungal networks. Science, 309: 1047.

Joner, E.J. and Johansen, A. 2000. Phosphatase activity of external hyphae of two arbuscular mycorrhizal fungi. Mycol. Res., 104: 81–86.

Joner, E.J., Leyval, C. and Colpaert, J.V. 2006. Ectomycorrhizas impede phytoremediation of poly aromatic hydrocarbons (PAHs) both within and beyond the rhizosphere. Environ. Poll., 142: 34–38.

Jones, H.E., Madeira, M., Herraez, L. et al. 1999. The effect of organic-matter management on the productivity of *Eucalyptus globulus* stands in Spain and Portugal: Tree growth and harvest residue decomposition in relation to site and treatment. For. Ecol. Manage., 122: 73–86.

Jones, M.D., Durall, D.M. and Tinker, P.B. 1990. Phosphorus relationships and production of extramatrical hyphae by two types of willow ectomycorrhizal at different soil phosphorus levels. New Phytol., 115: 259–267.

Jongbloed, R.H. and Borst Pauwels, G.W.F.H. 1988. Effects of Al^{3+} and NH_4^+ on growth and uptake of K^+ and $H_2PO_4^-$ by three ectomycorrhizal fungi in pure culture. pp. 47–52. *In*: Jansen, A.E., Dighton, J. and Bresser, A.H.M. (eds.). Ectomycorhiza and Acid Rain. Bilthoven, The Netherlands.

Jongbloed, R.H. and Borst Pauwels, G.W.F.H. 1989. Effects of ammonium and pH on growth and potassium uptake by the ectomycorrhizal fungus *Laccaria bicolor* in pure culture. Agric. Ecosyst. Environ., 28: 207–212.

Jongmans, A.G., van Breemen, N., Lundstrom, U. et al. 1997. Rock-eating fungi. Nature, 389: 682–683.

Jonsson, L., Dahlberg, A., Nilsson, M.-C., Karen, O. and Zackrisson, O. 1999b. Continuity of ectomycorrhizal fungi in self-regualting boreal *Pinus sylvestris* forests studied by comparing mycobiont diversity on seedlings and mature trees. New Phytol., 142: 151–162.

Jonsson, L., Dahlberg, A., Nilsson, M.-C., Zackrisson, O. and Karen, O. 1999a. Ectomycorrhizal fungal communities in late-successional Swedish boreal forests, and their composition following wildfire. Mol. Ecol., 8: 205–215.

Jonsson, L.M., Nilsson, M-C., Wardle, D.A. and Zackrisson, O. 2001. Context dependent effects of ectomycorhizal species richness on tree seedling productivity. Oikos, 93: 353–364.

Jumpponen, A. and Egerton-Warburton, L.M. 2005. Mycorrhizal fungi in successional environments: A community assembly model incorporating host plant, environmental and biotic filters. pp. 139–168. *In*: Dighton, J., White, J.F. and Oudemans, P. (eds.). The Fungal Community: Its Organization and Role in the Ecosystem, 3rd Edition, Taylor and Francis, Boca Raton.

Jumpponen, A., Mattson, K.G. and Trappe, J.M. 1998. Mycorrhizal functioning of *Phialocephala fortinii* with *Pinus contorta* on glacier forefront soil: Interactions with soil nitrogen and organic matter. Mycorrhiza, 7: 261–265.

Kårén, O. and Nylund, J.-E. 1997. Effects of ammonium sulphate on the community structure and biomass of ectomycorrhizal fungi in a Norway spruce stand in southwestern Sweden. Can. J. Bot., 75: 1628–1642.

Kerley, S.J. and Read, D.J. 1995. The biology of mycorrhizas in the Ericaceae XVIII. Chitin degradation by *Hymenoscyphus ericae* and transfer of chitin-nitrogen to the host plant. New Phytol., 131: 369–375.

Kinnear, J.E., Cockson, A., Christensen, P.E.S. and Main, A.R. 1979. The nutritional biology of the ruminants and ruminant-like mammals: A new approach. Comp. Biochem. Physiol., 64A: 357–365.

Koele, N. 2012. The role of the stonosphere for the interaction between mycorrhizal fungi and mycorrhizosphere bacteria during mineral weathering. pp. 172–180. *In*: Hock, B. (ed.). The Mycota IX Fungal Associatons, Springer-Verlag, Heidelberg.

Kong, F.X., Liu, Y., Hu, W., Shen, P.P., Zhou, C.L. and Wang, L.S. 2000. Biochemical responses of the mycorrhizae in *Pinus massoniana* to combined effects of Al, Ca and low pH. Chemosphere, 40: 311–318

Kotter, M.M. and Farentinos, R.C. 1984. Tassel-eared squirrels as spore dispersal agents of hypogeous mycorrhizal fungi. J. Mammal., 65: 684–687.

Kottke, I. and Oberwinkler, F. 1990. Pathways of elements in ectomycorrhizae in respect to Hartig net development and endodermis differentiation. *In*: Reisinger, A. and Bresinsky, A. (eds.). Abstracts of the Fourth International Mycological Congress, Regensburg, Germany.

Landeweert, R., Hoffland, E., Finlay, R.D., Kuyper, T.W. and van Breemen, N. 2001. Linking plants to rocks: ectomycorrhizal fungi mobilize nutrients from minerals. Trends Ecol. Evol., 16: 248–254.

Lapeyrie, F., Ranger, J. and Vairelles, D. 1991. Phosphate-solubilizing activity of ectomycorrhizal fungi in vitro. Can. J. Bot., 69: 342–346.

Last, F.T., Dighton, J. and Mason, P.A. 1987. Successions of sheathing mycorrhizal fungi. Tr. Ecol. Evol., 2: 157–161.

Leake, J.R. and Read, D.J. 1989. The biology of mycorrhiza in the Ericaceae. XIII. Some characteristics of the extracellular proteinase activity of the ericoid endophyte *Hymenoscyphus ericae*. New Phytol., 112: 69–76.

Leake, J.R. and Read. D.J. 1990a. Chitin as a nitrogen source for mycorrhizal fungi. Mycol. Res., 94: 993–995.

Leake, J.R. and Read, D.J. 1990b. Proteinase activity in mycorrhizal fungi. I. The effect of extracellular pH on the production and activity of proteinase by the ericoid endophytes of soils of contrasted pH. New Phytol., 115: 243–250.

Leake, J.R. and Miles, W. 1996. Phosphodiesters as mycorrhizal P sources. I. Phosphodiesterase production and utilization of DNA as a phosphorus source by the ericoid mycorrhizal fungus *Hymenoscyphus ericae*. New Phytol., 132: 435–443.

Levy, J.F. 1982. The place of basidiomycetes in the decay of wood in contact with the ground. pp. 161–178. *In*: Frankland, J.C., Hedger, J.N. and Swift, M.J. (eds.). Decomposer Basidiomycetes: Their Biology and Ecology. Cambridge University Press, Cambridge.

Leake, J.R. 2001. Is diversity of ectomycorrhizal fungi important for ecosystem function? New Phytol., 152: 1–8.

Leyval, C., Turnau, K. and Hasselwandter, K. 1997. Effect of heavy metal pollution on mycorrhizal colonization and function: Physiological, ecological and applied aspects. Mycorrhiza, 7: 139–153.

Li, Y., Bateman, C.C., Skelton, J., Justino, M.A., Nolen, Z.J., Simmons, D.R. and Hulcr, J. 2017. Wood decay fungus *Flavodon ambrosius* (Basidiomycota: Polyporales) is widely farmed by two genera of ambrosia beetles. Fungal Biol., 121: 984–989.

Lindahl, B. and Boberg, J. 2008. Distribution and function of litter basidiomycetes in coniferous forests. pp. 183–196. *In*: Boddy, L., Frankland, J.C. and van West, P. (eds.). Ecology of Saprotrophic Basidiomycetes, Elsevier Academic Press, Amsterdam.

Liss, B., Blaschke, H. and Schutt, P. 1984. Verleichende Feinwurzel untersuchungen an gesunden und erkrankter Altfichten auf zwei Standorten in Bayern - ein Beitrag zur Waldsterbenforschung. Eur. J. For. Path., 14: 90–102.

Lodge, D.J., McDowell, W.H., Macy, J., Ward, S.K., Leisso, R., Claudio-Campos, K. and Kühnert, K. 2008. Distribution and role of mat-forming saprobic basidiomycetes in a tropical forest. pp. 197–209. *In*: Boddy, L., Frankland, J. C. and van West, P. (eds.). Ecology of Saprotrophic Basidiomycetes, Elsevier Academic Press, Amsterdam.

MacFall, J.S. Slack, A. and Iyer, J. 1991. Effects of *Hebeloma arenosa* and phosphorus fertility on root acid phosphatase activity of red pine (*Pinus resinosus*) seedlings. Can. J. Bot., 69: 380–385.

Markkola, A.M. and Ohtonen, R. 1988. The effect of acid deposition on fungi in forest humus. pp. 122–126. *In*: Jansen, A.E., Dighton, J. and Bresser, A.H.M. (eds.). Ectomycorrhiza and Acid Rain. Bilthoven, The Netherlands.

Martino, E. and Perotto, S. 2010. Mineral transformations by mycorrhizal fungi. Geomicrobiol. J., 27: 609–623.

Marx, D.H. 1975. Mycorrhiza and establishment of trees on strip-mined land. Ohio J. Sci., 75: 288–297.

McLeod, A.R., Shaw, P.J.A. and Holland, M.R. 1992. The Liphook forest fumigation project: Studies of sulphur dioxide and ozone effects on coniferous trees. For. Ecol. Manage., 51: 121–127.

McNulty, S.G. and Aber, J.D. 1993. Effects of chronic nitrogen additions on nitrogen cycling in a high-elevation spruce-fir stand. Can. J. For. Res., 23: 1252–1263.

Melillo, J.M., Aber, J.D. and Muratore, J.F. 1982. Nitrogen and lignin control of hardwood leaf litter decomposition dynamics. Ecology, 63: 621–626.

Meyer, F.H. 1987. Das Wurzelsystem geschadigter Waldbestande. Allg. Forst Zeitscr., 27/28/29: 754–757.

Mietelski, J.W, Jasinska, M. Kubica, B. Kozak, K. and Macharski, P. 1994. Radioactive contamination of Polish mushrooms. The Science of the Total Environ. 157: 217–226.

Miller, S.L. 1995. Functional diversity in fungi. Can. J. Bot., 73 (Suppl 1): S50–S57.

Mitchell, D.T. and Read, D.J. 1981. Utilization of inorganic and organic phosphate by the mycorrhizal endophytes of *Vaccinium macrocarpon* and *Rhododendron ponticum*. Trans. Br. mycol. Soc., 76: 255–260.

Mousain, D. and Salsac, L. 1986. Utilisation du phytate et activites phosphatases acides chez *Pisolithus tinctorius*, basidiomycete mycorhizien. Physiol. Veg., 24: 193–200.

Mukerji, K.G. 1996. *Concepts in Mycorrhizal Research*. Dordrecht: Kluwer Academic Publishers.

Muramatsu, Y., Yoshida, S. and Sumia, M. 1991. Concentrations of radiocesium and potassium in basidiomycetes collected in Japan. Sci. Tot. Environ., 105: 29–39.

Myers, M.D. and Leake, J.R. 1996. Phosphodiesters as mycorrhizal P sources. II. Ericoid mycorrhiza and the utilization of nuclei as phosphorus and nitrogen source by *Vaccinium macrocarpon*. New Phytol., 132: 445–452.

Newbery, D. McC., I.J. Alexander and J.A. Rother. 1997. Phosphorus dynamics in a lowland African rain forest: The influence of the ectomycorrhizal trees. Ecol. Monogr., 67: 367–409.

Newell, K. 1984a. Interaction between two decomposer basidiomycetes and a collembolan under Sitka spruce: Distribution, abundance and selective grazing. Soil Biol. Biochem., 16: 227–233.

Newell, K. 1984b. Interactions between two decomposer basidiomycetes and a collembolan under Sitka spruce: Grazing and its potential effects on fungal distribution and litter decomposition. Soil Biol. Biochem., 16: 235–239.

Newman, E.I. and W.R. Eason. 1989. Cycling of nutrients from dying roots to living plants, including the role of mycorrhizas. Plant Soil, 115: 211–215.

Nylund, J.-E. 1989. Nitrogen, carbohydrate and ectomycorrhiza - The classical theories crumble. Agric. Ecosys. Environ., 28: 361–364.

Oberwinkler, F. 1993. Evolution of functional groups in basidiomycetes (fungi). pp. 143–163. *In*: Schulze, E.-D. and Mooney, H.A. (eds.). Biodiversity and Ecosystem Function, Springer, Berlin.

Oberwinkler, F. 2012. Basidiolichens. pp. 341–362. *In*: Hock. B. (ed.). Fungal Associations, The Mycota IX, 2nd Edition, Springer, Berlin.

Olsen, R.A., Joner, E. and Bakken, L.R. 1990. Soil fungi and the fate of radioceasium in the soil ecosystem - a discussion of possible mechanisms involved in the radiocaesium accumulation in fungi, and the role of fungi as a Cs-sink in the soil. pp. 657–663. *In*: Desmet, G., Nassimbeni, P. and Belli, M. (eds.). Transfer of Radionuclides in Natural and Semi-natural Environments. Elsevier Applied Science, London.

Olsson, S. 1995. Mycelial density profiles of fungi on heterogenous media and their interpretation in terms of nutrient reallocation patterns. Mycol. Res., 99: 143–153.

Olsson, S. and Jennings, D.H. 1991. Evidence for diffusion being the mechanism of translocation in the hyphae of three moulds. Exp. Mycol., 15: 302–309.

Orlovich, D.A., Draffin, S.J., Daly, R.L. and Stephenson, S.L. 2013. Piracy in the high trees: Ectomycorrhizal fungi form an aerial 'canopy soil' microhabitat. Mycologia, 105: 52–60.

Osono, T., Fukasawa, Y. and Takeda, H. 2003. Roles of diverse fungi in larch needle-litter decomposition. Mycologia, 95: 820–826.

Oulehle, F., Hofmeister, J., Cudlin, P. and Hruška, J. 2006. The effect or reduced atmospheric deposition on soil and soil solution chemistry at a site subjected to long-term acidification, Načetín, Czech Republic. Sci. Tot. Environ., 370: 532–544.

Palmer, J.G, Miller, O.K. and Gruhn, C. 1994. Fruiting of ectomycorrhizal basidiomycetes on unburned and prescribed burned hard-pine/hardwood plots after drought-breaking rainfalls on the Allegheny Mountains of southwestern Virginia. Mycorrhiza, 4: 93–104.

Parfitt, R.L., Salt, G.J. and Saggar, S. 2001. Post-harvest residue decomposition and nitrogen dynamics in *Pinus radiata* plantations of different N status. For. Ecol. Manage., 154: 55–67.

Parladé, J. and Alvarez, I.F. 1993. Coinoculation of aseptically grown Douglas fir with pairs of ectomycorrhizal fungi. Mycorrhiza, 3: 93.

Peay, K.G., Kennedy, P.G. and Bruns, T.D. 2011. Rethinking ectomycorrhizal succession: Are root density and hyphal exploration types drivers of spatial and temporal zonation? Fungal Ecol., 4: 233–240.

Pérez, F., Castillo-Guevara, C., Galindo-Flores, G., Cuautle, M. and Estrada-Torres, A. 2010. Effect of gut passage by two highland rodents on spore activity and mycorrhiza formation of two species of ectomycorrhizal fungi (*Laccaria trichodermophora* and *Suillus tomentosus*. Botany, 90: 1084–1092.

Perry, D.A., Margolis, H., Choquette, C., Molina, R., Marschner, H. and Trappe, J.M. 1989. Ectomycorrhizal mediation of competition between coniferous tree species. New Phytol., 112: 501–511.

Peterson, R.L., Massicott, H.B. and Melville, L.H. 2004. Mycorrhizas: Anatomy and Cell Biology. Ottowa, NRC-CNRC Research Press, CABI.

Piattoni, F., Amicucci, A., Iotti, M., Ori, F., Stocchi, V. and Zambonelli, A. 2014. Viability and morphology of *Tuber aestivum* spores after passage through the gut of *Sus scrofa*. Fung. Ecol., 9: 52–60.

Pietkäinen, A and Kytöviita, M.-M. 2007. Defoliation changes mycorrhizal benefit and competitive interactions between seedlings and adult plants. J. Ecol., 95: 639–647.

Ponge, J.F. 1990. Ecological study of a forest humus by observing a small volume. I. Penetration of pine litter by mycorrhizal fungi. Eur. J. For. Path., 20: 290–303.

Ponge, J.F. 1991. Succession of fungi and fauna during decomposition of needles in a small area of Scots pine litter. Plant Soil, 138: 99–113.

Ponge, J.F. 2005. Fungal communities: Relation to resource succession. pp. 169–180. *In*: Dighton, J., White, J.F. and Oudemans, P. (eds.). The Fungal Community: Its Composition and Role in the Ecosystem. Taylor and Francis, Boca Raton.

Porter, L.M. and Garber, P.A. 2010. Mycophagy and its influence on habitat use and ranging patterns in *Callimico goeldii*. Am. J. Phys. Anthropol., 142: 468–475.

Prenafeta-Boldu, F.X., Kuhn, A., Luykx, D.M.A.M., Anke, H., van Groenstijn, J.W. and de Bont, J.A.M. 2001. Isolation and characterization of fungi growing on volatile aromatic hydrocarbons as their sole source of carbon and energy source. Mycol. Res., 105: 477–484.

Qui, Z., Chappelka, A.H., Somers, G.L., Lockaby, B.G. and Meldahl, R.S. 1993. Effects of ozone and simulated acidic precipitation on ectomycorrhizal formation on loblolly pine seedlings. Environ. Exp. Bot., 33: 423–431.

Raskin, I. and Ensley, B.D. 2000. Phytoremediation of Toxic Metals: Using Plants to Clean Up the Environment. John Wiley & Sons, Inc., New York.

Rayner, A.D.M. 1978. Interactions between fungi colonizing hardwood stumps and their possible role in determining patterns of colonization and succession. Ann. Apppl. Biol., 89: 505–517.

Rayner, A.D.M. and Boddy, L. 1988. Fungal Decomposition of Wood. John Wiley, Chichester.

Rayner, A.D.M., Powell, K.A., Thompson, W. and Jennings, D.H. 1985. Morphogenesis of vegetative organs. pp. 249–279. *In*: Moore, D., Casselton, L.A., Wood, D.A. and Frankland, J.C. (eds.). Developmental Biology of Higher Fungi. Cambridge University Press, Cambridge.

Read, D.J., Francis, R. and Finlay, R.D. 1985. Mycorrhizal mycelia and nutrient cycling in plant communities. pp. 193–217. *In*: Fitter, A.H. et al. (eds.). Ecological Interactions in Soil, Plants, Microbes and Animals. Blackwell Scientific, Oxford.

Read, D.J., Leake, J.R. and Langdale, A.R. 1989. The nitrogen nutrition of mycorrhizal fungi and their host plants. pp. 181–204. *In*: Boddy, L., Marchant, R. and Read D.J. (eds.). Nitrogen, Phosphorus and Sulphur Cycling in Temperate Forest Ecosystems. Cambridge University Press, Cambridge.

Read, D.J. 1991a. Mycorrhizas in ecosystems. Experientia, 47: 376–391.

Read, D.J. 1991b. Mycorrhizas in ecosystems - nature's response to the "Law of the Minimum". pp. 101–130. *In*: Hawksworth, D.L. (ed.). Frontiers in Mycology. CAB International, Wallingford, UK.

Read, D.J., Lewis, D.H., Fitter, A. and Alexander, I.J. 1992. Mycorrhizas in Ecosystems. CAB International, Wallingford, UK.

Read, D.J. and Kerley, S. 1995. The status and function of ericoid mycorrhizal systems. pp. 499–520. *In*: Varma, A. and B. Hock (eds.). Mycorrhiza: Structure, Function, Molecular Biology and Biochemistry. Springer Verlag, Berlin.

Reddy, M.S. and Natarajan, K. 1997. Coinoculation efficiency of ectomycorrhizal fungi on *Pinus patula* seedlings in a nursery. Mycorrhiza, 7: 133–138.

Repáč, I. 1996. Effects of forest litter on mycorrhiza formation and growth of container-grown Norway spruce (*Picea abies* (L.) Karst.) seedlings. Lesnictvi Forestry, 42: 317–324.

Repáč, I. 1996. Inoculation of *Picea abies* (L.) Karst., seedlings with vegetative inocula of ectomycorrhizal fungi *Suillus bovinus* (L.: Fr.) O. Kuntze and *Inocybe lacera* (Fr.) Kumm. New Forests, 12: 41–54.

Rikkinen, J. and Vesala, R. 2017. Fungal diversity of Macrotermes – Termitomyces nests in Tsavo, Kenya. pp. 377–384. *In*: Dighton, J. and White, J.F. (eds.). The Fungal Community: Its Organization and Role in the Ecosystem, 4th Edition.CRC Press, Boca Raton.

Rizzo, D.M., Blanchette, R.A. and Palmer, M.A. 1992. Biosorption of metals by Armillaria rhizomorphs. Can. J. Bot., 70: 1515–1520.

Rommelt, R., Hiersche, L., Schaller, G. and Wirth, E. 1990. Influence of soil fungi (Basidiomycetes) on the migration of Cs134 + 137 and SR90 in coniferous forest soils. pp. 143–151. *In*: Desmet, G., Nassimbeni, P. and Belli, M. (eds.). Transfer of Radionuclides in Natural and Semi-natural Environments. Elsevier Applied Science, London.

Rosling, A., Roose, T., Herrmann, A.M., Davidison, F.A., Finlay, R.D. and Gadd G.M. 2009. Approaches to modelling mineral weathering by fungi. Fungal Biol. Rev., 23: 138–144.

Rozpadek, P., Domka, A. and Turneau, K. 2017. Mycorrhizal fungi and accompanying microorganisms in improving phytoremediation techniques. pp. 419–432. *In*: Dighton, J. and White, J.F. (eds.). The Fungal Community: Its Organization and Role in the Ecosystem, 4th Edition. CRC Press, Boca Raton.

Ruark, G.A., Thornton, F.C., Tiarks, A.E., Lockarby, B.G., Chappelka, A.H. and Meldahl, R.S. 1991. Exposing loblolly pine seedlings to acid precipitation and ozone: Effects on soil rhizosphere chemistry. J. Environ. Qual., 20: 828–832.

Rühling, A., Bååth, E., Nordgren, A. and Söderström, B. 1984. Fungi in metal-contaminated soil near the Gusum brass mill, Sweden. Ambio, 13: 34–36.

Rühling, A. and Söderström, B. 1990. Changes in fruitbody production of mycorrhizal and litter decomposing macromycetes in heavy metal polluted coniferous forests in North Sweden. Water, Air Soil Poll., 49: 375–387.

Rühling, A and Tyler, G. 1991. Effects of simulated nitrogen deposition to the forest floor on the macrofungal flora of a beech forest. Ambio, 20: 261–263.

Schickmann, S., Urban, A., Kräutler, K., Nopp-Mayr, U. and Hackländer, K. 2012. The interrelationship of mycophagous small mammals and ectomycorrhizal fungi in primeval, disturbed and managed Central European mountainous forests. Oecologia, 170: 395–409.

Schier, G.A. 1985. Response of red spruce and balsam fir seedlings to aluminium toxicity in nutrient solutions. Can. J. For. Res., 15: 29–33.

Senior, E., Smith, J.E., Watson-Craik, I.A. and Tosh, J.E. 1993. Ectomycorrhizae and landfill site reclamations: fungal selection criteria. Lett. Appl. Microbiol., 16: 142–146.

Shammas, K., O'Connell, A.M., Grove, T.S., McMurtrie, R., Damon, P. and Rance, S.J. 2003. Contribution of decomposing harvest residues to nutrient cycling in a second rotation *Eucalyptus globus* plantation in south western Australia. Biol. Fert. Soils, 38: 228–235.

Shaw, P.J.A., Dighton, J. and Poskitt, J.M. 1993. Studies on the mycorrhizal community infecting trees in the Liphook forest fumigation experiment. Agric. Ecosys. Environ., 47: 185–191.

Shaw, T.M., Dighton, J. and Sanders, F.E. 1995. Interactions between ectomycorrhizal and saprotrophic fungi on agar and in association with seedlings of lodgepole pine (*Pinus contorta*). Mycol. Res., 99: 159–165.

Silveira, M.L., Reddy, K.R. and Comerford, N.B. 2011. Litter decomposition and soluble carbon, nitrogen, and phosphorus release in a forest ecosystem. Eur. J. Soil Sci., 1: 86–96.

Simard, S.W., Jones, M.D., Durall, D.M., Perry, D.A., Myrold, D.D. and Molina, R. 1997a. Reciprocal transfer of carbon isotopes between ectomycorrhizal *Betula payrifrea* and *Pseudotsuga menziesii*. New Phytol., 137: 529–542.

Simard, S.W., Perry, D.A., Jones, M.D., Myrold, D.D., Durall, D.M. and Molina, R. 1997b. Net transfer of carbon between ectomycorrhizal tree species in the field. Nature, 338: 579–582.

Simard, S.W., Perry, D.A., Smith, J.E. and Molina, R. 1997c. Effects of soil trenching on occurrence of ectomycorrhizas of *Pseudotsuga menziesii* seedlings grown in mature forests of *Betula papyrifera* and *Pseudotsuga menziesii*. New Phytol., 136: 327–340.

Simard, S.W., Beiler, K.J., Bingham, M.A., Deslippe, J.R., Philip, L.J. and Teste, F.P. 2012. Mycorrhizal networks: Mechanisms, ecology and modelling. Fung. Biol. Rev., 26: 39–60.

Sinsabaugh, R.L. 2005. Fungal enzymes at the community scale. pp. 349–360. *In*: Dighton, J., Oudemans, P. and White, J.F. (eds.). The Fungal Community: Its Composition and Role in the Ecosystem. CRC Taylor and Francis, Boca Raton.

Sinsabaugh, R.L. and Liptak, M.A. 1997. Enzymatic conversion of plant biomass. pp. 347–357. *In*: Wicklow, D.T. and Soderstrom, B. (eds.). The Mycota IV. Springer-Verlag, Berlin Heidelberg.

Skeffington, R.A. and Brown, K.A. 1986. The effect of five years of acid treatment on leaching, soil chemistry and weathering of a humo-ferric podslol. Wat. Air Soil Poll., 31: 981–990.

Smith, S.E. and Read, D.J. 1997. Mycorrhizal Symbiosis. Academic Press, San Diego.

Sobotka, A. 1964. Effects of industrial exhalations on soil biology of Norway spruce stands in the Ore mountains. Lesnicky Casopis, 37: 987–1002.

Stribley, D.P. and Read, D.J. 1980. The biology of mycorhiza in the Ericaceae. VII. The relationship between mycorrhizal infection and the capacity to utilize simple and complex organic nitrogen sources. New Phytol., 86: 365–371.

Stroo, H.F. and Alexander, M. 1985. Effect of simulated acid rain on mycorrhizal infection of *Pinus strobus* L. Wat. Air Soil Poll., 25: 107–114.

Taylor, A.F. and Alexander, I. 2005. The ectomycorrhizal symbiosis: Life in the real world. Mycologist, 19: 102–112.

Tedersoo, L., Naadel, T., Bahram, M. et al. 2012. Enzymatic activities and stable isotope patterns of ectomycorrhizal fungi in relation to phylogeny and exploration types in an afrotropical forest. New Phytol., 195: 832–843.

Termorshuizen, A.J. and Schaffers, A.P. 1987. Occurrence of carpophores of ectomycorrhizal fungi in selected stands of *Pinus sylvestris* L. in the Netherlands in relation to stand vitality and air pollution. Plant Soil, 104: 209–217.

Termorshuizen, A.J. and Schaffers, A.P. 1989. The relation in the field between fruitbodies of mycorrhizal fungi and their mycorrhizas. Agric. Ecosyst. Environ., 28: 509–512.

Termorshuizen, A.J. and Schaffers, A.P. 1991. The decline of carpophores of ectomycorrhizal fungi in stands of *Pinus sylvestris* L. in the Netherlands: Possible causes. Nova Hedwigia, 53: 267–289.

Thomas, R.J. and Asakawa, N.M. 1993. Decomposition of leaf litter from tropical forage grasses and legumes. Soil Biol. Biochem., 25: 1351–1361.

Thompson, G.W. and Medve, R.J. 1984. Effects of aluminum and manganese on the growth of ectomycorrhizal fungi. Appl. Environ. Microbiol., 48: 556–560.

Thompson, R.M., Townsend, C.R., Craw, D., Frew, R. and Riley, R. 2001. (Further) links from rocks to plants. Trends Ecol. Evol., 16: 543.

Tibbett, M., Grantham, K., Sanders, F.E. and Cairney, J.W.G. 1998a. Induction of cold active acid phosphomonoesterase activity at low temperature in psychotrophic ectomycorrhizal *Hebeloma* spp. Mycol. Res., 102: 1533–1539.

Tibbett, M., Sanders, F.E. and Cairney, J.W.G. 1998b. The effect of temperature and inorganic phosphorus supply on growth and acid phosphatase production in arctic and temperate strains of ectomycorhizal *Hebeloma* spp., in axenic culture. Mycol. Res., 102: 129–135.

Tibbett, M., Sanders, F.E., Minto, S.J., Dowell, M. and Cairney, J.W.G. 1998c. Utilization of organic nitrogen by ectomycorrhizal fungi (*Hebeloma* spp.) of arctic and temperate origin. Mycol. Res., 102: 1525–1532.

Tibbett, M., Sanders, F.E., Cairney, J.W.G. and Leake, J.R. 1999. Temperature regulation of extracellular proteases in ectomycorrhizal fungi (*Hebeloma* spp.) grown in axenic culture. Mycol. Res., 103.

Tibbett, M. 2000. Roots, foraging and the exploitation of soil nutrient patches: The role of mycorrhizal symbionts. Func. Ecol., 14: 397–399.

Tietema, A., Riemer, L., Verstraten, J.M., van der Maas, M.P., van Wijk, A.J. and van Voorthuyzen, I. 1993. Nitrogen cycling in acid forest soils subject to increased atmospheric nitrogen input. For. Ecol. Manage., 57: 29–44.

Tlalka, M., Bebber, D., Darrah, P.R. and Watkinson, S.C. 2008. Mycelial networks: Nutrient uptake, translocation and role in ecosystems. pp. 43–62. *In*: Boddy, L., Frankland, J.C. and van West, P. (eds.). Ecology of Saprotrophic Basidiomycetes, Elsevier Academic Press, Amsterdam.

Trappe, J., Nicholls, A.O., Claridge, A.W. and Cork, S.J. 2006. Prescribed burning in a *Eucalyptus* woodland suppresses fruiting of hypogeous fungi, an important food source for mammals. Mycol. Res., 110: 1333–1339.

Tuininga, A.R., Dighton, J. and Gray, D.M. 2002. Burning, watering, litter quality and time effects on N, P, and K uptake by pitch pine (*Pinus rigida*) seedlings in a greenhouse study. Soil Biol. Biochem., 34: 865–873.

Tuininga, A.R. and Dighton, J. 2004. Changes in ectomycorrhizal communities and nutrient availability following prescribed burns in two upland pine-oak forests in the New Jersey pine barrens. Can. J. For. Res., 34: 1755–1765.

Turneau, K., Kottke, I., Dexheimer, J. and Botton, B. 1994. Element distribution in *Pisolithus tinctorius* mycelium treated with cadmium dust. Ann. Bot., 74: 137–142.

Tyler, G., Berggren, D., Bergkvist, B., Falkengren-Grerup, U., Folkeson, L. and Ruhling, A. 1987. Soil acidification and metal solubility in forests of southern Sweden. pp. 374–359. *In*: Hutchinson, T.C. and Meema, K.M. (eds.). Effects of Pollutants on Forests, Wetlands and Agricultural Ecosystems. Springer Verlag, Berlin.

Ulrich, B., Mayer, R. and Khanna, P.K. 1979. Deposition von Luftverunreinigungen und ihre Auswirkungen in Waldökosystemen im Solling. J.D. Sauerlanders Verlag, Frankfurt, Schriften aus der Forstlichen Fakultät der Üniversität Göttingen. 58.

Van Breemen, N. and Van Dijk, H.F.G. 1988. Ecosystem effects of atmospheric deposition of nitrogen in the Netherlands. Environ. Poll., 54: 249–274.

Varma, A. and Hock. B. 1995. Mycorrhiza: Structure, Function, Molecular Biology and Biotechnology. Springer-Verlag, Berlin.

Veerkamp, M.T., De Vries, B.W.L. and Kuyper, Th.W. 1997. Shifts in species composition of lignicolous macromycetes after application of lime in a pine forest. Mycol. Res., 101: 1251–1256.

Vinichuk, M., Rosén, K. and Dahlberg, A. 2013c. 137Cs in fungal sporocarps in relation to vegetation in a bog, pine swamp and forest along a transect. J. Environ. Radioact., 90: 713–720.

Visser, S. 1995. Ectomycorrhizal fungal succession in jack pine stands following wildfire. New Phytol., 129: 389–401.

Voříšková, J. and Baldrian, P. 2013. Fungal community on decomposing leaf litter undergoes rapid successional changes. ISME J., 7: 477–486.

Watling, R. 1982. Taxonomic status and ecological identities in the basidiomycete. pp. 1–32. *In*: Frankland, J.C., Hedger, J.N. and Swift, M.J. (eds.). Decomposer basidiomycetes: their biology and ecology. Cambridge University Press, Cambridge.

Watling, R. 2005. Fungal conservation: some impressions – a personal view. pp. 881–896. *In*: Dighton, J., White, J.F. and Oudemans, P. (eds.). The Fungal Community: Its Organization and Role in the Ecosystem, Taylor and Francis, Boca Raton.

Watling, R. 1999. Launch of the UK biodiversity action plan for lower plants. Mycologist, 13: 158.

Wells, J.M. and Boddy, L. 1990. Wood decay, and phosphorus and fungal biomass allocation, in mycelial cord systems. New Phytol., 116: 285–295.

Wells, J.M. and Boddy, L. 1995a. Phosphorus translocation by saprotrophic basidiomycete mycelial cord systems on the floor of a mixed deciduous woodland. Mycol. Res., 99: 977–999.

Wells, J.M and Boddy, L. 1995b. Effect of temperature on wood decay and translocation of soil-derived phosphorus in mycelial cord systems. New Phytol., 129: 289–297.

Wells, J.M., Thomas, J. and Boddy, L. 2001. Soil water potential shifts: Developmental responses and dependence on phosphorus translocation by the saprotrophic, cord-forming basidiomycete *Phanerochaete velutina*. Mycol. Res., 105: 859–867.

Went, F.W. and Stark, N. 1968. The biological and mechanical role of soil fungi. Proc. Natl. Acad. Sci. U.S.A., 60: 497–504.

Wilkins, D.A. 1991. The influence of sheathing (ecto-) mycorrhizas of trees on the uptake and toxicity of metals. Ag. Ecosys. Environ., 35: 245–260.

Woodward, S. and L. Boddy. 2008. Interactions between saprotrophic fungi. pp. 125–141. *In*: Boddy, L., Frankland, J.C. and van West, P. (eds.). Ecology of Saprotrophic Basidiomycetes. Elsevier Academic Press, Amsterdam.

Wu, B., Nara, K. and Hogetsu, T. 2002. Spatiotemporal transfer of carbon-14-labelled photosynthate from ectomycorrhizal *Pinus densiflora* seedlings to extraradical mycelia. Mycorrhiza, 12: 83–88.

Yamanaka, T. 1999. Utilization of inorganic and organic nitrogen in pure cultures by saprotrophic and ectomycorrhizal fungi producing sporophores on urea-treated forest floor. Mycol. Res., 103: 811–816.

Yoshida, S. and Muramatsu. Y. 1994. Accumulation of radiocesium in basidiomycetes collected from Japanese forests. Sci. Tot. Environ., 157: 197–205.

Zak, B. and Marx, D.H. 1964. Isolation of mycorrhizal fungi from roots on individual slash pines. For. Sci., 10: 214–222.

Zhu, H., Dancik, B.P. and Higginbotham, K.O. 1994. Regulation of extracellular proteinase production in an ectomycorrhizal fungus *Hebeloma crustuliniforme*. Mycologia, 82: 227–234.

6

Nutritional Attributes of Two Wild Mushrooms of Southwestern India

Sudeep D. Ghate and *Kandikere R. Sridhar**

INTRODUCTION

According to the World Health Organization (WHO) undernourishment is a level of food intake insufficient to meet dietary energy requirements needed for growth, maintenance and specific functions (FAO-IFAD-WFP, 2015). Malnutrition includes undernutrition, overnutrition and deficiency of micronutrients. An abnormal physiological condition is the result of inadequate, unbalanced or excessive consumption of macronutrients and micronutrients. Although considerable progress has been made over the last few decades in the field of nutrition, the task of solving malnutrition in the developing countries still persists. The United Nations Food and Agriculture Organization (FAO) has estimated up to 795 million people worldwide are suffering from undernourishment (FAO-IFAD-WFP, 2015). Global warming, urbanization and diminishing farmlands are the major constraints in providing a nutritionally balanced diet to the global population. About one-third of the population around the world suffers from deficiency of micronutrients (folate, iodine, iron, vitamin and zinc) (Allen et al., 2006; Miller and Welch, 2013). To meet the demand of nutritious food, exploration of indigenous and non-conventional sources is inevitable.

Macrofungi serve as a promising, alternative, non-conventional resource to meet the nutritional challenge. Although mushrooms are ranked lower than meat, they are superior to milk and other foods owing to their high protein content, which fulfils the dietary requirement of an adult human being (Chang and Miles, 1989; Afiukwa

---

Department of Biosciences, Mangalore University, Mangalagangotri, Mangalore 574 199, Karnataka, India.

* Corresponding author: kandikere@gmail.com

et al., 2015). Apart from proteins, mushrooms are a major source of components like carbohydrates, fibre, lipids, minerals and vitamins (Boa, 2004; Agrahar-Murugkar and Subbulakshmi, 2005). Mushrooms also serve as protein supplements to vegetarians and are almost equivalent to the protein quality found in meat (Gençcelep et al., 2009). Their consumption has been increasing worldwide in the recent past and there is a great demand, especially for wild mushrooms, due to their various nutraceutical qualities (Halpern and Miller, 2002; Barros et al., 2008).

The Indian subcontinent has been considered as one among 12 mega-biodiversity countries with 10 biogeographic zones, a vast coastline (13,000 km^2), several national parks and many wildlife sanctuaries (Singh and Chaturvedi, 2017). A wide variety of macrofungi have been reported from the Indian Peninsula, particularly in forests, mangroves, plantations and scrub jungles of the Western Ghats and western coastal region (Natarajan et al., 2005a, 2005b; Bhagwat et al., 2005; Brown et al., 2006; Swapna et al., 2008; Mohanan, 2011; Farook et al., 2013; Ghate et al., 2014; Senthilarasu, 2014; Karun and Sridhar, 2014; Karun et al., 2014, 2016; Usha and Janardhana, 2014; Pavithra et al., 2016; Greeshma et al., 2016; Ghate and Sridhar, 2016). The wild mushrooms, *Lentinus squarrosulus* and *Termitomyces clypeatus*, are common in the southwest coast of India (Karun et al., 2014; Greeshma et al., 2016). *Lentinus squarrosulus*, being edible, has been targeted as a candidate for isolating glucan and heteroglycon immunoenhancing properties (Bhunia et al., 2010, 2011). Similarly, the edible *T. clypeatus* is also endowed with heteroglycon, which has potential as an antioxidant (Ogunmdana and Fagade, 1982; Pattanayak et al., 2015). Literature on the nutritional qualities of wild *L. squarrosulus* and *T. clypeatus* is scanty, yet they are both consumed based on traditional knowledge. Thus, the objective of the present Chapter is to provide a comprehensive and comparative account of the nutritional and nutraceutical values of these mushrooms.

Mushrooms and Processing

Lentinus squarrosulus Mont. and *Termitomyces clypeatus* R. Heim were collected from the coastal sand dunes (12°47′N, 74°51′E; 14.5 m asl) and coconut plantations (12°49′N, 74°54′E; 27 m asl) of Southwest India during June–August 2014, respectively. The coastal sand dunes possess an average temperature of 26.2 ± 0.5°C, 81% humidity and 4 mm rainfall during the sampling period. The coconut plantations showed an average temperature of 25.1 ± 1°C, 70% humidity and 3 mm rainfall during sampling. The *L. squarrosulus* is usually gregarious on decomposing, standing, dead and fallen logs of coastal sand dunes and laterite scrub jungles, whereas *T. clypeatus* grows in and around the termite mounds in coconut plantations and scrub jungles. The fruit body of *L. squarrosulus* has cap, 2–8 cm diam.; stipe, 1–3 × 0.3–0.9 cm; wet weight, 14 ± 3 g. The fruit body of *T. clypeatus* has cap, 5–12 cm diam.; stipe, 5–12 × 0.5–1.5 cm; wet weight, 12.5 ± 2 g. They were identified based on their macro- and micro-morphological features using standard monographs (Pegler and Vanhaecke, 1994; Jordan, 2004; Phillips, 2006; Mohanan, 2011; Buczacki, 2012; Tibuhwa et al., 2010).

For assessment of proximal features, minerals, protein qualities and fatty acids, freshly collected mushroom samples (n = 5) were blotted and grouped into two portions in each of five replicates. One portion was oven dried (50–55°C) and another portion was cooked using a household pressure-cooker with finite water (6.5 L, Deluxe stainless steel, TTK Prestige™, Prestige Ltd., Hyderabad, India), this was followed by oven drying on aluminium foil (50–55°C). Dried mushroom samples were milled (Wiley Mill, mesh #30) and refrigerated (4°C) in airtight containers.

Proximal and Mineral Analysis

Proximal qualities, such as moisture content, crude protein, total lipids, crude fibre, ash, carbohydrates and energy value of cooked and uncooked mushroom flours were evaluated by standard methods. Moisture content was determined gravimetrically and expressed in per cent. Micro-Kjeldahl method was used to determine the crude protein content (N × 6.25) (Humphries, 1956). The Soxhlet method was employed for total lipid extraction using petroleum ether (60–80°C), while the quantity of crude fiber and ash was determined gravimetrically (AOAC, 1990). Total carbohydrates (%) content was calculated according to Müller and Tobin (1980): [Total carbohydrates (%) = 100 – (% Crude protein + % Total lipids + % Crude fiber + % Ash)]. Calorific value (kJ/100 g) was calculated based on Ekanayake et al. (1999): [Calorific value (kJ/100 g) = (Crude protein × 16.7) + (Total lipids × 37.7) + (Carbohydrate × 16.7)].

Mushroom flours were digested in a mixture of concentrated HNO_3, H_2SO_4 and $HClO_4$ (10:0.5:2 v/v) (AOAC, 1990). Quantities of minerals (Na, K, Ca, Mg, Fe, Cu and Zn) in digested samples were assessed by AAS (GBC Scientific Equipment Pty. Ltd., Serial # A 2826; Melbourne, Australia). To determine the total phosphorus content, vanadomolybdophosphoric acid method was followed by measuring the absorbance at 420 nm using KH_2PO_4 as standard (AOAC, 1990). The ratios of Na/K and Ca/P were calculated.

Amino Acid Analysis

The amino acids of mushroom flours were evaluated based on the method by Hofmann et al. (1997, 2003). As tryptophan is destroyed by acid hydrolysis, alkaline extraction was performed (Friedman and Finley, 1971; Allred and MacDonald, 1988), while oxidized samples were used in order to estimate sulfur amino acids (Allred and MacDonald, 1988). For other amino acids, known quantity of mushroom flour was hydrolyzed with HCl (6N, 15 ml) for 4 hr at 145°C. HCl was eliminated on cooling using a rotoevaporator (Büchi Laboratoriumstechnik AG RE121; Switzerland) attached to a diaphragm vacuum pump (MC2C; Vacuubrand GmbH, Germany). Trans-4-(Aminomethyl)-cyclohexanecarboxylic acid (Aldrich, purity, 97%) was added to each sample as the internal standard. The derivatization process of esterification with trifluoroacetylation was carried out (Brand et al., 1994).

The standard amino acids were weighed in reaction vials and dried using CH_2Cl_2 under a gentle stream of helium with slow heating in an oil bath (40–60°C) to eliminate traces of water. A 12 ml fresh acidified isopropanol (acetyl chloride: 2-propanol; 1:4) was added, followed by heating at 100°C for 1 hr. The reagent was eliminated from the samples after cooling by a gentle stream of helium at 60°C. Propanol and water contents were removed by evaporation in three successive aliquots of CH_2Cl_2. The leftover dry residue was trifluoroacetylated with trifluoroacetic anhydride (200 ml) overnight at room temperature. An aliquot of this solution without treatment was used for gas chromatography-combustion-isotope ratio mass spectrometry (GC-C-IRMS/MS).

The measurements of GC-C-IRMS/MS were conducted using a Hewlett-Packard 58590 II gas chromatograph connected via a split with a combustion interface to the IRMS system (GC-C-II to MAT 252, Finnigan MAT; Germany) for the isotopic determination of nitrogen and through a transfer line with a mass spectrometer (GCQ, Finnigan MAT; Germany) for determination of the amino acids. The capillary column of GC was a 50 m × 0.32 mm i.d. × 0.5 μm BPX5 (SGE), operating with the carrier gas flow of 1.5 ml/min with following temperature and pressure: initial 50°C (1 min), increased to 100°C at 10°C/min (10 min), increased to 175°C at 3°C/min (10 min), increased to 250°C/min (10 min); the head pressure, 13 psi (90 kpa).

Assessment of Protein Quality

Protein Digestibility

Estimation of *in vitro* protein digestibility (IVPD) was performed based on Akeson and Stahmann (1964). Defatted mushroom flours (100 mg) were treated at 37°C for 3 hr with pepsin (Sigma, 3165 units/mg protein) (1.5 mg 2.5/ml 0.1N HCl), followed by immediate inactivation (0.25 ml 1N NaOH). Incubation was continued for 24 hr at 37°C with trypsin (Sigma, 16,100 units/mg protein), followed by α-chymotrypsin (Sigma, 76 units/mg protein) (2 mg each 2.5/ml potassium phosphate buffer, pH 8.0, 0.1M) and inactivated immediately (0.7 ml, 100% TCA). Zero-time control was maintained by inactivation of enzyme prior to addition of mushroom flour. Supernatant was collected after centrifugation of inactivated mixture, the remaining residue was washed (10% TCA, 2 ml), followed by centrifugation. The supernatant was pooled twice by extraction with diethyl ether (10 ml) and the aqueous layer was retained by eliminating the ether layer. Aqueous layer was kept in boiling water bath for 15 min to remove traces of ether. On attaining room temperature, solution was made up to 25 ml in distilled water. Nitrogen content in 5 ml aliquots was determined using micro-Kjeldahl apparatus (Humphries, 1956) and protein in digest was estimated: [IVPD (%) = (Protein in digest/Protein in defatted flour) × 100].

EAA score, PDCAAS and PER

The essential amino acid score (EAAS) was determined according to FAO-WHO EAA requirement for adults (FAO-WHO, 1991): [EAAS (%) = (Amino acid content in the test protein in mg/g)/(FAO EAA reference pattern in mg/g) × 100]. The protein digestibility corrected amino acid score (PDCAAS) for adults (FAO-WHO, 1991) was calculated: [PDCAAS = (EAA in test protein in mg/g)/(FAO-WHO EAA reference pattern in mg/g) × IVPD (%)]. The protein efficiency ratio (PER) was calculated based on the amino acid composition of mushroom flours according to Alsmeyer et al. (1974): $PER_1 = -0.684 + 0.456 \times Leu - 0.047 \times Pro$; $PER_2 = -0.468 + 0.454 \times Leu - 0.105 \times Tyr$; $PER_3 = -1.816 + 0.435 \times Met + 0.78 \times Leu + 0.211 \times His - 0.944 \times Tyr$.

Fatty Acid Analysis

Total lipids of mushroom flours extracted by Soxhlet method were subjected to analysis of fatty acid methyl esters (FAMEs) based on the method outlined by Padua-Resurreccion and Benzon (1979). In brief, HCl (5%, 0.2 ml) and acetyl chloride (8.3 ml) was added drop by drop to absolute methanol (100 ml) kept in ice jacket. This mixture (0.2 ml) was added to extracted lipids of mushroom flour (200 mg) in screw-cap glass vials (15 ml capacity), mixed, vortexed, incubated at 70°C up to 10 hr, followed by cooling to room temperature. Later, it was suspended in distilled water (500 μl), n-hexane (HPLC grade, 100 μl) was added, vortexed and kept for separation. The top hexane layer was transferred into air-tight microcentrifuge tubes and preserved in a refrigerator (4°C) for assay of FAMEs.

Esterified samples (100 μl) in vials were diluted with n-hexane (HPLC grade, 900 μl). One μl aliquot was injected into the gas Chromatograph (GC-2010, Shimadzu, Japan) equipped with auto injector (AOI) and capillary column (BPX-70). The capillary column was conditioned for 10 hr prior to use. Elutants were detected on flame ionization detector (FID), amplified signals were transferred and followed by monitoring GC-Solutions software (http://www.shimadzu.eu/products/software/labsolutions/gcgcms/default.aspx). Analytical conditions of auto-sampler, injection port settings, column oven settings and column information of the gas chromatograph were followed according to Nareshkumar (2007). The FAMEs quantification in samples was performed based on the standard mixture (C_4–C_{24}) (Sigma, USA) run in similar conditions to the sample. Concentration and area of each peak of FAME was computed using the GC Post-run analysis software (http://www.umich.edu/~mssgroup/docs/GCinstructions.pdf). The ratios, such as total polyunsaturated to total saturated fatty acids (TUSFA/TSFA), total polyunsaturated to total monounsaturated fatty acids (TPUFA/TMUFSA), $C_{14:0}+C_{15:0}+(C_{16:0}/C_{18:0})$ and $C_{18:1}/C_{18:2}$, were calculated.

Data Analysis

Difference in proximal characteristics, minerals, protein fractions, amino acids, *in vitro* protein digestibility and fatty acids between uncooked and cooked mushroom flours was assessed by *t*-test using Statistica version # 8.0 (StatSoft Inc., 2008).

Discussion

Proximal Qualities

Protein content, ash content and calorific value in uncooked as well as cooked flours of *T. clypeatus* was higher than *L. squarrosulus* (Table 1), while it was opposite for crude fibre and total carbohydrates. Crude lipid content was significantly higher in uncooked flours than cooked flours in both mushrooms. Cooking significantly reduced crude protein, crude lipid, crude fibre and ash contents in both mushrooms. Total carbohydrates was significantly increased in cooked *L. squarrosulus,* while calorific value decreased in *T. clypeatus.* Such changes in the proximal contents could be due to pressure-cooking.

Proximal qualities of *L. squarrosulus* in this study considerably differed from Northern India (Gulati et al., 2011; Roy et al., 2017), China (Zhou et al., 2015) and Nigeria (Nwanze et al., 2006). Compared to the present study, the crude protein content was higher in *L. squarrosulus* sampled from Nigeria and China, whereas it was lower in samples of North-western India. Crude lipid content was higher than the reports from China as well as Northwest India, but corroborates with the report from Nigeria. Crude fibre content was higher than the samples from Nigeria, China and North-western India. Total carbohydrate content was more than the samples of China, but lower than North-eastern India as well as Nigeria. Such geographical differences in proximate composition in *L. squarrosulus* may be related to the abiotic factors prevailing in different geographic regions as well as the substrate quality. Besides wood or bark as substrates for cultivation, *L. squarrosulus* could be cultivated on non-woody substrates like leaf litter, paddy straw, rice bran/hull and grains (Fasidi and Kadiri, 1999; Adesina et al., 2011; De Leon et al., 2017). As this mushroom predominantly grows on logs and woody litter, dependence of proximate qualities on the type of woody material cannot be ruled out.

The protein content in *T. clypeatus* in the present study was comparable with *T. eurrhizus* (Singdevsachan et al., 2014), *T. globulus* (Sudheep and Sridhar, 2014) and *T. umkowaan* (Karun et al., 2016), but lower than other termitomycetes (*T. microcarpus, T. heimii* and *T. robustus*) (Aletor, 1995; Johnsy et al., 2011). Crude lipid as well as crude fibre contents in *T. clypeatus* in the present study was greater than other termitomycetes (*T. eurrhizus, T. globulus, T. heimii* and *T. microcarpus*). However, crude fibre content was comparable with *T. umkowaan* (Karun et al., 2016). The total carbohydrates in this study was also higher than *T. heimii*, but lower than other termitomycetes (*T. eurrhizus, T. globulus, T. microcapus* and *T. umkowaan*).

The two edible mushrooms studied, embodied with a high quantity of crude fibre as well as carbohydrates, are capable of imparting several health benefits to the consumer, in particular lowering the risk of diabetes, cardiovascular diseases and prevention of colorectal cancer (Cheung, 1997; Björck and Elmståhl, 2003; Chandalia et al., 2000; Murphy et al., 2012). Chitin and other polysaccharides in the mushrooms studied are responsible for the prevalence of high amounts of insoluble fibre, which are nutraceutically valuable (Kalač, 2009). The moderate quantity of carbohydrates in these mushrooms with considerable amounts of crude fibre has

Table 1. Proximate composition of uncooked and cooked mushrooms on dry weight basis (n = 5, mean ± SD; Significant difference between raw and cooked samples by *t*-test: $*p < 0.01$, $**p < 0.001$, $***p < 0.001$).

	Lentinus squarrosulus		*Termitomyces clypeatus*	
	Uncooked	**Cooked**	**Uncooked**	**Cooked**
Crude protein (g/100 g)	6.2 ± 0.1*	6.0 ± 0.1	25.8 ± 0.4***	21.7 ± 0.1
Crude lipid (g/100 g)	8.9 ± 0.5**	4.2 ± 0.2	8.2 ± 0.2**	6.1 ± 0.3
Crude fibre (g/100 g)	36.3 ± 1.3	39.6 ± 0.7**	16.0 ± 0.4	24.1 ± 1.2*
Ash (g/100 g)	5.6 ± 0.1**	4.3 ± 0.2	8.4 ± 0.2**	6.6 ± 0.2
Total carbohydrates (g/100 g)	43.1 ± 01.4	45.8 ± 0.5*	41.6 ± 0.7	41.4 ± 1.1
Calorific value (kJ/100 g)	1044.6 ± 29	1041.8 ± 8.6	1442.3 ± 2.9*	1319.4 ± 25

the capacity to combat intestinal cancers and induce low glycemic index, thereby reducing the risks of type II diabetes, obesity and dyslipoproteinemia (Venn and Mann, 2004; Hauner et al., 2012).

Mineral Profile

Six minerals evaluated (Na, K, Ca, Mg, P and Cu) were significantly higher in uncooked than cooked flours of *L. squarrosulus* (Table 2). Iron content was not significantly lowered on cooking, while zinc content showed a significant increase. Almost all minerals in *L. squarrosulus* were higher than *L. squarrosulus* from Northwestern India and *Lentinus tuberrigeum* in Southern India except for contents of copper, magnesium and zinc (Gulati et al., 2011; Manjunathan and Kaviyarasan, 2011). With exception of sodium and potassium contents, the remaining minerals in *L. squarrosulus* were higher than *Lentinus sajor-caju* and *L. connatus* (Gulati et al., 2011). All minerals except for zinc were significantly higher in uncooked than cooked flours of *T. clypeatus*. Sodium, potassium and calcium contents are higher than *T. globulus* (Sudheep and Sridhar, 2014) and phosphorus, iron, copper and zinc contents are comparable with *T. umkowaan* (Karun et al., 2016). Potassium and iron contents in *T. clypeatus* are higher in *T. eurrhizus* than *T. clypeatus* (Singdevsachan et al., 2014).

Mushrooms are generally known for having siginificant amounts of potassium (Dursun et al., 2006; Sudheep and Sridhar, 2014), but mushrooms in our study showed relatively low potassium content. Most of the minerals, including potassium, were drained in cooked flours. Thus, none of the minerals in both mushrooms (except iron in *L. squarrosulus*) were comparable to the recommended pattern of National Research Council-National Academy of Science (NRC-NAS) 1989. Thus, there is

Table 2. Mineral composition of uncooked and cooked mushrooms on dry weight basis (mg/100 g) (n = 5, mean ± SD; Significant difference between raw and cooked samples by *t*-test: *$p < 0.01$, **$p < 0.001$, ***$p < 0.001$; [a] NRC-NAS, 1989).

	Lentinus squarrosulus		*Termitomyces clypeatus*		Recommended Pattern[a]
	Uncooked	Cooked	Uncooked	Cooked	
Sodium	48.2 ± 1**	41.4 ± 0.5	36.7 ± 0.2***	30.2 ± 0.4	120–200
Potassium	146.3 ± 1.5**	132.4 ± 1.2	92.5 ± 0.3**	84.2 ± 0.4	500–700
Calcium	26.8 ± 0.2***	22.4 ± 0.2	15.2 ± 0.1***	12.3 ± 0.1	600
Magnesium	34.8 ± 0.6**	30.4 ± 0.2	29.1 ± 0.2*	28.4 ± 0.5	60
Phosphorus	85.2 ± 0.15***	40.5 ± 1.0	77.6 ± 0.8**	50.6 ± 1.3	500
Iron	9.5 ± 0.55	9.3 ± 0.3	8.1 ± 0.1**	6.1 ± 0.4	10
Copper	0.4 ± 0.01***	0.2 ± 0.01	0.3 ± 0.03*	0.1 ± 0.01	0.6–0.7
Zinc	2 ± 0.2	3.1 ± 0.2**	2.6 ± 0.2	4.5 ± 0.2**	5.0
Na/K ratio	0.32	0.31	0.44	0.32	
Ca/P ratio	0.26	0.66	0.30	0.16	

a need to develop a cooking strategy which helps in the retention of minerals. Iron content in *L. squarrosulus* uncooked as well as cooked flours is close to NRC-NAS (1989) pattern (10 mg/100 g), hence, it could be supplemented in food stuffs in order to combat anaemia (Gençcelep et al., 2009). Nevertheless, the Na/K ratio in both mushrooms are in a favourable range (< 1), which helps to control high blood pressure (Yusuf et al., 2007). The Na/K ratios were also in a favourable range in *T. globulus* and *T. umkowaan* (Sudheep and Sridhar, 2014; Karun et al., 2016). However, the Ca/P ratio in both mushrooms is not in a favourable range (> 1), which helps in prevention of calcium loss in urine as well as restoration of bone calcium (Shills and Young, 1988). The Ca/P ratio was also not favourable (< 1) in *T. umkowaan* (Karun et al., 2016), while it was favourable (> 1) in *T. globulus* (Sudheep and Sridhar, 2014).

Protein Bioavailability

Glycine is one the major amino acids in both mushrooms, which was followed by lysine and alanine in *L. squarrosulus*, and glutamine and alanine in *T. clypeatus* (Table 3). Cooking significantly decreased the presence of many amino acids with the exception of a few, while tryptophan was below detectable level. Unlike *T. globulus*, the quantity of EAA of *T. clypeatus* is comparable with *T. umkowaan* (Sudheep and Sridhar, 2014; Karun et al., 2016). Except for tryptophan, cystine and glutamine, most amino acids in uncooked and cooked mushrooms are higher or comparable with soybean, wheat and FAO-WHO (1991) recommended standards. The ratio of total EAA to total amino acids increased on cooking in both mushrooms suggesting nutritional advantage, even after cooking.

The IVPD analysis is an additional index which denotes protein quality in food stuffs along with EAA and EAAS (Table 4). Cooking significantly increased the IVPD in both mushrooms, indicating its novelty in the human diet. Cooking *T. globulus* also increased IVPD significantly, but it was lowered in *T. umkowaan* (Karun et al., 2016). The PDCAAS serves as another important variable in appraisal of protein quality in foods. The PDCAAS of both mushrooms followed a similar pattern as in EAAS, increasing in most amino acids on cooking, again ascertaining their suitability for consumption. However, as seen in our study, some EAAS were increased in cooked *T. globulus* (Sudheep and Sridhar, 2014), but decrease was seen in cooked *T. umkowaan* (Karun et al., 2016). The PER_{1-3} profile in uncooked and cooked mushrooms ranged from 2.1–3.9. The PER profile also showed higher range in *T. umkowaan* (Karun et al., 2016). If the PER is > 2 in foodstuffs, it is designated as high quality (1.5–2, moderate quality; < 1.5, poor quality) (Friedman, 1996). The PER_{1-3} are > 2 and cooking has increased PER in both mushrooms indicating their superiority in protein nourishment.

Fatty Acids Profile

Uncooked and cooked flours of *L. squarrosulus* consist of the highest amount of linoleic acid followed by palmitic, oleic and myristoleic acids (Table 5). The linoleic acid was highest in uncooked as well as cooked *T. clypeatus* followed by oleic, palmitic and stearic acids. Linoleic acid was significantly higher in cooked

Table 3. Amino acid composition of uncooked and cooked mushrooms in comparison with soybean, wheat and FAO-WHO pattern (g/100 g protein; n = 5 ± SD; Significant difference between raw and cooked samples by *t*-test: *$p < 0.01$; **$p < 0.001$; ***$p < 0.001$; [a]Bau et al., 1994; [b]USDA, 1999; [c]FAO-WHO, 1991 pattern; [d]Methionine + Cystine; [e]Phenylalanine + Tyrosine; [f]Total essential amino acids/Total amino acids; BDL, Below detectable level).

	Lentinus squarrosulus		*Termitomyces clypeatus*		Soybean[a]	Wheat[b]	FAO-WHO[c]
	Uncooked	Cooked	Uncooked	Cooked			
Essential amino acids							
His	2.61 ± 0.1**	1.72 ± 0.07	3.28 ± 0.11**	2.75 ± 0.08	2.50	1.9–2.6	1.90
Ile	5.09 ± 0.1	5.39 ± 0.14**	4.57 ± 0.11	5.27 ± 0.11*	4.62	3.4–4.1	2.80
Leu	7.31 ± 0.15	7.52 ± 0.1	6.8 ± 0.02	7.67 ± 0.2*	7.72	6.5–7.2	6.60
Lys	10.22 ± 0.09***	8.82 ± 0.1	7.67 ± 0.12	8.02 ± 0.06*	6.08	1.8–2.4	5.80
Met	1.51 ± 0.03	1.66 ± 0.04*	0.99 ± 0.12	1.4 ± 0.11*	1.22	0.9–1.5	2.50[d]
Cys	0.14 ± 0.03	0.08 ± 0.01	0.24 ± 0.02**	0.09 ± 0.001	1.70	1.6–2.6	
Phe	4.48 ± 0.06	4.70 ± 0.18	4.46 ± 0.1	4.41 ± 0.04	4.84	4.5–4.9	6.30[e]
Tyr	2.74 ± 0.08	2.95 ± 0.12**	3.16 ± 0.16	2.95 ± 0.13	1.24	1.8–3.2	
Thr	4.29 ± 0.11	4.13 ± 0.12	5.22 ± 0.12	4.99 ± 0.16	3.76	2.2–3.0	3.40
Trp	BDL	BDL	BDL	BDL	3.39	0.7–1.0	1.10
Val	6.81 ± 0.22	7.44 ± 0.14*	5.9 ± 0.11	6.16 ± 0.19	4.59	3.7–4.5	3.50
Non-essential amino acids							
Ala	8.77 ± 0.12	8.9 ± 0.12*	9.63 ± 0.07***	8.2 ± 0.05	4.23	2.8–3.0	
Arg	3.67 ± 0.12*	3.3 ± 0.09	4.12 ± 0.11	4.27 ± 0.08	7.13	3.1–3.8	
Asp	6.55 ± 0.21*	5.33 ± 0.17	6.48 ± 0.18	6.26 ± 0.12	11.30	3.7–4.2	
Glu	9.56 ± 0.19**	8.39 ± 0.16	9.89 ± 0.04**	8.59 ± 0.16	16.90	35.5–36.9	
Gly	10.24 ± 0.09	11.65 ± 0.15**	11.5 ± 0.12	11.81 ± 0.13	4.01	3.2–3.5	
Pro	5.24 ± 0.16	6.63 ± 0.15**	6.01 ± 0.12	6.36 ± 0.16	4.86	11.4–11.7	
Ser	6.81 ± 0.11	6.75 ± 0.13	7.41 ± 0.09	8.07 ± 0.06**	5.67	3.7–4.8	
TEAA/TAA ratio[f]	0.47	0.53	0.43	0.44	0.56	0.65–0.69	

mushrooms than in uncooked ones. Oleic acid was higher in cooked than uncooked flours in *L. squarrosulus*, whereas the opposite was observed in *T. clypeatus*. As seen in *T. clypeatus*, the linoleic acid was also highest in *T. umkowaan*, followed by oleic and palmitic acids (Karun et al., 2016), while in *T. globulus* the oleic acid was highest, followed by elaidic and palmitic acids (Sudheep and Sridhar, 2014). The overall fatty acid content in *T. clypeatus* was higher than *T. globulus*, but comparable

Table 4. *In vitro* protein digestibility (IVPD), essential amino acid score (EAAS), protein digestibility, corrected amino acid score (PDCAAS) and protein efficiency ratio (PER) of uncooked and cooked mushrooms (Significant difference in IVPD between raw and cooked samples by *t*-test: *$p < 0.01$).

	Lentinus squarrosulus		*Termitomyces clypeatus*	
	Uncooked	**Cooked**	**Uncooked**	**Cooked**
IVPD (%)	50.65 ± 1.31	70.33 ± 0.78*	33.25 ± 0.82	51.14 ± 1.1*
EAAS				
His	1.37	0.91	1.73	1.45
Ile	1.82	1.93	1.63	1.88
Leu	1.11	1.14	1.03	1.16
Lys	1.76	1.52	1.32	1.38
Met + Cys	0.66	0.70	0.49	0.60
Phe + Tyr	1.15	1.21	1.21	1.17
Thr	1.26	1.22	1.54	1.47
Val	1.95	2.13	1.69	1.76
PDCAAS				
His	0.37	0.34	0.30	0.39
Ile	0.33	0.48	0.19	0.34
Leu	0.09	0.12	0.05	0.09
Lys	0.15	0.18	0.08	0.12
Met + Cys	0.13	0.20	0.07	0.12
Phe + Tyr	0.09	0.14	0.06	0.10
Thr	0.19	0.25	0.15	0.22
Val	0.28	0.43	0.16	0.26
PER				
PER_1	2.40	2.43	2.13	2.52
PER_2	3.01	3.24	2.21	2.22
PER_3	3.36	3.88	2.16	2.24

to *T. umkowaan* (Sudheep and Sridhar, 2014; Karun et al., 2016). Presence of high linoleic and palmitic acids in mushrooms in our study followed a similar trend and is comparable with other wild mushrooms in India and other countries (Longvah and Deosthale, 1998; Kavishree et al., 2008; Lee et al., 2011; Marekov et al., 2012; Woldegiorgis et al., 2015). As seen in *T. clypeatus*, linoleic acid was also high in *T. umkowaan*, however it was not found in *T. globulus* (Sudheep and Sridhar, 2014; Karun et al., 2016). In addition, *T. clypeatus* also possesses another essential fatty acid, docosahexaenoic acid, which is comparable to *T. umkowaan* (Karun et al., 2016).

The two mushrooms studied are endowed with many essential fatty acids, along with a high content of linoleic acid, making these mushrooms an ideal food choice for those suffering from high cholesterol due to plasma cholesterol lowering effect, thereby reducing the possibilities of cardiovascular diseases (Chan et al., 1991; Kavishree et al., 2008). Palmitic and stearic acids are the saturated major fatty acids

Table 5. FAMEs of uncooked and cooked mushrooms in comparison with soybean and wheat (mg/100 g lipid) (n = 5, mean ± SD; Significant difference between raw and cooked samples by *t*-test: *p < 0.01, **p < 0.001, ***p < 0.001; [a]Wahnon et al., 1998; Cho, 1989; [b]Pomeramz, 1998; -Not detectable).

	Lentinus squarrosulus		*Termitomyces clypeatus*		Soybean[a]	Wheat[b]
	Uncooked	Cooked	Uncooked	Cooked		
Saturated fatty acid						
Myristic acid (C14:0)	0.44 ± 0.03	-	0.28 ± 0.01	0.30 ± 0.01*	Trace−45	
Pentadecanoic acid (C15:0)	0.83 ± 0.01*	0.80 ± 0.01	0.25 ± 0.01	0.26 ± 0.01		
Palmitic acid (C16:0)	18.89 ± 0.1	18.95 ± 0.01	16.03 ± 0.04	16.32 ± 0.07*	110−116	110−320
Heptadecanoic Acid (C17:0)	0.74 ± 0.01*	0.62 ± 0.03	0.36 ± 0.01*	0.32 ± 0.01		
Stearic acid (C18:0)	2.53 ± 0.09	2.70 ± 0.02	8.03 ± 0.006	7.95 ± 0.04	25−41	0−46
Arachidic acid (C20:0)	0.46 ± 0.01	-	0.58 ± 0.04	0.64 ± 0.01	Trace	
Heneicosanoic acid (C21:0)	-	-	0.34 ± 0.002*	0.33 ± 0.002		
Behenic acid (C22:0)	2.41 ± 0.01**	1.45 ± 0.09	0.76 ± 0.01[a]	0.84 ± 0.02*		
Lignoceric acid (C24:0)	0.62 ± 0.01	-	0.92 ± 0.01	0.92 ± 0.01		
Unsaturated fatty acid						
Myristoleic acid (C 14:1)	4.96 ± 0.01***	2.05 ± 0.02	-	-		
Cis-10-Pentadecanoic acid (C15:1)	0.56 ± 0.01	-	-	-		
Palmitoleic acid (C16:1)	2.43 ± 0.00***	1.76 ± 0.01	0.34 ± 0.01	0.38 ± 0.007*	Trace	
Oleic acid (C18:1)	6.75 ± 0.05	7.64 ± 0.06***	21.84 ± 0.03***	20.65 ± 0.01	211−220	110−290
Linoleic acid (C 18:2)	49.61 ± 0.09	59.26 ± 0.3***	49.07 ± 0.03	50.37 ± 0.07***	524−540	440−740
Linolenic (C18:3)	0.67 ± 0.01	0.84 ± 0.02**	-	-	71−75	7−44
Eicosenoic acid (C20:1)	-	1.76 ± 0.01	-	-		
Eicosatrienoic acid (C20:3)	0.45 ± 0.01	-	-	-		
Erucic acid (C22:1)	-	-	0.20 ± 0.01	-		
Docosadienoic acid (C22:6)	-	-	0.22 ± 0.01	-		
Docosahexaenoic acid (C22:6)	-	-	0.40 ± 0.01***	0.19 ± 0.01		

Table 5 contd. ...

...*Table 5 contd.*

	Lentinus squarrosulus		*Termitomyces clypeatus*		Soybean[a]	Wheat[b]
	Uncooked	**Cooked**	**Uncooked**	**Cooked**		
Nervonic acid (C24:1)	1.09 ± 0.01	-	-	-		
Total saturated fatty acids (TSFA)	26.92 ± 0.2***	24.51 ± 0.1	27.54 ± 0.08	27.88 ± 0.14*		
Total unsaturated fatty acids (TUFA)	66.53 ± 0.04	73.31 ± 0.28***	72.08 ± 0.007	72.52 ± 0.05**		
TUFA/TSFA	2.47	2.99	2.61	2.57		
TPUFA/TMUFA	3.45	4.55	2.22	2.40		
$C_{14:0}+C_{15:0}+(C_{16:0}/C_{18:0})$	8.73	7.81	2.52	2.62		
$C_{18:1}/C_{18:2}$	0.14	0.13	0.45	0.41		

in both mushrooms and these fatty acids are known to prevent as well as reverse liver damage in alcoholics (Nanji et al., 1995; Hayes, 2002). Although fatty acids of both mushrooms are not comparable with soybean and wheat, the total unsaturated fatty acids were higher than total saturated fatty acids, which again facilitates combating the risks of coronary heart diseases (Bentley, 2007). Other favourable qualities include higher TPUFA/TMUFA ratio in cooked flours of mushrooms, decreased ratio of $C_{14:0}+C_{15:0}+(C_{16:0}/C_{18:0})$ in cooked flours of *L. squarrosulus* and decreased ratio of $C_{18:1}/C_{18:2}$ in cooked flours of mushrooms.

Conclusions and Outlook

The wood-preferring mushroom *L. squarrosulus* consists of a considerable quantity of crude fibre, calorific value, iron content, low Na/K ratio, high quantity of EAA, improved IVPD, desired PER, high essential fatty acids and favourable fatty acid ratios. In addition to the above qualities, termite mound mushrooms *T. clypeatus* possess a protein content equivalent to many edible legumes. Thus, both mushrooms are of immense value in nutrition as well as health. Interestingly, based on traditional knowledge, these mushrooms are part and parcel of the diet of coastal dwellers during the wet season. To encourage growth of these mushrooms, retention of woody debris (standing dead and medium to large) and termite mounds takes place. Both mushrooms will erupt all of a sudden in large quantities in Southwest India. Once *L. squarrosulus* initiates growth on wood, it will be ready for harvest between 6–7 days (before cup turns upwards). Termite-mound mushroom *T. clypeatus* shoots up in and around termite mounds. This persists for only 3–5 hr, then the mushroom becomes spent or succumbs to insects. Besides immediate use as a food source, there is also a need to develop technical knowhow in order to collect, process and preserve them without loss of nutraceutical value. A close vigilance of locations supporting these mushrooms helps to harvest in bulk quantities and evaluation of their potential bioactive components pertaining to human health is one of the major needs of the hour.

Acknowledgements

Authors are grateful to Mangalore University for granting permission to carry out this study in the Department of Biosciences. SDG acknowledges the award of an INSPIRE Fellowship by the Department of Science and Technology, New Delhi. KRS is acknowledges the award of UGC-BSR Faculty Fellowship by the University Grants Commission, New Delhi.

References

Adesina, F.C., Fasidi, I.O. and Adenipekun, O.C. 2011. Cultivation and fruit body production of *Lentinus squarrosulus* Mont. (Singer) on bark and leaves of fruit trees supplemented with agricultural waste. Afr. J. Biotech., 10: 4608–4611.

Afiukwa, C.A., Ebem, E.C. and Igwe, D.O. 2015. Characterization of the proximate and amino acid composition of edible wild mushroom species in Abakaliki, Nigeria. J. Biosci., 1: 20–25.

Agrahar-Murugkar, D. and Subbulakshmi, G. 2005. Nutritional value of edible wild mushrooms collected from the Khasi hills of Meghalaya. Food Chem., 89: 599–603.

Akeson, W.R. and Stahmann, M.A. 1964. A pepsin pancreatin digest index of protein quality. J. Nutr., 83: 257–261.

Aletor, V.A. and Aladetimi, O.O. 1995. Compositional studies on edible tropical species of mushrooms. Food Chem., 54: 265–268.

Allen, L.H., De Benoist, B., Dary, O. and Hurrell, R. 2006. Guidelines on Food Fortification with Micronutrients, WHO and FAO of the United Nations, Rome.

Allred, M.C. and MacDonald, J.L. 1988. Determination of sulfur amino acids and tryptophan in foods and food and feed ingredients: Collaborative study. J. Assoc. Off. Anal. Chem., 71: 603–606.

Alsmeyer, R.H., Cunningham, A.E. and Happich, M.L. 1974. Equations predict PER from amino acid analysis. Food Technol., 28: 34–38.

AOAC. 1990. Official Methods of Analysis, Association of Official Analytical Chemists, Washington DC, 15th Edition.

Barros, L., Cruz, T., Baptista, P., Estevinho, L.M. and Ferreira, I.C. 2008. Wild and commercial mushrooms as source of nutrients and nutraceuticals. Food Chem. Toxicol., 46: 2742–2747.

Bau, H.M., Vallaume, C.F., Evard, F., Quemener, B., Nicolas, J.P. and Mejean, L. 1994. Effect of solid state fermentation using *Rhizophus oligosporus* sp. T-3 on elimination of antinutritional substances and modification of biochemical constituents of defatted rape seed meal. J. Sci. Food Agric., 65: 315–322.

Bentley, G. 2007. The health effects of dietary unsaturated fatty acids. Nutr. Bull., 32: 82–84.

Bhagwat, S., Kushalappa, C., Williams, P. and Brown, N. 2005. The role of informal protected areas in maintaining biodiversity in the Western Ghats of India. Ecol. Soc., 10: 1–40.

Bhunia, S.K., Dey, B., Maity, K.K., Patra, S., Mandal, S., Maiti, S., Maiti, T.K., Sikdar, S.R. and Islam, S.S. 2010. Structural characterization of an immunoenhancing heteroglycan isolated from an aqueous extract of an edible mushroom, *Lentinus squarrosulus* (Mont.) Singer. Carbohydr. Res., 345: 2542–2549.

Bhunia, S.K., Dey, B., Maity, K.K., Patra, S., Mandal, S., Maiti, S., Maiti, T.K., Sikdar, S.R. and Islam, S.S. 2011. Isolation and characterization of an immunoenhancing glucan from alkaline extract of an edible mushroom, *Lentinus squarrosulus* (Mont.) Singer. Carbohydr. Res., 346: 2039–2044.

Björck, I., and Elmståhl, H.L. 2003. The glycaemic index: Importance of dietary fibre and other food properties. Proc. Nutr. Soc., 62: 201–206.

Boa, E.R. 2004. Wild Edible Fungi: A Global Overview of Their Use and Importance to People, Food and Agricultural Organization, Rome.

Brand, W.A., Tegtmeyer, A.R. and Hilkert, A. 1994. Compound-specific isotope analysis, extending towards 15N/14N and 13C/12C. Org. Geochem., 21: 585–594.

Brown, N., Bhagwat, S. and Watkinson, S. 2006. Macrofungal diversity in fragmented and disturbed forests of the Western Ghats of India. J. Appl. Ecol., 43: 11−17.

Buczacki, S. 2012. Collins Fungi Guide, Harper-Collins Publishers, London.

Chan, J.K., Bruce, V.M. and McDonald, B.E. 1991. Dietary alpha-linolenic acid is as effective as oleic acid and linoleic acid in lowering blood cholesterol in normolipidemic men. Am. J. Clin. Nutr., 53: 1230−1234.

Chandalia, M., Garg, A., Lutjohann, D., von Bergmann, K., Grundy, S.M. and Brinkley, L.J. 2000. Beneficial effects of high dietary fiber intake in patients with type 2 diabetes mellitus. New Eng. J. Med., 342: 1392−1398.

Cheung, P.C.-K. 1997. Dietary fibre content and composition of some edible fungi determined by two methods of analysis. J. Sci. Food Agric., 73: 255−260.

Cho, B.H.S. 1989. Soybean oil: Its nutritional value and physical role related to polyunsaturated fatty acid metabolism. Am. Soybean Assoc. Tech. Bull. # 4HN6.

De Onis, M., Monteiro, C., Akré, J. and Clugston, G. 1993. The worldwide magnitude of protein-energy malnutrition: An overview from the WHO Global Database on Child Growth. Bull. World Health Org., 71: 703−712.

Dursun, N., Özcan, M.M., Kaşık, G. and Öztürk, C. 2006. Mineral contents of 34 species of edible mushrooms growing wild in Turkey. J. Sci. Food Agric. 86: 1087–1094.

Ekanayake, S., Jansz, E.R. and Nair, B.M. 1999. Proximate composition, mineral and amino acid content of mature *Canavalia gladiata* seeds. Food Chem. 66: 115–119.

FAO-IFAD-WFP. 2015. The State of Food Insecurity in the World. Meeting the 2015 international hunger targets: Taking stock of uneven progress, Food and Agriculture Organization, Rome.

FAO-WHO, 1991. Protein Quality Evaluation. Reports of a Joint FAO-WHO Expert Consultation, Food and Nutrition Paper # 51, pp. 1−66, Food and Agriculture Organization of the United Nations, FAO, Rome.

Farook, V.A., Khan, S.S. and Manimohan, P. 2013. A checklist of agarics (gilled mushrooms) of Kerala State, India. Mycosphere, 4: 97−131.

Friedman, M. and Finley, J.W. 1971. Methods of tryptophan analysis. J. Agric. Food Chem., 19: 626−631.

Friedman, M. 1996. Nutritional value of proteins from different food sources—A review. J. Agric. Food Chem., 44: 6–29.

Gençcelep, H., Uzun, Y., Tunçtürk, Y. and Demirel, K. 2009. Determination of mineral contents of wild-grown edible mushrooms. Food Chem., 113: 1033−1036.

Ghate, S.D. and Sridhar, K.R. 2016. Contribution to the knowledge on macrofungi in mangroves of the Southwest India. Plant Biosys., 150: 977−986.

Greeshma, A.A., Sridhar, K.R., Pavithra, M. and Ghate, S.D. 2016. Impact of fire on the macrofungal diversity of scrub jungles of Southwest India. Mycology, 7: 15−28

Gulati, A., Atri, N.S., Sharma, S.K. and Sharma, B.M. 2011. Nutritional studies on five wild Lentinus species from North-West India. World J. Dairy Food Sci., 6: 140−145.

Halpern, G.M. and Miller, A.H. 2002. Medicinal Mushrooms, M. Evans, New York.

Hauner, H., Bechthold, A., Boeing, H., Brönstrup, A., Buyken, A., Leschik-Bonnet, E., Linseisen, J., Schulze, M., Strohm, D. and Wolfram, G. 2012. Evidence-based guideline of the German Nutrition Society: Carbohydrate intake and prevention of nutrition-related diseases. Ann. Nutr. Met., 60: 1–58.

Hayes, K.C. 2002. Dietary fat and heart health: In search of the ideal fat. Asia Pac. J. Clin. Nutr., 11: 394–400.

Hofmann, D., Gehre, M. and Jung, K. 2003. Sample preparation techniques for the determination of natural 15N/14N variations in amino acids by gas chromatography-combustion-isotope ratio mass spectrometry (GC-C-IRMS). Isot. Environ. Health Stu., 39: 233–244.

Hofmann, D., Jung, K., Bender, J., Gehre, M. and Schürmann, G. 1997. Using natural isotope variations of nitrogen in plants an early indicator of air pollution stress. J. Mass Spectrom., 32: 855–863.

Humphries, E.C. 1956. Mineral composition and ash analysis. pp. 468–502. *In*: Peach, K. and Tracey, M.V. (eds.). Modern Methods of Plant Analysis, Volume #1. Springer, Berlin.

Johnsy, G., Sargunam, D., Dinesh, M.G. and Kaviyarasan, V. 2011. Nutritive Value of Edible Wild Mushrooms collected from the Western Ghats of Kanyakumari District. Bot. Res. Int., 4: 69−74.

Jordan, M. 2004. The Encyclopaedia of Fungi of Britain and Europe, Francis Lincoln Publishers Ltd., London.

Kalač, P. 2009. Chemical composition and nutritional value of European species of wild growing mushrooms: A review. Food Chem., 113: 9−16.

Karun, N.C. and Sridhar, K.R. 2014. A preliminary study on macrofungal diversity in an arboratum and three plantations of the Southwest coast of India. Cur. Res. Environ. Appl. Mycol., 4: 173–187.

Karun, N.C., Sridhar, K.R. and Appaiah, K.A.A. 2014. Diversity and distribution of macrofungi in Kodagu region (Western Ghats): A preliminary account. pp. 73−96. *In*: Pullaiah, T., Karuppusamy, S. and Rani, S.S., (eds.). Biodiversity in India, Volume # 7. Regency Publications, New Delhi.

Karun, N.C., Sridhar, K.R., Niveditha, V.R. and Ghate, S.D. 2016. Bioactive potential of two wild edible mushrooms of the Western Ghats of India. pp. 344−362. *In*: Watson, R.R. and Preedy, V.R., (eds.). Fruits, Vegetables, and Herbs. Elsevier Inc., Oxford.

Kavishree, S., Hemavathy, J., Lokesh, B.R., Shashirekha, M.N. and Rajarathnam, S. 2008. Fat and fatty acids of Indian edible mushrooms. Food Chem., 106: 597−602.

Lee, K.J., Yun, I.J., Kim, K.H., Lim, S.H., Ham, H.J., Eum, W.S. and Joo, J.H. 2011. Amino acid and fatty acid compositions of *Agrocybe chaxingu*, an edible mushroom. J. Food Comp. Anal., 24: 175−178.

Longvah, T. and Deosthale, Y.G. 1998. Compositional and nutritional studies on edible wild mushroom from Northeast India. Food Chem., 63: 331–334.

Manjunathan, J. and Kaviyarasan, V. 2011. Nutrient composition in wild and cultivated edible mushroom, *Lentinus tuberregium* (Fr.) Tamil Nadu., India. Int. Food Res. J., 18: 809−811.

Marekov, I., Momchilova, S., Grung, B. and Nikolova-Damyanova, B. 2012. Fatty acid composition of wild mushroom species of order Agaricales—Examination by gas chromatography-mass spectrometry and chemometrics. J. Chromatogr. B, 910: 54−60.

Miller, D.D. and Welch, R.M. 2013. Food system strategies for preventing micronutrient malnutrition. Food Policy, 42: 115−128.

Mohanan, C. 2011. Macrofungi of Kerala. Kerala, Kerala Forest Research Institute, Peechi, India, Handbook # 27.

Müller, H.G. and Tobin, G. 1980. Nutrition and Food Processing, Croom Helm Ltd., London.

Murphy, N., Norat, T., Ferrari, P., Jenab, M., Bueno-de-Mesquita, B., Skeie, G., Dahm, C.C., Overvad, K., Olsen, A., Tjonneland, A., Clavel-Chapelon, F., Boutron-Ruault, M.C., Racine, A., Kaaks, R., Teucher, B., Boeing, H., Bergmann, M.M., Trichopoulou, A., Trichopoulos, D., Lagiou, P., Palli, D., Pala, V., Panico, S., Tumino, R., Vineis, P., Siersema, P., van Duijnhoven, F., Peeters, P.H.M., Hjartaker, A., Engeset, D., Gonzalez, C.A., Sanchez, M.-J., Dorronsoro, M., Navarro, C., Ardanaz, E., Quiros, J. R., Sonestedt, E., Ericson, U., Nilsson, L., Palmqvist, R., Khaw, K.-T., Wareham, N., Key, T.J., Crowe, F.L., Fedirko, V., Wark, P.A., Chuang, S.-C. and Riboli, E. 2012. Dietary fibre intake and risks of cancers of the colon and rectum in the European prospective investigation into cancer and nutrition (EPIC). PLoS ONE, 7: e39361.

Nanji, A.A., Sadrzadeh, S.H., Yang, E.K., Fogt, F., Meydani, M., and Dannenberg, A.J. 1995. Dietary saturated fatty acids: A novel treatment for alcoholic liver disease. Gastroenterol., 109: 547−554.

Nareshkumar, S. 2007. Capillary gas chromatography method for fatty acid analysis of coconut oil. J. Pl. Crops, 35: 23−27.

Natarajan, K., Narayanan, K., Ravindran, C. and Kumaresan, V. 2005a. Biodiversity of agarics from Nilgiri Biosphere Reserve, Western Ghats, India. Curr. Sci., 88: 1890–1893.

Natarajan, K., Senthilarasu, G., Kumaresan, V. and Riviere, T. 2005b. Diversity in ectomycorrhizal fungi of a dipterocarp forest in Western Ghats. Curr. Sci., 88: 1893–1895.

NRC-NAS. 1989. Recommended dietary allowances, National Academy Press, Washington DC.

Nwanze, P.I., Jatto, W., Oranusi, S.U. and Josiah, J. 2006. Proximate analysis of Lentinus squarrosulus (Mont.) Singer and *Psathyrella atroumbonata* Pegler. Afr. J. Biotechnol., 5: 366−368.

Ogundana, S.K. and Fagade, O.E. 1982. Nutritive value of some Nigerian edible mushrooms. Food Chem., 8: 263−268.

Padua-Resurreccion, A.B. and Banzon, J.A. 1979. Fatty acid composition of the oil from progressively maturing bunches of coconut. Philip. J. Coconut Stud., 4: 1–15.

Pattanayak, M., Samanta, S., Maity, P., Sen, I.K., Nandi, A.K., Manna, D.K., Mitra, P., Acharya, K. and Islam, S.S. 2015. Heteroglycan of an edible mushroom *Termitomyces clypeatus*: Structure elucidation and antioxidant properties. Carbohydr. Res., 413: 30–36.

Pavithra, M., Sridhar, K.R., Greeshma, A.A. and Karun, N.C. 2016. Spatial and temporal heterogeneity of macrofungi in the protected forests of Southwestern India. J. Agric. Technol., 12: 105–124.

Pegler, D.N. and Vanhaecke, M. 1994. *Termitomyces* of Southeast Asia. Kew Bulletin # 49: 717–736.

Phillips, R. 2006. Mushrooms, Pan Macmillan, London.

Pomeramz, Y. 1998. Chemical composition of kernel structures. pp. 97–158. *In*: Pomeramz, Y. (ed.). Wheat Chemistry and Technology. American of Cereal Chemists, St. Paul, Minnesota.

Roy, D.A., Saha, A.K. and Das, P. 2017. Proximate composition and antimicrobial activity of three wild edible mushrooms consumed by ethnic inhabitants of Tripura in northeast India. Stu. Fungi, 2: 17–25.

Senthilarasu, G. 2014. Diversity of agarics (gilled mushrooms) of Maharashtra, India. Curr. Res. Environ. Appl. Mycol., 4: 58–78.

Shills, M.E.G. and Young, V.R. 1988. Modern Nutrition in Health and Disease. pp. 276–282. *In*: Neiman, D.C., Buthepodorth, D.E. and Nieman, C.N. (eds.). Nutrition. WmC Brown, Dubugue, USA.

Singdevsachan, S.K., Patra, J.K., Tayung, K., Sarangi, K. and Thatoi, K. 2014. Evaluation of nutritional and nutraceutical potentials of three wild edible mushrooms from Similipal Biosphere Reserve, Odisha, India. J. Verbr. Lebensm., 9: 111–120.

Singh, J.S. and Chaturvedi, R.K. 2017. Diversity of ecosystem types in India: A review. Proc. Indian Natn. Sci. Acad. 83: 569–594.

StatSoft. 2008. Statistica, version # 8. StatSoft Inc, Tulsa, Oklahoma, USA.

Sudheep, N.M. and Sridhar, K.R. 2014. Nutritional composition of two wild mushrooms consumed by the tribals of the Western Ghats of India. Mycology, 5: 64–72.

Swapna, S., Abrar, S. and Krishnappa, M. 2008. Diversity of macrofungi in semi-evergreen and moist deciduous forest of Shimoga District - Karnataka, India. J. Mycol. Pl. Pathol., 38: 21–26.

Tibuhwa, D.D., Kivaisi, A.K. and Magingo, F.S.S. 2010. Utility of the macro-micromorphological characteristics used in classifying the species of *Termitomyces*. Tanz. J. Sci., 36: 31–45.

USDA. 1999. Nutrient Data Base for Standard Reference Release # 13, Food Group 20: Cereal Grains and Pasta, U.S. Department of Agriculture, Agricultural Research Service, USA, Agriculture Handbook # 8–20.

Usha, N. and Janardhana, G.R. 2014. Diversity of macrofungi in the Western Ghats of Karnataka (India). Ind. Forest., 140: 531–536.

Venn, B.J. and Mann, J.I. 2004. Cereal grains, legumes and diabetes. Eur. J. Clin. Nutr., 58: 1443–1461.

Wahnon, R., Mokady, S. and Cogan, U. 1998. Proceedings of 19th World Congress, International Society for Fat Research, Tokyo.

Woldegiorgis, A.Z., Abate, D., Haki, G.D., Ziegler, G.R. and Harvatine, K.J. 2015. Fatty acid profile of wild and cultivated edible mushrooms collected from Ethiopia. J. Nutr. Food Sci., 5: 360.

Yusuf, A.A. Mofio, B.M. and Ahmed, A.B. 2007. Proximate and mineral composition of *Tamarindus indica* Linn 1753 seeds. Sci. World J., 2: 1–4.

Zhou, S., Tang, Q., Zhang, Z., Li, C. H., Cao, H., Yang, Y. and Zhang, J. 2015. Nutritional Composition of Three Domesticated Culinary-Medicinal Mushrooms: *Oudemansiella sudmusida*, *Lentinus squarrosulus*, and *Tremella aurantialba*. Int. J. Med. Mushroom, 17: 43–49.

7

Taxonomic and Domestication Studies on *Lentinus squarrosulus*

Atri, N.S.,[1, Rajinder Singh,[2] Mridu,[1] Lata[1] and Upadhyay, R.C.[3]*

INTRODUCTION

Collection of wild mushrooms and consumption in India has existed as a traditional practice since long ago. The most prized edible mushrooms include: *Termitomyces* R. Heim, *Lentinus* Fr., *Pleurotus* (Fr.) P. Kumm., *Podaxis* Desv., *Phellorinia* (Berk.), *Agaricus* L.: Fr. emend Karst., *Lepiota* (Pers.) Gray, *Volvariella* Speg. and *Calocybe* Kühner ex Donk. Among these mushrooms, genus *Lentinus* is a cosmopolitan saprobic basidiomycete belonging to the class Agaricomycetes and family Polyporaceae. According to Kirk et al. (2008), this genus has been represented by 40 species worldwide. Up to 20 valid species of *Lentinus* have been documented by various investigators from different parts of India (Sharma and Atri, 2015). *Lentinus* has been characterized by xeromorphic tough carpophores having lamellate hymenophore and growing in clusters (Pegler, 1977; Singer, 1986). Presence of dimitic hyphal system as well as firm and persistent texture differentiates it from other agarics. Fascicles of sterile hyphae known as hyphal pegs emerging from the hymenium surface is a common feature present in some genera of the family Polyporaceae, including *Lentinus*. The edible species of *Lentinus* possess distinct flavor and are rich in proteins, carbohydrates and dietary fibers with low lipid levels, affirming their edibility among

[1] Department of Botany, Punjabi University, Patiala 147 002, Punjab, India.
[2] Akal Group of Institutions, Mastuana Sahib, Sangrur 148 001, Punjab, India.
[3] Biotechnology Department, Bodoland University, Kokrajhar 783 370, Assam, India.
* Corresponding author: narinderatri04@gmail.com

other wild edible mushrooms (Atri et al., 2016). Besides, they are also reported to have many pharmacological benefits similar to *L. crinitus* (L.) Fr. in production of proteases, *L. tigrinus* (Bull.) Fr. in production of esterases and *L. strigosus* Fr. in synthesis of chemotherapeutic compounds with immunomodulatory activity against Chagas disease (Yang and Jong, 1989; Chang and Miles, 2004; Kirsch et al., 2011; Sabotič et al., 2007; Souza-Fagundes et al., 2010; Tahir et al., 2012). The enzyme complexes produced by *Lentinus* species are also known for their applications in bakery, food processing, beverage, cheese production and paper industries (Fonseca et al., 2014; Singhal et al., 2012; Sumantha et al., 2006). Being a white rot fungus, *Lentinus* degrades cellulose, hemicelluloses and lignin from wood. Owing to these properties, it can be cultivated on agro-industrial wastes or other lignocellulosic substrates which facilitate the ecofriendly disposal of agricultural waste (Ahmed et al., 2013; Dahmardeh, 2013). North Indian states such as Jammu-Kashmir, Himachal Pradesh, Punjab and Haryana favours the growth of a wide range of mushrooms owing to diverse topography, soil and climatic conditions. Among several wild edible mushrooms known in North India, *L. squarrosulus* has substantial edibility potential, which requires attention for domestication and popularization (Sharma and Atri, 2014; Mridu and Atri, 2017). In India, early attempts for domestication of this species were made by some workers (Natarajan and Manjula, 1978; Upadhyay and Rai, 1999). This Chapter discusses collection of *L. squarrosulus* from different locations of North West India and attempts at domestication.

Material and Methods

Mushroom

The fresh sporocarps of *L. squarrosulus* were collected during 2006 to 2017 in different habitats. Locations of collection include: Jammu-Kashmir, Himachal Pradesh, Punjab and Haryana. Earliest collection of *L. squarrosulus* (Accession # PUN 3414) was made from a dead stump of *Juglens regia* L. at Rajouri (J & K) in 2004; Kotla Barog in Sirmour District of Himachal Pradesh in 2006 (Accession # PUN 4127); *Juglens regia* rotten stump and stump of *Azadirachta indica* A. Juss. growing in caespitose clusters in the District Yamunanagar of Haryana in 2014 (Accession # PUN 3539). These samples were brought to the laboratory for identification based on morphological and anatomical characteristics. The samples were identified and hot air-dried to deposit in the herbarium of the Department of Botany, Punjabi University, Patiala (Punjab). Pure cultures were also raised from the fresh sporophores. The collections were air-dried and preserved in cellophane paper packets with a small amount of 1,4-dichlorobenzene in order to prevent microbial attack (Atri et al., 2005).

Taxonomy

For making taxonomic observations, macroscopic features of the sporophore were noted on the field key to mushroom collector (Atri et al., 2005). For studying microscopic details, freehand sections were cut and examined under oil emersion lens. Camera Lucida drawings of the microscopic details were drawn, and their measurements were taken. Microphotography of the internal details was carried out (Nikon Eclipse 80i microscope). The colour terminology of Kornerup and Wanscher (1978) was employed in order to record the colour of the spore deposit and various parts of the sporocarp. Identification of the material was carried out by comparing the details with the literature (Pegler, 1977, 1983).

Pure Culture

Pure culture of *L. squarrosulus* was raised from the pileus portion where the lamellae join the stipe. To raise the culture, a small piece of tissue was excised aseptically and sterilized by immersion in 0.02% mercuric chloride solution. Subsequently, it was inoculated into the pre-sterilized PDA slants and incubated at 27 ± 1°C. The total process of culturing was performed under aseptic conditions using laminar air flow. After 3–4 days of inoculation, white mycelia emerged and spread on the PDA slants. Purification of the culture for further maintenance and utilization in experiments was done by repeated sub-culturing. Cultures were preserved in refrigerator (+4°C) for future use.

Sporophore formation on Artificial Media

For sporophore formation, sterilized flasks containing yeast extract agar (YEA) and yeast glucose media (YGM) (Tuite, 1969) were inoculated with colonized mycelial discs on the agar plates and 1 ml of homogenate into the broth, incubated at 30 ± 1°C. Duration required for mycelial run and appearance of primordia were recorded.

Evaluation of Substrates for Spawn Production

The grains, as well as sawdust, were used for the preparation of spawn. The grains of wheat, maize, bajra, jowar and sawdust mixture were used for preparing spawn. In each case, healthy grains were taken and washed several times and soaked in water overnight. Later, the grains were boiled for 30 min, air dried for a while and mixed with 2% Gypsum ($CaSO_4$) and 4% Calcium carbonate ($CaCO_3$). Gypsum was used to separate the grains from one another and $CaCO_3$ was used to adjust the pH of the substrate. After thorough mixing, the grains were filled into test tubes (25 × 150 mm) up to three-fourths and plugged. The grain filled tubes were sterilized for 30 min. After cooling, the substrate filled tubes were inoculated with five uniform size mycelial discs, in all containing about 5–6 mg of mycelial load, and incubated (27 ± 1°C).

In sawdust, spawn was prepared by using sawdust mixture of *Populus deltoides* W. Bartram ex Marshall, *Vachellia nilotica* (L.) P.J.H. Hurter & Mabb., and *Eucalyptus tereticornis* Sm. The sawdust was hand mixed and soaked in water overnight and then filled into test tubes (25 × 150 mm). Later, the substrate was sterilized, cooled, inoculated and incubated. The linear growth (in cm) was recorded on daily basis. Subsequently, the mother spawn was prepared by using evaluated substrate (wheat grains), following the standard procedure as detailed above.

Evaluation of Substrates for Mycelial Colonization

The preliminary experiment was carried out for evaluation of natural substrates for vegetative growth. The substrates used include wheat straw, rice husk, sawdust and paddy straw in the test tubes. Three replicates of each substrate were soaked in water overnight in order to attain 80% moisture. Subsequently, the wet substrates were filled into the wide mouthed (25 × 150 mm) test tubes up to three-fourths of volume. The tubes were autoclaved and cooled, then the individual substrate was inoculated aseptically with actively growing (7 days) five mycelial discs and incubated at $30 \pm 1°C$.

Evaluation of Substrates for Cropping

Four different, locally available substrates including wheat straw, paddy straw, rice husk and sawdust were selected as substrates. The sawdust used was a mixture of *Mangifera indica* L., *Eucalyptus tereticornis* and *Dalbergia sissoo* Roxb. In addition, a mixture of four sawdust types in equal proportion was mixed and soaked in water overnight. Three replicates of each substrate were used in each bag containing 250–500 g of dry substrate. The water content in the substrate after soaking overnight and decanting was tested by squeezing the mixture in the hand. Pre-soaked substrate leaking out a few drops of water between the fingers upon squeezing was considered to be saturated to about 80% moisture level and was used for inoculation. No supplementation was made in the substrates during the evaluation studies. Preliminary experiments were performed by taking 250 g of dry substrate at $22 \pm 1°C$ and $30 \pm 1°C$. For final fruiting, experiment with half a kilogram of each substrate was carried out by incubating at $30 \pm 1°C$ and at $22 \pm 1°C$.

Bagging and Sterilization

The polypropylene plastic bags were used for packing the substrate constituents because they can withstand autoclaving. The substrates were packed into plastic bags and pressed to form a cylindrical cake. Three replicates of each substrate were filled to three-fourths of their capacity in polypropylene bags. The necks, prepared from plastic pipe, were fitted into the mouths of the filled bags and plugged with non-absorbent cotton plugs. The substrate filled bags were sterilized in an autoclave at 15 lb pressure for 30 min.

Inoculation

After cooling, the bags were inoculated with wheat grain spawn @ 3% by weight. The spawn inoculations were performed by mixing under aseptic conditions. The mother spawn was prepared much ahead of the inoculation.

Incubation and Cropping

After inoculation, the bags were kept in the incubation room at 30 ± 1°C in order to initiate and accelerate the spread of mycelium. After complete colonization, the bags were incubated (22 ± 1°C) for fruiting. The colonized cylinders were subjected to chilling treatment by dipping in ice chilled water (5–10 min). The time taken for protuberance emergence after spawn run as well as the yield was recorded. The morphological characteristics of different stages were also observed. During fruiting, exposure of colonized cylinders to diffused light and for aeration windows of the culture room were kept partially open. Humidity in the culture room was maintained at 85–95% by regular mist spray through humidifier.

Storage of Fruit Bodies

The freshly harvested fruit bodies were hot air dried at 45°C in an open mushroom drier and then packed in the airtight cellophane paper packets with a small amount of 1,4-dichlorobenzene in order to prevent insect infestation.

Biological Efficiency

The freshly harvested fruit bodies were weighed, then oven dried at 45°C for 24 hr and re-weighed. This was repeated twice before packing in the airtight cellophane paper packets. Biological efficiency of the mushroom (%) with individual substrate and mixture was calculated:

$$\text{Biological efficiency on fresh weight basis} = \frac{\text{Fresh weight of mushroom}}{\text{Dry weight of substrate}} \times 100$$

$$\text{Biological efficiency on dry weight basis} = \frac{\text{Dry weight of mushroom}}{\text{Dry weight of substrate}} \times 100$$

OBSERVATION AND RESULTS

Lentinus squarrosulus Mont. Annales des Sciences Naturelles Botanique 18:21 (1842)
(Fig. 1–3)

Taxonomic Account

Fructifications 4.5–7.5 cm in height. Pileus 9–9.5 cm in diameter, convex when young with depressed center, umbilicate to deeply infundibuliform at maturity, tough, coriaceous on drying; surface dry, pale yellow (2A$_3$) to yellowish white (4A$_2$), margin irregular, splitting at maturity, involute; scales recurved fibrillose to floccose fibrillose over the entire pileus surface; cuticle fully peeling; fleshy, up to 8 mm thick, unchanging, pileal veil scaly; taste sweet and odour mushroomy. Lamellae deeply decurrent, unequal, slightly interveined, crowded, up to 5 mm broad in the

Fig. 1. Field photograph of *Lentinus squarrosulus*.

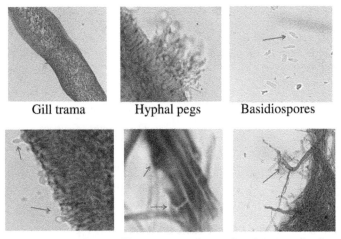

Gill trama   Hyphal pegs   Basidiospores

Pleurocystidia   Clamp connections   Bovista type hyphae

Fig. 2. Microphotographs of internal features of *Lentinus squarrosulus*.

centre, tough, orange white ($5A_2$); gill edges smooth; spore deposit pale yellow ($2A_3$) to creamish. Stipe central or eccentric, 4–4.6 cm long, 0.3–0.6 cm broad, concolorous with the pileus, fleshy, tough, cylindrical to slightly tapering downwards, solid; stipe surface covered by floccose fibrils over the entire surface.

Spores 5–8 × 2.4–3.2 µm (Q = 2.3 µm) thin walled, cylindrical with granular contents; inamyloid, cystidioles present on the sides of lamellae. Gill edges sterile with abundant, cystidiform hairs measuring 17.7–40.25 × 2.42–4 µm having clamp connections, sometimes branched, inamyloid. Pleurocystidia 16.1–37.8×2.42–6.44 µm in size. Clavate to fusiform and mucronate.

Pileus cuticle an epicutis of radially arranged generative 2.4–4.8 µm broad hyphae with clamp connections, bovista type binding hyphae measuring 1.6–7.2 µm

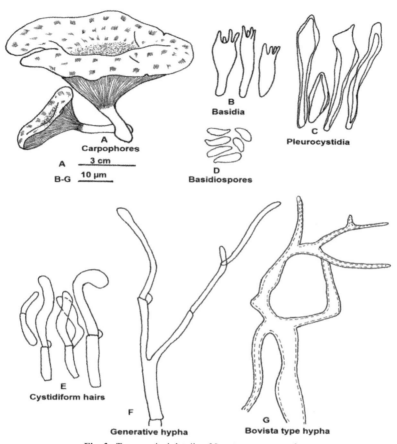

Fig. 3. Taxonomical details of *Lentinus squarrosulus.*

in width producing numerous lateral branches present in the context. Gill trama inamyloid, interwoven. Hyphal pegs numerous, extending beyond the basidial layer, made up of thin walled, hyaline hyphae, arising directly from the trama. Stipe formed of septate generative hyphae and aseptate binding hyphae.

Collections Examined (India)

Jammu, Rajouri (3400 m), growing gregariously on stump of *Juglens regia*, Harvinder Kaur, PUN 3414, May 12, 2004. Himachal Pradesh, Sirmour, Kotla Barog (3000 m), growing on dead stump of *Juglens regia,* N.S. Atri, PUN 4127, July 20, 2006; Haryana, Yamunangar (300 m) growing on a stump of *Azadirachta indica* in caespitose clusters, Mridu, PUN 3539, September 21, 2014.

Extralimital Collections (Africa)

Tanzania: Southern Highlands Province, Iringa District, Mufindi (Kigogo Valley) (1828 m) Coll. D. N. Pegler; T 772, 4 May, 1968, Kenya: Central Province, Font

Hall, Thika, Thika Falls (1463 m) Coll. D. N. Pegler K 98, 16 March, 1968. Tanzania: Southern Highlands Province. Iringa District, Kibau, Mufindi (Lumperu Tea Estate). Coll. D. N. Pegler T 780, 6 May, 1968. Tanzania: Northern Province. Arusha District, Arusha, N. P., Mt. Meru (1767 m) Coll. D. N. Pegler 1117, 28 May, 1968.

It is one of the commonly encountered species of *Lentinus*, which is quite close to *L. sajor-caju* (Fr.) Fr. (Pegler, 1977). It is characterized by caespitose habit, umbilicate coriaceous carpophores with sweet mushroomy taste, decurrent interveined lamellae, pale cream spore deposit, floccose fibrillar stipe and dimitic hyphal system with bovista type binding hyphae. These diagnostic features are in conformity with the details given by Pegler (1977) for *L. squarrosulus*. For the purpose of confirmation of identification, the Indian collection was compared with extralimital collections and was found to be in complete agreement. Earlier record of this fungus in India was by Montagne (1842) and Natarajan and Manjula (1978).

It is one of the commonly encountered species of *Lentinus*, which is quite close to *L. sajor-caju* (Fr.) Fr. (Pegler, 1977). It is characterized by caespitose habit, umbilicate coriaceous carpophores with sweet mushroomy taste, decurrent interveined lamellae, pale cream spore deposit, floccose fibrillar stipe and dimitic hyphal system with bovista type binding hyphae. These diagnostic features are in conformity with the details given by Pegler (1977) for *L. squarrosulus*. For the purpose of confirmation of identification, the Indian collection was compared with extralimital collections and was found to be in complete agreement. Earlier record of this fungus in India was by Montagne (1842) and Natarajan and Manjula (1978).

Evaluation of Substrates for Spawn Production

Locally available grains of wheat [*Triticum aestivum* L.], jowar [*Sorghum bicolor* (L.) Moench], bajra [*Pennisetum glaucum* (L.) R. Br.], maize [*Zea mays* L.] and sawdust were evaluated simultaneously for growth of *L. squarrosulus*. Prior to inoculation, the grains were boiled for different periods (10, 20 and 30 min.). In wheat grains boiled for 10 min, the linear growth was slow as compared to the growth in the grains boiled for 20 and 30 min. The best linear mycelial growth and grain colonization was observed in grains boiled for up to 30 min. At this boiling period, the mean mycelial growth was 1.62 cm/day. In Jowar and bajra grains the best linear growth of mycelium and grain colonization was recorded in grains boiled for 30 min. The average mycelial growth was 1.24–1.54 cm/day. Maize grains, being harder, showed a slightly different pattern of growth. The maximum growth was seen in grains boiled for up to 30 min. The mean mycelial growth was 1.44 cm/day, whereas growth was very slow in the maize grains boiled for 10 and 20 min. The mycelium was feebly spread, sparse and the colonization was poorer as compared to colonization on wheat, jowar and bajra grains, even after boiling for 30 min. Sawdust mixture, after boiling as above, was also used for preparing spawn, but it showed very poor growth as far as colonization is concerned. The best mycelial growth was seen in sawdust boiled for up to 30 min and the mean mycelial growth was 0.85/day.

Growth on all the grains was comparable, except for sawdust. However, maize grains, being harder, required more boiling time as compared to wheat, jowar and

Table 1. Evaluation of different substrates for production of spawn of *L. squarrosulus*.

Substrate	Days (for complete colonization)
Wheat Grains (200 g)	28
Maize Grains (200 g)	32
Jowar Grains (200 g)	36
Bajra Grains (200 g)	42
Sawdust	Not fully colonized even after 42 days

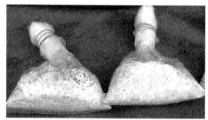

Fig. 4. Fully colonized wheat grain spawn bags.

bajra grains. Thus, wheat, jowar and bajra grains are suitable substrates for spawn production. Wheat grains, being readily available and less expensive, are the preferred substrate. Based on the growth on five different substrates, they were also tried for spawn production in polypropylene bags. Wheat grains were fully colonized after 28 days of incubation at 30 ± 1°C, followed by maize and jowar grains after 32 and 36 days of incubation, respectively. Bajra grains took 42 days, whereas sawdust remained partially colonized (Table 1). As has been the case for other mushrooms, wheat grains proved excellent substrate for spawn production in *L. squarrosulus* (Fig. 4).

Sporophore Production on Artificial Media

After 11 days of growth on artificial media, pigmentation began at the margins of the mycelial mat. Primordia appeared in the form of small protuberances after 16 days of inoculation. The majority of them were abortive and only two of them matured into full miniature sporophores. Initially the protuberances were white and on maturity these turned brownish towards the bases (Fig. 5).

Mycelial Colonization on Lingocellulosic Substrates

Linear growth of mycelium in the inoculated test tubes (25×150 mm) was recorded as a measure of colonization at regular time intervals (24 hr) on a daily basis up to 7 days (Table 2). In wheat straw, average mycelial growth was up to 1.13 cm/day. The

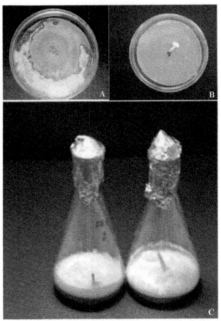

Fig. 5. Cultural characteristics on YEA (A), initiation of fruit body on YEA (B); cultural characteristics and initiation of fruit body on YGM (C).

Table 2. Mean linear growth of the mycelium recorded in different substrates (n = 3 ± SD).

Substrate	Growth of mycelium (cm/7 days)
Wheat Straw	7.90 ± 0.20
Rice Husk	7.16 ± 0.23
Sawdust	4.50 ± 0.10
Paddy Straw	4.10 ± 0.10

second-best growth, on a linear basis, was seen in rice husk with feeble and sparse spread (1.02 cm/day). In sawdust, the mean linear growth rate was 0.64 cm/day, while in paddy straw it was 0.58 cm/day.

According to the t-values, wheat straw supported maximum linear growth. As is apparent from the t-value table, the mycelial growth in wheat straw is significantly higher than growth in rice husk (t = 4.16, df = 3, $p < 0.05$), sawdust (t = 26.37, df = 3, $p < 0.01$) and paddy straw (t = 29.48, df = 3, $p < 0.01$) (Table 3).

Among the substrates evaluated, wheat straw proved as the better substrate over all others for mycelial colonization. In sawdust, in comparison to rice husk, the density of colonization was far better but not as good as it was in wheat straw.

Table 3. Matrix table showing t-values for evaluation of different substrates for mycelial growth (*Significant < 0.05; **Significant < 0.01).

Substrate	Wheat straw	Rice husk	Saw dust	Paddy straw
	7.9	7.16	4.5	4.1
Wheat Straw	-	4.16*	26.37**	29.48**
Rice Husk		-	18.11**	20.84**
Sawdust			-	4.92*
Paddy Straw				-

Sporophore Production on Lignocellulosic Substrates

Based on the preliminary experiment on the four lignocellulosic substrates (wheat straw, paddy straw, rice husk and sawdust), further experiment was performed by using 250 g and 500 g of these substrates individually and in combination (1:1:1:1) for sporophore production. Various steps for cropping, including substrate preparation, sterilization and spawning, colonization of substrate, chilling treatment, incubation and optimum conditions for fruiting, were maintained as detailed above.

Sporophore formation on Wheat Straw

On 250 g substrate bags incubated at 22 ± 1°C, primordia appeared after 20 days, as compared to 24 days in bags incubated at 30 ± 1°C. An average of 103 g of fresh weight of the sporophores was harvested from 250 g of dry wheat straw, giving 41.2% biological efficiency. In the bags incubated at 22 ± 1°C (Table 4), only 99 g of fresh weight of fruit bodies was harvested, with 39.6% biological efficiency on incubation at 30 ± 1°C (Table 5). The largest carpophores measured were 8 cm in

Table 4. Sporophore production and biological efficiency in 250 g substrate on incubation at 22 ± 1°C.

Substrates (250 g dry weight/bag)	Days taken for the appearance of primordia after inoculation	Number of sporophores/bag			Average fresh weight of sporophores/ bag (g)	Biological efficiency (%)
		I	II	III		
Wheat Straw	20	55	47	43	103	41.2
Sawdust	24	44	39	33	89	35.6
Rice Husk	35	12	10	8	24	9.6
Paddy Straw	29	-	-	-	-	0

Table 5. Sporophore production and biological efficiency 250 g substrate on incubation at 30 ± 1°C.

Substrates (250g dry weight/bag)	Days taken for the appearance of primordia after inoculation	Number of sporophores/bag			Average fresh weight of sporophores/bag (g)	Biological efficiency (%)
		I	II	III		
Wheat Straw	24	47	42	36	99	39.6
Sawdust	30	31	34	29	84	33.6
Rice Husk	40	9	7	4	19	7.6
Paddy Straw	32	-	-	-	-	0

Fig. 6. Growth and sporophore formation on wheat straw.

height. The pileus size of larger carpophores varied from 7.5–8.5 cm in diameter and the stipe measured from 4.3–7.5 × 0.5–0.9 cm. Incubation at 22 ± 1°C proved better for sporophore production. The sporophores obtained were natural in appearance with respect to their size, color and consistency (Table 4, 5, Fig. 6).

The bags containing 500 g of wheat straw also showed early colonization after 13 days of incubation. After 19 days, appearance of brownish pigmentation was seen. In comparison to other substrates, appearance of brown pigmentation took 23–25 days after inoculation. Simultaneously, protuberance emergence began, among them some of the primordia were abortive, while 53–60 primordia matured into full-fledged sporophores in all the three replicate bags. Fruiting appeared in 4–5 flushes in 5–7 weeks. Overall, an average of 277.6 g fresh weight of the sporophores were harvested from 500 g of dry wheat straw, giving 55.5% biological efficiency. The largest sporophores measured 8.5 cm in height, the pileus size varied between 8.5 and 9.0 cm in diameter with stipe size ranging from 4.0–7.5 × 0.5–1.0 cm. The fresh weight of the individual sporophores was between 0.29 g and 10.1 g, while their dry weight varied from 0.027–1.49 g. The sporophores obtained were creamish white and fleshy in consistency. Net moisture percentage in the sporophores was 86.7%, which contributes towards the soft texture of the fruit body. By taking total dry weight of the harvested sporophores into consideration, the biological efficiency was 7.4%. In comparison to other substrates, the conversion efficiency of *L. squarrosulus* in wheat straw turned out to be far better than rice husk and sawdust (Table 6, Fig. 6).

Sporophore formation on Sawdust

In comparison to wheat straw, in the 250 g substrate bags with sawdust mixture incubated at 22 ± 1°C, primordia appeared after 24 days of inoculation as compared to 30 days in bags incubated at 30 ± 1°C. On an average, 89 g fresh weight of the fruit bodies were harvested from 250 g of dry sawdust mixture, giving 35.6% biological efficiency (Table 4). In bags incubated at 30 ± 1°C, 84 g of fresh weight of sporophores were obtained with 33.6% biological efficiency (Table 5). On the sawdust, the largest sized sporophores measured were up to 5.4 cm in height with stipe 5 × 1 cm, whereas pileus measured up to 6.5 cm in diameter. The colour of the sporophores obtained was light creamish with brownish tinge as compared to sporophores obtained on other substrates. Largely sporophores were fleshy to coriaceous in texture.

The 500 g sawdust mixture also took 24 days for colonization. Compared to the wheat straw, in sawdust an average of 96.1 g fresh weight of the mushroom sporophores were harvested giving 19.21% biological efficiency (Table 6). Fruiting appeared in 3–4 flushes spread over a period of 5 weeks. On the sawdust mixture, the largest sized carpophores measured were up to 5.2 cm in height with stipe measuring 3.4–5 × 0.9–1 cm, whereas pileus measured up to 4.5–6.5 cm in diameter. The fresh weight of the individual sporophores varied from 0.957–6.326 g, while their dry weight varied from 0.105–0.577 g. The sporophores were creamish white with brownish tinge compared to sporophores obtained on the other substrates. Largely sporophores were soft and fleshy in texture. The net moisture content of the sporophores was 84.22%, which is almost comparable to the moisture content of sporophores obtained from wheat straw. This may be the reason for soft and fleshy texture of sporophores. By considering the dry weight of the harvested sporophores, the biological efficiency on sawdust mixture was only 3.03%, which is far less as compared to the biological efficiency obtained in wheat straw (Table 6, Fig. 7).

Sporophore formation on Rice Husk

In comparison to sporophore production on the wheat straw and mixture of sawdust, the rice husk substrate proved to be a poorer substrate. In this, the substrate bags containing 250 g rice husk incubated at 22 ± 1°C began showing primordia formation after 35 days of inoculation as compared to 40 days in case of bags incubated at 30 ± 1°C. An average of 24 g of fresh weight of the sporophores was harvested from

Table 6. Evaluation of different substrates for cultivation of *L. squarrosulu.*

Substrate	Net weight of dry substrate/ bag (g)	Number of days taken after inoculation for complete colonization of bags	Number of days taken after inoculation for appearance of first primordium	Number of sporophores transformed into fruit bodies/bag			Average number of sporophores/bag	Fresh weight of mushroom (g)			Total fresh weight (g)	Total dry weight (g)	Net moisture (%)	Biological efficiency on fresh weight basis (%)	Biological efficiency on dry weight basis (%)
				Bag I	Bag II	Bag III		Bag I	Bag II	Bag III					
Wheat Straw	500	13	19	60	53	55	56	275.59	284.24	272.94	277.59	37.02	86.66	55.52	7.40
Paddy Straw	500	30	34	–	–	–	–	–	–	–	–	–	–	–	–
Rice Husk	500	19	29	28	26	26	27	38.45	30.13	36.42	35.00	13.21	63.46	7	2.64
Sawdust	500	24	27	39	36	36	37	97.56	92.31	98.37	96.08	15.16	84.22	19.21	3.03
Mixture (1:1:1)	500	27	33	72	77	73	74	345.37	340.29	338.42	341.36	46.82	86.28	68.27	9.36

Fig. 7. Growth and sporophore formation on sawdust mixture.

250 g of dry rice husk substrate, giving 9.6% biological efficiency from the bags incubated at 22 ± 1°C (Table 4), in comparison to 19 g of fresh weight of sporophores obtained with 7.6% biological efficiency in the bags incubated at 30 ± 1°C (Table 5). The largest sporophores measured were up to 5.5 cm in height. The pileus size varied between 4.5–6 cm in diameter and the stipe size was between 4.3–5.2 × 0.5–1 cm. Sporophore formation in rice husk took more time as compared to sporophore formation on other substrates.

Rice husk (500 g) bags took 19 days for colonization. Relatively fewer and smaller sporophores were harvested from rice husk than other substrates. The sporophore formation was limited to three flushes spread over a period of 5 weeks. In comparison to fruiting on the wheat straw and the sawdust mixture, rice husk proved as a poorer substrate. An average of 35 g fresh weight of the sporophores were harvested, giving 7% biological efficiency (Table 6). The largest sporophores measured were up to 4.5 cm in height, the pileus size varied from 5–6.5 cm in diameter and the stipe size was 2.9–4.2 × 0.5–0.8 cm. The fresh weight of the individual sporophores varied from 0.65–3.12 g, while their dry weight ranged from 0.095–0.25 g. Sporophore formation in rice husk took more duration compared to fruiting on other substrates. The sporophores were creamish with a slight brownish tinge and tough and leathery texture. The total moisture content of harvested mushrooms was much lower (63.46%) as compared to sporophores harvested from wheat straw as well as sawdust mixture. The biological efficiency on net dry weight basis was 2.64%, which is less than in wheat straw (7.40%) but it was quite close to the biological efficiency obtained in sawdust substrate (3.03%) (Table 6, Fig. 8).

Sporophore formation on Paddy Straw

The paddy straw bags colonized after 29 days of inoculation at 30 ± 1°C compared to 32 days at 22 ± 1°C. Within the next few days of colonization, some abortive primordia appeared, which never progressed further into sporophores (Table 4, 5).

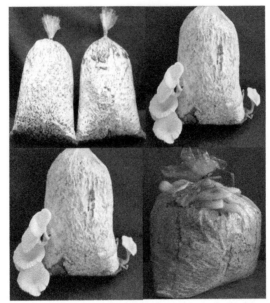

Fig. 8. Growth and sporophore formation on rice husk.

The colonization efficiency was also less in paddy straw even after 30 days of inoculation. Only a few abortive primordia were formed without formation of sporophores (Table 6).

Sporophore formation on Mixed Substrate

The bags with mixed substrates took 27 days for colonization, where the primordia numbered between 72 and 77, matured into sporophores giving 4–5 flushes within 3–5 weeks. Amongst the harvested sporophores, the largest sporophores measured 4.5–8.5 cm in height, the pileus size ranged from 5.25–7.9 cm in diameter and the stipe size was 3.5–8.0 × 0.4–0.9 cm. The fresh weight of the individual sporophores varied between 0.3 and 13.1 g, while their dry weight 0.029 and 1.62 g. The sporophores were pale yellow to creamish white and fleshy in consistency. The biological efficiency on the mixture was as high as 68.27% on wet weight basis. The sporophores harvested were comparable to those obtained on wheat straw. The moisture content was 86.28%, which is comparable to the moisture content in the sporophores obtained from wheat straw (86.66%). On dry weight basis, biological efficiency of the mushroom on the mixture was better (9.36%) than on wheat straw (7.40%), sawdust (3.03%) and rice husk (2.64%) (Table 6, Fig. 9).

Fig. 9. Sporophore formation on mixture of wheat straw + sawdust + rice husk + paddy straw (1:1:1:1).

Biological Efficiency

It is apparent from the experiments performed by using 250 g of each substrate that *L. squarrosulus* gave variable biological efficiency with different substrates. Maximum biological efficiency ranging between 39.6–41.2% was achieved in wheat straw at two ranges of temperatures (22 ± 1°C and 30 ± 1°C) followed by 33.6–35.6% in sawdust (Table 4, 5). Based on the present findings, rice husk and rice straw proved to be inferior in nourishing the mushroom. Among the two rice substrates, on rice husk, biological efficiency was between 7.6–9.6%, whereas there was no sporophore formation in paddy straw. Based on incubation temperature, although fruiting took place at wider temperatures, sporophore formation was better at 22 ± 1°C (Fig. 10). Using 500 g substrate, yield of mushroom on fresh weight basis was maximum in the mixture of four substrates followed by wheat straw, sawdust and rice husk. Use of paddy straw individually did not yield any sporophore.

The most favourable temperature for substrate colonization was 30 ± 1°C while for sporophore production it was 22 ± 1°C. Maximum biological efficiency of up to 68.27% was achieved in the substrate mixture followed by wheat straw (55.52%), sawdust mixture (19.21%) and rice husk (7%) (Table 6, Fig. 11). Comparison of the fresh weight of the mushrooms obtained in 250 g and 500 g dry mass of each substrates and their mixture have also been evaluated (Fig. 12).

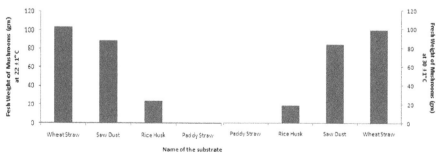

Fig. 10. Comparison of fresh weight of fruit body on different substrates at 22 ± 1°C and 30 ± 1°C.

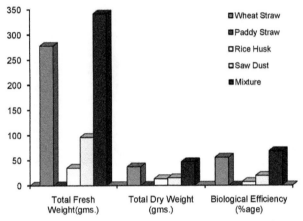

Fig. 11. Fresh weight, dry weight and biological efficiency on different substrates (500 g).

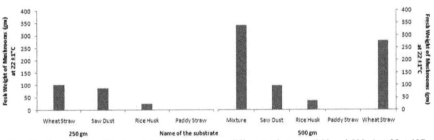

Fig. 12. Comparison of fresh weight of fruit body on different substrates (250 and 500 g) at 22 ± 1°C.

Discussion

Lentinus squarrosulus is an economically viable mushroom, commonly consumed in Nigeria under the traditional name 'Tifa' (Oso, 1975; Akpaja et al., 2003). In Thailand it is locally known as 'Hed Khon Khao' (Petcharat, 1995) and in China it is also known as "shiitake with tilted scales" (Zhou et al., 2015). This mushroom is common in occurrence under dense vegetation as well as in open habitats exposed to large temperature variations throughout equatorial Africa, South-East Asia, the Pacific Islands and Australia (Neda and Doi, 1998; Castillo et al., 2004). In India, it has been reported from Southern (Butler and Bisby, 1960; Manimohan et al., 2004; Natarajan and Manjula, 1978; Venkatachalapathi and Paulsamy, 2016) as well as from the Northern parts (Sharma and Atri, 2014, 2015; Mridu and Atri, 2017); however, there are a few reports about its edibility and consumption in India. Workers in Tamil Nadu (Natarajan and Manjula, 1978; Venkatachalapathi and Paulsamy, 2016) and Bhutan (Upadhyay and Rai, 1999) documented its utilization in human consumption. However, *L. squarrosulus* is still an underutilized mushroom in India as well as the world over (Lau and Abdullah, 2017). This study has proved its cultivation for future domestication.

On evaluation of substrate (wheat, jowar and bajra) for spawn production, boiling substrates up to 30 min is better than boiling up to 10 and 20 min. Linear growth on all the grains boiled up to 30 min was comparable, except for the sawdust mixture. However, maize grains, being harder, required more boiling time than wheat, jowar and bajra grains. On wheat grains, after boiling for 30 min, the mycelial colonization was found to be quite fast and extensive. In wheat grains (200 g) complete colonization was accomplished in 28 days compared to maize and jowar grains which took 32 and 36 days, respectively. Bajra grains took 42 days for colonization, whereas sawdust (100 g) remained partially colonized even after 42 days of inoculation and incubation. As in other edible mushrooms, wheat grains boiled for 30 min has been recommended for spawn production in *L. squarrosulus*. This inference correlates with the observations of others (e.g., Nwanze et al., 2008; Das et al., 2015) who also recommended the use of wheat grains for spawn preparation. Abbot et al. (2008) and Ayodele et al. (2007) used sorghum [*Sorghum bicolor* (L.) Moench] grains as substrates for spawn production. Okhuoya et al. (2005) raised the spawn on a sorghum-based material, while De Leon et al. (2017) found *Sorghum bicolor* seeds and cracked maize grains as good substrates. In Punjab, wheat could be fetched more easily and also cheaper compared to grains of jowar and bajra.

Substrate selection for raising successful crop similar to spawn preparation is a matter of great interest. Lignocellulosic agricultural wastes are available in Punjab could be suitable substrate for the cultivation of *L. squarrosulus*. In our study, for substrate analysis, four commonly available non-composted substrates (wheat straw, paddy straw, rice husk, sawdust and mixture of these in the ratio 1:1:1:1) were considered for evaluation. It was seen that mushroom cultivation on fresh weight basis was maximum in mixture of four substrates followed by wheat straw, sawdust and rice husk. Biological efficiency was maximum (68.27%) in mixture of substrates followed by wheat straw (55.52%), sawdust (19.21%) and rice husk (7%). Vegetative growth of mushroom *L. squarrosulus* was best achieved at 30 ± 1°C, while sporophore formation was better at 22 ± 1°C. The crop produced 3 to 5 flushes of mushrooms within 3-5 weeks.

In Nigeria, Nwanze et al. (2008) used various non-composted substrates including sawdust, animal bedding, rice and lime supplemented with different lipid sources to study the effect of lipids on sporophore production. Abbot et al. (2008) used sawdust of five economically important tropical tree species along with supplementation as substrate for the cultivation of *L. squarrosulus*. Ayodele et al. (2007) used sawdust from seven economically important tree species in cultivation of *L. squarrosulus*. Petcharat (1995) experimented with para-rubber sawdust supplemented with rice bran and sticky rice flour for cultivation of *L. squarrosulus*. Ediriweera et al. (2015) reported initiation of primordial formation in 50–60 days after inoculation on rubber sawdust-based substrate which took 15–20 days to complete one cycle of mushroom production. Okhuoya et al. (2005) also cultivated *L. squarrosulus* on the sawdust of five economically important tropical tree species. Gbolagade et al. (2006) evaluated different bacteria and their effects on the decomposition of compost (this decomposition makes the compost suitable for the growth of *L. squarrosulus*). Kadiri and Arzai (2004) cultivated *L. subnudus* on composted agricultural wastes and wooden logs of tropical hardwood plants. De Leon et al. (2017) cultivated *L. squarrosulus* on rice straw supplemented with rice bran and rice hull with varying concentration. An attempt was made to utilize sawdust mixture of common trees in Punjab (e.g., *Mangifera indica*, *Eucalyptus tereticornis* and *Dalbergia sissoo*), however the biological efficiency achieved was 19.21% as compared to 68.27% biological efficiency achieved in equal mixture of wheat straw, paddy straw, rice husk and sawdust.

Nwanze et al. (2008) incubated polypropylene bags containing substrates in darkness (30 ± 2°C) up to three weeks followed by aeration in light. Abbot et al. (2008) incubated spawned bags for thirty days in shaded condition. Upadhyay and Rai (1999) incubated the bags for colonization at 24–28°C with relative humidity between 75 and 80% in Bhutan and recommended composted wheat and paddy straw for cultivation of *L. squarrosulus*. Natarajan and Manjula (1978) produced abortive sporophores of this mushroom on sawdust medium. Presently, the colonized bags were never exposed to direct sunlight, instead during the day time after inoculation they were exposed to diffused light and for aeration a small slit was kept in the windows of the culture room. Relative humidity in the culture room was maintained between 85–95% and the temperature was reduced to 22 ± 1°C during cropping. However, for colonization of the substrate, temperature of the cropping room was maintained at 30 ± 1°C. Abbot et al. (2008) observed first primordial emergence after 20.6 ± 0.06 days. Ayodele et al. (2007) observed first primordial emergence after 40 days. Okhuoya et al. (2005) reported primordial emergence after 20.6 ± 0.16 days. In our study, primordial emergence was achieved after 19 (in wheat straw) to 33 days (in mixture of substrates).

Lentinus squarrosulus sporophores obtained during cultivation were as good as other edible mushrooms in their nutritional profile (e.g., button mushroom, shiitake and dhingri). Biochemical analysis of sporophores of *L. squarrosulus* confirmed the presence of nutritionally important components. As reported by Kaul (1983) the quality and quantity of protein present in this mushroom is comparable to other edible mushrooms or even much higher than commonly consumed vegetables like cabbage (1.5%), peas (2.6%) and cauliflower (2.4%). Nwanze et al. (2006) studied the proximate composition of the stipe and pileus of *L. squarrosulus* and reported that *L. squarrosulus* has significantly lower protein efficiency ratio (PER) values than the standard casein diet. Mushrooms possess a wide range of fat content varying from 1.2–20.6% (Adriano and Cruz, 1933). The nutritional and nutraceutical profile of Indian strain of *L. squarrosulus* was also found to be comparable to other wild edible mushrooms (Atri et al., 2016; Mridu and Atri, 2017).

Conclusions

It is apparent from this study that *L. squarrosulus* is an important mushroom and fits into the diet of the modern calorie conscious society. In view of its nutritional and nutraceutical profile it serves as a suitable alternative for mushroom consumers. This mushroom has a potential to provide equally potent culinary option as seen in commercial edible mushrooms like *Agaricus*, *Lentinula*, *Pleurotus* and *Volvariella*. Efforts should be focused for its commercialization and popularization amongst the consumers. It is quite easy to culture its vegetative mycelia for large scale production with locally available inexpensive lignocellulosic substrates (e.g., wheat straw, paddy straw, rice husk and sawdust). Cultivated fresh fruit bodies of this mushroom are fleshy with, creamish, broad depressed pileus and stout stipe. Owing to these characteristic features, *L. squarrosulus* is yet another addition to the list of Indian specialty mushrooms, comprising of shiitake, dhingri, milky mushroom, paddy straw mushroom, Jew's ear mushroom and more. It has immense potential for large scale production, popularization and consumption throughout the Indian subcontinent.

Acknowledgements

Authors are thankful to the Head, Department of Botany, Punjabi University, Patiala for providing necessary laboratory facilities. We are grateful to the UGC, New Delhi for financial support under BSR fellowship scheme and DBT for technical support under IPLS project.

References

Abbot, O., Okhuoya, J.A. and Akpaja, E.O. 2008. Growth of *Lentinus squarrosulus* (Mont.) Singer on sawdust of different tropical tree species. Afr. J. Food Sci., 3: 7–10.

Adriano, F.T. and Cruz, R.A. 1933. The chemical composition of Philippines mushrooms. Philipp. J. Agric., 4: 1–11.

Ahmed, M., Abdullah, N., Ahmed, K.U. and Bhuyan, M.H.M.B. 2013. Yield and nutritional composition of oyster mushroom strains newly introduced in Bangladesh. Pesq. Agropec. Bras., 2: 197–202.

Akpaja, E.O., Isikhuemhen, O.S. and Okhuoya, J.A. 2003. Ethnomycology and usage of edible and medicinal mushrooms among the Igbo people of Nigeria. Int. J. Med. Mush., 5: 313–319.

Atri, N.S., Kaur, A. and Kour, H. 2005. Wild Mushrooms—Collection and identification. pp. 9–26. *In*: Rai, R.D., Upadhyay, R.C. and Sharma, S.R. (eds.). Frontiers in Mushroom Biotechnology, National Research Center for Mushrooms, Chambaghat, Solan.

Atri, N.S., Babita, K., Sapan, K., Upadhyay, R.C., Ashu, G., Lata and Arvind, G. 2016. Nutritional profile of wild edible mushrooms of India. pp. 372–395. *In*: Deshmukh, S.K., Misra, J.K., Tiwari, J.P. and Papp, T. (eds.). Fungi: Applications and Management Strategies, CRC Press, Taylor and Francis.

Ayodele, S.M., Akpaja, E.O. and Anyiador, F. 2007. Evaluation of the Yield of *Lentinus squarrosulus* (Mont.) Singer on selected economic species. Pak. J. Biol. Sci., 10: 4243–4286.

Butler, E.J. and Bisby, G.R. 1960. The Fungi of India. Indian Council of Agricultural Research, New Delhi.

Castillo, G., Nihoul, A. and Demoulin, V. 2004. Correlation between *in vitro* growth response to temperature and the habitat of some lignicolous fungi from Papua New Guinea coastal forests. Mycologia, 25: 57–81.

Chang, S.T. and Miles, G. 2004. Mushroom: Cultivation, Nutritional Value, Medicinal Effects and Environmental Impact, 2nd Edition, CRC Press, Boca Raton, Florida.

Dahmardeh, M., 2013. Use of oyster mushroom (*Pleurotus ostreatus*) grown on different substrates (Wheat and barley straw) and supplemented at various levels of spawn to change the nutritional quality forage. Int. J. Agric. For., 4: 138–140.

Das, A.R., Borthakur, M., Saha, A.K., Joshi, S.R. and Das, P. 2015. Growth of mycelial biomass and fruit body cultivation of *Lentinus squarrosulus* collected from home garden of Tripura in Northeast India. J. Appl. Biol. Biotech., 3: 017–019.

De Leon, A.M., Guinto, L.J.Z.G., De Ramos, P.D.V. and Kalaw, S.P. 2017. Enriched cultivation of *Lentinus squarrosulus* (Mont.) Singer: A newly domesticated wild edible mushroom in the Philippines. Mycosphere, 8: 615–629.

Ediriweera, S.S., Wijesundera, R.L.C., Nanayakkara, C. M., and Weerasena, O.V.D.S.J. 2015. Comparative study of growth and yield of edible mushrooms, *Schizophyllum commune* Fr., *Auricularia polytricha*

(Mont.) Sacc. and *Lentinus squarrosulus* Mont. on lignocellulosic substrates. Mycosphere, 6: 760–765.

Fonseca, T.R.B., Barroncas, J.F. and Teixeira, M.F.S. 2014. Production in solid matrix and partial characterization of proteases of edible mushroom in the Amazon rainforest. Rev. Bra. de Tec. Agroindust., 1: 1227–1236.

Gbolagade, J.S., Fasidi, I.O., Ajayi, E.J. and Sobowale, A.A. 2006. Effect of physico-chemical factors and semi-synthetic media on vegetative growth of *Lentinus subnudus* Berk., an edible mushroom from Nigeria. Food Chemistry, 99: 742–747.

Kadiri, M. and Arzai, A.H. 2004. Cultivation of *Lentinus subnudus* Berk. (Polyporales: Polyporaceae) on woodlogs. Biores. Technol., 94: 65–67.

Kaul, T.N. 1983. Cultivated Edible Mushrooms. CSIR, New Delhi.

Kirk, P.M., Cannon, P.F., Minter, P.F. and Stalpers, J.A. 2008. Ainsworth Bisby's Dictionary of Fungi, 10th Edition, CAB International Wallingford, Oxon, UK.

Kirsch, L. de S., Pinto, A.C., Porto, T.S., Porto, A.L. and Teixeira, M.F.S. 2011. The influence of different submerged cultivation conditions on mycelial biomass and protease production by *Lentinus citrinus* Walleynet Rammeloo DPUA 1535 (Agaricomycetideae). Int. J. Med. Mushr., 2: 185–192.

Kornerup, A. and Wansche, J.H. 1978. Methuen Handbook of Colour, 3rd Edition. Eyre Methuen, London.

Lau, B.F. and Abdullah, N. 2017. Bioprospecting of *Lentinus squarrosulus* Mont., an underutilized wild edible mushroom, as a potential source of functional ingredients: A review. Tr. Food Sci. Technol., 61: 116–131.

Manimohan, P., Divya, N., Kumar, T.K.A., Vrinda, K.B. and Pradeep, C.K. 2004. The genus *Lentinus* in Kerala state, India. Mycotaxon, 90: 311–314.

Montagne, J.F.C. 1842. Cryptogamae Nilgherenses. Ann. Sci. Nat. (II Ser.), 18: 12–23.

Mridu and Atri, N.S. 2017. Nutritional and Nutraceutical Characterization of Three Wild Edible Mushrooms from Haryana, India. Mycosphere, 8: 1035–1043.

Natarajan, K. and Manjula, M. 1978. Studies on *Lentinus polychorus* Lév., and *L. squarrosulus* Mont. Ind. Mush. Sci., 1: 451–453.

Neda, H. and Doi, Y. 1998. Notes on Agaricales in Kyushu District. Memoirs of the National Science Museum, # 31, Tokyo, 89–95.

Nwanze, P.I., Ameh, J.B. and Umoh, V.J. 2008. The effect of the interaction of various spawn grains and oil types on carpophore dry weight, stipe length and stipe and pileus diameters of *Lentinus squarrosulus* (Mont.) Singer. Afr. J. Food Agric Nutr. Develop., 8: 1–12.

Nwanze, P.I., Khan, A.U., Ameh, J.B. and Umoh, V.J. 2006. Nutritional studies with *Lentinus squarrosulus* (Mont.) Singer and *Psathyrella atroumbonata* Pegler: Animal assay. Afr. J. Biotechnol., 5: 457–460.

Okhuoya, J.A., Akpaja, E.O. and Abbot, O. 2005. Cultivation of *Lentinus squarrosulus* (Mont.) Singer on sawdust of selected tropical tree species. Int. J. Mush. Sci., 2: 41–46.

Oso, B.A. 1975. Mushroom and Yoruba people of Nigeria. Mycologia, 67: 311–319.

Pegler, D. N. 1983. The genus *Lentinus*—A World Monograph. Kew Bulletin, Additional Series X. Royal Botanic Garden, Kew.

Pegler, D.N. 1977. A preliminary Agaric flora of East Africa. Kew Bulletin Additional Series VI. Royal Botanic Garden, Kew.

Petcharat, V. 1995. Cultivation of wild mushroom - I. Hed Khon Khao (*Lentinus squarrosulus* Mont.). Songklanakarin J. Sci. Technol. (Thailand), 17: 44–56.

Sabotič, J., Trček, T., Popovič, T. and Brzin, J. 2007. Basidiomycetes harbour a hidden treasure of proteolytic diversity. J. Biotechnol., 2: 297–307.

Sharma, S.K. and Atri, N.S. 2014. Nutraceutical composition of wild species of genus *Lentinus* Fr. from Nothern India. Curr. Res. Environ. Appl. Mycol., 4: 11–32.

Sharma, S.K. and Atri, N.S. 2015. The genus *Lentinus* (Basidiomycetes) from India - an annotated checklist. J. Threat. Taxa, 7: 7843–7848.

Singer, R. 1986. The Agaricales in Modern Taxonomy, 4th Edition, Koeltz. Koenigstein, Germany.

Singhal, P., Nigam, V.K. and Vidyarthi, A.S. 2012. Studies on production, characterization and applications of microbial alkaline proteases. Int. J. Adv. Biotechnol. Res., 3: 653–669.

Souza-Fagundes, E.M., Cota, B.B., Rosa, L.H., Romanha, A.J., Corrêa-Oliveira, R., Rosa, C.R. et al. 2010. *In vitro* activity of hypnophilin from *Lentinus strigosus*: A potential prototype for Chagas disease and leishmaniasis chemotherapy. Braz. J. Med. Biol. Res., 11: 1054–1061.

Sumantha, A., Larroche, C. and Pandey, A. 2006. Microbiology and industrial biotechnology of food-grade proteases: A perspective. Food Technol. Biotechnol., 2: 11–220.

Tahir, L., Ali, M.I., Zia, M., Atiq, N., Hasan, F. and Ahmed, S. 2012. Production and characterization of esterase in *Lentinus tigrinus* for degradation of polystyrene. Pol. J. Microbiol., 1: 101–108.

Tuite, J. 1969. Plant Pathological Methods-Fungi, and Bacteria. Burges Publishing Company, USA.

Upadhyay, R.C. and Rai, R.D. 1999. Cultivation and nutritive value of *Lentinus squarrosulus*. Mush. Res., 8: 35–38.

Venkatachalapathi, A. and Paulsamy, S. 2016. Exploration of wild medicinal mushroom species in Walayar valley, the Southern Western Ghats of Coimbatore District Tamil Nadu. Mycosphere, 7: 118–130.

Yang, Q.Y. and Jong, S.C. 1989. Medicinal mushrooms in China. Mushroom Science XII (Part 1), 631–643.

Zhou, S., Tang, Q.J., Zhang, Z., Li, C.H., Cao, H., Yang, Y. et al. 2015. Nutritional composition of three domesticated culinary-medicinal mushrooms: *Oudemansiella submucida, Lentinus squarrosulus,* and *Tremella aurantialba*. Int. J. Med. Mush., 17: 43–49.

8

Some Edible, Toxic and Medicinal Mushrooms from Temperate Forests in the North of Mexico

Fortunato Garza Ocañas,[1,] Miroslava Quiñónez Martínez,[2] Lourdes Garza Ocañas,[3] Artemio Carrillo Parra,[4] Horacio Villalón Mendoza,[1] Humberto Gonzalez Rodríguez,[1] Ricardo Valenzuela Garza,[5] Gonzalo Guevara Guerrero,[6] Jesús García Jiménez[6] and Mario García Aranda[7]*

INTRODUCTION

Mexico is considered as a biologically mega diverse country. It is located between the Neartic and Neotropical biogeographic regions and, as a consequence, has a high diversity of topographic and climatic conditions (Challenger, 1998; González et al., 2015; Salinas et al., 2017; Mororne et al., 2017). Mexico has a high diversity of ecosystems which are ecologically complex and heterogeneous. Semiarid vegetation

---

[1] Facultad de Ciencias Forestales, Carretera Nacional km 145, Linares, N.L. Mexico, A.P. 41, C.P. 67700, México.

[2] Universidad Autónoma de Ciudad Juárez, Instituto de Ciencias Biomédicas, Avenida Plutarco Elías Calles 1210, Fovissste Chamizal Ciudad Juárez, Chihuahua, México C.P. 32310, México.

[3] Facultad de Medicina, Departamento de Farmacología y Toxicología, Universidad Autónoma de Nuevo León, México.

[4] Instituto de Silvicultura e Industria de la Madera, Universidad Juárez del Estado de Durango, México.

[5] Escuela Nacional de Ciencias Biológicas Carpio y Plan de Ayala, Santo Tomás Instituto Politécnico Nacional México, D.F., México.

[6] Emilio Portés Gil, 1301 Poniente, Apartado Postal 175, Ciudad Victoria, Tamaulipas, México, 87010, México.

[7] Mario García Aranda Mar Caspio, 8212, Loma Linda, Monterrey, N.L. C.P. 64120, México.

* Corresponding author: fortunatofgo@gmail.com

and grassland are found in the high plateau in the North and temperate forests are mainly located in the Sierra Madre Occidental in the west and Sierra Madre Oriental in the east and tropical rain forests in the south (González et al., 2012). Different vegetation types including grassland and oak chapparral sometimes merge with forests in some areas forming very important and highly diverse fungal communities. Mushrooms are widely distributed in the world and in Mexico and they are intimately linked to forest ecosystems functioning as they grow and live, degrading organic matter through different enzymatic activities, functioning either as saprotrophic, parasitic, pathogenic or mycorrhizal symbionts (Schmidt and Müller, 2007). The relationship between plants and fungi has a very long history, stretching back to the Devonian period, and big fossil sporocarps of some species like *Prototaxites* sp. or *Gondwanagaricites magnificus* from lower Cretaceous crato formations are the evidence of such long-lasting coexistence (Retallack and Landing, 2014; Heads et al., 2017). Thus, forests around the world evolved over time and important groups of plants formed intimate relationships with fungi as a strategy to survive. Nowadays, they are widely dispersed through the world and some have migrated from the Arctic southwards into Tropical regions (e.g., *Pinus* spp.) in America. Conifers and Oaks are widely distributed in North America, Mexico and Central America, and mushroom species are intimately related to these species (Halling and Müller, 2002; Müller et al., 2006; Binion et al., 2008). Mycorrhizal fungi played an important role in migration of this group of plants as they help with nutrient uptake from different soils including poor soils, sapotrophic, parasitic and pathogenic fungal species also established nutrition relationships with Oaks and Conifer trees and contribute significantly to the cycle of nutrients in forests (Hybbett, 2006). In America, fungi have interesting relationships with Oak and Conifer forests and contribute to their development and growth (Singer et al., 1983; North, 2002; Allen, 2005). Current distribution of some plants is the result of their simbiotic associations. Spore dispersion mechanisms contribute significantly to the survival of fungi and mycophagia, i.e., consumption of mushrooms by animals is a very important ecological mechanism wherein animals such as slugs, mites, insects, as well as vertebrates are involved (turkeys, squirrels, mice and rats, rabbits, turtles, peccaries, deers and cervids in general as well as foxes, bears and wolves) (Frank et al., 2007; Binion et al., 2008). Air and rain water are also important spore dispersion factors and millions of spores are carried away through the forests every year. Spore germination processes *in vitro* occur with different degrees of difficulty and parasites, pathogens and sapotrophic fungal species have evolved to germinate their spores more easily in comparison to some mycorrhizal species. Thus, water and some enzymes in the animal digestive system promote germination and these mechanisms have evolved over millions of years. Some squirrels and mice search for mushrooms during the rainy season and eat the fertile hymenophore of epigeous and hypogeous mushrooms, thereby dispersing spores in their excreta (Kendrick, 1994). These spores are ready to germinate and grow, producing new mycelium of the species involved. They will start searching for food in the vicinity; some will find it in fallen leaves, others in tree branches and trunks and some others in living roots depending on their growth habit. All together mushrooms are an important part of the forests and they are alive using different survival strategies during the dry season. Mexico is known for the great diversity of organisms living in the different climatic and edaphic conditions occurring throughout its territory (Sarukhán et al., 2015). Mushrooms are abundant and it is believed that there are ca. 8000 fungi species studied so far out of the 200,000 probable species occurring in the different vegetation types of the country (Guzmán, 1998; Hawksworth, 2001). The North of Mexico has an influence of Artic and Neotropical biogeographic regions and temperate forests have a high diversity of hosts for fungi such as Pines, Fir, Douglas fir, and Oaks and a high variety of edible, toxic, and medicinal fungal species are associated (Peinado et al., 1994; Müller et al., 2006). In the states of Chihuahua and Durango, toxic mushroom species have caused casualties, mostly due to missidentification of edible species like *Amanita caesarea* group confused with old specimens of toxic species, e.g., *A. muscaria* var. *flavovolvata* (Garza et al., 1985; Quiñónez and Garza, 2015). In these states ethnic groups and people in general may eat wild edible mushrooms and there are no records of people eating wild edible mushrooms from the states of Coahuila, Nuevo Leon and Tamaulipas. Some of the edible wild species frequently used as food when in season include *Amanita jacksonii, A. cochiseana, A. rubescens* (this one is edible only after cooking), *Agaricus campestris, A. bitorquis, Hypomyces lactifluorum, Lactarius deliciosus, L. indigo, Russula delica, Cantharellus cibarius, Boletus pinophilus, B. rubriceps, B. barrowsii, Harrya chromapes* and *Aureoboletus russellii*. Also, some medicinal mushrooms occur in temperate forests in the north of Mexico, such as *Coriolus versicolor, Ganoderma* spp., *Pycnoporus sanguineus* and others.

This paper aims to provide information regarding mushroom diversity, as well as identifying those species reported as edible, toxic or medicinal. This, together with a list of some of the main Conifers, Fagaceae and Ericaceae tree hosts from temperate forests in the North of Mexico. It is also intended to generate relevant information regarding the potential that many mushroom species have for research as well as for social and economic development for people in rural forestry conditions in the North of Mexico.

Materials and Methods

Field collection of mushrooms was carried out during the months of February and March in the state of Baja California; July and August for the states of Chihuahua, Sonora and Durango; and from September-October in the states of Coahuila, Nuevo León and Tamaulipas. These states are located in the North of Mexico in the Neartic and Neotropical regions and they include the biogeographic provinces known as California, Sonora, Sierra Madre Occidental, Sierra Madre Oriental and Tamaulipas (Morrone et al., 2017). Collection of mushrooms was carried out in very many localities in the last 25 years at several altitudes ranging from 360–3650 m. The main vegetation types studied were temperate Forests with Conifers, i.e., *Picea, Pseudotsuga, Abies,* and *Pinus*; Oaks, i.e., *Quercus* spp., as well as *Arbutus* spp. and *Arctostaphyllos* spp. Trees from these genera host many mycorrhizal fungi species both in the Basidiomycetes and Ascomycetes. Mushrooms were collected following traditional methods recording macroscopic characteristics, photographs were taken in the field and specimens were transported to the laboratory, specimens were dehydrated and microscopic characteristics recorded for identification using specilized keys for the different groups (Leonard, 2010). Some edible and toxic species of fungi associated to temperate forests in the North of Mexico have been reported (Quiñónez and Garza, 2015; García et al., 2014; Guevara et al., 2014). For toxic species, the identification criteria used was based on several authors (Lyncoff and Mitchell, 1977; Groves, 1979; Hallen and Adams, 2002; Hall et al., 2003; Lindequist et al., 2005; Duffy, 2008). An artificial classification of species depending on the degree of danger for health was made as follows:

1) Species which can be fatal

2) Species causing serious intoxications requiring hospital attention but are not fatal

3) Species causing light poisoning not requiring hospital attention and are not life threatening.

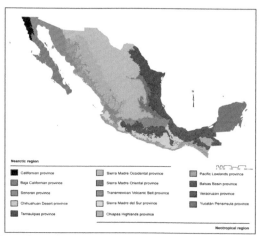

Fig. 1. Biogeographic provinces of Mexico (Morrone et al., 2017) (http://mexicanmap.atlasbiogeografico. com/MexicanBiogeographicProvinces.tif).

Results

In this study, 282 species were identified from the states of Baja California Norte, Chihuahua, Sonora, Coahuila, Nuevo Leon and Tamaulipas (Tables 1–3; Figs. 5–12). A total of 272 species belong to the Phyllum Basidiomycota and 10 to the Phyllum Ascomycetes, 54 families and 120 genera were studied (Table 1). They are associated with temperate forests and 147 tree hosts are widely distributed in these states. Most edible and toxic species occur in the Sierra Madre Occidental and are closely related to species found in the south states of North America. Species from the Sierra Madre Oriental are more diverse and there are edible species in the forests but are not very abundant, many toxic species are abundant in this Sierra. Oak forests and mixed oak-pine forests have a higher diversity of fungal species throughout the North, followed by pine-oak forests.

Results of diversity and distribution of fungal species show that in the state of Nuevo León more collections have been made with 225 species, followed by the

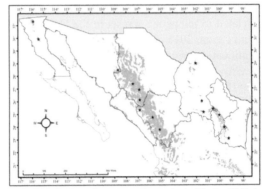

Fig. 2. Distribution of temperate forests in the North of México.

Fig. 3. Forests in the North of México. Top left Pine forest; Top right Oak forest.

Fig. 4. Forests in the North of Mexico. Top left 1. Conifer forests at 3200 m.a.s.l.; Top right 2. inside look of conifer forest; Central left 3. *Quercus* spp. forests at 2300 m.a.s.l.; Central right 4. Inside look of *Quercus* forest; Below left 5. *Pinus-Quercus* forests; Below right 6. Inside look of the *Pinus-Quercus* forests.

Table 1. List of tree species in temperate forests in the North of Mexico.

Family Pinaceae
P. pseudostrobus var. *apulcensis* (Lndl.) Shaw
Pinus teocote Schltdl. & Cham.
Pinus nelsonii Shaw
Pinus greggii Engelm. ex Parl. var. *gregii*
P. culminicola Andresen & Beaman
P. cembroides Zucc.
P. monophylla Torr. & Frém.
P. quadrifolia Parl. ex Sudw.
P. coulterii D. Don
P. engelmannii Carr.
P. lambertiana Douglas
*P. montezumae Lamb.*var. *montezumae*
P. johannis (Rob.) Pass.
P. hartwegii Lindl.
P. leiophylla Schiede ex Schltdl. & Cham.
P. arizona var. *cooperii* (C.E. Blanco) Farjon
P. duranguensis Martínez
P. chihuahuana Engelm.
P. devoniana Lindl.
P. discolor Bailey & Hawksw.
P. douglasiana Martinez
P. patula Schiede ex Schltdl. & Cham.
P. lumholtzii B.L. Rob. & Fernald
P. ayacahuite Ehrenb. ex Schltdl.var. brachyptera
P. flexilis E. James
P. herrerae Martinez
Pinus strobiformis Engelm.
P. yecorensis Debreczy & I. Rácz
P. arizonica var. *arizonica* Engelm.
P. arizonica var. *stormiae* Martínez
P. jeffreyi Balf.
Abies duranguensis Martinez
A. vejarii Martinez
A. concolor (Gordon & Glend) Lindl.
Abies guatemalensis Rehd.
Picea chihuahuana Martinez
Picea engelmanii ssp. *mexicana* (Martínez) P. Schmidt
P. engelmanii Parry ex Engelm.
Pseudotsuga mensiezii (Mirb.) Franco

Table 1 cont. ...

...Table 1 cont.

Family Fagaceae
Q. affinis Scheidw.
Quercus emoryi Torr.
Q. canbyi Trel.
Q. cupreata Trel. et C.H. Mull.
Q. castanea Née
Q. conspersa Benth.
Q. crassifolia Bonpl.
Q. crassipes Bonpl.
Q. delgadoana Nixon & L.M. Kelly
Q. alpescens Trel.
Q. depressa Bonpl.
Q. flocculenta C.H. Mull.
Q. fulva Liebm.
Q. furfuracea Liebm.
Q. galeanensis C.H. Mull.
Q. gentry C.H. Mull.
Q. graciliramis C.H. Mull.
Q. gravesii Sudw.
Q. hintoniorum Nixon et C.H. Mull.
Q. hirtifolia M.L. Vázquez
Q. hypoleucoides A. Camus
Q. eduardii Trel.
Q. hypoxantha Trel.
Q. jonesii Trel.
Q. laurina Bonpl.
Q. mexicana Bonpl.
Q. miquihuanensis Nixon et. C.H. Mull.
Q. ocoteifolia Liebm.
Q. pinnativenulosa C.H. Mull.
Q. runcinatifolia Trel. et C.H. Mull.
Q. rysophylla Weath.
Q. saliicifolia Née
Q. saltillensis Trel.
Q. sapotifolia Liebm.
Q. sartorii Liebm.,
Q. sideroxyla Bonpl.
Q. skinneri Benth.
Q. tenuiloba C.H. Mull.
Q. viminea Trel.
Q. xalapensis Bonpl.
Q. chihuahuensis Trel

Table 1 cont. ...

...Table 1 cont.

Family Fagaceae
Q. convallata Trel.
Q. diversifolia Née
Q. edwardsae C.H. Mull.
Q. fusiformis Small
Q. germana Bonpl.
Q. glaucoides M. Mart et Gal.
Q. gregii (A. DC.) Trel.
Q. intrincata Trel.
Q. invaginata Trel.
Q. laceyi Small.
Q. laeta Liebm.
Q. lancifolia Schltdl. et Cham.
Q. magnolifolia Née
Q. microlepis Trel.
Q. microphylla Née
Q. muehlenbergii Engelm.
Q. oblongifolia Torr.
Q. obtusata Bonpl.
Q. oleoides Schltdl. et Cham.
Q. opaca Trel.
Q. pastorensis C.H. Muller
Q. pendicularis Née
Q. polymorpha Schltdl. et Cham.
Q. praeco Trel.
Q. pringlei Seemen ex Loes.
Q. pungens Liebm.
Q. rugosa Trel.
Q. sebifera Trel.
Q. sinuata Walt var. *breviloba* (Torr.) C. H. Mull.
Q. splendens Née
Q. striatula Trel.
Q. supranitida C.H. Mull.
Q. thinkhamii C.H. Mull.
Q. toxicodendrifolia Trel.
Q. monterreyensis Trel. et C.H. Mull.
Q. vaseyana Buckl.
Q. verde C.H. Mull.
Q. mohriana Buckl. ex Rydb.
Q. toumeyei Sarg.
Q. tarahumara Spellenb., Bacon & Breedlove
Q. basaseachicensis C.H. Mull.

Table 1 cont. ...

...Table 1 cont.

Family Fagaceae
Q. chrysolepis Liebm.
Q. arizonica Sarg.
Q. durifolia Seem.
Q. gambelli Nutt.
Q. grisea Liebm.
Q. potosina Trel.
Q. tuberculata Liebm.
Q. mcvaughii Spellenb.
Q. scytophylla Liebm.
Q. subpathulata Trel.
Q. deppressipes Trel.
Q. intricata Trel.
Q. tardifolia C.H. Mull.
Q. clivicola Trel. & C.H. Mull.
Q. sartorii Liebm.
Q. agrifolia Née
Q. dumosa Nutt.
Q. urbanii Trel.
Q. virginiana Mill.
Q. coahuilensis Nixon & C.H. Müll.
Q. carmenensis C.H. Müll.
Family Ericaceae
Arbutus xalapensis var. *texana* (Buckley) A. Grey
Arbutus arizonica (A. Gray) Sarg.
Arctostaphylos pungens KBK
A. glauca Lindl.
A. peninsularis Wells
A. platyphylla (A. Gray) Kuntze

states of Chihuahua, Durango, Tamaulipas, Sonora, Coahuila and Baja California Norte with 189, 188, 186, 104, 106, and 70 respectively.

Classification Depending on the Degree of Danger of Species

1) From the species studied in the North of Mexico, 180 have been reported as toxic and 20 of these can be deadly poisonous, including *Amanita verna, A. virosa, A. bisporigera, A. phalloides, Boletus satanas* and *Hypholoma fasciculare* (Figs. 8–11). They have different toxins and some of them destroy liver cells and cause irreversible liver failure and death. These species represent a real threat or risk for humans and should be avoided.

2) Some species that should not be ingested raw include *Amanita rubescens, Boletus subvelutipes, Suillelus luridus* and *Boletus amigdalinus*. The first species is used

Table 2. List of edible, toxic and medicinal mushroom species from temperate forests in the North of Mexico (xeranquical taxonomic arrangement as in Kirk et al., 2008).

Family	Genus	Species	Authors	Distribution	Habit	Edibility
Leotiaceae	*Leotia*	*lubrica*	(Scop.) Pers.	Wide	Saprotrophic	Toxic
Leotiaceae	*Leotia*	*viscosa*	Fr.	Low	Saprotrophic	Toxic
Discinaceae	*Gyromitra*	*esculenta*	(Person:Fries) Fr.	Low	Saprotrophic	Toxic
Discinaceae	*Hydnotrya*	*cerebriformis*	Harkn.	Low	Mycorrhizal	Toxic
Helvellaceae	*Helvella*	*crispa*	(Scop.).	Wide	Saprotrophic	Toxic
Helvellaceae	*Helvella*	*elastica*	Bull.	Wide	Saprotrophic	Toxic
Helvellaceae	*Helvella*	*macropus*	(Pers.) P. Karst.	Low	Mycorrhizal	Toxic
Pezizaceae	*Hydnobolites*	*cerebriformis*	Tul. & C. Tul.	Low	Mycorrhizal	Toxic
Pezizaceae	*Pachyploeus*	*carneus*	Karkn.	Low	Mycorrhizal	Toxic
Pezizaceae	*Pachyploeus*	*virescens*	Gilkey	Low	Mycorrhizal	Toxic
Ophiocordycipitaceae	*Elaphocordyceps*	*capitata*	(Holmsk.) G.H. Sung, J.M. Sung & Spataphora	Medium	Mycorrhizal	Toxic
Morchellaceae	*Morchella*	*esculenta*	(L.) Pers.	Medium	Mycorrhizal	Edible
Tuberaceae	*Tuber*	*lyonii*	Butters	Medium	Mycorrhizal	Edible
Tuberaceae	*Tuber*	*regimontanum*	G. Guevara, Bonito & Julio Rodr.	Low	Mycorrhizal	Edible
Agaricaceae	*Agaricus*	*xanthodermus*	Genev.	Wide	Saprotrophic	Toxic

Table 2 cont. ....

...Table 2 cont.

Family	Genus	Species	Authors	Distribution	Habit	Edibility
Agaricaceae	Agaricus	campestris	L.	Wide	Saprotrophic	Edible
Agaricaceae	Agaricus	silvaticus	Schaeff.	Medium	Saprotrophic	Toxic
Agaricaceae	Agaricus	silvicolae-similis	Bohus & Locsmándi	Low	Saprotrophic	Toxic
Agaricaceae	Agaricus	arvensis	Schaeff.	Medium	Saprotrophic	Edible
Agaricaceae	Agaricus	bitorquis	(Quél.) Sacc.	Low	Saprotrophic	Edible
Agaricaceae	Calvatia	bovista	(L.) Murrill	Wide	Saprotrophic	Edible
Agaricaceae	Calvatia	cyathiformis	(Bosc) Morgan	Wide	Saprotrophic	Edible
Agaricaceae	Chlorophyllum	molybdites	(G. Mey.) Masse	Wide	Saprotrophic	Toxic
Agaricaceae	Lepiota	cristata	(Bolton) P. Kumm.	Wide	Saprotrophic	Toxic
Agaricaceae	Lepiota	naucina	(Fr.) P. Kumm.	Wide	Saprotrophic	Toxic
Agaricaceae	Lepiota	clypeolaria	(Bull. ex Fr.) P. Kumm.	Wide	Saprotrophic	Toxic
Agaricaceae	Leucoagaricus	rubrotinctus	(Peck) Singer	Midium	Saprotrophic	Toxic
Agaricaceae	Leucocoprinus	birnbaumii	(Corda) Singer	Medium	Saprotrophic	Toxic
Agaricaceae	Leucocoprinus	fragilissimus	(Berk. & Rav.) Pat.	Medium	Saprotrophic	Toxic
Agaricaceae	Leucocoprinus	cepistipes	(Sow. ex Fr.) Pat.	Medium	Saprotrophic	Toxic
Agaricaceae	Lycoperdon	perlatum	Pers.	Wide	Saprotrophic	Edible

Family	Genus	species	Author	Distribution	Nutrition	Edibility
Agaricaceae	Lycoperdon	pyriforme	Schaeff.	Wide	Saprotrophic	Edible
Agaricaceae	Lycoperdon	mammaeforme	Pers.	Low	Saprotrophic	Edible
Agaricaceae	Lycoperdon	marginatum	Vitt.	Medium	Saprotrophic	Toxic
Agaricaceae	Cystoderma	amianthinum	(Scop.) Fayod	Low	Saprotrophic	Toxic
Agaricaceae	Cystodermella	cinnabarina	(Alb. & Schwein.) Harmaja	Medium	Saprotrophic	Toxic
Agaricaceae	Cystodermella	granulosa	(Batsch) Harmaja	Medium	Saprotrophic	Toxic
Agaricaceae	Panaeolus	papilionaceus	(Bull.) Quél.	Wide	Saprotrophic	Toxic
Agaricaceae	Panaeolus	semiovatus	(Sowerby) S. Lundell y Nannf.	Medium	Saprotrophic	Toxic
Agaricaceae	Panaeolus	antillarum	(Fr.) Dennis	Wide	Saprotrophic	Toxic
Agaricaceae	Panaeolus	cinctulus	(Bolton) Sacc.	Medium	Saprotrophic	Toxic
Agaricaceae	Coprinus	comatus	(OF Müll.) Pers.	Wide	Saprotrophic	Edible
Agaricaceae	Phaeolepiota	aurea	(Matt.) Maire	Low	Saprotrophic	Toxic
Amanitaceae	Amanita	cochiseana	Sanchez et al., 2015	Medium	Mycorrhizal	Edible
Amanitaceae	Amanita	basii	Guzmán & Ram.-Guill.	Medium	Mycorrhizal	Edible
Amanitaceae	Amanita	phalloides	(Vaill. Ex Fr.) Link	Medium	Mycorrhizal	Toxic
Amanitaceae	Amanita	jacksonii	Pomerl.	Medium	Mycorrhizal	Edible
Amanitaceae	Amanita	abrupta	Peck	Low	Mycorrhizal	Toxic

Table 2 cont. ...

...*Table 2 cont.*

Family	Genus	Species	Authors	Distribution	Habit	Edibility
Amanitaceae	Amanita	crocea	(Quél.) Singer	Low	Mycorrhizal	Toxic
Amanitaceae	Amanita	fulva	Fr.	Medium	Mycorrhizal	Toxic
Amanitaceae	Amanita	muscaria	(L.) Lam.	Wide	Mycorrhizal	Toxic
Amanitaceae	Amanita	citrina	Pers.	Wide	Mycorrhizal	Toxic
Amanitaceae	Amanita	flavoconia	G.F. Atk.	Wide	Mycorrhizal	Toxic
Amanitaceae	Amanita	gemmata	(Fr.) Bertill.	Wide	Mycorrhizal	Toxic
Amanitaceae	Amanita	pantherina	(DC.) Krombh.	Wide	Mycorrhizal	Toxic
Amanitaceae	Amanita	polypyramis	(Berk. & M.A. Curtis) Sacc.	Wide	Mycorrhizal	Toxic
Amanitaceae	Amanita	rubescens	Pers.	Wide	Mycorrhizal	Toxic
Amanitaceae	Amanita	flavorubescens	G.F. Atk.	Medium	Mycorrhizal	Toxic
Amanitaceae	Amanita	virosa	Bertill.	Medium	Mycorrhizal	Toxic
Amanitaceae	Amanita	verna	(Bull.) Lam.	Medium	Mycorrhizal	Toxic
Amanitaceae	Amanita	bisporigera	G.F. Atk.	Medium	Mycorrhizal	Toxic
Amanitaceae	Amanita	strobiliformis	(Paulet ex Vittad.) Bertill.	Medium	Mycorrhizal	Toxic
Amanitaceae	Amanita	calyptroderma	G.F. Atk. & VG Ballen	Low	Mycorrhizal	Toxic
Amanitaceae	Amanita	cokeri	E.-J. Gilbert & Kühner ex E.-J. Gilbert	Low	Mycorrhizal	Toxic

Amanitaceae	Amanita	magniverrucata	Thiers & Ammirati	Low	Mycorrhizal	Toxic
Amanitaceae	Amanita	onusta	(Howe) Sacc.	Low	Mycorrhizal	Toxic
Amanitaceae	Amanita	perpasta	Corner & Bas	Low	Mycorrhizal	Toxic
Amanitaceae	Amanita	peckiana	Kauffman	Low	Mycorrhizal	Toxic
Amanitaceae	Amanita	porphyria	Alb. & Schwein.	Low	Mycorrhizal	Toxic
Amanitaceae	Amanita	smithiana	Bas	Medium	Mycorrhizal	Toxic
Amanitaceae	Amanita	vaginata	(Bull.) Lam.	Medium	Mycorrhizal	Toxic
Amanitaceae	Amanita	variabilis	E.-J. Gilbert y Cleland	Low	Mycorrhizal	Toxic
Amanitaceae	Amanita	tuza	Guzmán	Medium	Mycorrhizal	Toxic
Amanitaceae	Amanita	spreta	(Peck) Sacc.	Low	Mycorrhizal	Toxic
Amanitaceae	Limacella	illinita	(Fr.) Maire	Low	Mycorrhizal	Toxic
Bolbitiaceae	Conocybe	lateritia	(Fr.) Kühner	Medium	Saprotrophic	Toxic
Bolbitiaceae	Conocybe	apala	(Fr.) Arnolds	Low	Saprotrophic	Toxic
Bolbitiaceae	Bolbitius	titubans	(Bull.) Fr.	Low	Saprotrophic	Toxic
Cortinariaceae	Cortinarius	sanguineus	(Wulfen) Gray	Wide	Mycorrhizal	Toxic
Cortinariaceae	Cortinarius	semisanguineus	(Fr.) Gillet	Medium	Mycorrhizal	Toxic
Cortinariaceae	Cortinarius	elegantissimus	Rob. Henry	Low	Mycorrhizal	Toxic

Table 2 cont. ...

...*Table 2 cont.*

Family	Genus	Species	Authors	Distribution	Habit	Edibility
Cortinariaceae	*Cortinarius*	*violaceus*	(L.) Gray	Wide	Mycorrhizal	Toxic
Cortinariaceae	*Cortinarius*	*collinitus*	(Sowery) Gray	Medium	Mycorrhizal32	Toxic
Cortinariaceae	*Cortinarius*	*corrugatus*	Peck	Low	Mycorrhizal	Toxic
Cortinariaceae	*Cortinarius*	*purpureus*	(Bull.) Bidaud, Moënne-Locc. & Reumaux	Medium	Mycorrhizal	Toxic
Cortinariaceae	*Cortinarius*	*traganus*	(Fr.:Fr.) Fr.	Medium	Mycorrhizal	Toxic
Cortinariaceae	*Cortinarius*	*magnivelatus*	Dearness ex Fogel	Low	Mycorrhizal	Toxic
Cortinariaceae	*Cortinarius*	*smithii*	Amirati, Niskaenen & Liimat	Low	Mycorrhizal	Toxic
Cortinariaceae	*Cortinarius*	*pinetorum*	Fr. Kauffman	Low	Mycorrhizal	Toxic
Cortinariaceae	*Cortinarius*	*paleaceus*	(Weinm.) Fr.	Low	Mycorrhizal	Toxic
Entolomataceae	*Entoloma*	*mexicanum*	(Murrill) Hesler	Low	Mycorrhizal	Toxic
Entolomataceae	*Entoloma*	*sinuatum*	(Bull.) P. Kum.	Wide	Saprotrophic42	Toxic
Entolomataceae	*Entoloma*	*incanum*	(Fr.:Fr.) Hesler	Low	Saprotrophic	Toxic
Entolomataceae	*Entoloma*	*abortivum*	(Berk. & MA Curtis) Donk	Low	Saprotrophic	Toxic
Hydnangiaceae	*Laccaria*	*bicolor*	(Maire) PD Orton	Medium	Mycorrhizal	Edible
Hydnangiaceae	*Laccaria*	*laccata*	(Scop.) Cooke	Wide	Mycorrhizal	Edible
Hydnangiaceae	*Laccaria*	*proxima*	(Boud.) Pat.	Low	Mycorrhizal	Edible

Hygrophoraceae	*Hygrocybe*	*acutoconica*	(Clem.) Singer	Medium	Saprotrophic	Toxic
Hygrophoraceae	*Hygrophorus*	*erubescens*	(Fr.) Fr.	Medium	Mycorrhizal	Toxic
Hygrophoraceae	*Hygrophorus*	*russula*	(Schaeff. Ex Fr.) Kauffman	Wide	Saprotrophic	Edible
Hygrophoraceae	*Hygrocybe*	*conica*	(Schaeff.) P. Kumm.	Medium	Saprotrophic	Toxic
Hygrophoraceae	*Gliophorus*	*psittacinus*	(Schaeff.) Herink	Low	Saprotrophic	Toxic
Inocybaceae	*Crepidotus*	*mollis*	(Schaeff.) Staude	Wide	Saprotrophic	Toxic
Inocybaceae	*Inocybe*	*erubescens*	A. Blytt	Medium	Mycorrhizal	Toxic
Inocybaceae	*Inocybe*	*geophylla*	(Bull.) P. Kumm.	Medium	Mycorrhizal	Toxic
Inocybaceae	*Inocybe*	*histrix*	(Fr.) P. Karst.	Low	Mycorrhizal	Toxic
Inocybaceae	*Inocybe*	*rimosa*	(Bull.)P.Kumm.	Medium	Mycorrhizal	Toxic
Inocybaceae	*Inocybe*	*lacera*	(Fr.) Kumm.	Medium	Mycorrhizal	Toxic
Inocybaceae	*Inocybe*	*confusa*	P. Karst.	Low	Mycorrhizal	Toxic
Inocybaceae	*Inocybe*	*maculata*	Boud.	Low	Mycorrhizal	Toxic
Lyophyllaceae	*Lyophyllum*	*decastes*	(Fr.) Singer	Low	Saprotrophic	Edible
Mycenaceae	*Mycena*	*pura*	(Pers.) P. Kumm	Medium	Saprotrophic	Toxic
Mycenaceae	*Panellus*	*pusillus*	(Pers. ex Lév.) Burds. & O.K. Mill.	Medium	Saprotrophic	Toxic
Mycenaceae	*Panellus*	*stipticus*	(Bull.) P. Karst.	Medium	Saprotrophic	Toxic

Table 2 cont. ...

...*Table 2 cont.*

Family	Genus	Species	Authors	Distribution	Habit	Edibility
Mycenaceae	Xeromphalina	cauticinalis	(With.) Kühner & Maire	Medium	Saprotrophic	Toxic
Mycenaceae	Xeromphalina	tenuipes	(Schwein.) A.H. Sm.	Medium	Saprotrophic	Toxic
Physalacriaceae	Armillaria	mellea	(Vahl) P. Kumm.	Wedium	Pathogenic	Toxic
Physalacriaceae	Desarmillaria	tabescens	(Scop.) RA Koch y Aime	Medium	Pathogenic	Toxic
Physalacriaceae	Cryptotrama	chrysopeplum	(Berk. & M. A. Curtis) Singer	Low	Saprophitic	Toxic
Physalacriaceae	Flammulina	velutipes	(Curtis) Singer	Low	Saprotrophic	Edible
Physalacriaceae	Oudemansiella	canarii	(Jungh) Höhn.	Low	Saprotrophic	Edible
Pleurotaceae	Pleurotus	dryinus	(Pers.) P. Kumm.	Low	Saprotrophic	Edible
Pluteaceae	Pluteus	petasatus	(Fr.) Gillet	Low	Saprotrophic	Toxic
Pluteaceae	Pluteus	cervinus	(Schaeff.) P. Kumm.	Wide	Saprotrophic	Toxic
Pluteaceae	Volvariella	volvacea	(Bull.) Singer	Low	Saprotrophic	Edible
Pluteaceae	Volvariella	bombycina	(Schaeff.) Cantante	Low	Saprotrophic	Edible
Pluteaceae	Volvariella	gloiocephala	(DC.) Boekhout y Enderle	Medium	Saprotrophic	Edible
Psathyrellaceae	Psathyrella	candolleana	(Fr.) Maire.	Wide	Saprotrophic	Toxic
Psathyrellaceae	Coprinellus	atramentarius	(Bull.) Redhead, Vilgalys & Monclavo	Wide	Saprotrophic	Toxic
Schizophyllaceae	Schizophyllum	commune	P.	Wide	Saprotrophic	Edible

Strophariaceae	*Pholiota*	*adiposa*	(Batsch: Fr.) P. Kumm.	Medium	Saprotrophic	Toxic
Strophariaceae	*Pholiota*	*squarrosa*	(Batsch.) P.Kumm.	Low	Saprotrophic	Toxic
Strophariaceae	*Protostropharia*	*semiglobata*	(Batsch ex Fr.) Quél.	Wide	Saprotrophic	Toxic
Tapinellaceae	*Tapinella*	*panuoides*	(Fr.) Fr.	Wide	Saprotrophic	Toxic
Tapinellaceae	*Tapinella*	*atrotomentosa*	(Batsch) Šutara	Wide	Saprotrophic	Toxic
Omphalotaceae	*Gymnopus*	*alkalivirens*	(Singer) Halling	Low	Saprotrophic	Toxic
Omphalotaceae	*Gymnopus*	*confluens*	(Pers.) Antonín, Halling & Noordel.	Medium	Saprotrophic	Toxic
Omphalotaceae	*Gymnopus*	*fusipes*	(Bull.) Gray	Medium	Saprotrophic	Toxic
Omphalotaceae	*Omphalotus*	*olivascens*	H.E. Bigelow, OK Mill. & Thiers	Low	Saprotrophic	Toxic
Omphalotaceae	*Omphalotus*	*subilludens*	(Murrill) HE Bigelow	Medium	Saprotrophic	Toxic
Tricholomataceae	*Leucopaxillus*	*gentianeus*	(Quél.) Kotl.	Medium	Mycorrhizal	Toxic
Tricholomataceae	*Leucopaxillus*	*albissimus*	(Peck) Singer	Low	Mycorrhizal	Toxic
Tricholomataceae	*Resupinatus*	*applicatus*	(Batsch) Gray	Medium	Saprotrophic	Toxic
Tricholomataceae	*Clitocybe*	*dealbata*	(Sow.:Fr.) Kumm.	Medium	Saprotrophic	Toxic
Tricholomataceae	*Clitocybe*	*gibba*	(Pers.) P. Kumm.	Medium	Mycorrhizal	Edible
Tricholomataceae	*Lepista*	*nuda*	(Bull.) Cooke	Medium	Saprotrophic	Edible
Tricholomataceae	*Lepista*	*sordida*	(Schumach.) Cantante	Low	Saprotrophic	Edible

Table 2 cont. ...

...*Table 2 cont.*

Family	Genus	Species	Authors	Distribution	Habit	Edibility
Tricholomataceae	*Tricholoma*	*sejunctum*	(Sow. ex Fr.) Quél	Medium	Mycorrhizal	Toxic
Tricholomataceae	*Tricholoma*	*sulphureum*	(Bull.:Ffr.) Kumm.	Medium	Mycorrhizal	Toxic
Tricholomataceae	*Tricholoma*	*virgatum*	(Fr.:Fr.) Kumm.	Medium	Mycorrhizal	Toxic
Tricholomataceae	*Tricholoma*	*vaccinum*	(Pers.:Fr.)Kumm.	Medium	Mycorrhizal	Toxic
Tricholomataecee	*Tricholoma*	*flavovirens*	(Pers.) S. Lundell	Low	Mycorrhizal	Edible
Tricholomataecee	*Tricholoma*	*magnivelare*	(Peck) Redhead	Low	Mycorrhizal	Edible
Tricholomataceae	*Tricholomopsis*	*rutilans*	(Schaeff.) Singer	Medium	Saprotrophic	Toxic
Tricholomataceae	*Tricholomopsis*	*decora*	(Fr.) Singer	Low	Saprotrophic	Toxic
Auriculariaceae	*Auricularia*	*nigricans*	(Sw.) Birkebak, Looney & Sánchez-García	Medium	Saprotrophic	Edible
Auriculariaceae	*Exidia*	*glandulosa*	(Bull.) Fr.	Wide	Saprotrophic	Toxic
Auriculariaceae	*Exidia*	*recisa*	(Ditmar) Fr.	Low	Saprotrophic	Toxic
Boletaceae	*Caloboletus*	*inedulis*	(Murrill) Vizzini	Medium	Mycorrhizal	Toxic
Boletaceae	*Boletellus*	*coccineus*	(Sacc.) Singer	Low	Mycorrhizal	Edible
Boletaceae	*Boletellus*	*ananas*	(MA Curtis) Murrill	Medium	Mycorrhizal	Edible
Boletaceae	*Hortiboletus*	*rubellus*	(Krombh.) Simonini, Vizzini & Gelardi	Wide	Mycorrhizal	Edible
Boletaceae	*Hortiboletus*	*campestris*	(AH Sm. & Thiers) Biketova & Wasser	Medium	Mycorrhizal	Edible

Boletaceae	*Boletus*	*variipes*	Peck	Medium	Mycorrhizal	Toxic
Boletaceae	*Boletus*	*amigdalinus*	(Thiers) Thiers	Low	Mycorrhizal	Toxic
Boletaceae	*Boletus*	*paulae*	J. García, Singer y F. Garza-Ocañas	Low	Mycorrhizal	Edible
Boletaceae	*Boletus*	*subluridellus*	A.H. Sm. & Thiers	Medium	Mycorrhizal	Toxic
Boletaceae	*Boletus*	*subvelutipes*	Peck	Medium	Mycorrhizal	Toxic
Boletaceae	*Boletus*	*luridellus*	(Murrill) Murrill	Low	Mycorrhizal	Toxic
Boletaceae	*Boletus*	*barrowsi*	Thiers & AH Sm.	Low	Mycorrhizal	Toxic
Boletaceae	*Boletus*	*pinophilus*	Pilát & Dermek	Wide	Mycorrhizal	Edible
Boletaceae	*Porphyrellus*	*cyaneotinctus*	(A.H. Sm. & Thiers) Cantante	Medium	Mycorrhizal	Toxic
Boletaceae	*Sutorius*	*eximius*	(Peck) Halling, Nuhn & Osmundson	Wide	Mycorrhizal	Toxic
Boletaceae	*Aureoboletus*	*russellii*	(Escarcha) G. Wu y Zhu L. Yang	Wide	Mycorrhizal	Edible
Boletaceae	*Butyriboletus*	*frostii*	(JL Russell) G. Wu, Kuan Zhao y Zhu L. Yang	Wide	Mycorrhizal	Edible
Boletaceae	*Suillellus*	*luridus*	(Schaeff.) Murrill	Wide	Mycorrhizal	Toxic
Boletaceae	*Leccinum*	*manzanitae*	Thiers	Medium	Mycorrhizal	Edible
Boletaceae	*Leccinum*	*aurantiacum*	(Bull.) Gray	Low	Mycorrhizal	Edible
Boletaceae	*Harrya*	*chromapes*	(Frost) Halling, Nuhn, Osmundson y Manfr. Binder	Medium	Mycorrhizal	Edible
Boletaceae	*Tylopilus*	*felleus*	(Bull.) P. Karst.	Low	Mycorrhizal	Toxic

Table 2 cont. ...

...*Table 2 cont.*

Family	Genus	Species	Authors	Distribution	Habit	Edibility
Boletaceae	*Tylopilus*	*plumbeoviolaceus*	(Snell y EA Dick) Snell y EA Dick	Medium	Mycorrhizal	Toxic
Boletaceae	*Tylopilus*	*tabacinus*	(Peck) Singer	Low	Mycorrhizal	Toxic
Boletaceae	*Tylopilus*	*alboater*	(Schwein) Murrill	Low	Mycorrhizal	Toxic
Boletaceae	*Strobilomyces*	*confusus*	Singer	Medium	Mycorrhizal	Edible
Boletaceae	*Strobilomyces*	*strobilaceus*	(Scop.) Berk.	Wide	Mycorrhizal	Edible
Boletinellaceae	*Boletinellus*	*merulioides*	(Schwein.) Murrill	Medium	Mycorrhizal	Edible
Boletinellaceae	*Phlebopus*	*portentosus*	(Berk. & Broome) Boedijn	Low	Mycorrhizal	Edible
Boletinellaceae	*Phlebopus*	*brassiliensis*	Singer	Low	Saprotrophic	Edible
Gyroporaceae	*Gyroporus*	*castaneus*	(Bull.) Quél.	Medium	Mycorrhizal	Edible
Gyroporaceae	*Gyroporus*	*subalbellus*	Murrill	Low	Mycorrhizal	Edible
Gyroporaceae	*Gyroporus*	*castaneus*	(Bull.) Quél.	Medium	Mycorrhizal	Edible
Hygrophoropsidaceae	*Hygrophoropsis*	*aurantiaca*	(Wulf. ex Fr.) Maire	Medium	Saprotrophic	Toxic
Rhizopogonaceae	*Rhizopogon*	*occidentalis*	Zeller & CW Dodge	Low	Mycorrhizal	Edible
Rhizopogonaceae	*Rhizopogon*	*luteolus*	Fr.	Medium	Mycorrhizal	Edible
Sclerodermataceae	*Scleroderma*	*areolatum*	Ehrenb.	Wide	Mycorrhizal	Toxic
Sclerodermataceae	*Scleroderma*	*verrucosum*	(Bull.) Pers.	Wide	Mycorrhizal	Toxic

Sclerodermataceae	*Scleroderma*	*texense*	Berk.	Medium	Mycorrhizal	Toxic
Sclerodermataceae	*Scleroderma*	*cepa*	Pers.	Medium	Mycorrhizal	Toxic
Sclerodermataceae	*Scleroderma*	*citrinum*	Pers.	Medium	Mycorrhizal	Toxic
Sclerodermataceae	*Pisolithus*	*arhizus*	(Scop.) Rauschert	Wide	Mycorrhizal	Edible
Suillaceae	*Suillus*	*brevipes*	(Peck) Kuntze	Wide	Mycorrhizal	Edible
Suillaceae	*Suillus*	*cothurnatus*	Cantante	Medium	Mycorrhizal	Edible
Suillaceae	*Suillus*	*granulatus*	(L.) Roussel	Wide	Mycorrhizal	Edible
Suillaceae	*Suillus*	*spraguei*	(Berk. & MA Curtis) Kuntze	Medium	Mycorrhizal	Edible
Suillaceae	*Suillus*	*pseudobrevipes*	A.H. Sm. Y Thiers	Medium	Mycorrhizal	Edible
Suillaceae	*Suillus*	*tomentosus*	Singer	Wide	Mycorrhizal	Edible
Suillaceae	*Suillus*	*luteus*	(L.) Roussel	Medium	Mycorrhizal	Edible
Cantharellaceae	*Cantharellus*	*cibarius*	Fr.	Wide	Mycorrhizal	Edible
Cantharellaceae	*Cantharellus*	*lateritius*	(Berk.) Cantante	Medium	Mycorrhizal	Edible
Cantharellaceae	*Cantharellus*	*cinnabarinus*	(Schwein.) Schwein.	Low	Mycorrhizal	Edible
Cantharellaceae	*Craterellus*	*fallax*	A.H. Smith	Low	Mycorrhizal	Edible
Cantharellaceae	*Craterellus*	*cornucupioides*	(L.) Pers.	Low	Mycorrhizal	Edible
Hydnaceae	*Hydnum*	*repandum*	L.	Medium	Mycorrhizal	Edible

Table 2 cont. ...

...*Table 2 cont.*

Family	Genus	Species	Authors	Distribution	Habit	Edibility
Clavariadelphaceae	*Clavariadelphus*	*truncatus*	Donk	Wide	Saprotrophic	Edible
Clavariadelphaceae	*Clavariadelphus*	*ligula*	(Schaeff.) Donk	Low	Saprotrophic	Edible
Gomphaceae	*Gomphus*	*clavatus*	(Pers.) Gray	Medium	Mycorrhizal	Edible
Gomphaceae	*Turbinellus*	*floccosus*	(Schwein.) Earle & Giachini & Castellano	Low	Mycorrhizal	Toxic
Gomphaceae	*Ramaria*	*formosa*	(Pers.:Fr.) Quél.	Medium	Mycorrhizal	Toxic
Gomphaceae	*Ramaria*	*flava*	(Schaeff.) Quél.	Wide	Mycorrhizal	Toxic
Gomphaceae	*Ramaria*	*botrytoides*	(Peck) Corner	Medium	Mycorrhizal	Toxic
Tremellaceae	*Tremella*	*foliacea*	Pers.	Medium	Saprotrophic	Edible
Tremellaceae	*Tremella*	*lutescens*	Lloyd	Wide	Saprotrophic	Edible
Tremellaceae	*Tremella*	*fuciformis*	Berk.	Medium	Saprotrophic	Edible
Hymenogastraceae	*Hebeloma*	*crustuliniforme*	(Bull.) Quél.	Medium	Mycorrhizal	Toxic
Hymenogastraceae	*Galerina*	*marginata*	(Batsch) Kühner	Low	Saprotrophic	Toxic
Hymenogastraceae	*Gymnopilus*	*fulvosquamulosus*	Hesler	Low	Saprotrophic	Toxic
Hymenogastraceae	*Gymnopilus*	*penetrans*	(Fr.) Murrill	Medium	Saprotrophic	Toxic
Hymenogastraceae	*Hypholoma*	*capnoides*	(Fr.) P. Kumm.	Medium	Saprotrophic	Toxic
Hymenogastraceae	*Hypholoma*	*fasciculare*	(Huds. ex Fr.) P. Karst.	Wide	Saprotrophic	Toxic

Hymenogastraceae	Psilocybe	mexicana	R. Heim.	Low	Saprotrophic	Toxic
Hymenogastraceae	Deconica	coprophila	(Bull.) P. Kumm.	Wide	Saprotrophic	Toxic
Hymenogastraceae	Gymnopilus	aeruginosus	(Peck) Singer	Wide	Saprotrophic	Toxic
Fomitopsidaceae	Phaeolus	schweinitzii	(Fr.) Pat.	Wide	Pathogenic	Toxic
Ganodermataceae	Ganoderma	applanatum	(Pers.) Pat.	Wide	Pathogenic	Medicinal
Ganodermataceae	Ganoderma	curtisii	(Berk.) Murrill	Medium	Pathogenic	Medicinal
Ganodermataceae	Ganoderma	resinaceum	Boud.	Medium	Pathogenic	Medicinal
Ganodermataceae	Ganoderma	lobatum	(Cooke) GF Atk.	Medium	Pathogenic	Medicinal
Ganodermataceae	Ganoderma	colossus	(Fr.) Baker	Low	Pathogenic	Medicinal
Ganodermataceae	Ganoderma	oerstedii	(Fr.) Torrend	Medium	Pathogenic	Medicinal
Ganodermataceae	Ganoderma	brownii	(Murrill) Gilb.	Medium	Pathogenic	Medicinal
Ganodermataceae	Humphreya	coffeata	(Berk.) Stacyaert	Low	Pathogenic	Medicinal
Meruliaceae	Cymatoderma	caperatum	(Berk. & Mont.) D.A.	Medium	Saprotrophic	Medicinal
Meruliaceae	Phlebia	tremellosa	(Schrad.) Nakasone & Burds.	Medium	Saprotrophic	Medicinal
Phanerochaetaceae	Byssomerulius	incarnatus	(Schwein.) Gilb.	Wide	Pathogenic	Medicinal
Polyporaceae	Hexagonia	hydnoides	(Sw.) M. Fidalgo	Wide	Saprotrophic	Medicinal
Polyporaceae	Pycnoporus	sanguineus	(L.) Murrill	Wide	Saprotrophic	Medicinal

Table 2 cont. ...

...*Table 2 cont.*

Family	Genus	Species	Authors	Distribution	Habit	Edibility
Polyporaceae	*Trametes*	*versicolor*	(L.) Lloyd	Medium	Saprotrophic	Medicinal
Polyporaceae	*Lentinus*	*crinitus*	(L.) Fr.	Wide	Saprotrophic	Edible
Polyporaceae	*Neolentinus*	*lepideus*	(Fr.) Redhead & Ginns	Low	Pathogenic	Edible
Albatrellaceae	*Albatrellus*	*ellisi*	(Berk.) Pouzar	Low	Pathogenic	Edible
Fomitopsidaceae	*Laetiporus*	*sulphureus*	(Bull.) Murrill	Low	Saprophitic	Edible
Bondarzewiaceae	*Amylosporus*	*campbellii*	(Berk.) Ryvarden	Low	Saprotrophic	Medicinal
Bondarzewiaceae	*Heterobasidion*	*annosum*	(Fr.) Bref.	Low	Pathogenic	Toxic
Hericiaceae	*Hericium*	*erinaceus*	(Bull.) Pers.	Wide	Saprotrophic	Edible
Hericiaceae	*Hericium*	*coralloides*	(Scop.) Pers.	Low	Saprotrophic	Edible
Russulaceae	*Lactarius*	*rufus*	(Scop.) Fr.	Medium	Mycorrhizal	Toxic
Russulaceae	*Lactarius*	*scrobiculatus*	(Scop.) Fr.	Low	Mycorrhizal76	Toxic
Russulaceae	*Lactarius*	*torminosus*	(Schaeff.) Gray	Low	Mycorrhizal	Toxic
Russulaceae	*Lactarius*	*vellereus*	(Fr.) Fr.	Low	Mycorrhizal	Toxic
Russulaceae	*Lactarius*	*chrysorrheus*	Fr.	Low	Mycorrhizal	Toxic
Russulaceae	*Lactarius*	*piperatus*	(L.) Pers.	Low	Mycorrhizal	Toxic
Russulaceae	*Lactarius*	*uvidus*	(Fr.) Fr.	Low	Mycorrhizal	Toxic

Family	Genus	species	author			
Russulaceae	Lactarius	deliciosus	(L.) Gray	Medium	Mycorrhizal	Edible
Russulaceae	Lactarius	indigo	(Schwein.) P.	Low	Mycorrhizal	Edible
Russulaceae	Lactarius	volemus	(Fr.) Fr.	Low	Mycorrhizal	Edible
Russulaceae	Lactarius	zonarius	(Bull.) Fr.	Low	Mycorrhizal	Toxic
Russulaceae	Russula	emetica	(Schaeff.) Pers.	Wide	Mycorrhizal	Toxic
Russulaceae	Russula	nigricans	Fr.	Medium	Mycorrhizal	Toxic
Russulaceae	Russula	albonigra	Fr.	Low	Mycorrhizal	Toxic
Russulaceae	Russula	cyanoxantha	(Schaeff.).	Medium	Mycorrhizal	Edible
Russulaceae	Russula	brevipes	Peck	Wide	Mycorrhizal	Edible
Russulaceae	Russula	virescens	(Schaeff.).	Medium	Mycorrhizal	Edible
Russulaceae	Russula	rosea	Pers.	Low	Mycorrhizal	Toxic
Russulaceae	Russula	foetens	(Pers.) Fr.	Medium	Mycorrhizal	Toxic
Stereaceae	Stereum	ostrea	(Blume & T. Nees)	Wide	Saprotrophic	Toxic
Sebacinaceae	Tremellodendron	schweinitzii	(Peck) G.F. Atk.	Medium	Mycorrhizal	Toxic
Bankeraceae	Hydnellum	scrobiculatum	(Fr.) P. Karst.	Medium	Mycorrhizal	Toxic
Bankeraceae	Phellodon	niger	(Fr.) P. Karst.	Medium	Mycorrhizal	Toxic
Bankeraceae	Sarcodon	scabrosus	(Fr.) P. Karst.	Medium	Mycorrhizal	Toxic

Table 2 cont. ...

...Table 2 cont.

Family	Genus	Species	Authors	Distribution	Habit	Edibility
Bankeraceae	*Sarcodon*	*imbricatus*	(L.) P. Karst.	Medium	Mycorrhizal	Toxic
Hypocreaceae	*Hypomyces*	*lactifluorum*	(Schwein.) Tul. Y C. Tul.	Wide	Pathogenic	Edible
Pyronemataceae	*Aleuria*	*aurantia*	(Pers.) Fuckel	Wide	Saprotrophic	Toxic

Table 3. Distribution of species in the Northern States of Mexico.

Genus	Species	BCN	Sonora	Chih.	Durango	Coahuila	Nuevo León	Tam.
Leotia	*lubrica*			x	x	x	x	x
Leotia	*viscosa*			x	x			x
Gyromitra	*esculenta*			x	x		x	
Hydnotrya	*cerebriformis*						x	x
Helvella	*crispa*	x	x	x	x	x	x	x
Helvella	*elastica*			x	x	x	x	x
Helvella	*macropus*	x	x	x	x	x	x	x
Hydnobolites	*cerebriformis*						x	x
Pachyploeus	*carneus*						x	x
Pachyploeus	*virescens*						x	x
Elaphocordyceps	*capitata*			x	x		x	x
Morchella	*esculenta*			x		x	x	x
Tuber	*lyonii*					x	x	x
Tuber	*regimontanum*						x	
Agaricus	*xanthodermus*	x		x	x	x	x	x
Agaricus	*campestris*	x	x	x	x	x	x	x

Table 3 cont. ...

...Table 3 cont.

Genus	Species	BCN	Sonora	Chih.	Durango	Coahuila	Nuevo León	Tam.
Agaricus	silvaticus	x		x	x		x	
Agaricus	silvicolae-similis			x	x		x	
Agaricus	arvensis	x		x	x	x	x	
Agaricus	bitorquis	x	x	x	x			
Calvatia	bovista		x	x	x	x	x	x
Calvatia	cyathiformis	x		x	x	x	x	x
Chlorophyllum	molybdites	x	x	x	x	x	x	x
Lepiota	cristata	x	x	x	x	x	x	x
Lepiota	naucina		x	x	x		x	
Lepiota	clypeolaria		x	x	x	x	x	x
Leucoagaricus	rubrotinctus				x		x	x
Leucocoprinus	birnbaumii	x		x	x	x	x	x
Leucocoprinus	fragilissimus			x	x		x	x
Leucocoprinus	cepistipes		x	x	x	x	x	x
Lycoperdon	perlatum	x	x	x	x	x	x	x
Lycoperdon	pyriforme	x	x	x	x	x	x	x

Lycoperdon	*mammaeforme*					x	x	x
Lycoperdon	*marginatum*			x	x	x	x	x
Cystoderma	*amianthinum*	x		x	x	x	x	x
Cystodermella	*cinnabarina*		x	x	x	x	x	x
Cystodermella	*granulosa*		x	x	x	x		x
Panaeolus	*papilionaceus*	x	x	x	x	x	x	x
Panaeolus	*semiovatus*			x		x		x
Panaeolus	*antillarum*	x	x	x	x	x		x
Panaeolus	*cinctulus*			x		x		
Coprinus	*comatus*	x	x	x	x	x	x	x
Phaeolepiota	*aurea*					x		x
Amanita	*cochiseana*		x	x	x	x	x	x
Amanita	*basii*		x	x	x	x		
Amanita	*phalloides*			x	x			
Amanita	*jacksonii*				x			x
Amanita	*abrupta*		x	x				
Amanita	*crocea*		x	x	x		x	x

Table 3 cont. ...

...Table 3 cont.

Genus	Species	BCN	Sonora	Chih.	Durango	Coahuila	Nuevo León	Tam.
Amanita	*fulva*		x	x	x	x	x	x
Amanita	*muscaria*	x	x	x	x	x	x	x
Amanita	*citrina*		x	x	x	x		
Amanita	*flavoconia*	x	x	x	x	x	x	x
Amanita	*gemmata*	x	x	x	x	x	x	x
Amanita	*pantherina*	x	x	x	x	x		x
Amanita	*polypyramis*	x	x	x	x	x		x
Amanita	*rubescens*	x	x	x	x	x	x	x
Amanita	*flavorubescens*			x	x			
Amanita	*virosa*		x	x	x	x		x
Amanita	*verna*	x		x	x	x		x
Amanita	*bisporigera*	x	x	x	x	x		x
Amanita	*strobiliformis*	x		x	x			
Amanita	*calyptroderma*		x	x	x			
Amanita	*cokeri*	x	x	x	x			
Amanita	*magniverrucata*	x	x	x	x			
Amanita	*onusta*			x	x	x		x

Amanita	*perpasta*			X				
Amanita	*peckiana*			X	X			
Amanita	*porphyria*		X	X	X	X		X
Amanita	*smithiana*		X	X	X			
Amanita	*vaginata*	X	X	X	X	X	X	X
Amanita	*variabilis*		X	X	X			
Amanita	*tuza*			X	X			
Amanita	*spreta*		X	X	X			
Limacella	*illinita*					X		X
Conocybe	*lateritia*			X	X	X		X
Conocybe	*apala*		X	X	X	X		X
Bolbitius	*titubans*		X	X		X		X
Cortinarius	*sanguineus*		X	X	X	X	X	
Cortinarius	*semisanguineus*		X	X	X	X	X	
Cortinarius	*elegantissimus*				X			X
Cortinarius	*violaceus*		X	X	X	X	X	
Cortinarius	*collinitus*				X	X		X

Table 3 cont. ...

...Table 3 cont.

Genus	Species	BCN	Sonora	Chih.	Durango	Coahuila	Nuevo León	Tam.
Cortinarius	corrugatus					x		x
Cortinarius	purpureus			x				
Cortinarius	traganus				x			
Cortinarius	magnivelatus			x				
Cortinarius	smithii		x	x				
Cortinarius	pinetorum					x		
Cortinarius	paleaceus					x		
Entoloma	mexicanum					x		
Entoloma	sinuatum			x		x		x
Entoloma	incanum					x		
Entoloma	abortivum				x	x		
Laccaria	bicolor	x		x	x	x		
Laccaria	laccata	x	x	x	x	x	x	x
Laccaria	proxima	x				x		x
Hygrocybe	acutoconica			x				
Hygrophorus	erubescens		x	x	x			

		1	2	3	4	5	6	7	8
Hygrophorus	*russula*			x	x	x	x		x
Hygrocybe	*conica*	x	x	x	x	x	x	x	x
Gliophorus	*psittacinus*				x	x	x		
Crepidotus	*mollis*	x	x	x	x	x	x	x	x
Inocybe	*erubescens*								
Inocybe	*geophylla*					x	x		x
Inocybe	*histrix*				x		x		x
Inocybe	*rimosa*		x	x	x	x	x		
Inocybe	*lacera*				x	x	x		
Inocybe	*confusa*				x				
Inocybe	*maculata*					x			
Lyophyllum	*decastes*				x				
Mycena	*pura*	x	x	x	x	x	x	x	x
Panellus	*pusillus*	x	x	x	x	x	x	x	x
Panellus	*stipticus*					x	x		x
Xeromphalina	*cauticinalis*	x	x	x	x	x			x
Xeromphalina	*temuipes*		x	x	x	x	x		x

Table 3 cont. ...

...Table 3 cont.

Genus	Species	BCN	Sonora	Chih.	Durango	Coahuila	Nuevo León	Tam.
Armillaria	mellea	x	x	x	x	x	x	x
Desarmillaria	tabescens	x	x	x	x	x		x
Cryptotrama	chrysopeplum				x	x		x
Flammulina	velutipes		x	x		x		x
Oudemansiella	canarii					x		x
Pleurotus	dryinus		x	x	x	x		
Pluteus	petasatus					x		x
Pluteus	cervinus	x	x	x	x	x		x
Volvariella	volvacea					x		x
Volvariella	bombycina					x		x
Volvariella	gloiocephala	x				x		
Psathyrella	candolleana	x	x	x	x	x		x
Coprinellus	atramentarius	x	x	x	x	x	x	x
Schizophyllum	commune	x	x	x	x	x	x	x
Pholiota	adiposa					x		
Pholiota	squarrosa			x		x		x

Protostropharia	*semiglobata*	x		x	x	x		x
Tapinella	*panuoides*			x	x	x		x
Tapinella	*atrotomentosa*			x	x			x
Gymnopus	*alkalivirens*				x			
Gymnopus	*confluens*		x	x	x	x		x
Gymnopus	*fusipes*		x	x	x	x	x	
Omphalotus	*olivascens*	x	x	x	x	x		
Omphalotus	*subilludens*			x	x	x	x	x
Leucopaxillus	*gentianeus*	x		x	x	x		x
Leucopaxillus	*albissimus*			x	x	x		
Resupinatus	*applicatus*	x	x	x	x	x		x
Clitocybe	*dealbata*			x	x	x		
Clitocybe	*gibba*		x	x	x	x	x	x
Lepista	*nuda*	x	x	x	x		x	x
Lepista	*sordida*					x		
Tricholoma	*sejunctum*			x		x		x
Tricholoma	*sulphureum*			x	x	x	x	x

Table 3 cont. ....

...Table 3 cont.

Genus	Species	BCN	Sonora	Chih.	Durango	Coahuila	Nuevo León	Tam.
Tricholoma	virgatum					x	x	x
Tricholoma	vaccinum					x	x	x
Tricholoma	flavovirens					x		x
Tricholoma	magnivelare			x	x	x		x
Tricholomopsis	rutilans					x		x
Tricholomopsis	decora					x		
Auricularia	nigricans			x	x	x	x	x
Exidia	glandulosa		x			x	x	x
Exidia	recisa		x	x	x	x		x
Caloboletus	inedulis					x		
Boletellus	coccineus			x	x	x		x
Boletellus	ananas		x					x
Hortiboletus	rubellus			x	x	x		x
Boletus	variipes			x	x	x		x
Hortiboletus	campestris				x	x		
Boletus	paulae					x		x
Boletus	subluridellus			x		x		x

Boletus	subvelutipes	x			x	x	x	x	x
Boletus	amigdalinus		x	x					
Boletus	luridellus				x		x		x
Boletus	barrowsi	x	x	x	x		x	x	
Boletus	pinophilus			x	x	x	x	x	x
Porphyrellus	cyaneotinctus				x	x	x		x
Sutorius	eximius			x	x	x	x	x	x
Aureoboletus	russellii			x	x	x	x	x	x
Butyriboletus	frostii			x	x	x	x		x
Suillellus	luridus				x	x	x	x	x
Leccinum	manzanitae		x		x	x			
Leccinum	aurantiacum			x	x	x	x		x
Harrya	chromapes			x	x	x	x		x
Tylopilus	felleus			x	x				
Tylopilus	plumbeoviolaceus			x	x	x	x		x
Tylopilus	tabacinus				x		x		x
Tylopilus	alboater			x					

Table 3 cont. ...

...Table 3 cont.

Genus	Species	BCN	Sonora	Chih.	Durango	Coahuila	Nuevo León	Tam.
Strobilomyces	confusus				x	x	x	x
Strobilomyces	strobilaceus	x	x	x	x	x	x	x
Boletinellus	merulioides					x		x
Phlebopus	portentosus					x		x
Phlebopus	brassiliensis					x		
Gyroporus	castaneus		x	x	x	x	x	x
Gyroporus	subalbellus					x		x
Gyroporus	castaneus		x	x	x	x		x
Hygrophoropsis	aurantiaca	x	x	x		x		x
Rhizopogon	occidentalis	x		x				
Rhizopogon	luteolus					x	x	
Scleroderma	areolatum				x	x		
Scleroderma	verrucosum			x		x		x
Scleroderma	texense					x		
Scleroderma	cepa			x	x	x		
Scleroderma	citrinum		x			x		
Pisolithus	arhizus			x	x	x	x	x

Suillus	brevipes			x	x	x		x
Suillus	cothurnatus			x		x		x
Suillus	granulatus		x	x	x	x	x	x
Suillus	spraguei	x		x	x	x		
Suillus	pseudobrevipes	x	x	x	x	x		
Suillus	tomentosus		x	x	x	x	x	x
Suillus	luteus				x	x	x	x
Cantharellus	cibarius	x	x	x	x	x	x	x
Cantharellus	lateritius				x	x		x
Cantharellus	cinnabarinus			x	x	x	x	x
Craterellus	fallax			x	x	x		
Craterellus	cornucupioides	x	x	x		x		x
Hydnum	repandum	x	x	x	x	x	x	x
Clavariadelphus	truncatus		x	x	x	x	x	x
Clavariadelphus	ligula			x				
Gomphus	clavatus				x	x		x
Turbinellus	floccosus	x	x		x	x	x	x

Table 3 cont. ...

...Table 3 cont.

Genus	Species	BCN	Sonora	Chih.	Durango	Coahuila	Nuevo León	Tam.
Ramaria	formosa			x				
Ramaria	flava				x			x
Ramaria	botrytoides		x			x		x
Tremella	foliacea			x	x	x		x
Tremella	lutescens		x	x	x	x	x	x
Tremella	fuciformis	x		x	x	x		x
Hebeloma	crustuliniforme			x		x		
Galerina	marginata				x	x		
Gymnopilus	fulvosquamulosus					x		
Gymnopilus	penetrans			x				
Hypholoma	capnoides					x		
Hypholoma	fasciculare	x	x	x	x	x		x
Psilocybe	mexicana			x		x		
Deconica	coprophila	x	x	x	x	x	x	x
Gymnopilus	aeruginosus		x	x	x	x	x	
Phaeolus	schweinitzii	x	x	x	x	x	x	x
Ganoderma	applanatum		x	x	x	x	x	x

Ganoderma	curtisii	x				x	
Ganoderma	resinaceum	x			x	x	
Ganoderma	lobatum				x	x	
Ganoderma	colossus					x	
Ganoderma	oerstedii					x	
Ganoderma	brownii			x	x	x	
Humphreya	coffeata	x				x	
Cymatoderma	caperatum			x			x
Phlebia	tremellosa			x	x	x	x
Byssomerulius	incarnatus			x	x	x	x
Hexagonia	hydnoides	x				x	
Pycnoporus	sanguineus	x				x	
Trametes	versicolor	x	x	x	x	x	
Lentinus	crinitus	x	x		x	x	
Neolentinus	lepideus				x	x	
Albatrellus	ellisi				x		
Laetiporus	sulphureus	x			x		
Amylosporus	campbellii					x	

Table 3 cont. ....

...Table 3 cont.

Genus	Species	BCN	Sonora	Chih.	Durango	Coahuila	Nuevo León	Tam.
Heterobasidion	annosum					x		
Hericium	erinaceus	x	x	x	x	x		x
Hericium	coralloides			x				
Lactarius	rufus			x	x	x		x
Lactarius	scrobiculatus			x	x	x		x
Lactarius	torminosus				x	x		x
Lactarius	vellereus	x	x	x				
Lactarius	chrysorrheus			x	x	x	x	x
Lactarius	piperatus					x		x
Lactarius	uvidus				x	x		x
Lactarius	deliciosus			x	x	x	x	x
Lactarius	indigo			x	x	x	x	x
Lactarius	volemus			x		x		x
Lactarius	zonarius		x	x	x	x	x	x
Russula	emetica	x	x	x	x	x	x	x
Russula	nigricans			x	x	x		x

Russula	*albonigra*				x		
Russula	*cyanoxantha*		x	x	x	x	x
Russula	*brevipes*	x	x	x	x	x	x
Russula	*virescens*			x	x	x	x
Russula	*rosea*			x			
Russula	*foetens*		x	x	x		x
Stereum	*ostrea*	x	x	x	x	x	x
Tremellodendron	*schweinitzii*		x	x	x		x
Hydnellum	*scrobiculatum*			x	x		x
Phellodon	*niger*			x	x		
Sarcodon	*scabrosus*		x	x	x		x
Sarcodon	*imbricatus*			x	x		x
Hypomyces	*lactifluorum*	x	x	x	x	x	x
Aleuria	*aurantia*		x	x	x		x

as food after fruiting bodies are boiled several times, water should be discarded every time. After this procedure, people in the Sierra Madre Occidental in the state of Chihuahua use this species as food, as the toxins are degraded in this way.

3) Some species, such as *Deconica coprophila*, *Panaeolus antillarum*, *Panaeolus cinctulus* and *Protostropharia semiglobata*, can cause gastrointestinal upsets. Other species, like *Corpinellus atramentarius*, should never be used as food if mixed with alcoholic drinks as their toxins interact with alcohol and cause intoxications.

Species with Unknown Edibility

There are very many mushroom species in temperate forests that are little known regarding their edibility (e.g., *Russula* spp., *Lactarius* spp., *Amanita* spp., *Boletus* spp., *Laccaria* spp.). Many other species are ecologically important and abundant; some are mycorrhizal with many hosts and others may be saprotrophic or parasites of some trees but are not recommended as some or many species belong to known toxic groups, including *Cortinarius* spp., *Tricholoma* spp., *Inocybe* spp., *Hypholoma* spp., *Clitocybe* spp., *Conocybe* spp., *Tapinella* spp., *Omphalotus* spp. and *Turbinellus* spp.

Small Edible Species

Some species are known to be edible but they are very small and it is often difficult to collect a sufficient amount for them to be used as food (e.g., *Laccaria laccata*); others are edible but their stem is rather fibrous in consistency and only the pileus can be used for food (*Neolentinus lepideus*, *Aureoboletus russelli*) (Figs. 5–7). Some species have been considered as edible for a long time but there are some recent reports regarding the presence of toxins in *Armillaria mellea* and *Desarmillaria tabescens*. Some species are considered edible in some countries, such as Brasil, even if they have a hard leather or corky consistency (*Lentinus crinitus*, *L. tigrinus*). Others are rather thin, soft or leathery (e.g., *Hexagonia papyracea*). A few species have a woody consistency and some people cut them into small pieces and grind them to prepare infusions (e.g., *Inonotus obliquus* in the USA), others are edible but not very popular due to their cartilaginous consistency, e.g., *Auricularia mesenterica*, *A. nigricans* and *Tremella foliacea*.

Medicinal Species

There are very many scientific reports regarding some species' medicinal usage, including *Coriolus versicolor*, *Picnoporus sanguineus*, *Ganoderma curtisii*, *G. oerstedii*, *G. applanatum*, *G. titans*, *G. resinossum*, *G. lobatum*, *Neolentinus lepideus* and *Calvatia cyathiformis* (Fig. 12). All these species are abundant in the temperate forests in the North of Mexico and some strains of these species have been obtained and studied in order to reveal the effects of their secondary metabolism on cancerous liver cells *in vitro* as well as in laboratory mice with good results, thus becoming a very promising research line (Ramírez et al., 2006).

Fig. 5. Edible species: 1. *Amanita jacksonii;* 2. *A. cochiseana;* 3. *A. bassi;* 4. *A. rubescens* group; 5. *Boletus barrowsii;* 6. *Butiriboletus frostii;* 7. *Boletus chipewaensis;* 8. *B. pseudopinophilus;* 9. *Boletus* group a *edulis;* 10. B. group b *edulis;* 11. *Boletus rubriceps;* 12. *Boletinellus merulioides.*

Forests and Mushrooms

Results showed that most mycorrhizal species were associated either with pines, oaks, spruce, fir, *Arbutus* spp. or *Arctostaphyllos* spp. in temperate forests, in altitudes ranging from 550–3650 m, saprotrophic, parasitic and pathogenic species are also present. There is a high diversity of mushrooms species at lower altitudes, i.e., 500–700 m, most of them are saprotrophic (*Leucocoprinus birnbaumii, Lepiota rubrotincta, Psathyrella* spp., *Agaricus xanthodermus, Chlorophyllum molybdites*) but some mycorrhizal species also grow associated to oaks, pines and elms (*Hortiboletus rubellus, Pisolithus tinctorius, Inocybe rimosa, I. geophylla, Scleroderma areolatum. S. cepa* and *S. verrucosum*) amongst many others. These are also present in piedmont forests associated with Oaks and many other hosts.

Distribution of Species

1) Most edible or poisonous species are located in altitudes from 1500–3000 m. Many of them form mycorrhizas with a wide range of hosts (refer to hosts list) and some are saprotrophic or parasitic species (Figs. 5–7; Table 2). Some of the main edible species are: *Amanita cochiseana* (from the caesarea group),

Fig. 6. Edible species: 1. *Lactarius indigo;* 2. *Lactarius deliciosus;* 3. *Flammulina velutipes;* 4. *Harrya chromapes;* 5. *Hypomyces lactifluorum growing on Russula brevipes;* 6. *Gomphus clavatus;* 7. *Hydnum repandum;* 8. *Hortiboletus rubellus;* 9. *Russula virescens;*10. *Morchella conica;* 11. *Aureoboletus russellii;* 12. *Tuber* sp.; 13. *Boletus edulis group;* 14. *Strobilomyces strobilaceus;*15. *Polyporus tenuicolus;* 16. *Lycoperdon piriforme.*

Fig. 7. Edible species: 1. *Hericium erinaceus*; 2. *Russula brevipes*; 3. *Hericium coraloides*; 4. *Lycoperdon perlatum*; 5. *Calvatia cyathiformis*; 6. *Hortiboletus campestris*; 7. *Lactarius volemus*; 8. *Albatrellus ellisii*; 9. *Suillus granulatus*; 10. *Cantharellus cibarius.*

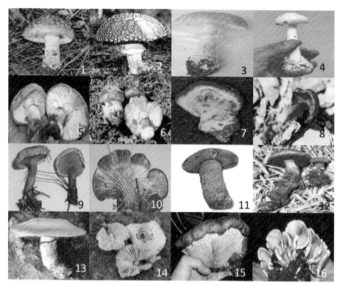

Fig. 8. Toxic species: 1. *Amanita pantherina*; 2. *Amanita muscaria*; 3. *Amanita verna*; 4. *Amanita bisporigera*; 5. *Amanita flavorubescens*; 6. *Amanita flavoconia*; 7. *Cortinarius* sp., 8. *Cortinarius sanguineus*; 9. *Hygrophoropsis aurantiaca*; 10. *Tapinella panuoides*; 11. *Sutorius eximius*; 12. *Boletus subvelutipes*; 13. *Agaricus xanthodermus*; 14. *Chlorophyllum molybdites*; 15. *Tapinella atrotomentosus*; 16. *Hypholoma fasciculare*.

Fig. 9. Toxic species: 1. *Cortinarius violaceus;* 2. *Cortinarius brunneus*; 3. *Cortinarius aureoturviantus;* 4. *Cortinarius semisanguineus;* 5. *Cortinarius* sp.; 6. *Leucocoprinus birmbaumii*; 7. *Scleroderma areolatum*; 8. *Scleroderma verrucosum*; 9. *Omphalotus subilludens*; 10. *Collybia alkalivirens*; 11. *Stropharia semiglobata*; 12. *Paneolus papilonaceus*; 13. *Paneolus cinctulus*; 14. *Amanita plumbea*; 15. *Gymnopilus aeruginacens*; 16. *Cortinarius pinetorum*.

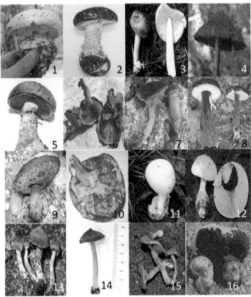

Fig. 10. Toxic species: 1. *Amanita abrupta;* 2. *Amanita muscaria var. flavivolvata;* 3. *Amanita vaginata;* 4. *Cortinarius phoeniceus;* 5. *Cortinarius violaceus;* 6. *Omphalotus olivacens;* 7. *Omphalotus subilludens;* 8. *Boletus plumbeoviolaceus;* 9. *Tylopilus alboater;* 10. *Tylopilus felleus;* 11. *Amanita phalloides;* 12. *Amanita perspasta;* 13. *Cortinarius paleaceus;* 14. *Inocybe gerardii;* 15. *Inocybe calamistrata;* 16. *Boletus satanas.*

A. jacksonii, A. calyptrodema, Cantharellus cibarius, Craterellus cornucupiodes, C. fallax, Tricholoma magnivelare, Lactarius deliciosus, L. indigo, Russula delica, R. brevipes, Boletus pseudopinophilus, B. barrowsi, B. chippewaensis, B. appendiculatus, Leccinum aurantiacum, L. manzanitae, Hydnum repandum, Gomphus clavatus, Hortiboletus rubellus, Suillus tomentosus, S. granulatus, S. luteus, S. brevipes, S. pseudobrevipes, Lepista nuda, Albatrellus ellisii, Harrya chromapes, Hericium erinaceus, H. coraloides, Agaricus campestris and *A. bitorquis.* Native truffles have been found, *Tuber regiomontanum* for example, and considering the diversity of Oaks and Pines occurring in the North of Mexico it is possible that many species have not yet been discovered. Many other interesting hypogeous species from different genera have been reported recently (Cázares et al., 1992; Guevara et al., 2014).

2) There are quite a number of toxic species and many are mycorrhizal with oaks and conifers, i.e., *Amanita verna, A. bisporigera, A. virosa, A. muscaria, var. flavivolvata, A. phalloides* (Figs. 8–11; Table 2). Also, there are saprotrophic or parasitic toxic species: *Hypholoma fasiculare, H. capnoides, Mycena pura, Gymnopilus aeruginacens., Cortinarius sanguineus, C. semisanguineus, C. violaceus, Hebeloma crustuliniforme, Inocybe fastigiata, I. geophylla, I. rimosa* and *I. lacera.*

3) Medicinal species are mostly associated to Oak forests and they are either saprotrophic or parasitic: *Coriolus versicolor y Ganoderma curtissi, G. oerstedii, G. resinaceum* and *G. applanatum* (Fig. 12; Table 2).

Fig. 11. Toxic species: 1. *Amanita magniverrucata;* 2. *Amanita novinupta;* 3. *Amanita rubescens;* 4. *Amanita verna;* 5. *Amanita bisporigera;* 6. *Amanita fulva;* 7. *Tricholoma flavovirens;* 8. *Entoloma lividum;* 9. *Entoloma mexicana;* 10. *Scleroderma citrinum;* 11. *Boletus luridellus;* 12. *Paneolus antillarum;* 13. *Suillelus luridus;* 14. *Boletus amigdalinus;* 15. *Cortinarius magnivelarum;* 16. *Sarcodon imbricatus;* 17. *Xerocomus* sp.; 18. *Lepiota cristata;* 19. *Russula emetica;* 20. *Pluteus petasatus.*

Risks of Intoxication by Toadstools and Mushrooms in City Gardens and Nearby Forests

It is interesting to mention that eating poisonous species may occur and is mainly associated with species of some genera, i.e., *Amanita, Cortinarius, Russula, Leucocoprinus, Lepiota, Chlorophyllum* and *Scleroderma*. In the field, some toxic species like *Amanita verna* have be confused with edible species like *Agaricus campestris or A. bitorquis* as they grow in lawns placed very close to pine and oak forests where the toxic species *Amanita verna, A. virosa and A. bisporigera* grow. Also, oak species (*Quercus fusiformis, Quercus rysophylla* and *Q. polymorpha*) have recently been planted extensively in many city gardens in the north of Mexico. These trees form mycorrhizal associations in nurseries with many ectomycorrhizal fungi and, once established in their final destination in the citygardens, they produce fruiting bodies that may cause intoxications (e.g., *Scleroderma texense, S. areolatum,*

Fig. 12. Medicinal species: 1. *Ganoderma oerstedii;* 2. *Auricularia nigricans*; 3. *Ganoderma brownii*; 4. *Ganoderma lobatum*; 5. *Ganoderma curtisii*; 6. *Ganoderma collosus*; 7. *Calvatia craneiiformis*; 8. *Pycnoporus sanguineus*; 9. *Coriolus versicolor*; 10. *Ganoderma resinaceus*; 11. *Lycoperdon piryforme*; 12. *Calvatia cyathiformis*.

S. cepa, Inocybe fastigiata, I. calamistrata, Cortinarius violaceus, Amanita muscaria, Boletus luridellus and *Suillelus luridus*). Some saprotrophyc species, namely *Chlorophyllum molybdites, Lepiota rubrotincta, Leucocoprinus birmbaumii* and *Agarcius xanthodermus*, have also been collected.

Altitudinal Distribution

The species identified in this study are located in altitudes from 500 to 3500m in the east, while in central and west states, most species are located in altitudes from 1500–2500m in Oak forests. Most toxic fatal species like *Amanita verna* grow in high altitude forests and are associated with *Quercus* spp. and *Pinus* spp., forming mycorrhizas as well as some saprotrophic species or pathogens (*Hypholoma fasciculare, Armillaria mellea* and *Desarmillaria tabescens*). Some toxic saprotrophic species grow in lower altitudes ca. 500–1500 (*Chlorophyllum molybdites*). Medicinal species follow the same pattern, with some species growing in *Quercus-Pinus* forests (e.g., *Ganoderma applanatum* and *Neolentinus lepideus* and *Coriolus versicolor*) and other species growing in lower altitudes (500–2000) (*Ganoderma resinosum*). Thus, most toxic species are located in altitudes from 500 to 3200 m (*Amanita bisporigera, A. phalloides, A. virosa, A. rubescens, Tricholomopsis rutilans, Inocybe fastigiata, I.*

calamistrata, Pholliota squarrosa, P. adiposa, Cortinarius violaceus, C. paleaceus, C. pinetorum, Mycena pura and *Hypholoma fasciculare*).

Importance of Species

All species are ecologically important but there was a lack of information regarding toxic species (*Amanita phalloides, A. verna, A. bisporigera* and *A. virosa*) occurring in these extensive temperate forests. They should be known in order to avoid poisoning with possible fatal events in people living in rural forestry communities. Identification of edible, toxic and medicinal species is very important as many research lines can be established from each one of these groups (Garza et al., 2014). Mycorrhizal species have relevance both because they promote seedling growth and because many species are edible and can be collected and sold (e.g., *Amanita caesarea* group, now called *cochiseana*) (Sánchez et al., 2015), *A. jacksonii, Boletus edulis* group, *Tricholoma magnivelare, Lactarius deliciosus, Cantharellus cibarius* and some others (Burk, 1983; Barros et al., 2008; Garza et al., 2012). In the case of some of the medicinal species, we have carried out research searching for secondary metabolites with antimicrobial and anticancer activity in liver cells *in vitro* with very promising results (Garza et al., 2006).

Discussion

This study confirms the presence of a high diversity of edible, toxic and medicinal mushroom species in temperate forests in the North of Mexico. Oak and mixed oak pine forests had the higher diversity of species, as reported previously by (Pérez et al., 1986; Lafferriere and Gilbertson, 1992; Moreno et al., 1994; Quiñónez et al., 1999; Garza et al., 2002; North, 2002; García et al., 2014; Quiñonez and Garza 2015; Sánchez et al., 2015; García et al., 2017) and some species had been reported from Oak forests in several states. Many of the species found in this study have also been reported in Oak forests in Northeastern North America (Bessette et al., 1997; Bessette et al., 2007; Binion et al., 2008). Many of the fungal species have evolved from the Artic regions towards the south, those species from the West in the Sierra Madre Occidental are related to those species found in Southern parts of North America (e.g., California). According to Salinas et al. (2017), characteristics such as location, climate, physiography, soil types, geological age, karstic landscape, and wide range of elevations are important attributes for plant endemisms, including species in the Fagaceae in the Sierra Madre Oriental. These field characteristics and those from forests and trees themselves, together with management and conservation activities, are very important for the development of fungal growth in temperate forests in the North of Mexico (Garza et al., 1985; Garza et al., 2014; Quiñónez et al., 2014; Quiñonez and Garza, 2015). Regarding mushroom intoxications, individual health and susceptibility or immune response can play an important role in mushroom intoxications. Some individuals can develop allergic reactions even when eating edible cultivated species like the common champignon *Agaricus campestris* or *Pleurotus ostreatus*. When eating edible mushrooms, it is always recommended to eat only a little bit and when wanting to eat wild edible mushrooms one should always count on the identification help of an expert, otherwise it is never recommended to risk a fatal reaction for a few mushrooms. Toxic species grow beside edible species as they can share tree hosts in the forests.

References

Allen, M.F., Egerton-Warburton, I., Tresede, K., Cario, C., Lnidahl, A., Lansing, J., Querejeta, J., Karen O., Harney, S. and Zink, T. 2005. Biodiversity of mycorrhizal fungi in Southern California, USDA forest Service Gen. Tech. Rep. PSW-GTR-195.

Barros, L., Ferreira, M.J., Queiros, B., Ferreira, I.C.F.R. and Baptista, P. 2008. Wild and commercial mushrooms a source of nutrients and nutraceuticals, Food Chem. Toxicol., 46: 2742–2747.

Bessette, A.E., Bessette, A.R. and Fischer, D.W. 1997. Mushrooms of Northern United States. Syracuse University Press, Syracuse, Beutler.

Bessette, A.E. Roody, W.C. Bessette, A.R. Dunaway, D.C. 2007. Mushrooms of southern United States. Syracuse University Press, Syracuse, Beutler.

Binion, D., Stephenson, S., Roody, W., Burdsall, H., Vasilyeva, L. and Miller, O.K. 2008. Mushrooms associated with oaks of Eastern North America. West Virginia University Press, Morgantown.

Boa, E. 2004. Wild Edible Fungi a Global Overview of Their Use and Importance to People, Food and Agricultural Organization, Rome.

Burk, W.R. 1983. Puffball usages among North American Indian. J. Ethnobiol., 3: 55–62.

Cázares, E., García, J. and Castillo, J. 1992. Hypogeous fungi from northern México. Mycologia, 84: 341–359.

Challenger, A. 1998. Utilización y conservación de los ecosistemas terrestres de México. Pasado, Presente y Futuro. Conabio.

Duffy, T. 2008. Toxic Fungi of Western North America. Published by Mykoweb.

Fischer, D. and Bassette, A. 1992. Edible wild mushrooms of North America. University of Texas Press, Austin.

Frank, J., Barry, S., Madden, J. and Southwork, D. 2007. Oaks belowground: Mycorrhizas, truffles and small mammals. USDA, Forest Service, Tech. Rep. PSW-GTR_217.

García, G., Valenzuela, R., Raymundo, T., García, L., Guevara, L., Garza, F., Cázares, E. and Ruiz, E. 2014. Macrohongos asociados a encinares (*Quercus* spp.) en algunas localidades del estado de Tamaulipas, México. Capitulo 5. En Biodiversidad Tamaulipeca 2, SEP, Instituto Tecnológico de Ciudad Victoria, Tam., Volye, 1: 103–140.

García, J. and Garza, F. 2001. Conocimiento de los hongos de la Familia Boletaceae de México. Revista Ciencia UANL Volume IV, # 3 Julio–Septiembre 2001, 336–343.

Garza, L., Ramírez, X.S., Garza, F., Salinas, M.C., Waksman, N., Alcaraz, Y. and Torres, O. 2006. Evaluación de la actividad biológica de extractos acuosos de macromicetos del Noreste de México. Revista Ciencia UANL, Volume IX, # 2.

Garza, F., García, J. and Castillo, J. 1985. Macromicetos asociados al bosque de *Quercus rysophylla* en algunas localidades del centro del estado de Nuevo León. Rev. Mex. Mic., 1: 423–437.

Garza, F. 1986. Hongos ectomicorrícicos en el estado de Nuevo León. Rev. Mex. Mic., 2: 197–205.

Garza, F., García, J., Estrada E. and Villalón, H. 2002. Macromicetos, ectomicorrizas y cultivos de *Pinus culminicola* en Nuevo León. Revista Ciencia UANL Volume V, # 2, Abril-Junio, 204–210.

Garza, F., Cázares González, E., Carrillo Parra, A., Garza Ocañas, L. and Quiñónez Martínez, M. 2012. Economía micológica en bosques templados y plantaciones: Un enfoque educativo-productivo para el desarrollo socioeconómico rural, el manejo sustentable y la conservación de los hongos silvestres comestibles de México. Capitulo 4 En: Economía en el manejo sustentable de los recursos naturales. U.A.N.L., 77–111.

Garza, F., Carrillo, A., Garza, L., Quiñónez, M., García, J. and Guevara, G. 2014. Técnicas para el manejo de hongos ectomicorrícicos: Del bosque al laboratorio y viceversa. En Técnicas en el manejo sustentable de los recursos naturales, UANL, FCF, 71–119.

González, S., González, M., Tena, J.A., Raucho, I. and López, I. 2012. Vegetación de la Sierra Madre Occidental, México: una síntesis. Acta Bot. Mex., 100: 351–403.

Groves, J.W. 1979. Edible and poisonous mushrooms of Canada. Research Branch Agriculture, Canada.

Guevara, G., Garza, F. and Cázares, E. 2004. Estudio del ITS nuclear de algunas especies del género Cantharellus de México. Revista Ciencia, UANL, 371–378.

Guevara, G. and Garza, F. 2005. Estudio de la subunidad mayor del ADN ribosomal nuclear de algunas especies del género *Cantharellus* de México. Rev. Mex. Micol., 20: 21–26.

Guevara, G., Cázares, E., Bonito, G., Healy, R., Stielow, B., Garcóa, J., Garza, F., Castellano, M. and Trappe, J. 2014. Hongos hipogeos de Tamaulipas, México. Capitulo 4. En Biodiversidad Tamaulipeca 2, SEP, Instituto Tecnológico de Ciudad Victoria, Tam., 1: 87–101.

Guzmán, G. 1998. Inventorying the fungi of México. Biodiver. Conserv., 7: 369–384.

Hall, I.R., Stephenson, S.L., Buchanan, P.K., Yun, W. and Cole, A.L.J. 2003. Edible and poisonous mushrooms of the world. Timber press.

Hallen, H. and Adams, G. 2002. Don´t pick poison when gathering mushrooms for food in Michigan. Michigan State University, Department of Plant Pathology. Bulletin # E-1080.

Halling, R. and Muller, G. 2002. Agarcis and Boletes of Neotropical Oakwoods. pp. 1–10. *In*: Watling, R., Frankling, J.C., Ainsworth, A.M., Isaac, S. and Robinson, C.H. (eds.). Tropical Mycology.

Hawksworth, D.L. 2001. The magnitude of Fungal Diversity: 1.5 million species estimates revisited. Mycol. Res., 105: 1422–1432.

Heads, S., Miller, A., Crane, J., Thomas, M., Ruffatto, D., Methven, A., Raudabaugh, D. and Wang, Y. 2017. The oldest fossil mushroom. Plos. One 12: e0178327.

Hybbett, D. 2006. A phylogenetic overview of Agaricomycotina. Mycologia, 98: 917–925.

Kendrick, B. 1994. Evolution in action from mushrooms to truffles. Mc Ilvanea, 11: 34– 37.

Laferriere, E.J. and Gilbertson, R.L. 1992. Fungi of Nabogame, Chihuahua, México. Mycotaxon, 44: 73–87.

Leonard, P.L. 2010. A Guide to Collecting and Preserving Fungal Specimens for the Queensland Herbarium. Queensland Government. Department of environmental and research management. Brisbane.

Lindequist, U., Niedermeyer, T.H.J. and Jülich, W.D. 2005. The pharmacological potential of mushrooms, Evid. Based Complement. Altern. Med., 2: 285–299.

Lyncoff, G. and Mitchell, D.H. 1977. Toxic and Hallucinogenic Mushroom Poisoning—A Handbook for Physicians and Mushrooms Hunters. Van Nostrand Reinhold Company, New York.

Moreno-Fuentes, Aguirre-Acosta, A.E., Villegas, M. and Cifuentes, J. 1994. Estudio fungístico de los macromicetos en el municipio de Bocoyna, Chihuahua, México. Rev. Mex. Mic., 10: 63–76.

Morrone, J., Escalante, T. and Rodriguez, G. 2017. Mexican biogeographic provinces: Map and shapefiles. Zootaxa, 4277: 277–279.

Müller, G., Haling, R., Carranza, J., Mata, M. and Schmidt, J.P. 2006. Saprotrophic and ectomycorrhizal macrofungi of Costa Rican Oak forests. pp. 55–68. *In*: Kappelle, M. (ed.). Ecology and Conservatrion of Neotropical Montane Oak Forests. Springer Verlag, Berlin.

North, M.P. 2002. Seasonality and abundance of truffles from Oak woodlands to red Fir Forests. USDA Forest Service Tech. Rep. PSW-GTR-183.

Olimpia Mariana García-Guzmán, O., Garibay-Orijel, R., Hernández, E., Arellano-Torres, E. and Oyama, K. 2017. Word-wide meta-analysis of *Quercus* forests ectomycorrhizal fungal diversity reveals southwestern Mexico as a hotspot. Mycorrhiza, 27: 811–822.

Peinado, M., Bartolome, C., Delgadillo, J. and Aguado, I. 1994. Pisos de vegetación de la Sierra de San Pedro Martir, Baja California, México. Acta Bot. Mex., 29: 1–30.

Pérez-Silva, E. and Aguirre-Acosta, E. 1986. Flora Micológica del estado de Chihuahua, México I. Inst. Biol., 57: 17–32.

Quiñonez, M., Garza, F., García, J., Saenz, J. and Mendoza, J. 1999. Guía de Hongos de la región de Bosque Modelo. Universidad Autónoma de Chihuahua, Facultad de Zootecnia, Gobierno del Estado de Chihuahua.

Quiñónez, M., Garza, F., Sosa, M., Lebgue, T., Lavin, P. and Bernal, S. 2008. Índices de diversidad y similitud de hongos ectomicorrizógenos en bosques de Bocoyna, Chihuahua, México. Revista Ciencia Forestal en México. Volyme 33(103).

Quiñonez, M., Ruan, F., Aguilar, I., Garza, F., Lebgue, T., Lavin, P.A. and Enriquez, I. 2014. Knowledge and use of edible mushrooms in two municipalities of the Sierra Tarahumara, Chihuhua, México. J. Ethnobiol. Ethnomed., 10: 1–13.

Quiñonez, M. and Garza, F. 2015. Hongos silvestres comestibles de la Sierra Tarahumara de Chihuahua. Fomix-Conacyt, Universidad Autónoma de Ciudad Juárez.

Ramírez Gómez, XS., Garza Ocañas, L., Garza Ocañas, F. and Salinas Carmona, M.C. 2006. Evaluación de la citotoxicidad selectiva, actividad antioxidante e immunomoduladora de macromicetos del Noreste de México. Revista Ciencia UANL. Volume IX (1).

Retallack, G. and Landing, E. 2014. Afinities and arquitectura of Devonian trunks of *Prototaxites loganii*. Mycologia, 106: 1143−1158.

Salinas, M.M., Estrada, A.E. and Villarreal, J.A. 2017. Endemic vascular plants of the Sierra Madre Oriental, México Phytotaxa, 328: 1–52.

Sánchez, N., Soria, I., Romero, L., López, M., Rico, R. and Portillo, A. 2015. Los hongos Agaricales de las áreas de encino del estado de Baja California, México. Estudio de Diversidad Volumen IUniversity of Nebraska, Lincoln, 215−226.

Sánchez, S., Tulloss, R.E. Guzman, L., Valos, N-Da, Cifuentes, J., Valenzuela, R., Estrada, A., Ruan, F., Diaz, R., Hernández, N., Torres, M., Leo, H. and Moncalvo, J.M. 2015. In and out of refugia: historical patterns of diversity and demography in the North American Caesar's mushroom species complex. Mol. Ecol., 24: 5938–5956.

Sarukhán, J., Urquiza-Haas, T., Koleff, P., Carabias, J., Dirzo, R., Ezcurra, E., Cerdeira-Estrada, S., and Soberón, J. 2015. Strategic actions to value, conserve, and restore the natural capital of Megadiversity Countries: The case of Mexico. *In*: Bioscience A Forum for Integrating the Life Sciences. Am. Inst. Biol. Sci., 65(2).

Singer, R., Araujo, I. and Ivory, M. 1983. The ectotrophycally mycorrhizal fungi of the Neotropical Lowlands, Especially Central Amazonia Forests 2, J. Cramer, Vaduz.

Villarreal, J.A., Encina, J.A. and Carranza, M.A. 2008. Los encinos (*Quercus*: Fagaceae) de Coahuila, México. J. Bot. Res. Inst. Texas, 2: 1235–1278.

9

Biotechnological Exploitation of Macrofungi for the Production of Food, Pharmaceuticals and Cosmeceuticals

Susanna M. Badalyan[1],* and Alessandra Zambonelli[2]

INTRODUCTION

Macrofungi (from the Greek "makros" meaning large) are ascomycetous and basidiomycetous mushrooms that fruit above or below the ground with the word mushroom used in its widest sense as "any fungus with a distinctive fruiting body that is large enough to be ... picked by hand" (Chang and Wasser, 2017). All of them form a thallus, a vegetative mycelium composed from undifferentiated false tissue and it is from this that their fruiting bodies develop. Unlike plants they lack chlorophyll and so are unable to use light in photosynthesis to produce organic molecules. Instead, they are heterotrophic, gaining their energy by colonising and decomposing carbon compounds present in the substrates they colonize. They do this by secreting enzymes into the substrates they colonize and decomposing organic material extracellularly. The macrofungi are found almost everywhere in natural ecosystems, as well as industrial and agriculture wastes.

The fungi functionally and evolutionary are more closely related to animals than plants and are placed in a separate kingdom—Fungi (Feeney et al., 2014).

[1] Laboratory of Fungal Biology and Biotechnology, Institute of Pharmacy, Department of Biomedicine, Yerevan State University, 1 Aleg Manoogian St., 0025, Yerevan, Armenia.

[2] Dipartimento di Scienze e Tecnologie Agro-Alimentari, University of Bologna, via Fanin 46, 40127 Bologna, Italy.

* Corresponding author: s.badalyan@ysu.am

For example, their cell-wall contains the polysaccharide chitin which is also found in arthropods. The macrofungi are placed in two phyla, the Basidiomycota (class Agaricomycetes) and Ascomycota (class Pezizomycetes) in the sub-kingdom Dikarya and have a huge morphological range. Some, like the truffles, form below the ground but the majority are epigeous and include agaricoid, gasteroid, bracket and jelly mushrooms (Chang and Miles, 1992; Hibbett, 2007; Schoch et al., 2009; Kües and Navarro-Gonzalés, 2015). They are classified into different ecological groups. The saprotrophs live on dead organic matter; xylotrophs live on wood or other substrates containing lignin and coprotrophs that grow on dung (dung mushrooms). Others form very close associations with other organisms and include the mycorrhizal fungi that form symbiotic relationships with the roots of the vast majority of green plants; the entomogenous fungi that grow inside insects and parasitic species that grow on or in living plants or animals. Some of these are obligate pathogens and can only live within the tissues of the host whilst others are facultative and may, for example, kill their host and continue to live on the dead tissues afterwards.

Macrofungi have been used since ancient times not only as food, but also as medicines (Wasser, 2010; De Silva et al., 2013; Grienke et al., 2014; Guzmán, 2015; de Mattos-Shipley et al., 2016; Karun and Sridhar, 2017) and styptics such as the one found in Ice Man's posessions (http://bioweb.uwlax.edu/bio203/2011/manske_bria/). Those which are eaten are often regarded as gourmet foods as they have excellent culinary properties such as color, texture and flavour, as well as pungent, garlicky, cheesy, musky and fruity aromas. Good examples of a group of prized edible macrofungi are the truffles, ascomyceteous fungi characterized by hypogeous ascomata with spores sequestered in a more or less spherical mass. Some species of the genus *Tuber* (also referred to as the "true truffles") (Zambonelli et al., 2016) and the genera *Tefezia* and *Tirmania* (the desert truffles) (Kagan-Zur et al., 2014) have a considerable economic value for the excellent gastronomic and medicinal (Badalyan, 2012) proprieties of their ascomata.

Wild and cultivable mushroom resources can also be sources of healthy functional food (nutraceuticals and nutriceuticals), medicine (pharmaceuticals) and cosmetic products (cosmeceuticals, nutricosmetics) (Chang and Buswell, 2001; Badalyan et al., 2007; Wasser and Weiss, 1999; Wasser, 2011; De Silva et al., 2012a; Bishop et al., 2015; Wu et al., 2016; Taofiq et al., 2016a,c; 2017a). For example, basidiomycetous *Ganoderma* species complex are widely used to develop different health-enhancing food additives or nutriceuticals that are used as tonics, teas, soups, and alcoholic beverages; pharmaceuticals, and organic cosmetic products (cosmeceuticals) applied to the skin, and nutricosmetics that are administered orally (Chang and Buswell 2001; De Silva et al., 2012a; Bishop et al., 2015; Wu et al., 2016; Taofiq et al., 2016a,c, 2017a).

Some macrofungi are widely used in bioremediation (mycoremediation) processes. They posess great potential for generating environmental and socio-economic impacts in human welfare at local, national and global levels. Macrofungi are important in development of agriculture industry and are an only group of fungal organisms allowing realization of "a non-green revolution" in the world (Chang and Miles, 2004; Stamets, 2005). In our Chapter we examine the development of the above industries based on macrofungi and examine those species which can be exploited for their food value or biotech products including pharmaceuticals, nutriceuticals, nutraceuticals and cosmeceuticals.

Taxonomy and Distribution of Macrofungi

Fungi are morphologically extremely diverse and have a nomenclature and taxonomy to match. They are found in almost all habitats with estimates of total numbers ranging between 0.5 and 5.1 million with current working estimates in the region of 1.5 to 3.0 million (Blackwell, 2011; Hawksworth, 2012; Hibbet and Taylor, 2013; Tedersoo et al., 2014; Dai et al., 2015; Peay et al., 2016). Of these, about 150,000 species are macrofungi, of which 10% (14,000–16,000) are known to science (Hawksworth, 1991; Chang, 2001; Mueller et al., 2007; Kirk et al., 2008; Wasser, 2010; Blackwell, 2011).

In recent molecular surveys of soil mycobiomes from different continents, latitudes and ecosystems, about half of all detected fungal species (80,000–81,000) were Agaricomycetes (phylum Basidiomycota) and about 2.0% Pezizomycetes (phylum Ascomycota) (Tedersoo et al., 2014; Peay et al., 2016). About 7,000 known mushroom species are considered to be edible and about half of these from 231 genera are highly regarded (Hawksworth, 1991; Wasser and Weiss, 1999; Chang, 2001; Wasser, 2002, 2010). In contrast, only 3% of the known species are considered poisonous (Boa, 2004). The current number of poisonous mushrooms approximates around 500 species (Chang and Wasser, 2017) which is considerably more than those previously listed by Bresinsky and Besl (1990). However, this number could be even higher considering that at least 200 poisonous species are listed in China alone (Li et al., 2003; Chen et al., 2012).

Many fungi produce chemicals that have pharmacological properties, and many are used in traditional Chinese medicine. *Ganoderma* spp. has been used there for at least 2000 years and there is currently considerable research on the lanostane triterpenes they produce and their bioactivity (Wang et al., 2015). *Ophiocordyceps sinensis* (dong cheong xia cao) is another highly sought-after species (Winkler, 2017) and is the world's most expensive with rumours of prices going as high as US$ 100,000/kg in the USA. Other medicinal macrofungi include *Trametes versicolor, Schizophyllum commune, Lentinula edodes, Grifola frondosa, Fomes fomentarius, Fomitopsis pinicola, Pleurotus ostreatus, Cantharellus cibarius, Boletus edulis* and *Tuber borchii*. There may be as many as 700 medicinal species but a lack of knowledge is limiting their use (Wasser and Weiss, 1999; Hawksworth, 2001; Wasser, 2002, 2011; Lindequist et al., 2005; Badalyan, 2012; De Silva et al., 2012a,b; Kües and Badalyan, 2017). Generally, they are viewed to be safe and possess around 130 different therapeutic effects (Wang et al., 2002; Wasser, 2002, 2010).

Recent molecular analysis of the localities from which fungi new to science have been described over the past 10 years has revealed that about 60% were from the tropics. This is also the case for mushrooms, especially ectomycorrhizal species, and in some tropical locations as many as 75% of species have yet to be described (Hawksworth, 2012). However, following the development of next generating sequencing (NGS) and large-scale DNA-sequence datasets have become available that have revised estimates of fungal diversity and provided a new understanding of how fungal diversity shapes, and is shaped by, other ecosystem components (Peay et al., 2016). Currently, about 1200 new species of fungi are being described annually and while this is impressive at the current rate it would take another 4000 years to process the 5 million species that await study (Chang and Wasser, 2017). We can assume that approximate 2% of world fungal biota and around 10% of world mushroom biodiversity have been discovered by mycologists to date so the bulk of fungal biodiversity still remains hidden.

Nutritional, Medicinal and Cosmetic Potential for Macrofungi

Macrofungi have long been appreciated all over the world for their nutritional properties and medicinal value (Moore and Chiu, 2001; Chang and Miles, 2004; Barros et al., 2008; Wasser, 2011; Valverde et al., 2015; Chang and Wasser, 2017; Bandara et al., 2015, 2017). The early civilizations of Greeks, Egyptians and Romans, Chinese, Japanese and Mexican people prized them for their therapeutic value and as treasures in religious rites where hallucinogenic species were employed (Hobbs, 1995; Wasser, 2010; Guzmán, 2015). There are rich sources of ethno-mycological and ethno-medicinal information on wild mushrooms in different countries of the world and some of this still awaits exploitation in western medicine (Boa, 2004; Grinke et al., 2014; Karun and Sridhar, 2017).

Asian population use mushrooms (fresh or dried) and mushroom products (teas, tablets, capsules and extracts) to prevent and cure different diseases (Wasser and Weiss, 1999; Wasser, 2010). Applications of medicinal mushrooms in Western societies are limited in forms of nutraceuticals (functional foods) and nutriceuticals (dietary supplements) (Wasser, 2002, 2011; Cheung, 2008; Chen et al., 2017) although there is a move to include them in natural cosmetic products (cosmeceuticals, nutricosmetics) (Wu et al., 2016; Taofiq et al., 2016a,c; 2017a,b).

Mushrooms have a high protein content containing all the essential amino acids, vitamins (B_1, B_2, B_3, B_5, B_9, C, D), high levels of dietary fiber, unsaturated fatty acids, high contents of minerals particularly phosphorus and iron and almost no cholesterol. They are also excellent sources of beta-glucan and selenium (Manzi et al., 1999; Mattila et al., 2000; Poucheret et al., 2006; Ribeiro et al., 2009; Cheung, 2010, 2013; Ulziijargal and Mau, 2011; Badalyan, 2012; Ahmad et al., 2013; Kalăc, 2013; Toafiq et al., 2017b). Mushroom nucleic acid also plays a part in the regulation of various physiological processes in the human body (Phan et al., 2017). As a consequence the composition of mushrooms plus the biotech products obtained from them represent low energy healthy foods which are excellent additions to human diets (Chung, 2006; Dai et al., 2009; Ferreira et al., 2010; Bishop et al., 2015; Chaiyasut and Sivamaruthi, 2017).

Mushroom biotech products developed from fruiting bodies and/or mycelia are consumed worldwide. These are not strictly pharmaceutical products (medicines) but represent a novel class of functional food additives—nutraceuticals and nutraceuticals consisting of dried mycelial biomass or fruiting bodies or aqueous-alcohol extracts.

Among the recognized 7000 edible mushrooms species more than 3,000 species from 231 genera are highly regarded and include the highly prized mycorrhizal species *Tuber melanosporum, Tuber magnatum, Tricholoma matsutake, C. cibarius* and *B. edulis* as well as the saprobic morels in particular *Morchella sextelata* and *Morchella importuna* which are being widely cultivated in Sichuan, China (Peng et al., 2015). However, only about 200 species of mushrooms are grown, either commercially or experimentally. Of these only half arecommercially cultivated with just 10 taking the lion's share of the market worldwide (Badalyan, 2012; Ajmal et al., 2015; Chang and Wasser, 2017) (Table 1) (Fig. 1).

Fig. 1. Commercially cultivated species: the medicinal fungus *Ganoderma lucidum* (a); the edible and medicinal *Coprinus comatus* (b), the gourmet fungi *Morchella* sp. (c), and *Tuber melanosporum* (d).

The truffles (phylum Ascomycota) are the most expensive edible fungi. In Europe the retail price of the *T. magnatum* ascomata can be up to 5.000 €/kg. *T. melanosporum* commands lower prices which are however higher (600–1.200 €/kg) than those of any mushroom. The less valuable truffles, like *T. borchii* or *T. aestivum* usualy sell for 300–600 €/kg. Outside of the truffle growing countries of Europe prices can reach even dizzier heights particularly for *T. magnatum* (Fig. 2).

Of course, such prices have no relevance to their nutritional value but for their flavor (Crisan and Sands, 1978) where just tiny amounts are grated over pasta, eggs and white meat or in wafer thin slices by the head chef in the most exclusive of restaurants. What makes truffles so special a food is their unique aroma which is made of a mixture of hydrocarbons containing alcohol, ketone, aldehyde, and ester functional groups but particularly sulfur containing molecules (Splivallo and Culleré, 2016). To date, more than 300 volatiles have been described from about eleven species (Splivallo et al., 2011). The aromatic profile of a single species is distinctive and can contain up more than 100 volatile compounds (Gioacchini et al.,

Fig. 2. (a) *Tuber magnatum* sold in Harrods, London at 6,500 £/kg on 25 September 2010 (Ian Hall picture). (b) *Tuber melanosporum* sold in Harrods, London at 1600 pound/kg on 18 February 2018 (Susanna Badalyan picture).

2005; March et al., 2006; Vita el al., 2015). However, when it really comes to the nitty gritty it is the occasion and the realization that the food on the plate in front of you contains the ultimate in foods and which may not be repeated in a lifetime.

In China in the 1970s, Chinese species of truffle was used as pig food but is now regarded as having medicinal value (http://chinatruffles.com/14-0-health.htm). Recently, fermentation techniques were tested in order to produce extracellular and intracellular polysaccharide by *Tuber sinense* and *T. melanosporum* (Liu et al., 2008, 2009).

The first use of the desert truffles *Terfezia* spp. and *Tirmania nivea* in cuisine dates back to the third millennium BCE in the Bronze Age by the nomadic Amorites in the area of middle Euphrates (Shavit, 2014). Later they became gourmet foods highly appreciated by Egyptians, Greeks and Romans. The desert truffles have also been also used in Middle Eastern and North African traditional medicines (El Enshasy et al., 2013). They are used for the treatment of several diseases and in particular for skin and eye ailments. Like in Europe, they are also considered as aphrodisiacs (Hall et al., 2007; Shavit and Shavit, 2014). It has also been reported that they have immune-modulating, hepatoprotective, antidepressant, antibacterial, antifungal, antiviral, antioxidant and antiradical properties from, for example, their phenolic, carotenoid, anthocyanin, ascorbic acid, flavonoid, tannin, glycoside, and ergosterol content (Owaid, 2017). Several recent studies have also shown them to have antimicrobial properties against a wide range of human pathogens (Shavit and Shavit, 2014). The basidiomycetous desert truffles of Australia *Mycoclelandia bulundari* and *Elderia arenivaga* are also used by Australian aborigines for medicinal (Trappe et al., 2008b) and cosmetic (Shavit and Shavit, 2014) purposes, whereas in the Africa Kalahari the local species *Kalaharituber pfeilii* is used not only as food and for its medicinal proprieties but also as antidote against the poison they put on arrows (Trappe et al., 2008a).

Another well-known highly prized edible ascomycetous is *Morchella esculenta* (true morel). Due to unique flavor, taste and texture it has high culinary value. Total world production of true morel is 150 tons dry weight which is about 1.5 million tons fresh weight. The price of dry morels ranges from 200–750 USD/kg. *M. esculenta* possesses a wide range of bioactive constituents (vitamins, minerals, tocopherols, carotenoid, organic and phenolic acids) with nutritional and pharmacological (antioxidant, antitumor, antimicrobial, anti-inflammatory, immune-stimulating) properties (Wasser and Weiss, 1999; Ferreira et al., 2009, 2010; Alves et al., 2012; Heleno et al., 2013; Ajmal et al., 2015). In traditional medicine, true morel is used as a laxative, purgative, emollient, tonic, wound-healing, and to control general weakness and stomach problems (Hobbs, 1995).

Different bioactive compounds, such as proteins (aegerolysins, lectins, ribosome-inactivating proteins and protein inhibitors), ribosomal and non-ribosomal peptides, polysaccharides, peptidoglycans, phenolics, different types of terpenoids (sesquiterpenes, diterpenes, triterpenes), lipids, steroids, alkaloids, lectines, phenolics, polyketides, and others were isolated from wild and cultivated macrofungi (Mizuno et al., 1995; Wasser and Weiss, 1999; Wasser, 2002; Cerigini et al., 2007; Ferreira et al., 2010; De Silva et al., 2013; Grienke et al., 2014; Duru and Çayan, 2015; Kües and Badalyan, 2017). They possess different therapeutic effects, such as antiviral, antibacterial, antifungal, antioxidative, immunomodulatory and immunosuppressive, anti-inflammatory, cytotoxic, hypotensive, hepatoprotective, hypoglycemic and hypocholesterolemic and mitogenic/regenerative (Wasser and Weiss, 1999; Wasser, 2002, 2010, 2011; Poucheret et al., 2006; Stanikunaite et al., 2007; Baggio et al., 2010; Palaciois et al., 2011; Badalyan, 2012; Chang and Wasser, 2012; Patel, 2012; Saltarelli et al., 2009, 2015; Bandara et al., 2015, 2017; Fu et al., 2016; Phan et al., 2017; Kolundžić et al., 2017; Chen et al., 2017; Shen et al., 2017; Chaiyasut and Sivamaruthi, 2017) (Table 1). However, further clinical survey to substantiate the revealed therapeutic effects is required (Money, 2016; Wasser, 2017; Taofiq et al., 2017a).

About 80–85% of all medicinal mushroom products, such as (1-3)-β-D glucan Lentinan extracted from *L. edodes,* polysaccharide PSK and Krestin extracted from *T. versicolor* are derived from the fruiting bodies either commercially produced or collected from the wild. A smaller percent of mushroom products are obtained from broth cultures, e.g., (1-3)-, (1-6)-β-D-glucan Schizophyllan from *S. commune*, a protein bound polysachcharide complex PSPC from *Tricholoma laboyense*, polysaccharides from fruiting bodies and mycelium of *Hericium erinaceus* and so on.

Mycelia and cultural broth of edible mushrooms *Pleurotus eryngii* and *Suillus bellinii,* as potential alternative sources of bioactive compounds with antioxidative, anti-inflammatory or cytotoxic activities, was recently evaluated (Souilem et al., 2017). However, biotechnological cultivation of mycelia is advancing and production of mycelial biotech products from macrofungi is constantly being improved (Moore and Chiu, 2001; Chang and Miles, 2004; Asatiani et al., 2007; Elisashvili, 2012; He et al., 2017).

Extracts from sclerotia of several edible medicinal species, such as *Polyporus umbellatus* and *Pleurotus tuberregium* are also used for manufacturing pharmaceutical products, food supplements, as well as cosmetics and beverages (Badalyan et al., 2008; Bandara et al., 2015). Extractable compounds from inedible species also possess biotechnological potential and are also being studied in the development of health care products with different formulations (Barros et al., 2008; Dai et al., 2009; Reis et al., 2011, 2012; Heleno et al., 2011, 2013, 2015a,b).

Nowadays, there is a growing demand for healthier biotech and safe cosmetics products containing natural and/or organic ingredients. The use of natural bioactive molecules, such as vitamins, peptides and phenolics obtained from plant and mushrooms, as cosmeceutical ingredients are receiving increased attention because of their multifunctional benefits on skin (nutritive, antioxidative, antityrosinase, anti-hyaluronidase, anticollagenase, anti-elastase, anti-pigmentaion and photoprotective, antimicrobial and anti-inflammatory). The anti-inflammatory, regenerative, moisturizing and antioxidative properties of mushrooms and their extracts (and/or ingredients) make prospective their application in manufacturing of cosmeceuticals and nutricosmetics to inhibit collagenase, elastase and tyrosinase enzymes associated with inflammatory diseases, wrinkles, aging and hyperpigmentation (Wu et al., 2016; Taofiq et al., 2016a, 2017a; Chandrawanshi et al., 2017).

Although mushrooms have been studied in terms of nutritional value and medicinal properties, there is still slow progress in their exploitation as natural cosmeceutical ingredients (Taofiq et al., 2015; 2016a–c; 2017a; Wu el al., 2016). However, revealed medicinal properties of mushrooms reinforce the interest to their further usage as cosmetic ingredients. Phenolics, polyphenolics, terpenoids, selenium, polysaccharides, vitamins and other fungal bioactive molecules with antioxidative, anti-aging, anti-wrinkle, anti-pigmentation, skin whitening, and moisturizing effects are particularly beneficial in formulation of skin and hair care bioproducts (Wu et al., 2016; Taofiq et al., 2017a). Other potential mycochemicals, such asceramides, lentinan, schizophyllan, omega 3, 6, and 9 fatty acids, carotenoids, resveratrol, alkaloids, glycosides, steroids, saponins, tannins and flavonoids have also potential to be used in formulations of different cosmetic products (Chiu et al, 2000; Badalyan et al., 2007; Hyde et al., 2010; Camassola, 2013; Yildiz et al., 2015; Wu et al, 2016).

In recent years, use of mushroom extracts as a source of multifunctional ingredients in the production of organic cosmetic products for topical (cosmeceuticals) and oral (nutricosmetics) applications is increasing. Mycocosmetics can be used to control skin hyperpigmentation, suppress inflammation and prevent photoaging processes (Taofiq et al., 2016a,c; Wu et al., 2016). Extracts and bioactive compounds obtained from several species (e.g., *L. edodes, G. lucidum, F. fomentarius, Inonotus obliquus, Tremella fuciformis, P. ostreatus* and *S. commune*) as organic ingredients are either used or patented to be used in mycocosmetology (Liu et al., 2016; Wu et al., 2016).

Natural ingredients for skin and hair care are also becoming popular due to their protective role to reduce production of different enzymes (collagenase, elastase, tyrosinase and oxidative), which cause degradation of skin and catalyzes the most important step in melanin biosynthesis (Tamsyn et al., 2009; Lee et al., 2013; Yan et al., 2014; Chandrawanshi et al., 2017). The cosmetic industry is in search of natural inhibitors of these enzymes used as effective ingredients in anti-aging cosmetic products (creams, lotions, or ointments) without allergic/toxic side effects (Soto et al., 2015). Mycochemical analysis, evaluation of bioactivity and cosmetic potential of macrofungi from different ecological groups showed that phenolic acids and ergosterol particularly possess anti-tyrosinase, anti-hyaluronidase, anti-collagenase and anti-elastase activity. These compounds can be used to design cosmetic formulations and products with antioxidative, anti-aging, antibacterial, anti-inflammatory and anti-pigmentation effects (Liao et al., 2014; Wu et al., 2016; Taofiq et al., 2015; 2017a,b). Several proteins, L-ergothioneine, polysaccharides, phenolic acids, ergosterol, vitamins, and minerals with antioxidative (Cheah and Halliwell, 2012; Zembron-Lacny et al., 2013; Stefan et al., 2015), anti-inflammatory (Mizuno et al., 2009; Fernandes et al., 2015; Taofiq et al., 2015) immunomodulatory, antimicrobial, and antitumor effects (Facchini et al., 2014) have been reported in edible medicinal mushrooms *Agaricus bisporus, P. ostreatus,* and *L. edodes.* Antioxidative activity, due to phenolic compounds, was detected in *A. bisporus, B. edulis, Calocybe gambosa, C. cibarius, Craterellus cornucopioides, Hygrophorus marzuolus, Lactarius deliciosus* and *P. ostreatus* (Palaciois et al., 2011). As a natural source of bioactive ingredients, these species can be incorporated in the production of organic cosmetic products. For example, edible and inedible mushrooms, such as *L. edodes, G. frondosa, G. lucidum, Wolfiporia extensa, O. sinensis, Sparassis latifolia,*and jelly *Tremella* spp. (*T. fuciformis* and *T. mesenterica*) are used in topical creams, lotions, serums and other products in Eastern and Western countries, while *A. bisporus, P. ostreatus, Hypsizygus ulmarius* and *F. fomentarius*, are more recognized in Western countries (van Griensven, 2009; Wani et al., 2010; Choi et al., 2014). Other popular mushrooms used in cosmetic products are *Agaricus subrufescens, C. comatus, M. esculenta, H. erinaceus, Mycoleptodonoides aitchisonii, Phellinus linteus, S. commune* and *Volvareilla volvacea* (Bernás et al., 2006; Wani et al., 2010; Meng et al., 2010; Deepalakshmi and Mirunalini, 2011; Chandrasekaran et al., 2012; Mortimer et al., 2012; Llarena-Hernández et al., 2015).

Thus, fruiting bodies and mycelia of wild and cultivatable mushrooms, as well as their extracts, are representing excellent natural sources of bioactive compounds to produce safe cosmetic products against skin aging, inflammation, and hyperpigmentation.

Nowadays, many mushroom-based skin and hair care organic products of American (Dr. Weil For Origins™ Mega-Mushroom, Revlon, Aveeno and Osmia), European (Yves Saint Laurent, Yves Rocher; Estée Lauder, One Love Organics and La Bella Figura) and Asian (Sulwhasoo, Hankook Sansum and Sekkisei) companies containing extracts and compounds from *L. edodes, O. sinensis, G. lucidum* and other macrofungi are commercially available in the market (Wu et al., 2016; Taofiq et al., 2016a). Several species (*L. edodes, Lyophyllum decastes, Ophiocordyceps bassiana, P. eryngii, H. ulmarius* and *G. frondosa)* are used in skin renewal and against atopic dermatitis, Reishi mushrooms (*Ganoderma* spp.) and *S. commune* in anti-aging and skin lifting, *O. sinensis, Fomes officinalis, Flammulina velutipes* and *Tremella* species in skin moisturizing and anti-melanin cosmetic products (Wu et al., 2016). Further advances in fungal biology, genomics, proteomics, metabolomics and system pharmacology will open new perspectives of development of mycocosmetology.

Edible and medicinal mushrooms are also used to develop organic food supplements for pets which contains not only medicinal plants and other ingredients but also dietary fibers and polysacharides isolated from mycelial biomass. These products obtained from *L. edodes, G. frondosa, H. erinaceus, P. eryngii, T. versicolor, A. subrufescens, Agaricus blazei, O. sinensis, I. obliquus* and *G. lucidum* were reported as stimulating immune activity in animals (http://www.alohamedicinals. com/).

The standardization of mushroom biotech-products is still in its early stages. The most important problems of analyzing and manufacturing such products are the lack of set standards and specific protocols that would assure the quality of mushroom products, insufficient understanding of their bioactivity effects and others (Wasser et al., 2000; Wasser, 2011).

Advances in the Cultivation of Edible and Medicinal Macrofungi

The first macrofungus to be cultivated was *Auricularia auricula-judae,* in China in 600 AD. The next to be cultivated were *L. edodes* and *F. velutipes* in 1100 AD and *A. bisporus* in 1650 AD. (Miles and Chang, 1997; Chang and Miles, 1989, 2004). In the past, the mushroom-based industry only involved the production of fresh or dried fruiting bodies as food. However, during the last three decades, there has been a dramatic increase in the interest in production of mushrooms also as pharmaceuticals and cosmeceuticals (Fig. 3a).

In 2012 the world's total edible and medicinal mushrooms production was estimated at over 31 million tonnes, which was valued at 20 billion US dollars (Chang and Wasser, 2012). The cultivation methods can involve a relatively simple farming activity, as with *V. volvacea* and *Pleurotus pulmonarius* var. *stechangii* (=*P. sajorcaju*), or a high-technology industry, as with *A. bisporus, F. velutipes*, and *Hypsizygus marmoreus*. In China, for example, mushroom production is an integrative farming activity to promote sustainable development in rural communities. Mushroom production is often obtained in open space or in rudimental greenhouses using as substrate local agricultural or forest waste (Zhang et al., 2014) (Fig. 3b). On the other

Fig. 3. Prof. Wei Ping Xiong is showing a *Ganoderma leucocontextum* cultivation in Tibet (China) (Ian Hall picture) (a). Field cultivation of *Pleurotus pulmonarius* var. *stechangii* in the rural area of Panzhihua (Sichuan province, China) (b). Highly technological cultivation of *Agaricus bisporus* in Holland (c and d).

hand, there are highly technological mushroom farms, in China as well in many other developed countries, where mushrooms are cultivated in climate-controlled rooms and were most of the processes are automatized (Fig. 3c,d). In each case, however, continuous production of successful crops requires both practical experience and scientific knowledge.

Although the nutritional value of edible mushrooms has been well established and medicinal properties are gradually obtaining scientific value, the existing problems in biotechnological cultivation of fruiting bodies and mycelia, particularly mycorrhiza-forming ascomycetous and basidiomycetous species, is still limiting their practical usage (Chang, 2006).

Current trends in biotechnological cultivation of macrofungi are:

a) fruiting bodies (mushrooms) production

b) mycelium cultivation.

The submerged cultivation of mycelium has significant industrial potential and it is the best technique for obtaining consistent and safe biotech products. However, there are other advantages, as well. The submerged cultivation of mycelia takes several days, whereas mushrooms production is a long-term (1–2 months) process; mycelial cultures can be established for almost all known species, while only a limited (about 50) number of species is possible to cultivate as fruiting bodies. Widely used medicinal mushrooms (*G. lucidum, G. frondosa, T. versicolor, S. commune* and *P. ostreatus*) are saprotrophs and cultivated in large scale on wood, straw and other plant residues for mushroom production (Chang and Miles, 2004; Chang, 2006). However, mycelia of these species may also be produced in submerged fermentation to obtain certain therapeutic compounds, such as anti-tumor β-glucans (Asatiani et al., 2007; Bisen et al., 2010; Elisashvili, 2012).

Research into devising methods for the artificial or semi-artificial cultivation of highly prized edible mycorrhizal mushrooms is gathering momentum because due to the uncontrolled harvesting their abundance in natural environments is being affected. However, few species of ectomycorrhizal fungi have been successfully cultivated in the field till now (Wang and Hall, 2004; Masaphy, 2010). Fermentation technologies are an alternative method for obtaining mycelial biomass for production of functional compounds (Tang et al., 2008). However, it should be mentioned that growth on artificial media of symbiotic species is poor (Iotti et al., 2005; Barros et al., 2006) and large-scale production of bioactive compounds cannot be fully expected (Clericuzio et al., 2006).

Mycorrhization with their host trees ("domestication") has growing interest for production of ectomycorrhizal species to ensure more reliable and enhanced yields of highly valuable edible and medicinal mushrooms. Such approach will prevent commercial overharvest of mycorrhizal species in order to protect their natural resources (Hall et al., 2003; Stamets, 2005; Badalyan et al., 2005; Kües and Martin, 2011; Savoie and Largeteau, 2011). Currently, *T. melanosporum* and *T. aestivum* are commercially cultivated with success in several countries. Some progress has also been made in the cultivation of other ascomycetes, such as *T. borchii, Tuber macrosporum, Tuber indicum, Terfezia claveryi* and related species and of basidiomycetes including *L. deliciosus, Lyophyllum shimeji, Rhizopogon rubescens,*

Table 1. The most important biotechnologically cultivated and exploited macrofungi (*: ECM, ectomycorrhizal; SH, saprobic humicolous; SL, saprobic litter decaying; SW, saprobic wood decaying; **, Reference in bold indicates attempts for cultivation).

Species	Trophic category*	Used parts	Prevalent biotechnological interests	Bioactive compounds	Medicinal or cosmetic effects	References**
Agaricus bisporus (J.E. Lange) Imbach	SH	Fruiting bodies, mycelium	Food, medicines, cosmetics	Alkaloids, polysaccharides, phenols and polyphenols, proteins and aminoacids, saponins, tannins	Antimicrobial, anti-inflammatory, antioxidant, anticancer	Sánchez, 2004; Palaciois et al., 2011; Taofiq et al., 2015; Wu et al., 2016
Agaricus blazei Murrill	SH	Fruiting bodies	Food, medicines	Polysaccharides	Anticancer, anti-inflammatory, immune-modulating, hypoglycemic, hepatoprotective,	Padilha et al., 2009; González et al., 2010; Lin et al., 2012; Førland et al., 2011
Auricularia auricula-judae (Bull.) Quél.	SW	Fruiting bodies	Food, medicines, cosmetics	Polysaccharides, minerals, amino acids	Anti-tumor, blood anti-coagualting, hypoglycemic, hypocholesterolemic, anti-inflammatory, anti-fatigue, antioxidant, antiviral	Sanchez, 2004; Nguyen et al., 2012; Ohiri and Bassey, 2017
Coprinus comatus (O.F. Müll.) Pers.	SH	Fruiting bodies, mycelium	Food, medicines, cosmetics	Polysaccharides, alkaloids, phenolic compounds, proteins, amino acids, unsaturated fatty acids	Antitumor, antimicrobial, anti-inflammatory, antioxidant, hypoglycemic, hypocholesterolemic, anti-obesity, antiandrogenic modulator	Yang et al., 2003; Sánchez, 2004; Dotan et al., 2010; Ren et al., 2012; Wu et al., 2016; Badalyan, 2016; Tešanović et al., 2017
Flammulina velutipes (Curtis) Singer	SW	Fruiting bodies, mycelium	Food, medicines, cosmetics	glycosides	Anti-inflammatory, antioxidant	Sánchez, 2004; Wu et al., 2016
Ganoderma lucidum complex	SW	Fruiting bodies, mycelium	Medicines, cosmetics	Polysaccharides, terpenoids, lanostanetriterpenes, saponins, phenols and polyphenols	Anticancer, antioxidant, anti-inflammatory, hepatoprotective, immune-modulating, collagen elasticity, skin anti-aging, hair protection, anti-fade	Saltarelli et al., 2009; Wang et al., 2015; Deepalakshmi and Mirunalini, 2011; Ferreira et al., 2015 ; Wu et al., 2016; Liu et al., 2016 ; Zhou, 2017

Species				Compounds	Bioactivities	References
Ganoderma tsugae Murrill	SW	Fruiting bodies, mycelium	Medicines	Polysaccharides, terpenoids	Antitumor, immune-stimulating, wound healing	Zhang et al., 1994; Mizuno et al., 1995; Su et al., 1999
Grifola frondosa (Dicks.) Gray	SW	Fruiting bodies	Medicines, cosmetics	Polysaccharides, glycosides	Anticancer, immune-modulating, anti-inflammatory, antioxidant, hypoglycemic, hypocholesterolemic, anti-obesity, hypotensive	Chen, 1999; Wu et al., 2016
Hericium erinaceus (Bull.) Pers.	SW	Fruiting bodies, mycelium	Medicines, cosmetics	Polysaccharadies, triterpens hericenone and erinacine, steroids	Anti-inflammatory, antioxidative,immune-stimulating, anti-mutagenic, hypoglycemic, anticarcinogenic, antidiabetic, antifatigue, antihypertensive, antihyperlipodemic, cardioprotective, hepatoprotective, nephroprotective, neuroprotective	Oei, 2003; Friedman, 2015
Hypsizygus marmoreus (Peck) H.E. Bigelow	SW	Fruiting bodies	Medicines, cosmetics	Glycosides	Anti-inflammatoy, antioxidant, antitumor, hypocholesterolemic	Nakamura, 2006; Wu et al., 2016
Lactarius deliciosus (L.) Gray	ECM	Fruiting bodies	Food, medicines	Flavonoids, sesquiterpenoids, lectins, phenolic compounds	Antibacterial, antifungal, cytotoxic, anti-inflammatory, insecticidal, nematocidal, antioxidant	Chaumont and Simeray, 1982; Wang et al., 2002; Ferreira et al., 2007; Queirós et al., 2009; Ferreira et al., 2009; Palacios et al., 2011; Badalyan, 2012; Guerin-Laguette et al., 2014; Wu et al., 2016

Table 1 cont. ...

...Table 1 cont.

Species	Trophic category*	Used parts	Prevalent biotechnological interests	Bioactive compounds	Medicinal or cosmetic effects	References**
Lentinula edodes (Berk.) Pegler	SW	Fruiting bodies, mycelium	Food, medicines, cosmetics	Polysaccharides, glucan lentinan, L-ergothioneine, flavonoids, glycosides, phenols and polyphenols, proteins and amino acids, tannins phenolic acids, ergosterol	Antitumor, immune-modulatory, antimicrobial, anti-inflammatory, antioxidant, moisturizer, anti-aging, skin brightness	Sánchez, 2004; Mizuno et al., 2009; Tepwong et al., 2011; Cheah and Halliwell, 2012; Wu et al., 2016
Lyophyllum shimeji (Kawam.) Hongo	ECM	Fruiting bodies	Gourmet food, medicines	Polysaccharides	Antimicrobial	Yamanaka, 2008; Wu et al., 2016
Morchella spp.	SL/ECM	Fruiting bodies	Gourmet food, medicines	Galactomannan (a-D-glucan), phenolic compounds, polysaccharides, organic acids, tocopherols, carotenoid	Immune-modulating, antioxidant, antiallergenic, antimicrobial anti-inflammatory, antitumor, laxative, purgative, emollient, tonic, wound-healing, for general weakness and stomach problems	Hobbs, 1995; Duncan et al., 2002; Ferreira et al., 2009; 2010; Alves et al., 2012; Heleno et al., 2013; Ajmal et al., 2015; Liu et al., 2017; Tietel and Masaphy, 2017
Pholiota nameko (T. Itô) S. Ito & S. Imai	SW	Fruiting bodies	Food, medicines, cosmetics	Glycosides, polysaccharides	Anti-inflammatory, antioxidant, hypolipidemic	Gizaw, 2015; Li et al., 2010; Wu et al., 2016

Pleurotus eryngii (DC.) Quél.	SW	Fruiting bodies, mycelium and cultural broth	Food, medicines	Steroids, polysaccharides, terpenoids, polyphenols	Anticancer, antimicrobial, antiviral, anti-inflammatory,anti-allergic,immune-modulating, hypolipidemic antioxidant, cytotoxic	Zervakis, 2005; Wu et al., 2016; Fu et al., 2016; Souilem et al., 2017
Pleurotus ostreatus (Jacq.) P. Kumm.	SW	Fruiting bodies	Food, medicines, cosmetics	Alkaloids, polysaccharides, phenols and polyphenols, saponins	Antimicrobial, anti-inflammatory, antioxidant, antitumor	Sánchez, 2004; Palaciois et al., 2011; Facchini et al., 2014; Stefan et al., 2015;Taofiq et al., 2015; Taofiq et al., 2016c; Wu et al., 2016
Pleurotus pulmunarius (Fr.) Quél.	SW	Fruiting bodies	Food, medicines	Polysaccharides	Analgesic or antinociceptive activity	Baggio et al., 2010; Liang et al., 2011
Pleurotus sajor-caju (Fr.) Singer	SW	Fruiting bodies	Food, Medicines, cosmetics	Polysaccharides pleurane and β-glucan, phenolic acids, tocopherols, polyunsaturated fatty acids, organic acids, proteoglycans, phenolic acids, terpenes, proteins,sterols	Antioxidant,antibacterial,cytotoxic, anti-inflammatory, antiviralantifungal,anti-parasitic, antitumor, antihypertensive,antidiabetic	Chang and Wasser, 2012; Kanagasabapathy et al., 2012; Pokhrel et al., 2013; Alves et al., 2013; Teplyakova and Kosogova, 2016; Ademola and Odeniran, 2017; Finimundy et al., 2018
Polyporus umbellatus (Pers.) Fr.	SW	Fruiting bodies, mycelium, sclerotia	Medicines, cosmetics	Polysaccharides, glucans, Ergone, polyporusterones, phenolics, steroids, fattyacids	Antioxidant, diuretic, anti-inflammatory, antitumor, anticancer, immune-modulating, hair growth, antiviral, antibacterial, anti- protozoal	Bandara et al., 2015

Table 1 cont. ...

...Table 1 cont.

Species	Trophic category*	Used parts	Prevalent biotechnological interests	Bioactive compounds	Medicinal or cosmetic effects	References**
Schizophyllum commune Fr.	SW	Fruiting bodies	Medicines, cosmetics	Glucan schizophyllan	Antimicrobial, antitumor, immune-modulating, antioxidant, anti-aging	Wasser, 2002; Mirfat, et al., 2014; Wu et al., 2016; Dasanayaka and Wijeyaratne, 2017; Chandrawanshi et al., 2017
Stropharia rugosoannulata Farl. ex Murrill	SW	Fruiting bodies, mycelium	Food, medicines	Steroids, strophasterols, polysaccharides	Nematocitic, antifungal, anti-inflammatory, growth regulatory, hypoglycemic, hypocholesterolemic	Oei, 2003; Luo et al., 2006; Wu et al., 2013; Zhai et al., 2013
Terfezia boudieri Chatin	ECM	Fruiting bodies	Gourmetfood, medicines	Phenolic compounds	Antioxidant, antimicrobial	Slama et al., 2010; Dundar et al., 2012; Shavit and Shavit, 2014
Terfezia claveryi Chatin	ECM	Fruiting bodies	Gourmet food, medicines	Peptide antibiotic	Antibacterial, antimutagenic, anticarcinogenic, antioxidant	Janakat et al., 2005; Badalyan, 2012; Morte et al., 2012
Trametes versicolor (L.) Lloyd	SW	Fruiting bodies, mycelium	Medicines	Peptitoglucans, polysaccharides PSK and Krestin, polypeptides	Antimicrobial, immune-modulating, antitumor, anticancer	Wasser and Weiss, 1999; Chu et al, 2002; Veena and Pandey, 2012
Tuber aestivum Vittad.	ECM	Fruiting bodies, mycelium	Gourmetfood, medicines	Phenolic acids (hydroxycinnamic acidderivatives, o- and p-coumaric acids), flavonoids, ergosteryl ester	Antioxidant, mutagenic and antimutagenic properties	Hall et al., 2007; Fratianni et al., 2007; Villares et al., 2012; Patel, 2012
Tuber borchii Vittad.	ECM	Fruiting bodies, mycelium	Gourmetfood, medicines	Lectins cyanovirin-N (CVN)and TBF-1	Antiviral	Percudani et al., 2005; Cerigini et al., 2007; Hall et al., 2007; Patel, 2012; Iotti et al., 2016

Tuber indicum complex	ECM	Fruiting bodies, mycelium	Gourmetfood, medicines	Polysaccharides, phenolics, flavonoids, ergosterol, tuberoside, polyhydroxylated ergosterol glucoside, androstenol, ceramides	Antioxidant and radical scavenging activities, immune-modulating, antitumor,	Luo et al., 2011; Jinming et al, 2001; Tang et al., 2008; Liu et al., 2008; Villares et al., 2012; Patel, 2012; Wang, 2012
Tuber melanosporum Vittad.	ECM	Fruiting bodies, mycelium	Gourmetfood, medicines	Phenolic acids, ergosterol, polysaccharides	Antioxidant, neurotropic	Tardif, 2000; Hall et al., 2007; Liu et al., 2009; Villares et al., 2012; Patel, 2012
Volvariella volvacea (Bull.) Singer	SW	Fruiting bodies	Food, medicines, cosmetics	Alkaloids, carbohydrates, protein and amino acids, tannins	Antimicrobial, anti-inflammatory, antioxidant	Sánchez, 2004; Wu et al., 2016

and *T. matsutake* and allies (Hall et al., 2003; Wang and Hall, 2004; Yamada et al., 2001; 2006; Slama et al., 2010).

Cultivation of Truffles – from the Past to an Industry

The economic value of the truffles has stimulated the researchers to find reliable methods of cultivation since the XVIII century. However only in the XIX century did truffle cultivation become successful when Joseph Talon decided to sow acorns at the foot of truffle-producing oak trees that cultivate the prized fungi. The mycelium of the truffle infecting the roots of the producing oak can spread out on the roots of the seedling which could become mycorrhized with truffle and when transplanted in new sites are able to produce truffles (Hall et al., 2007). Talon's technique is quite empiric and its success depends on the ability of *Tuber* mycelium to infect the roots of the seedlings before the other ectomycorrhizal fungi in the soil. However, Talon's technique remained the only method for cultivating truffles for 170 years and resulted in large harvests of truffles across France.

Only at the end of last century has modern truffle cultivation been introduced elsewhere in Europe. It consists of producing seedlings mycorrhized with truffle in a greenhouse, which are then planted in suitable sites. Modern truffle cultivation was developed in Italy and Spain to begin with, then it was introduced into countries outside of Europe starting with the USA, followed by Israel, New Zealand, Australia, Canada, Argentina, Chile, Morocco, China and South Africa (Kagan-Zur et al. 2001; Hall et al., 2007; Garcia-Montero et al., 2007; Wang, 2012; Berch and Bonito, 2014; Reyna and Garcia-Barreda, 2014; Hall et al., 2017). Although, *Tuber* mycorrhized plants could be produced using different inoculation techniques nowadays all the plants are inoculated with truffle spores. This technique is simple and consists of inoculating seedlings in the nursery with a spore suspension or a spore powder obtained by grinding fresh or preserved ascomata (Hall et al., 2007). It provides good results with most of the edible *Tuber* species (*T. melanosporum, T. aestivum, T. borchii, T. macrosporum* and *T. indicum*) and the inoculated plants are usually well mycorrhized four-six months after inoculation, but not with *T. magnatum*. Plants inoculated with *T. magnatum* spore suspension form few mycorrhizas which often disappear soon after.

Other limiting factors with this method are the high cost of truffles used to produce the inoculum and the risk of introducing different *Tuber* species in the inoculum. In order to avoid this last problem all the truffles have to be morphologically or molecularly checked. Morphological identification includes the microscopic observation of the spore features. Alternatively, molecular techniques can be used with *Tuber* specific primers in multiplex PCR which allows the simultaneous detection of contaminating species (Rubini et al., 1998; Amicucci et al., 2000; Zambonelli and Bonito, 2012).

Another technique is "the mother plant technique" which is a modification of Talon's technique. In this, a plant already mycorrhized with the desired *Tuber* species is placed inside a jar filled with sterile soil, then surrounded by sterile seedlings (Hall et al., 2007). The mycelium of the *Tuber* then grows out from the mother plant and colonizes the seedlings. The limit of this technique is the risk of contamination. The mycelium of ectomycorrhizal species, which often contaminate *Tuber* mycorrhized plants, develops more quickly than the mycelium of *Tuber* spp. which grows very slowly *in vitro* and in soil conditions (Iotti et al., 2002; Gryndler et al., 2015). Thus, these contaminants can take over rendering all the plants contaminated.

A third technique, mycelial inoculation, uses mycelia in pure culture to inoculate the plants (Hall et al., 2007). The difficulties associated with the isolation of truffle mycelium and the slow growth rate of *Tuber* spp. cultures has limited its use commercially. Moreover, after discovering that truffles are heterothallic (Paolocci et al., 2006; Martin et al., 2010), it was assumed that plants individually inoculated with only one mating type would never produce (Zambonelli et al., 2008). For this reason, over the past 20 years *T. borchii* plants raised under controlled conditions in Italian laboratories have focused on plant-fungus molecular interactions (Giomaro et al., 2005). However, in spring 2007, seedlings of *Pinus pinea*, *Quercus pubescens*, *Corlylus avellana* and *Quercus robur* were experimentally inoculated with the *T. borchii* strains Tb98, 2352, 2292, 1Bo and planted later that year (Iotti et al., 2016). The first truffles were harvested in 2016, confirming that the technique was a distinct option for the future. In this plantation the two *T. borchii* mating types were able to cross from one block to the other allowing fertilization and fruiting body production (Leonardi et al., unpub. obs.).

The use of mycelial inoculation techniques commercially would eliminate contaminants being introduced during the inoculation process. The application of this method will also allow for the selection of pairs of cultures carrying compatible mating types tailored to specific combinations of host, soils, climate, and yield (Zambonelli et al., 2015). Indeed, it is already known that some *T. borchii* strains can be more susceptible to high temperatures (Leonardi et al., 2017).

Conclusions and Future Prospects

Mushrooms include around 16,000 known species distributed worldwide (Chang and Wasser, 2017), although many species are yet to be discovered. They represent an important source of foods and bioactive compounds which is largely unexplored. Fewer than 200 species are cultivated commercially, mostly as food resource. However, during the last three decades, there has been a dramatic increase in the interest in mushroom cultivation for their biotechnological exploitation as pharmaceuticals and cosmeceuticals or as gourmet food products. This has not always been a simple task, particularly for the ectomycorrhizal species, like the truffles, which require one to establish a symbiotic relationship with the roots of a suitable host plant in order to produce their fruiting bodies. Significant progresses were obtained from the first empirical attempt of their cultivation (Zambonelli et al., 2015) and, thanks to the recent discoveries regarding truffle biology and genetics, truffle cultivation can be realized by planting mycorrhized plants obtained through mycelial inoculation (Iotti et al., 2016).

Biotechnological exploitation of mushrooms as pharmaceuticals and cosmeceuticals involves not only fruiting body production but also submerged cultivation of mycelia or production in *in vitro* specialized vegetative structures like sclerotia. For large scale biotechnological cultivation of macrofungi, studies of biological, genetic and growth characteristics of mycelia are required (Badalyan et al., 2015; Kües et al., 2016). In this regard establishment and maintenance of cultures collections of different groups of basidiomycetous and ascomycetous macrofungi are of great importance not only to study their biodiversity and distribution, but also for the conservation of their genetic resources, their cultivation, and exploitation (Badalyan et al., 2012; Bisko et al., 2012, 2016; Piattoni et al., 2017; Badalyan and Gharibyan, 2017).

Further research in the biology and biotechnology of edible and medicinal mushrooms, development of new cultivation techniques, as well as advances in genomics, proteomics and metabolomics will assist in the future use of macrofungi with the ultimate aim of producing new mushroom-based biotech products.

Acknowledgements

This chapter is a result of long-standing cooperation between the two authors on mushroom research supported by the University of Bologna, Italy, and Yerevan State University, Armenia, as well as SCS RA research grants (13RF-110 and 15RF-064). The authors are particularly grateful to Ian Hall, for editing the manuscript. Thanks also go to our colleagues and collaborators who contributed to the development of macrofungi research, their biotechnological cultivation and exploitation as sources of valuable and gourmet food, pharmaceuticals and cosmeceuticals.

References

Ademola, I.O. and Odeniran, P.O. 2017. Novel trypanocide from an extract of *Pleurotus sajor-caju* against *Trypanosoma congolense*. Pharmac. Biol., 55: 132–138.

Ahmad, M.F., Ahmad, F.A., Azad, Z., Ahmad, A., Alam, M.I., Ansari, J.A. and Panda, B.P. 2013. Edible mushrooms as health promoting agent. Adv. Sci. Focus, 1: 189–196.

Ajmal, M., Akram, A., Ara, A., Akhund, S. and Nayyar, B.G. 2015. *Morchella esculenta*: An edible and health beneficial mushroom. Pak. J. Food Sci., 25: 71–78.

Alves, M.J., Ferreira, I.C.F.R., Dias, J., Teixeira, V., Martins, A. and Pintado, M.A. 2012. Review on antimicrobial activity of mushroom (Basidiomycetes) extracts and isolated compounds. Planta Med., 78: 1707–1718.

Alves, M.J., Ferreira, I.C.F.R., Dias, J., Teixeira, V., Martins, A. and Pintado, M. 2013. A review on antifungal activity of mushroom (Basidiomycetes) extracts and isolated compounds. Curr. Topics Med. Chem., 13: 2648–2659.

Amicucci, A., Zambonelli, A., Potenza, L. and Stocchi, V. 2000. Multiplex PCR for the identification of white *Tuber* species. FEMS Microbiol. Lett., 189: 265–269.

Asatiani, M.D. Elisashvili, E.I., Wasser, S.P., Reznick, A.Z. and Nevo, E. 2007. Free-radical scavenging activity of submerged mycelium extracts from higher basidiomycetes mushrooms. Biosci. Biotechnol. Biochem., 71: 3090–3092.

Badalyan, S.M., Hovsepyan, R.A., Iotti, M. and Zambonelli, A. 2005. On the presence of truffles in Armenia. Flora Mediterrenea, 15: 683–692.

Badalyan, S.M., Gharibyan, N.G. and Kocharyan, A.E. 2007. Perspectives of the usage of bioactive substances of medicinal mushrooms in pharmaceutical and cosmetic industries. Int. J. Med. Mushrooms, 9: 275–276.

Badalyan, S.M., Isikhuemhen, O.S. and Gharibyan, N.G. 2008. Antagonistic/antifungal activity of *Pleurotus tuberregium* (Fr.) Singer against selected fungal pathogens. Int. J. Med. Mushrooms, 10: 155–162.

Badalyan, S.M. 2012. Medicinal aspects of edible ectomycorrhizal mushrooms. pp. 317–334. *In*: Zambonelli, A. and Bonito, G. (eds.). Edible Ectomycorrhizal Mushrooms, Current Knowledge and Future Prospects, Volume 34. Springer-Verlag, Berlin/Heidelberg.

Badalyan, S.M., Gharibyan, N.G., Iotti, M. and Zambonelli, A. 2012. Morphological and genetic characteristics of different collections of *Ganoderma* P. Karst. species. Proceedings of the 18th Congr. ISMS, Beijing, China, 247–254.

Badalyan, S.M., Shnyreva, A.V., Iotti, M. and Zambonelli, A. 2015. Genetic resources and mycelial characteristics of several medicinal polypore mushrooms (Higher Basidiomycetes). Int. J. Med. Mushrooms, 17: 371–384.

Badalyan, S.M. 2016. Fatty acid composition of different collections of coprinoid mushrooms (Agaricomycetes) and their nutritional and medicinal values. Int. J. Med. Mushrooms, 18: 883–893.

Badalyan, S.M. and Gharibyan, N.G. 2017. Characteristics of Mycelial Structures of Different Fungal Collections. YSU Press, Yerevan.

Baggio, C.H., Freitas, C.S., Martins, D.F., Mazzardo, L., Smiderle, F.R., Sassaki, G.L., Lacomini M., Marques, M.C. and Santos, A.R. 2010. Antinociceptive effects of $(1\to3),(1\to6)$-linked β-glucan isolated from *Pleurotus pulmonarius* in models of acute and neuropathic pain in mice: Evidence for a role for glutamatergic receptors and cytokine pathways. Journal of Pain, 11: 965–971.

Bandara, A.R., Rapior, S., Bhat, D.J., Kakumyan, P., Chamyuang, S., Xu J. and Hyde, K.D. 2015. *Polyporus umbellatus,* an edible-medicinal cultivated mushroom with multiple developed health-care products as food, medicine and cosmetics: A review. Cryptogam. Mycol., 36: 3–42.

Bandara, A.R., Karunarathna, S.C., Mortimer, P.E., Hyde K.D., Khan, S., Kakumyan, P. and Xu, J. 2017. First successful domestication and determination of nutritional and antioxidant properties of the red ear mushroom *Auricularia thailandica* (Auriculariales, Basidiomycota). Mycol. Progress, 16: 1029–1039.

Barros, L., Calhelha, R.C., Vaz, J.A., Ferreira, I.C.F.R., Baptista, P. and Estevinho, L.M. 2006. Antimicrobial activity and bioactive compounds of Portuguese wild edible mushrooms methanolic extracts. Eur. Food Res. Technol., 225: 151–156.

Barros, L., Cruz, T., Baptista, P., Estevinho, L.M. and Ferreira, I.C.F.R. 2008. Wild and commercial mushrooms as source of nutrients and nutraceuticals. Food Chem. Toxicol., 46: 2742–2747.

Berch, S.M. and Bonito, G. 2014. Cultivation of Mediterranean species of *Tuber* (Tuberaceae) in British Columbia, Canada. Mycorrhiza, 24: 473–479.

Bernás, E., Jaworska, G. and Lisiewska, Z. 2006. Edible mushrooms as a source of valuable nutritive constituents. Acta Sci. Pol. Technol. Aliment., 5: 5–20.

Bisen, P.S., Baghel, R.K., Sanodiya, B.S., Thakur, G.S. and Prasad, G.B. 2010. *Lentinus edodes*: A macrofungus with pharmacological activities. Curr. Med. Chem., 17: 2419–2430.

Bishop, K.S., Kao, C.H.J., Xu, Y., Glucinac, M.P., Paterson, R.R.M. and Ferguson, L.R. 2015. From 2000 years of *Ganoderma lucidum* to recent developments in nutraceuticals. Phytochemistry, 114: 56–65.

Bisko, N.A., Babitskaya, V.G., Bukhalo, A.S., Krupoderova, T.A., Lomberg, M.L., Mikhaylova, O. B., Puchkova, T.A., Solomko E.F. and Shcherba, V.V. 2012. Biological characteristics of medicinal macrofungi in culture. Volume 1 and Volume 2. M.G. Kholodny Inst. Botany NAS of the Ukraine. Kyiv.

Bisko, N.A., Lomberg, M.L., Mytropolska, N.Y. and Mykchaylova, O.B. 2016. The IBK Mushroom Culture Collection. M.G. Kholodny Institute of Botany, NAS of the Ukraine, Alterpress, Kyiv.

Blackwell, M. 2011. The fungi: 1, 2, 3, 5.1 million species? Am. J. Bot., 98: 426–438.

Boa, E.R. 2004. Wild edible fungi a global overview of their use and importance to people. Food and Agriculture Organization of the United Nations, Rome. (http://www.fao.org/docrep/007/ y5489e/ y5489e00.htm).

Bresinsky, A. and Besl, H. 1990. A Colour Atlas of Poisonous Fungi. Wurzburg: Wolfe Publishing Ltd., Germany.

Camassola, M. 2013. Mushrooms - The incredible factory for enzymes and metabolites productions. Ferment. Technol., 2: e117.

Cerigini, E., Palma, F., Buffalini, M., Amicucci, A., Ceccaroli, P., Saltarelli, R. and Stocchi, V. 2007. Identification of a novel lectin from the Ascomycetes fungus *Tuber borchii*. Int. J. Med. Mushrooms, 9: 287.

Chaiyasut, C. and Sivamaruthi, B.S. 2017. Anti-hyperglycemic property of *Hericium erinaceus* – A mini review. Asian Pac. J. Trop. Biomed., 7: 1036–1040.

Chandrasekaran, G., Oh, D.S. and Shin, H.J. 2012. Versatile applications of the culinary-medicinal mushroom *Mycoleptodonoides aitchisonii* (Berk.) Maas G. (Higher Basidiomycetes): A review. Int. J. Med. Mushrooms, 14: 395–401.

Chandrawanshi, K.N., Tandia, D.K. and Jadhav, S.K. 2017. Nutraceutical properties evaluation of *Schizophyllum commune*. Indian J. Sci. Res., 13: 57–62.

Chang, S.T. and Miles, P.G. 1989. Edible mushrooms and their cultivation. Boca Raton, FL: CRC Press.

Chang, S.T. and Miles, P.G. 1992. Mushroom biology - a new discipline. Mycologist, 6: 64–65.

Chang, S.T. 2001. Mushroom production. pp. 74–93. *In*: Doelle, H.W., Rokem, S., Berovic, M. (eds) Encyclopedia of Life Support Systems (EOLSS), Biotechnology, Vol VII. EOLSS Publishers, Oxford.

Chang, S.T. and Buswell, J.A. 2001. Nutriceuticals from mushrooms. pp. 53–73. *In*: Doelle, H.W., Rokem, S., Berovic, M. (eds.). Encyclopedia of Life Support Systems (EOLSS), Biotechnology. Volume VII. EOLSS Publishers, Oxford.

Chang, S.T. and Miles, P.G. 2004. Mushroom: Cultivation, nutritional value, medicinal effect, and environmental impact, 2nd Edition. CRC Press. Boca Raton, FL.

Chang, S.T. 2006. Development of the culinary-medicinal mushrooms industry in China: Past, present, and future. Int. J. Med. Mushrooms, 8: 1–12.

Chang, S.T. and Wasser, S.P. 2012. The role of culinary-medicinal mushrooms on human welfare with a pyramid model for human health. Int. J. Med. Mushrooms, 14: 95–134.

Chang, S.T. and Wasser, S.P. 2017. The Cultivation and Environmental Impact of Mushrooms. *In*: Oxford Research Encyclopedia. Environmental Science - Agriculture and the Environment, Oxford University press USA, 2016, pp. 1–39.

Chaumont, J.P. and Simeray, J. 1982. Les propriétés antifongiques des 225 Basidiomycetes et Ascomycetes vis-á-vis des 7 champignons pathogenes cultivés *in vitro*. Cryptogam. Mycol., 3: 249–259.

Cheah, I.K. and Halliwell, B. 2012. Ergothioneine: Antioxidant potential, physiological function and role in disease. Biochim. Biophys. Acta, 1822: 784–793.

Chen, A.W. 1999. A practical guide for synthetic-log cultivation of the medicinal mushroom *Grifola frondosa* (Dich.: Fr.) S. Fr. Gray (Maitake). Int. J. Med. Mushrooms, 1: 153–168.

Chen, B., Tian, J., Zhang, J., Wang, K., Liu, L., Yang, B., Bao, L. and Liu, H. 2017. Triterpenes and meroterpenes from *Ganoderma lucidum* with inhibitory activity against HMGs reductase, aldose reductase and α-glucosidase. Fitoterapia, 120: 6–16.

Chen, Z., Zhang, P. and Zhang, Z. 2012. Investigation and analysis of 102 mushroom poisoning cases in Southern China from 1994 to 2012. Fungal Divers., 64: 123–131.

Cheung, P.C. 2008. Mushrooms as Functional Foods; John Wiley & Sons: Hoboken, NJ, USA, 1–34.

Cheung, P.C.K. 2010. The nutritional and health benefits of mushrooms. Nutrition Bulletin, 35: 292–299.

Cheung, P.C.K. 2013. Mini-review on edible mushrooms as source of dietary fiber: Preparation and health benefits. Food Science and Human Wellness, 2: 162–166.

Chiu, S.W., Law, S.C., Ching, M.L., Cheung, K.W. and Chen, M.J. 2000. Themes for mushroom exploitation in the 21st century: Sustainability, waste management, and conservation. J. Gen. Appl. Microbiol., 46: 269–282.

Choi, M.H., Han, H.K., Lee, Y.J., Jo, H.G. and Shin, H.J. 2014. *In vitro* anti-cancer activity of hydrophobic fractions of *Sparassis latifolia* extract using AGS, A529, and HepG2 cell lines. J. Mushroom, 12: 304–310.

Chu, K.K., Ho, S.S. and Chow, A.H. 2002. *Coriolus versicolor*: A medicinal mushroom with promising immunotherapeutic values. J. Clin. Pharmacol., 42: 976–984.

Chung, S.T. 2006. The need for scientific validation of culinary-medicinal mushroom products. Int. J. Med. Mushrooms, 8: 187–195.

Clericuzio, M., Tabasso, S., Bianco, M.A., Pratesi, G., Beretta, G., Tinelli, S., Zunino, F. and Vidari, G. 2006. Cucurbitane triterpenes from the fruiting bodies and cultivated mycelia of *Leucopaxillus gentianeus*. J. Nat. Prod., 69: 1796–1799.

Crisan, E.V. and Sands, A. 1978. Nutritional value. pp. 137–165. *In*: Chang, S.T., Hayes, W.A. (eds.). The Biology and Cultivation of Edible Mushrooms. New York Academic Press, USA.

Dai, Y.C., Yang, Z.L., Cui, B.K., Yu, C.J. and Zhou, L.W. 2009. Species diversity and utilization of medicinal mushrooms and fungi in China (Review). Int. J. Med. Mushrooms, 11: 287–302.

Dai, Y.C, Cui, B.K., Si, J., He, S.H., Hyde, K.D., Yuan, H.S., Liu, X.Y. and Zhou, L.W. 2015. Dynamics of the worldwide number of fungi with emphasis on fungal diversity in China. Mycol. Prog., 14: 62.

Dasanayaka, P.N. and Wijeyaratne, S.C. 2017. Cultivation of *Schizophyllum commune* mushroom on different wood substrates. J. Trop. For. Environ., 07: 65–73.

de Mattos-Shipley, K.M.J., Ford, K.L., Alberti, F., Banks, A.M., Bailey, A.M. and Foster, G.D. 2016. The good, the bad and the tasty: the many roles of mushrooms. Stud. Mycol., 85: 125–157.

De Silva, D.D., Rapior, S., Fons, F., Bahkali, A.H. and Hyde, K.D. 2012a. Medicinal mushrooms in supportive cancer therapies: An approach to anti-cancer effects and putative mechanisms of action. Fungal Divers., 55: 1–35.

De Silva, D.D., Rapior, S., Hyde, K.D. and Bahkali A.H. 2012b. Medicinal mushrooms in prevention and control of diabetes mellitus. Fungal Divers., 56: 1–29.

De Silva D.D., Rapior, S., Sudarman, E., Stadler, M., Xu, J., Alias, S.A. and Hyde, K.D. 2013. Bioactive metabolites from macrofungi: ethnopharmacology, biological activities and chemistry. Fungal Divers., 62: 1–40.

Deepalakshmi, K. and Mirunalini, S. 2011. Therapeutic properties and current medical usage of medicinal mushroom: *Ganoderma lucidum*. Int. J. Pharm. Sci. Res., 2: 1922–1929.

Dotan, N., Wasser, S.P. and Mahajna, J. 2010. The culinary-medicinal mushroom *Coprinus comatus,* as a natural antiandrogenic modulator. Integrative Cancer Therapies., 20: 1–12.

Duncan, C.J., Pugh, N., Pasco, D.S. and Ross, S.A. 2002. Isolation of a galactomannan that enhances macrophage activation from the edible fungus *Morchella esculenta*. J. Agric. Food Chem., 50: 5683–5685.

Dundar, A., Yesil, O.F., Acay, H., Okumus, V., Ozdemir, S. and Yildiz, A. 2012. Antioxidant properties, chemical composition and nutritional value of *Terfezia boudieri* (Chatin) from Turkey. Food Sci. Technol. Int., 18: 317–328.

Duru, M.E. and Çayan, G.T. 2015. Biologically active terpenoids from mushroom origin: A review. Rec. Nat. Prod., 9: 456–483.

El Enshasy, H., Elsayed, E.A., Aziz, R. and Wadaan, M.A. 2013. Mushrooms and truffles: Historical biofactories for complementary medicine in Africa and in the Middle East. Evidence-Based Compl. Alt. Med., 620451.

Elisashvili, V. 2012. Submerged cultivation of medicinal mushrooms: Bioprocesses and products (review). Int. J. Med. Mushrooms, 14: 211–239.

Facchini, J.M., Alves, E.P., Aguilera, C., Gern, R.M.M., Silveira, M.L.L., Wisbeck, E. and Furlan, S.A. 2014. Antitumor activity of *Pleurotus ostreatus* polysaccharide fractions on Ehrlich tumor and Sarcoma 180. Int. J. Biol. Macromol., 68: 72–77.

Feeney, M.J., Miller, A.M. and Roupas, P. 2014. Mushrooms-biologically distinct and nutritionally unique: Exploring a "Third Food Kingdom". Nature Today, 49: 301–307.

Fernandes, A., Barros, L., Martins, A., Herbert, P. and Ferreira, I.C.F.R. 2015. Nutritional characterisation of *Pleurotus ostreatus* (Jacq. ex Fr.) P. Kumm. Produced using paper scraps as substrate. Food Chem., 169: 396–400.

Ferreira, I.C.F.R., Baptista, P., Vilas-Boas, M. and Barros, L. 2007. Free-radical scavenging capacity and reducing power of wild edible mushrooms from Northeast Portugal: Individual cap and stipe activity. Food Chem., 100: 1511–1516.

Ferreira, I.C.F.R., Barros, L. and Abreu, R.M.V. 2009. Antioxidants in wild mushrooms. Cur. Med. Chem., 16: 1543–1560.

Ferreira, I.C.F.R., Ferreira, I.C., Vaz, J.A., Vasconcelos, M.H. and Martins, A. 2010. Compounds from wild mushrooms with antitumor potential. Anticancer Agents Med. Chem., 10: 424–436.

Ferreira, I.C.F.R., Heleno, S.A., Reis, F.S., Stojkovic, D., Queiroz, M.J.R.P., Vasconcelos, M.H. et al. 2015. Chemical features of *Ganoderma* polysaccharides with antioxidant, antitumor and antimicrobial activities. Phytochemistry, 114: 38–55.

Finimundy, T.C., Barros, L., Calhelha, R.C., Alves, M.J., Prieto, M.A., Abreu, R.M.V., Dillon, A.J. P., Henriques, J.A.P., Roesch-Ely, M. and Ferreira, I.C.F.R. 2018. Multifunctions of *Pleurotus sajor-caju* (Fr.) Singer: A highly nutritious food and a source for bioactive compounds. Food Chem., 245: 150–158.

Førland, D.T., Johnson, E., Saetre, L., Lyberg, T., Lygren, I. and Hetland, G. 2011. Effect of an extract based on the medicinal mushroom *Agaricus blazei* Murill on expression of cytokines and calprotectin in patients with ulcerative colitis and Crohn's disease. Scand. J. Immunol., 73: 66–75.

Fratianni, F., Luccia A.D., Coppola, R. and Nazzaro, F. 2007. Mutagenic and antimutagenic properties of aqueous and ethanolic extracts from fresh and irradiated *Tuber aestivum* black truffle: A preliminary study. Food Chem., 102: 471–474.

Friedman, M. 2015. Chemistry, nutrition, and health-promoting properties of *Hericium erinaceus* (Lion's Mane) mushroom fruiting bodies and mycelia and their bioactive compounds. J. Agric. Food Chem., 63: 7108–7123.

Fu, Z., Liu, Y. and Zhang, Q. 2016. A potent pharmacological mushroom: *Pleurotus eryngii*. Fungal Genom. Biol., 6: 139.

García-Montero, L.G., Pascual, C., García-Cañete, J., Grande, M.A. and Diaz, P. 2007. Historical summary and reference list of Spanish truffles (*Tuber* spp.) research (1962–2006). N. Z. J. Crop Hortic. Sci., 35: 129–138.

Gioacchini, A.M., Menotta, M., Bertini, L., Rossi, I., Zeppa, S. Zambonelli, A., Piccoli, G. and Stocchi, V. 2005. Solid-phase microextraction gas chromatography/mass spectrometry: A new method for species identification of truffles. Rapid.Commun. Mass Spectrom., 19: 2365–2370.

Giomaro, G.M., Sisti, D. and Zambonelli, A. 2005. Cultivation of Edible ectomycorrhizal fungi by *in vitro* mycorrhizal syntesis. pp. 253–267. *In*: Declerck, S., Strullu, D.G., Fortin, J.A. (eds.). *In vitro* Culture of Mycorrhizas, Soil Biology Volume 4, Springer, Berlin, Heidelberg.

Gizaw, B. 2015. Cultivation and yield performance of *Pholiota nameko* on different agro-industrial wastes. Acad. J. Food. Res., 3: 32–42.

González Matute, R., Figlas, D. and Curvetto, N. 2010. Sunflower seed hull-based compost for *Agaricus blazei* Murrill cultivation. Int. Biodet. Biodegr., 64: 742–747.

Grienke, U., Zöll, M., Peintner, U. and Rollinger J.M. 2014. European medicinal polypores. A modern view on traditional uses. J. Ethnopharmacol., 154: 564–583.

Gryndler, M., Beskid, O., Hršelová. H., Bukovská. P., Hujslová, M., Gryndlerová, H., Konvalinková, T., Schnepf, A., Sochorová, L. and Jansa, J. 2015. Mutabilis in mutabili: Spatio-temporal dynamics of a truffle colony in soil. Soil Biol. Biochem., 90: 62–70.

Guerin-Laguette, A., Cummings, N., Butler, R.C., Willows, A., Hesom-Williams, N., Li, S. and Wang, Y. 2014. *Lactarius deliciosus* and *Pinus radiata* in New Zealand: Towards the development of innovative gourmet mushroom orchards. Mycorrhiza, 24: 511–523.

Guzmán, G. 2015. New studies on hallucinogenic mushrooms: History, diversity, and applications in psychiatry. Int. J. Med. Mushrooms, 17: 1019–1030.

Hall, I.R., Wang, Y. and Amicucci, A. 2003. Cultivation of edible ectomycorrhizal mushrooms. Trends Biotechnol., 21: 433–438.

Hall, I.R., Brown, G.T. and Zambonelli, A. 2007. Taming the truffle—the history, lore and science of the ultimate mushroom. Timber, Portland.

Hall, I.R., Fitzpatrick N., Miros P. and Zambonelli, A. 2017. Counter-season cultivation of truffles in the Southern Hemisphere: an update. Italian Journal of Mycology, 46: 21–36.

Hawksworth, D.L. 1991. The fungal dimension of biodiversity magnitude, significance and conservation. Mycol. Res., 95: 641–655.

Hawksworth, D.L. 2001. Mushrooms: The extent of the unexplored potential. Int. J. Med. Mushrooms, 3: 333–337.

Hawskworth, D.L. 2012. Global species numbers of fungi: Are tropical studies and molecular approaches contributing to a more robust estimate? Biodivers. Conserv., 21: 2425–2433.

He, X., Wang, X., Fang, J., Chang, Y., Ning, N., Guo, H., Huang, L., Huang, X. and Zhao, Z. 2017. Structures, biological activities, and industrial applications of the polysaccharides from *Hericium erinaceus* (Lion's Mane) mushroom: A review. Int. J. Biol. Macromol., 97: 228–237.

Heleno, S.A., Battos, L., Sousa, M.J., Martins, A., Santos-Buelga, C. and Ferreira, I.C.F.R. 2011. Targeted metabolites analysis in wild *Boletus* species. LWT - Food Science and Technology, 44: 1343–1348.

Heleno, S.A., Stojkovic, D., Barros, L., Glamoclija, J., Sokovic, M., Martins, A., Queiroz, M.J.R. and Ferreira, I.C.F.R. 2013. A comparative study of chemical composition, antioxidant and antimicrobial properties of *Morchella esculenta* (L.) Pers. from Portugal and Serbia. Food Res. Int., 51: 236–243.

Heleno, S.A., Ferreira, R.C., Antonio, A.L., Queiroz, M.J.R.P., Barros, L. and Ferreira, I.C.F.R. 2015a. Nutritional value, bioactive compounds and antioxidant properties of three edible mushrooms from Poland. Food Biosci., 11: 48–55.

Heleno, S.A., Barros, L., Martins, A., Morales, P., Fernandez-Ruiz, V., Glamoclija, J. et al. 2015b. Nutritional value, bioactive compounds, antimicrobial activity and bioaccessibility studies with wild edible mushrooms. LWT-Food Science and Technology, 63: 799–806.

Hibbett, D.S. 2007. After the gold rush, or before the flood? Evolutionary morphology of mushroom-forming fungi (Agaricomycetes) in the early 21st century. Mycol. Res., 111: 1003–1020.

Hibbet, D.S. and Taylor, J.W. 2013. Fungal systematics: Is a new age of enlightenment at hand? Nat. Rev. Microbiol., 11: 129–133.

Hobbs, C.R. 1995. Medicinal mushrooms: An exploration of tradition, healing and culture, 2nd Edn. Botanical, Santa Cruz, CA.

Hyde, K.D., Bahkali, A.H. and Moslem, M.A. 2010. Fungi - an unusual source for cosmetics. Fungal Divers., 43: 1–9.

Iotti, M., Amicucci, A., Stocchi, V. and Zambonelli, A., 2002. Morphological and molecular characterisation of mycelia of some *Tuber* species in pure culture. New Phytol., 155: 499–505.

Iotti, M., Barbieri, E., Stocchi, V. and Zambonelli, A. 2005. Morphological and molecular characterisation of mycelia of ectomycorrhizal fungi in pure culture. Fungal Divers., 19: 51–68.

Iotti, M., Piattoni, F., Leonardi, P., Hall, I.R. and Zambonelli, A. 2016. First evidence for truffle production from plants inoculated with mycelial pure cultures. Mycorrhiza, 26: 793–798.

Janakat, S., Al-Fakhiri, S. and Sallal, A.-K. 2005. Aqueous extract of the truffle *Terfezia claveryi* contains a potent antimicrobial agent that is protein in nature and may be used in the treatment of eye infections caused by *P. aeruginosa*. Saudi Med. J., 26: 952–955.

Jinming, G., Lin, H. and Jikai, L. 2001. A novel sterol from Chinese truffles *Tuber indicum*. Steroid, 66: 771–775.

Kagan-Zur, V., Freeman, S., Luzzati, Y., Roth-Bejerano, N. and Shabi, E. 2001. Survival of introduced *Tuber melanosporum* at two sites in Israel as measured by occurrence of mycorrhizas. Plant Soil, 229: 159–166.

Kagan-Zur, V., Roth-Bejerano, N., Sitrit, Y. and Morte, A. 2014. Desert truffles. Soil Biology 38. Dordrecht, Springer.

Kaláč, P. 2013. A review of chemical composition and nutritional value of wild-growing and cultivated mushrooms. J. Sci. Food Agric., 93: 209–218.

Kanagasabapathy, G., Kuppusamy, U.R., Abd Malek, S.N., Abdulla, M.A., Chua, K.-H. and Sabaratnam, V. 2012. Glucan-rich polysaccharides from *Pleurotus sajor-caju* (Fr.) Singer prevents glucose intolerance, insulin resistance and inflammation in C57BL/6Jmice fed a high-fat diet. BMC Complem. Altern. Med., 12: 261–261.

Karun, N.C. and Sridhar, K.R. 2017. Edible wild mushrooms of the Western Ghats: Data on the ethnic knowledge. Data in Brief, 14: 320–328.

Kirk, P.M., Cannon, P.F., Minter, D.W. and Stalpers, J.A. 2008. Ainsworth and Bisby's Dictionary of the Fungi, 10th ed.; Centre for Agriculture and Biosciences International: Wallingford, UK.

Kolundžić, M., Stanojković, T., Radović, J., Tacić, A., Dodevska M., Milenković, M., Sisto, F., Masia, C., Farronato, G., Nikolić, V. and Kundaković, T. 2017. Cytotoxic and antimicrobial activities of *Cantharellus cibarius* Fr. (Cantarellaceae). J. Med. Food, 20: 790–796.

Kües, U. and Martin, F. 2011. On the road to understanding truffles in the underground. Fungal Genet. Biol., 48: 555–560.

Kües, U. and Navarro-González, M. 2015. How do Agaricomycetes shape their fruiting bodies? 1. Morphological aspects of development. Fungal Biol. Rev., 29: 63–97.

Kües, U., Badalyan, S.M., Gießler, A. and Dörnte, B. 2016. Asexual sporulation in Agaricomycetes. The Mycota. Growth, Differentiation and Sexuality, Volume I, 3rd Edition, Jürgen Wendland (ed.). Springer-Verlag, 269–328.

Kües, U. and Badalyan, S.M. 2017. Making use of genomic information to explore the biotechnological potential of medicinal mushrooms. pp. 397–458. *In*: Book series Medicinal and Aromatic Plants of the World, Vol. 4. Medicinal Plants and Fungi - Recent Advances in Research and Development, Volume 2, Springer Nature Singapore.

Lee, J.S., Shin, D.B., Lee, S.M., Kim, S.H., Lee, T.S. and Jung, D.C. 2013. Melanogenesis inhibitory and antioxidant activities of *Phellinus baumii* methanol extract. Korean J. Mycol., 41: 104–111.

Leonardi, P., Iotti, M., Donati Zeppa, S., Lancellotti, E., Amicucci, A. and Zambonelli, A. 2017. Morphological and functional changes in mycelium and mycorrhizas of *Tuber borchii* due to heat stress. Fungal Ecol., 29: 20–29.

Li, H., Zhang, M. and Ma, G. 2010. Hypolipidemic effect of the polysaccharide from *Pholiota nameko*. Nutrition., 26: 556–562.

Li, T., Yang, Z., Chen, Z., Song, B. and Deng, W. 2003. Poisonous mushrooms known from China - species resources and distribution. Fifth International conference on mushroom biology and mushroom products. http://wsmbmp.org/proceedings/5th%20international%20conference/pdf/chapter%2062. pdf.

Liang, Z.C., Wu, K.J., Wang, J.C., Lin, C.H. and Wu, C.Y. 2011. Cultivation of the culinary-medicinal lung oyster mushroom, *Pleurotus pulmonarius* (Fr.) Quél. (Agaricomycetideae) on grass plants in Taiwan. Int. J. Med. Mushrooms, 13: 193–199.

Liao, W.C., Hsueh, C.Y. and Chan, C.F. 2014. Antioxidative activity, moisture retention, film formation, and viscosity stability of *Auricularia fuscosuccinea*, white strain water extract. Biosci. Biotechnol. Biochem., 78: 1029–1036.

Lin, J.G., Fan, M.J., Tang, N.Y., Yang, J.S, Hsia, T.C., Lin, J.J., Lai, K.C., Wu, R.S., Ma, C.Y., Wood, W.G. and Chung, J.G. 2012. An extract of *Agaricus blazei* Murill administered orally promotes immune responses in murine leukemia BALB/c mice *in vivo*. Integr. Cancer Ther., 11: 29–36.

Lindequist, U., Niedermeyer, T.H.J. and Jülich, W.D. 2005. The pharmacological potential of mushrooms. eCAM, 2: 285–299.

Liu, Q.N., Liu, R.S., Wang, Y.H., Mi, Z.Y., Li, D.S., Zhong, J.J. and Tang Y.J. 2009. Fed-batch fermentation of *Tuber melanosporum* for the hyperproduction of mycelia and bioactive *Tuber* polysaccharides. Bioresour Technol., 100: 3644–3649.

Liu, Q., Ma, H., Zhang, Y. and Dong, C. 2017. Artificial cultivation of true morels: Current state, issues and perspectives. Crit. Rev. Biotechnol., 6: 1–13.

Liu, R.S., Li, D.S., Li, H.M. and Tang, Y.J. 2008. Response surface modeling the significance of nitrogen source on the submerged cultivation of Chinese truffle *Tuber sinense*. Process Biochem., 43: 868–876.

Liu, Y., Zhang, J., Tang, Q., Yang, Y., Xia, Y., Zhou, S., Wu, D., Zhang, Z., Dong, L. and Cui, S.W. 2016. Rheological properties of b-D-glucan from the fruiting bodies of *Ganoderma lucidum*. Food Hydrocoll., 58: 120e125.

Llarena-Hernández, R.C., Xavier Vitrac, E.R., Mérillon, J.-M. and Savoie, J.-M. 2015. Antioxidant activities and metabolites in edible fungi, a focus on the almond mushroom *Agaricus subrufescens*. Mérillon, J.-M. and K.G. Ramawat (eds.). Fungal Metabolites: doi:10.1007/978-3-319-19456-1_35-1.

Luo H., Li, X., Li, G., Pan Y. and Zhang, K. 2006. Acanthocytes of *Stropharia rugosoannulata* function as a nematode-attacking device. Appl. Environ. Microbiol., 72: 2982–2987.

Luo, Q., Zhang, J., Yan, L., Tang, Y., Ding, X., Yang, Z. and Sun, Q. 2011. Composition and antioxidant activity of water-soluble polysaccharides from *Tuber indicum*. J. Med. Food, 14: 1609–1616.

Manzi, P., Gambelli, L., Marconi, S., Vivanti, V. and Pizzoferrato, L. 1999. Nutrients in edible mushrooms: an inter-species comparative study. Food Chem., 65: 477–482.

March, R.E., Richards, D.S. and Ryan, R.W. 2006. Volatile compounds from six species of truffle-head-space analysis and vapor analysis at high mass resolution. Int. J. Mass Spectrom., 249: 60–67.

Martin, F., Kohler, A., Murat, C., Balestrini, R., Coutinho, P.M., Jaillon, O., Montanini, B., Morin, E., Noel, B. et al. 2010. Périgord black truffle genome uncovers evolutionary origins and mechanisms of symbiosis. Nature, 464: 1033–1038.

Masaphy, S. 2010. Biotechnology of morel mushrooms: Successful fruiting body formation and development in a soilless system. Biotechnol. Lett., 32: 1523–1527.

Mattila, P., Suonpa, K. and Piironen, V. 2000. Functional properties of edible mushrooms. Nutrition, 6: 694–696.

Meng, F., Zhou, B., Lin, R., Jia, L., Liu, X., Deng, P., Fan, K., Wang, G., Wang L. and Zhang, J. 2010. Extraction optimization and *in vivo* antioxidant activities of exopolysaccharide by *Morchella esculenta* SO-01. Bioresour. Technol., 101: 4564–4569.

Miles, P.G. and Chang, S.T. 1997. Mushroom Biology: Concise Basics and Current Development. World Scientific, Singapore.

Mirfat, A.H.S., Noorlidah, A. and Vikineswary, S. 2014. Antimicrobial activities of split gill mushroom *Schizophyllum commune*. Fr. Amer. J. Res. Commun., 2: 113–124.

Mizuno, M., Nishitani, Y., Hashimoto, T. and Kanazawa, K. 2009. Different suppressive effects of fucoidan and lentinan on IL-8 mRNA expression in *in vitro* gut inflammation. Biosci. Biotechnol. Biochem., 3: 2324–2325.

Mizuno, T., Wang, G., Zhang, J. et al. 1995. Reishi, *Ganoderma lucidum* and *Ganoderma tsugae*: bioactive substances and medicinal effects. Food Rev. Intl., 11: 151–166.

Money, N.P. 2016. Are mushrooms medicinal? Fungal Biol., 120: 449–453.

Moore, D. and Chiu, S.W. 2001. Fungal products as food. pp. 223–251. *In*: Pointing, S.B., Hyde K.D. (eds.). Bio-Exploitation of Filamentous Fungi. Fungal Diversity Press: Hong Kong.

Morte, A., Andrino, A., Honrubia, M. and Navarro-Ródenas, A. 2012. *Terfezia* cultivation in arid and semiarid soils. pp. 241–263. *In*: Zambonelli, A. and Bonito, G.M. (eds.). Edible Ectomycorrhizal Mushrooms, Soil biology, Volume 34. Berlin, Springer-Verlag, Heidelberg.

Mortimer, P.E., Karunarathna, S.C., Li, Q., Gui, H., Yang, X. Yang, X. and Hyde, K.D. 2012. Prized edible Asian mushrooms: Ecology, conservation and sustainability. Fungal Divers., 56: 31–47.

Mueller, G.M., Schmit, J.P., Leacock, P.R., Buyck, B. et al. 2007. Global diversity and distribution of macrofungi. Biodivers. Conserv., 16: 37–48.

Nakamura, K. 2006. Bottle cultivation of culinary-medicinal buna-shimeji mushroom *Hypsizygus marmoreus* (Peck) Bigel. (Agaricomycetideae) in Nagano Prefecture (Japan). Int. J. Med. Mushrooms, 8: 179–186.

Nguyen, T.L., Chen, J., Hu, Y., Wang, D., Fan, Y., Wang, J., Abula, S., Zhang, J., Qin, T., Chen, X., Xiaolan Chen, X., Kakaikhakamec, S. and Dang, B.K. 2012. *In vitro* antiviral activity of sulfated *Auricularia auricula* polysaccharides. Carbohydrate Polymers, 90: 1254–1258.

Oei, P. 2003. Mushroom cultivation, appropriate technology for mushroom growers. Leiden: Backhugs Publishers.

Ohiri, R.C. and Bassey, E.E. 2017. Evaluation and characterization of nutritive properties of the jelly ear culinary-medicinal mushroom *Auricularia auricula-judae* (Agaricomycetes) from Nigeria. Int. J. Med. Mushrooms, 19: 173–177.

Owaid, M.N. 2018. Bioecology and uses of desert truffles (Pezizales) in Middle Eastern. WJST, 15(3): 179–188.

Padilha, M.M., Avila, A.A., Sousa, P.J., Cardoso, L.G., Perazzo, F.F. and Carvalho, J.C. 2009. Anti-inflammatory activity of aqueous and alkalin extracts from mushrooms (*Agaricus blazei* Murill). J. Med. Food., 2: 359–364.

Palaciois, I., Lozano, M., Moro, C., D'Arrigo, M., Rostagno, M.A., Martínez, J.A., García-Lafuente, A., Guillamón, E. and Villares, A. 2011. Antioxidant properties of phenolic compounds occurring in edible mushrooms. Food Chem., 128: 674–678.

Paolocci, F., Rubini, A., Riccioni, C. and Arcioni, S. 2006. Reevaluation of the life cycle of *Tuber magnatum*. Appl. Environ. Microbiol., 72: 2390–2393.

Patel, S. 2012. Food, Health and Agricultural Importance of Truffles: A Review of Current Scientific Literature. Curr.Trends Biotechnol. Pharm., 6: 15–27.

Peay, K.G., Kennedy, P.G. and Talbot, J.M. 2016. Dimension of Biodiversity in the Earth mycobiome. Nat. Rev. Microbiol., 14: 434–447.

Peng, W., Chen, Y., Tan, H., Tang, J. and Gan, B. 2015. Artificial Cultivation of Morels is Blooming in Sichuan. WSMBMP Bulletin Number 13.http://wsmbmp.org/Bol13/5.html.

Percudani, R., Montanini, B. and Ottonello, S. 2005. The anti-HIV cyanovirin-N domain is evolutionarily conserved and occurs as a protein module in eukaryotes. Proteins, 60: 670–678.

Phan, C.W., David, P. and Sabaratnam, V. 2017. Edible and medicinal mushrooms: Emerging brain food for the mitigation of neurodegenerative diseases. J. Med. Food, 20: 1–10.

Piattoni, F., Leonardi, P., Siham, B., Iotti, M. and Zambonelli, A. 2017. Viability and infectivity of *Tuber borchii* after cryopreservation. Cryo Letters, 38: 58–64.

Pokhrel, C.P., Kalyan, N., Budathoki, U. and Yadav, R.K.P. 2013. Cultivation of *Pleurotus sajor-caju* using different agricultural residues. Int. J. Agricult. Pol. Res., 1: 19023.

Poucheret, P., Fons, F. and Rapior, S. 2006. Biological and pharmacological activity of higher fungi: 20-year retrospective analysis. Cryptogam. Mycol., 27: 311–333.

Queirós, B., Barreira, J.C.M., Sarmento, A.C. and Ferreira, I.C.F.R. 2009. In search of synergistic effects in antioxidant capacity of combined edible mushrooms. Int. J. Food Sci. Nutr., 60: 160–172.

Reis, F.S., Heleno, S.A., Barros, L., Sousa, M.J., Martins, A., Santos-Buelga, C. and Ferreira, I.C. F.R. 2011. Toward the antioxidant and chemical characterization of mycorrhizal mushrooms from Northeast Portugal. J. Food Sci., 76: 824–830.

Reis, F.S., Martins, A., Barros, L. and Ferreira, I.C.F.R. 2012. Antioxidant properties and phenolic profile of the most widely appreciated cultivated mushrooms: A comparative study between in vivo and in vitro samples. Food Chem. Toxicol., 50: 1201–1207.

Ren, J., Shi, J.-L., Han, Ch.-Ch., Liu, Z.-Q. and Guo, J.-Y. 2012. Isolation and biological activity of triglycerides of the fermented mushroom of *Coprinus comatus*. BMC Complem. Alternat. Medicine, 12: 52–56.

Reyna, S. and Garcia-Barreda, S. 2014. Black Truffle cultivation: a global reality. For. Syst., 23(2): 317–328.

Ribeiro, B., de Pinho, P.G., Andrade, P.B., Baptista, P. and Valentão, P. 2009. Fatty acid composition of wild edible mushrooms species: A comparative study. Microchem. J., 93: 29–35.

Rubini, A., Paolocci, F., Granetti, B. and Arcioni, S. 1998. Single step molecular characterization of morphologically similar black truffle species. FEMS Microbiol. Lett., 164: 7–12.

Saltarelli, R., Ceccaroli, P., Iotti, M., Zambonelli, A., Buffalini, M. Casadei, L., Vallorani, L. and Stocchi, V. 2009. Biochemical characterisation and antioxidant activity of mycelium of *Ganoderma lucidum* from Central Italy. Food Chem., 116: 143–151.

Saltarelli, R., Ceccaroli, P., Buffalini, M., Vallorani, L., Casadei, L., Zambonelli, A., Iotti, M., Badalyan, S.M. and Stocchi, V. 2015. Biochemical characterization, antioxidant and antiproliferative activities of different *Ganoderma* collections. J. Mol. Microbiol. Biotechnol., 25: 16–25.

Sánchez, C. 2004. Modern aspects of mushrooms culture technology. Appl. Microbiol. Biotechnol., 64: 756–762.

Savoie, J.M. and Largeteau, M.L. 2011. Production of edible mushrooms in forests: trends in development of a mycosilviculture. Appl. Microbiol. Biotechnol., 89: 971–979.

Schoch, C.L., Sung, G.H., López-Giráldez, F., Townsend, J.P., Miadlikowska, J. et al. 2009. The Ascomycota tree of life: A phylum-wide phylogeny clarifies the origin and evolution of fundamental reproductive and ecological traits. Syst. Biol., 58: 224–239.

Shavit, E. 2014. The history of desert truffle use. pp. 217–241. *In*: Kagan, Zur, V., Roth Bejerano, N., Sirit, Y. and Morte, A. (eds.). Desert Truffle. Phylogeny, Physiology, Distribution and Domestication. Soil Biology, Volume 38. Springer-Verlag, Berlin, Heidelberg.

Shavit, E. and Shavit, E. 2014. The medicinal value of desert truffles. pp. 323–340. *In*: Kagan Zur, V., Roth Bejerano, N., Sirit, Y., Morte, A. (eds.). Desert Truffle - Phylogeny, Physiology, Distribution and Domestication. Soil Biology, Volume 38. Springer-Verlag, Berlin, Heidelberg.

Shen, H.-S., Shao, S., Chen, J.-C. and Zhou, T. 2017. Antimicrobials from mushrooms for assuring food safety. Compr. Rev. Food Sci. Food. Saf., 16: 316–329.

Slama, A., Fortas, Z., Boudabous, A. and Neffati, M. 2010. Cultivation of an edible desert truffle (*Terfezia boudieri* Chatin). Afr. J. Microbiol. Res., 4: 2350–2356.

Soto, M.L., Falqué, E. and Domínguez, H. 2015. Relevance of natural phenolics from grape and derivative products in the formulation of cosmetics - Review. Cosmetics, 2: 259–276.

Souilem, F., Fernandes, A., Calhelha, R.C., Barreira, J.C.M., Barros, L., Skhiri, F., Martins, A. and Ferreira, I.C.F.R. 2017. Wild mushrooms and their mycelia as sources of bioactive compounds: Antioxidant, anti-inflammatory and cytotoxic properties. Food Chem., 230: 40–48.

Splivallo, R., Ottonello, S., Mello, A. and Karlovsky, P. 2011. Truffle volatiles: From chemical ecology to aroma biosynthesis. New Phytol., 189: 688–699.

Splivallo, R. and Culleré, L. 2016. The smell of truffles: from aroma biosynthesis to product quality. pp. 393–407. *In*: Zambonelli, A., M. Iotti and C. Murat (eds.). True Truffle (*Tuber* spp.) in the

World: Soil Ecology, Systematics and Biochemistry. Soil Biology, Volume 47. Springer-Verlag, Berlin, Heidelberg.

Stamets, P. 2005. Mycelium Running: How mushrooms can help save the world. Ten Speed Press, Berkeley, CA.

Stanikunaite, R., Trappe, J.M., Khan, S.I. and Ross, S.A. 2007. Evaluation of therapeutic activity of hypogeous Ascomycetes and Basidiomycetes from North America. Int. J. Med. Mushrooms, 9: 7–14.

Stefan, R.I., Vamanu, E. and Angelescu, G.C. 2015. Antioxidant activity of crude methanolic extracts from *Pleurotus ostreatus*. Res. J. Phytochem., 9: 25–32.

Su, C.H., Sun, C.S., Juan, S.W., Ho, H.O., Hu, C.H. and Sheu, M.T. 1999. Development of fungal mycelia as skin substitutes: Effects on wound healing and fibroblast. Biomaterials, 20: 61–68.

Tamsyn, S.A.T., Pauline, H. and Declan, P.N. 2009. Anti-collagenase, anti-elastase and anti-oxidant activities of extracts from 21 plants. BMC Complement.Altern. Med., 9: 1–11.

Tang, Y.J., Zhu, L.L., Li, D.S., Mi, Z.Y. and Li, H.M. 2008. Significance of inoculation density and carbon source on the mycelial growth and *Tuber* polysaccharides production by submerged fermentation of Chinese truffle *Tuber sinense*.Process. Biochem., 43: 576–586.

Taofiq, O., Calhelha, R.C., Heleno, S.A., Barros, L., Martins, A., Santos-Buelga, C., Queiroz, M.J. R.P. and Ferreira, I.C.F.R. 2015. The contribution of phenolic acids to the anti-inflammatory activity of mushrooms: Screening in phenolic extracts, individual parent molecules and synthesized glucuronated and methylated derivatives. Food Res. Int., 76: 821–827.

Taofiq, O., González-Paramás, A.M., Martins, A., Barreiro, M.F. and Ferreira, I.C.F.R. 2016a. Mushrooms extracts and compounds in cosmetics, cosmeceuticals and nutricosmetics—A review. Ind. Crops Prod., 90: 38–48.

Taofiq, O., Martins, A., Barreiro, M. F. and Ferreira, I.C.F.R. 2016b. Anti-inflammatory potential of mushroom extracts and isolated metabolites. Trend. Food Sci. Technol., 50: 193–210.

Taofiq, O., Heleno, S., Calhelha, R., Alves, M., Barros, L., Barreiro, M., Gonzalez-Paramas, A. and Ferreira, I.C.F.R. 2016c. Development of mushroom-based cosmeceutical formulations with anti-inflammatory, anti-tyrosinase, antioxidant, and antibacterial properties. Molecules, 21: 1372; doi:10.3390/molecules21101372.

Taofiq, O., Heleno, S.A., Calhelha, R.C., Alves, M.J., Barros, L., Gonzalez-Paramas, A.M., Barreiro, M.F. and Ferreira, I.C.F.R. 2017a. The potential of *Ganoderma lucidum* extracts as bioactive ingredients in topical formulations, beyond its nutritional benefits. Food Chem. Toxicol., 108: 139–147.

Taofiq, O., Fernandes, A., Barros, L., Barreiro, M.F. and Ferreira, I.C.F.R. 2017b. UV-irradiated mushrooms as a source of vitamin D2: A review. Trends Food Sci. Technol., 70: 82–94.

Tardif, A. 2000. La Mycothérapie ou les Propriéteés Médicinales des Champignons. Le Courrier du Livre, Paris.

Tedersoo, L. Bahram, M. Põlme, S., Kõljalg, U., Yorou, N.S., Wijesundera, R., Ruiz, V. et al. 2014. Global diversity and geography of soil fungi. Science, 346: 1256688.

Teplyakova, T.V. and Kosogova, T.A. 2016. Antiviral effect of Agaricomycetes mushrooms (Review). Int. J. Med. Mushrooms, 18: 375–386.

Tepwong, P., Giri, A. and Ohshima, T. 2011. Effect of mycelial morphology on ergothioneine production during liquid fermentation of *Lentinula edodes*. Mycoscience, 53:102–112.

Tešanović, K., Pejin, B., Šibul, P., Matavulj, M., Rašeta, M., Janjuševik, R. and Karaman, M. 2017. A comparative overview of antioxidative properties and phenolic profiles of different fungal origins: Fruiting bodies and submerged cultures of *Coprinus comatus* and *Coprinellus truncorum*. J. Food Sci. Technol., 54: 430–438.

Tietel, Z. and Masaphy, S. 2017. True morels (*Morchella*) - nutritional and phytochemical composition, health benefits and flavor: A review. Crit. Rev. Food Sci. Nutr., 28: 1–14.

Trappe, J.M., Claridge, A.W., Arora, D. and Smit, W.A. 2008a. Desert truffles of the African Kalahari: ecology, ethnomycology, and taxonomy. Econ Bot., 62: 521–529.

Trappe, J.M., Claridge, A.W., Claridge, D.L. and Liddle, L. 2008b. Desert truffles of the Australian outback: Ecology, ethnomycology and taxonomy. Econ Bot., 62: 497–506.

Ulziijargal, E. and Mau, J.L. 2011. Nutrient compositions of culinary-medicinal mushroom fruiting bodies and mycelia. Int. J. Med. Mushrooms, 13: 343–349.

Valverde, M.E., Hernández-Pérez, T. and Paredes-López, O. 2015. Edible mushrooms: Improving human health and promoting quality life. Int. J. Microbiol., Article ID 376387.

van Griensven, L.J. 2009. Culinary-medicinal mushrooms: Must action be taken? Int. J. Med. Mushrooms, 11: 281–286.

Veena, S.S. and Pandey M. 2012. Physiological and cultivation requirements of *Trametes versicolor*, a medicinal mushroom to diversify Indian mushroom industry. Ind. J. Agricult. Sci., 82: 672–675.

Villares, A., García-Lafuente, A., Guillamón, E. and Ramos, A. 2012. Identification and quantification of ergosterol and phenolic compounds occurring in *Tuber* spp. Truffles J. Food Compos. Anal., 26: 177–182.

Vita, F., Taiti, C., Pompeiano, A., Bazihizina, N., Lucarotti, V., Mancuso, S. and Alpi, A. 2015. Volatile organic compounds in truffle (*Tuber magnatum* Pico): Comparison of samples from different regions of Italy and from different seasons. Scientific Reports, 5: 12629.

Wang, K., Bao, L., Xiong, W., Ma, K., Han, J., Wang, W., Yin, W. and Liu, H. 2015. Lanostane triterpenes from the Tibetan medicinal mushroom *Ganoderma leucocontextum* and their inhibitory effects on HMG-CoA reductase and alpha-glucosidase. J. Nat. Prod., 78: 1977–1989.

Wang, X. 2012. Truffle Cultivation in China. pp. 227–240. *In*: A. Zambonelli and G.M. Bonito (eds.). Edible Ectomycorrhizal Mushrooms: Current Knowledge and Future Prospects. Soil Biology Volume 34. Berlin, Heidelberg: Springer-Verlag.

Wang, Y., Bunchanan, P. and Hall, I.R. 2002. A list of edible ectomycorrhizal mushrooms. Edible Mycorrhizal Mushrooms and their Cultivation. *In*: Hall, I.R., W. Yun, A. Zambonelli and E. Danell (eds.). Proceedings of the 2nd Int. Conf. Edible Mycorrhizal Mushrooms, New Zealand Institute for Crop & Food Research Limited. Christchurch, New Zealand. CD-ROM.

Wang, Y., Hall, I.R., Dixon, C., Hance-halloy, M., Strong, G. and Brass, P. 2002. The cultivation of *Lactarius deliciosus* (saffron milk cap) and *Rhizopogon rubescens* (shoro) in New Zealand. Edible Mycorrhizal Mushrooms and Their Cultivation. Proceedings of the 2nd Int. Conf. Edible Mycorrhizal Mushrooms. New Zealand Institute for Crop & Food Research Limited, Christchurch, New Zealand. CD-ROM.

Wang, Y. and Hall, I.R. 2004. Edible ectomycorrhizal mushrooms: Challenges and achievements. Can. J. Bot., 82: 1063–1073.

Wani, B.A., Bodha, R.H. and Wani, A.H. 2010. Nutritional and medicinal importance of mushrooms. J. Med. Plants Res., 4: 2598–2604.

Wasser, S.P. and Weiss, A.L. 1999. Therapeutic effects of substances occurring in higher Basidiomycetes mushrooms: A modern perspective. Crit. Rev. Immunol., 19: 65–96.

Wasser, S.P., Nevo, E., Sokolov, D., Reshetnikov, S. and Timor-Tismenetsky, M. 2000. Dietary supplements from medicinal mushrooms: Diversity of types and variety of regulations. Int. J. Med. Mushrooms, 2: 1–19.

Wasser, S.P. 2002. Medicinal mushrooms as a source of antitumor and immunomodulating polysaccharides. Appl. Microbiol. Biotechnol., 60: 258–274.

Wasser, S.P. 2010. Medicinal mushroom science: History, current status, future trends, and unsolved problems. Int. J. Med. Mushrooms, 12: 1–16.

Wasser, S.P. 2011. Current findings, future trends, and unsolved problems in studies of medicinal mushrooms. Appl. Microbiol. Biotechnol., 89: 1323–1332.

Wasser, S.P. 2017. Medicinal mushrooms in human clinical studies. Part I. anticancer, oncoimmunological, and immunomodulatory activities: A review. Int. J. Med. Mushrooms, 19: 279–317.

Winkler, D. 2017. The Wild Life of Yartsa Gunbu (*Ophiocordyceps sinensis*) on the Tibetan Plateau. *In*: Fungi. Volume 10:1, Spring, pp. 53–64.

Wu, J., Kobori, H., Kawaide, M., Suzuki, T., Choi, J.-H., Yasuda, N., Noguchi, K., Matsumoto, T., Hirai, H. and Kawagishi, H. 2013. Isolation of bioactive steroids from the *Stropharia rugosoanulata* mushroom and absolute configuration of strophasterol B. Biosci. Biotechnol. Biochem., 77: 1779–1781.

Wu, Y., Choi, M.H, Li, J., Yang, H. and Shin, H.J. 2016. Mushroom cosmetics: the present and future. Cosmetics, 3: 22.

Yamada, A., Ogura, T. and Ohmasa, M. 2001. Cultivation of mushrooms of edible ectomycorrhizal fungi associated with *Pinus densiflora* by *in vitro* mycorrhizal synthesis. I. Primordium and basidiocarp formation in open-pot culture. Mycorrhiza, 11: 59–66.

Yamada, A., Maeda, K., Kobayashi, H. and Murata, H. 2006. Ectomycorrhizal symbiosis *in vitro* between *Tricholoma matsutake* and *Pinus densiflora* seedlings that resembles naturally occurring "shiro". Mycorrhiza, 16: 111–116.

Yamanaka, K. 2008. Commercial cultivation of *Lyophyllum shimeji*. pp. 197–202. *In*: Lelley, J. I. and Buswell, J.A. (eds). Proceedings of the Sixth International Conference on Mushroom Biology and Mushroom Products 2008, Bonn, Germany.

Yan, Z.F., Yang, Y., Tian, F.H., Mao, X.X., Li, Y. and Li, C.T. 2014. Inhibitory and acceleratory effects of *Inonotus obliquus* on tyrosinase activity and melanin formation in B16 melanoma cells. Ev.-Bas. Compl. Alt. Med., 1–11.

Yang, X., Wan, M., Mi K., Feng, H., Chan, D.K.O. and Yang, Q. 2003. The quantification of (1, 3)-β-glucan in edible and medicinal mushroom polysaccharides by using limulus G test. Mycosystema, 22: 296–302.

Yildiz, O., Can, Z., Laghari, A.Q., Sahin, H. and Malkoç, M. 2015. Wild edible mushrooms as a natural source of phenolics and antioxidants. J. Food Biochem., 39: 148–154.

Zambonelli, A., Iotti, M. and Piattoni, F. 2008. Problems and perspectives in the production of *Tuber* infected plants. pp. 263–271. *In*: Lelley, J.I. and Buswell, J.A. (eds.). Mushroom Biology and Mushroom Products, Proceedings of the Sixth International Conference on Mushroom Biology and Mushroom Products 2008, Bonn, Germany.

Zambonelli, A. and Bonito, G.M. 2012. Edible ectomycorrhizal mushrooms: Current knowledge and future prospects, Soil Biology, Volume 34. Springer-Verlag, Berlin, Heidelberg.

Zambonelli, A., Iotti, M. and Hall, I. 2015. Current status of truffle cultivation: Recent results and future perspectives. Micologia Italiana, 44: 31–40.

Zambonelli, A., Murat, C. and Iotti, M. 2016. True truffle (*Tuber* spp.) in the world. Soil Biology.Volume 47. Springer-Verlag, Berlin, Heidelberg.

Zembron-Lacny, A., Gajewski, M., Naczk, M. and Siatkowski, I. 2013. Effect of shiitake (*Lentinus edodes*) extract on antioxidant and inflammatory response to prolonged eccentric exercise. J. Physiol. Pharmacol., 64: 249–254.

Zervakis, G. 2005. Cultivation of the king-oyster mushroom *Pleurotus eryngii* (DC. : Fr.) Quél. on substrates deriving from the olive-oil industry Int. J. Med. Mushrooms, 7: 486–487.

Zhai, X., Zhao, A., Geng, L. and Xu, C. 2013. Fermentation characteristics and hypoglycemic activity of an exopolysaccharide produced by submerged culture of *Stropharia rugosoannulata*. Ann. Microbiol., 63: 1013–1020.

Zhang, J., Wang, G., Li, H., Zhuang, C., Mizuno, T., Ito, H., Mayuzumi, I., Okamoto, H. and Li, J. 1994. Antitumor active protein-containing glycans from the Chinese mushroom songshan lingzhi, *Ganoderma tsugae* mycelium. Biosci. Biotechnol. Biochem., 58: 1202–1205.

Zhang, Y., Geng, W., Shen, Y., Wang, Y. and Dai, Y.C. 2014. Edible mushroom cultivation for food security and rural development in China: Bio-innovation, technological dissemination and marketing. Sustainability, 6: 2961–2973.

Zhou, X.W. 2017. Cultivation of *Ganoderma lucidum*. pp. 385–414. *In*: Diego, C.Z., Pardo-Giménez A. (eds.). Edible and Medicinal Mushrooms: Technology and Applications. John Wiley & Sons, Ltd, Chichester, UK.

10

Bioactive Metabolites from Basidiomycetes

Christian Agyare* and *Theresa Appiah Agana*

INTRODUCTION

The kingdom Fungi is divided into four different phyla: Basidiomycota (club fungi), Deuteromycota (imperfect fungi), Ascomycota (sac fungi) and Zygomycota (conjugation fungi). Species in the phylum Basidiomycota include the Basidiomycetes class that have macroscopic fruiting bodies, large enough to be seen with the naked eye. Many mushrooms look like umbrellas growing from the ground. Among the more famous genera are *Agaricus*, *Amanita* (including species that are deadly, delicious and hallucinogenic) and *Cantherellus*. The phylum Ascomycota includes the morel and truffle mushrooms. The phylum Zygomycota includes the *Penicillium* (Penicillin) and *Trichophyton* (Alexopoulos et al., 1996; Margulis and Schwartz, 1988).

Mushrooms (Basidiomycetes) are macrofungi (higher fungi) with distinctive fruiting bodies and reproductive structures (Prasad and Wesely, 2008). From a taxonomic point of view, many fungi in the class Basidiomycetes are considered mushrooms however some species belong to the Ascomycetes. About 10,000 of the known mushroom species belong to the class Basidiomycetes of which about 5,000 species are edible, while over 1,800 species are considered to have medicinal properties (Diyabalanage et al., 2008). Mushrooms are an unlimited source of therapeutically useful bioactive agents (Lindequist et al., 2005; Barros et al., 2007).

Department of Pharmaceutics, Faculty of Pharmacy And Pharmaceutical Sciences, College of Health Sciences, Kwame Nkrumah University of Science and Technology, Kumasi, Ghana.
* Corresponding author: cagyare.pharm@knust.edu.gh

A bioactive compound is a compound which has the ability to interact with one or more component(s) of living tissues by presenting a wide range of probable effects. The origin of these substances can be natural (that is terrestrial or aquatic; plant, animal or other source such as microorganisms) or synthetic (Ribeiro et al., 2011). Typical bioactive mushroom compounds are produced as secondary metabolites that are not necessary for normal functioning of the mushroom (such as growth) (Aksel, 2010), but play an important role in competition, defense (Dudareva and Pichersky, 2000). A number of bioactive molecules including polysaccharides, lectins, terpenoids and phenolics have been isolated from various mushrooms (Maiti et al., 2008; Patel and Goya, 2012).

Bioactive Polysaccharides from Basidiomycetes

Polysaccharides are biopolymers made up of monosaccharides linked together through glycosidic bonds. These structures can be linear or contain branched side chains. Polysaccharides have a general formula of $C_x(H_2O)_y$ where x is usually a large number between 200 and 2500. Considering that the repeating units in the polymer backbone are often six-carbon monosaccharides, the general formula is also represented as $(C_6H_{10}O_5)$ n where $40 \leq n \leq 3000$ (Zong et al., 2012).

Based on the monosaccharide composition, polysaccharides can be grouped into two main categories; *homopolysaccharides* and *heteropolysaccharides*. Both homo and heteropolysaccharides may possess homolinkages or heterolinkages in regard to configuration and/or linkage position. Again, heteropolysaccharides have various types and sequences of glycosidic linkages, resulting in practically limitless structural diversity (Ruthes et al., 2016).

Among the bioactive compounds found in mushrooms, polysaccharides are the main component responsible for the bioactivities of some mushroom species. A number of polysaccharides with diverse activities such as antitumor, immunomodulatory, anti-inflammatory, antiviral, antioxidative, antinociceptive, hypoglycemic and hepatoprotective effects have been reported (Xu et al., 2014; Yan et al., 2014; Ferreira et al., 2015; Ruthes et al., 2016; Singdevsachana et al., 2016; He et al., 2017) (Table 1).

Bioactive Phenolic Compounds from Basidiomycetes

Phenolic compounds constitute about 8000 different phenolic structures. They have one or more aromatic rings with one or more hydroxyl groups, including different subclasses such as tannins, flavonoids, phenolic acids, lignans, stilbenes and oxidized polyphenols (Barros et al., 2009; Fraga et al., 2010; Carocho and Ferreira, 2013).

Cinnamic acid and its derivatives such as o-coumaric, p-coumaric, caffeic, ferulic and chlorogenic acids have been found in several mushrooms, including *Agaricus arvensis, Amanita caesarea, Amanita muscaria, Amanita pantherina,*

Table 1. Bioactive polysaccharide from basidiomycetes.

Mushroom	Polysaccharide	Biological property
Agaricus bisporus, *Agaricus brasiliensis,* *Phellinus linteus* and *Ganoderma lucidum*	Varying amounts of both α- and β-glucans	Antioxidative and immunomodulating activities (Kozarski et al., 2011)
Agaricus blazei	Glucomannan	Antitumour activity (Mizuno et al., 1999)
Agaricus blazei	Heteroglucans, glycoprotein	Remarkable antitumor activity (Mizuno et al., 1990)
Agaricus blazei	α-1,6- and α-1,4-glucan	Stimulate lymphocyte T-cell subsets in mice (Mizuno et al., 1998)
Calocybe indica	Glucan	Stimulates the splenocytes and thymocytes (Mandal et al., 2012)
Collybia dryophila	β-D-glucans	Anti-inflammatory activity (Pacheco-Sanchez et al., 2006)
Dictyophora indusiata	Branched (1→3)-β-D-glucan	Anti-inflammatory effects (Hara et al., 1982)
Grifola frondosa (Maitake)	Polysaccharide	Antitumor activity (Nanba et al., 1987)
Hybrid mushroom *Pfle1r* of *Pleurotus florida* and *Lentinula edodes*	Glucan consisting of terminal, (1→3,6)-linked, and (1→6)-linked β-D-glucopyranosyl moieties	Macrophages, splenocytes, and thymocytes activation (Maji et al., 2012)
Hybrid mushroom (backcross mating between PfloVv12 and *Volvariella volvacea*)	Two glucans (watersoluble PS-I, water-insoluble PS-II) were isolated. PS-I was found to consist of only (1→6)-linked β-D-glucopyranose PS-II was composed of terminal, (1→3,4)-linked, and (1→3)-linked β-D-glucopyranosyl moieties	PS-I showed macrophages, splenocytes, and thymocytes activation as well as antioxidant property (Sarkar et al., 2012)
Lentinula edodes	2 polysaccharide fractions (IA-a and IA-b). IA-a is a 1,4-α-glucan exhibiting 1,6-branching. This branching structure suggests that IA-a has a glycogen-like structure. The IA-b component appears to be an arabinoxylan-like polysaccharide, mainly consisting of arabinose and xylose	Potently stimulated cytokine production stimulates phagocytosis (Kojima et al., 2010)

Table 1 contd. ...

...Table 1 contd.

Mushroom	Polysaccharide	Biological property
Lentinula edodes	Polysaccharide LT1 which has a backbone chain composed of 1→4-linked and 1→3-linked glucopyranosyl residues and has branches of single glucosyl stubs at C-6 of β-(1→4)-linked glucopyranosyl	Natural antitumor drug (Yu et al., 2010)
Lentinus edodes	D-glucopyranose	Immunomodulating activities (Zheng et al., 2005)
Lentinus edodes	Lentinan, LC-11, LC-12, LC-13, EC-11 and EC-14	Antitumor effect; inhibited the growth of Sarcoma 180 implanted s.c. in mice (Chihara et al., 1970)
Lentinus edodes	β-D-glucan	Antitumor effect (Sasaki and Takasuka, 1976)
Lentinus squarrosulus	Glucan	Optimum activation of macrophages as well as splenocytes and thymocytes (Bhunia et al., 2011)
Lentinus squarrosulus	Heteroglycan	Immunoenhancing (Bhunia et al., 2010)
Pleurotus florida	(1→3)-,(1→6)-branched glucan, a heteroglycan 9 (Fr. II) consisting of D-mannose, D-glucose, and D-galactose, a (1→6)-a-glucan and a (1→3)-, (1→6)-β-D-glucan	Immunoenhancing, stimulates macrophages, splenocytes and thymocytes (Roy et al., 2009)
Pleurotus florida	Glucan	Immunomodulating (Rout et al., 2004)
Pleurotus florida	Glucans (PS-I, PS-II and PS-III)	Antitumor and immunomodulating properties (Ojha et al., 2010)
Pleurotus florida	Polysaccharide consisting of D-glucose and D-galactose	Immunoenhancing (Dey et al., 2010)
Pleurotus florida	Two glucans (PS-I and PS-II) isolated	PS-I showed macrophage, splenocyte and thymocyte activations (Dey et al., 2012)
Pleurotus ostreatus	Heteropolysaccharide	Stimulates macrophages, splenocytes, and thymocytes (Maity et al., 2011)
Russula albonigra	Glucans	Excellent activations of macrophages as well as splenocytes and thymocytes *in vitro* (Nandi et al., 2012)
Somatic hybrid (PfloVv5FB) of *Pleurotus florida* and *Volvariella volvacea*	β-glucan	Strong immunoactivation of macrophages, splenocytes as well as Thymocytes (Maity et al., 2013)

Table 1 contd. ...

...Table 1 contd.

Mushroom	Polysaccharide	Biological property
Somatic hybrid mushroom PCH9FB of *Pleurotus florida* and *Calocybe indica* var.	Glucan	Immunoenhancing and antioxidant properties with immune activation of macrophage, splenocyte, and thymocyte (Maity et al., 2011)
Somatic hybrid mushroom (PfloVv1aFB), raised through protoplast fusion between the strains of *Pleorutus florida* and *Volverilla volvacea*	Heteroglycan consisting of (1→3)-, (1→6)-, (1→3,4)-linked, and terminal β-D-Glcp along with (1→2,6)-α-D-Galp and terminal α-D-Manp	Strong immunostimulating activity of macrophages as well as splenocytes and thymocytes (Bhunia et al., 2012)
Somatic hybrid mushroom of *Pleurotus florida* and *Calocybe indica* var. APK2	Novel polysaccharide consisting of galactose, fucose and glucose	Macrophage, splenocyte, thymocyte activation as well as antioxidant property (Maity et al., 2011)
Termitomyces robustus	β-glucans	Immunostimulating, significant macrophage, splenocyte, and thymocyte activation (Bhanja et al., 2012)
Trametes gibbosa	β-glucans	Anti-tumor (Ooi and Liu, 2000)
Tricholoma crissum	Glucan	Macrophageactivation *in vitro* by NO production in a dose dependent manner and strong splenocyte and thymocyte immunostimulation in mouse cell culture Medium; also exhibited good inhibition activity toward lipid peroxidation (Samanta et al., 2013)
Tricholoma crissum	Heteropolysaccharide	Splenocyte, thymocyte as well as macrophage activations (Patra et al., 2012)

Amanita rubescens, Armillaria mellea, Auricularia auricula-judae, Boletus aereus, Boletus edulis, Calocybe gambosa, Cantharellus cibarius, Chroogomphus fulmineus, Citocybe odora, Coprinus comatus, Lepista nuda, Lentinus edodes, Lycoperdon molle, Phellinus linteus, Pleurotus eryngii, Pleurotus ostreatus, Ramaria botrytis, Tricholoma acerbum, Tricholoma equestre, etc., to exhibit antimicrobial activities (Mattila et al., 2001; Valentao et al., 2005; Puttaraju et al., 2006; Kim et al., 2008; Barros et al., 2009; Heleno et al., 2011; Reis et al., 2011; Vaz et al., 2011a,b; Heleno et al., 2012).

Phenolic acids including benzoic and cinnamic acid derivatives are the most common phenolic compounds obtained from mushrooms. Among benzoic acid derivatives, p-hydroxybenzoic, gallic, protocatechuic, syringic and vanillic acids were identified in different mushroom species such as *Agaricus arvensis, Amanita pantherina, Boletus edulis, Boletus reticulatus, Cantharellus cibarius, Chroogomphus fulmineus, Citocybe odora, Hygrophorus marzuolus, Hygrophorus olivaceo-albus, Lactarius salmonicolor, Lactarius volemus, Lepista nuda, Lentinus edodes, Pleurotus eryngii, Pleurotus ostreatus, Ramaria botrytis, Russula cyanoxantha, Sparassis crispa,* etc., to exhibit antimicrobial activities (Puttaraju et al., 2006; Kim et al., 2008; Barros et al., 2009; Heleno et al., 2011, 2012; Reis et al., 2011; Vaz et al., 2011a,b).

Flavonoids such as quercetin, rutin and chrysin (Valentao et al., 2005; Ribeiro et al., 2006; Kim et al., 2008; Jayakumar et al., 2009; Yaltirak et al., 2009) and tannins like ellagic acid (Ribeiro et al., 2007) have been isolated from mushrooms such as *Fistulina hepatica, Agaricus blazei, Ionotus obliquus, Flammulina velutipes, Ganoderma lucidum, Sparassis crispa, Suillus luteus, Suillus granulatus Cantharellus cibarius, Pleurotus ostreatus,* etc., have been found to exhibit antimicrobial activities.

Bioactive Proteins and Peptides from Basidiomycetes

Mushrooms produce bioactive proteins and peptides such as lectins, ribosome-inactivating proteins (RIPs), fungal immunomodulatory proteins (FIPs) and laccases (Sánchez, 2017). Lectins are carbohydrate-binding proteins with specific binding capacities found in several organisms including fungi (Li et al., 2008). More than 400 edible mushroom species have been screened and about 50% of them found to contain lectins (Pemberton, 1994). Lectins have biological activities such as antitumor, hemagglutinating, antiproliferative and insecticidal activities (Wang et al., 1995; Wang et al., 1997; Wang et al., 2000; Trigueros et al., 2003; Yang et al., 2005; Xu et al., 2014). They have been isolated from several mushrooms including *Volvariella volvacea* (She et al., 1998), *Pleurotus citrinopileatus* (Wang et al., 1997), *Pleurotus ostreatus* (Wang et al., 2000), *Tricholoma mongolicum* (Li et al., 2008), *Tricholoma mongolicum* (Wang et al., 1997), *Agrocybe cylindracea* (Wang et al., 2002) and *Agrocybe aegerita* (Yang et al., 2005a, 2005b).

Ribosome-inactivating proteins (RIPs) are enzymes that inactivate ribosomes by eliminating adenosine residues from rRNA. RIPs from mushrooms have been reported to exhibit antitumor activity (Wong et al. 2008). Laccases are phenol oxidases which fungi use to degrade lignocellulosic substrates. Laccases with antiviral (Wang and Ng, 2006; El Fakharany et al., 2010) and antitumor activities (Zhang et al., 2010a) have been purified from some mushrooms.

Fungal immunomodulatory proteins (FIPs) which target immune cells have been isolated from mushrooms with potential application for tumor immunotherapy (Ding et al., 2009; Chang et al., 2010). Detailed information on their structures and biological activities proteins and peptides isolated from the basidiomycetes has been provided in Table 2 (Singh et al., 2015).

Table 2. Bioactive proteins and peptides from Mushrooms.

Mushroom	Structural property	Biological property
Agaricus arvensis	Lectin (30.4-kDa, Homodimeric)	Antiproliferative effects against HepG2 and MCF7 tumor cells (Zhao et al., 2011)
Agrocybe cylindracea	Lectin (31.5-kDa, Heterodimeric [15.3-kDa and 16.1-kDa subunits])	Potent mitogenic activity toward mouse splenocytes (Yagi et al., 1997; Wang et al., 2002; Hu et al., 2013)
Agrocybe aegerita	Lectin (43-kDa, Monomeric)	Induces cell apoptosis *in vitro* (Jiang et al., 2012)
Agrocybe aegerita	Lectin (15.8-kDa Homodimeric and a member of the galectin family)	Tumor-suppressing function via apoptosis-inducing activity in cancer cells (Sun et al., 2003; Zhao et al., 2003; Yang et al., 2005a; Yang et al., 2005b; Yang et al., 2009; Ren 2013)
Auricularia polytricha	Lectin (23-kDa, Monomeric)	Able to agglutinate only trypsinized human erythrocytes (Yagi et al., 1988)
Armillaria luteo-virens	Lectin (29.4-kDa, Dimeric, fairly thermostable)	Potent mitogenic activity toward splenocytes and antiproliferative activity toward tumor cells (Feng et al., 2006)
Aleuria aurantia	Lectin (72-kDa, Homodimeric Non-glycosylated, composed of two identical 312-amino acid subunits)	Able to agglutinate all types of human blood erythrocytes when treated with alpha (1 leads to 2)-fucosidase (Kochibe and Furukawa, 1980; Olausson et al., 2008)
Boletus edulis	Lectin (Homodimeric, 16.3-kDa subunits)	Stimulating effect on mitogenic response of mouse splenocytes and able to inhibit HIV-1 reverse transcriptase enzyme *in vitro*. Antineoplastic or antitumor properties Zheng et al., 2007; Bovi et al., 2013)
Boletopsis leucomelas	Lectin (15-kDa, Monomeric)	Apoptosis-inducing activity just like mistletoe lectins (Koyama et al., 2002)
Clavaria purpurea	Lectin (16-kDa, Monomeric)	Potential interest for detection and characterization of glycoconjugates containing Galα1-4Gal and other α-galactosyl sugars on the cell surfaces (Lyimo et al., 2012)
Clitocybe nebularis	Lectin (15.9-kDa)	Induces maturation and activation of dendritic cells via the toll-like receptor 4 pathway. Also has an immunomodulatory property on leukaemic T-cell lines. Insecticidal and anti-nutritional properties (Svajger et al., 2011; Pohleven et al., 2012)

Table 2 contd. ...

Table 2 contd. ...

Mushroom	Structural property	Biological property
Cordyceps militaris	Lectin (Monomeric, 31-kDa comprised of 27% α-helix, 12% β-sheets, 29% β-turns, and 32% random coils)	Exhibits mitogenic activity against mouse splenocytes (Jung et al., 2007)
Flammulina velutipes	Lectin (12-kDa)	Inhibits proliferation of leukemia L1210 cells (Yatohgo et al., 1988; Ng et al., 2006)
Gymnopilus spectabilis	Lectin (52.1-kDa and 64.4-kDa subunits Glycoprotein)	Inhibits *in vitro* the growth of *Staphylococcus aureus* and *Aspergillus niger* (Alborés et al., 2014)
Grifola frondosa	Lectin (68-kDa, Homodimeric, high content of acidic and hydroxyl amino acids and low content of methionine and histidine)	Cytotoxic against HeLa cells (Kawagishi et al., 1990; Stepanova et al., 2007)
Ganoderma lucidum (GLL-M and GLL-F)	Lectin (GLL-M:18-kDa, GLL-F:12-kDa)	Health-promoting and therapeutic effects (Kawagishi et al., 1997)
Ganoderma capense	Lectin (18-kDa fairly heat stable)	Potent mitogenic activity toward mouse splenocytes, and antiproliferative activity toward leukemia (L1210 and M1) cells and hepatoma (HepG2) cells (Ngai and Ng, 2004)
Hericium erinaceum	Lectin (54-kDa, Heterodimeric with 15-kDa and 16-kDa subunits)	Used in Chinese medicine (Kawagishi et al., 1994)
Hygrophorus russula	Lectin [18.5-kDa Subunits (Homotetrameric)]	Shows mitogenic activity against spleen lymph cells (F344 rat) and strong binding of to HIV-1 gp120 (Suzuki et al., 2012)
Lactarius flavidulus	Lectin (29.8-kDa, Dimeric)	Suppresses the proliferation of hepatoma (HepG2) and leukemic (L1210) cells. Inhibits the activity of HIV-1RT enzyme (Wu et al., 2011)
Laetiporus sulfureus	Lectin (35-kDa, Hexameric Non-glycoprotein)	Hemolytic property by pore forming towards blood cells (Mancheño et al., 2005)
Lactarius rufus	Lectin [98-kDa (containing six subunits)]	The lectin agglutinates human etrythrocytes without any marked group specificity (Panchak et al., 2007)
Lentinus edodes	Lectin (43-kDa, Monomeric)	Mitogenic towards murine splenic lymphocytes (Wang et al., 1999)
Lyophyllum decastes	Lectin (10-kDa, Homodimeric)	The lectin shares carbohydrate binding preference with verocytoxin of bacteria *Shigella dysenteriae* and *E. coli* 0157:H7 (Goldstein et al., 2007)

Table 2 contd. ...

Table 2 contd. ...

Mushroom	Structural property	Biological property
Macrolepiota procera	Lectin (16-kDa Monomeric)	Has toxic effects towards the nematode indicating a protecting role against predators and parasites (Žurga et al., 2014)
Marasmius oreades	Lectin [(Consists of an intact (33-kDa) and truncated (23-kDa) subunit in addition to a small polypeptide (10-kDa)]	Has proteolytic activity and inhibits protein and DNA synthesis in NIH/3T3 cells. May induce BAX-mediated apoptosis (Cordara et al., 2014)
Pholiota adiposa	Lectin (32-kDa, Homodimeric)	Antiproliferative activity toward hepatoma Hep G2 cells and breast cancer MCF7 cells. It also exhibits HIV-1 reverse transcriptase inhibitory activity (Zhang et al., 2009)
Pleurotus citrinopileatus	Lectin (32.4-kDa subunits, Homodimeric)	Potent antitumor, mitogenic and HIV-1 reverse transcriptase inhibitory activities (Li et al., 2008)
Pholiota squarrosa	Lectin (4.5-kDa)	Able to differentiate between primary and metastatic colon cancer tissues in the expression of α1-6 fucosylation (Kobayashi et al., 2004)
Pleurotus ostreatus	Lectin (40- and 41-kDa subunits, Heterodimeric)	Potent antitumor activity in sarcoma S-180 bearing and hepatoma H-22 bearing mice. Enhances immunogenicity of some vaccines in transgenic mice. Possesses anti-inflammatory activities (Wang et al., 2000; Jedinak et al., 2011; Gao et al., 2013)
Pleurotus ferulae	Lectin (35-kDa, Homodimeric)	Highly potent hemagglutinating and proliferative activities toward mouse splenocytes (Xu et al., 2014)
Pleurotus tuber-regium	Lectin (32-kDa)	Exhibits hemagglutinating activity toward trypsinized rabbit erythrocytes but not toward untrypsinized rabbit erythrocytes (Wang and Ng, 2003)
Polyporus adusta	Lectin (12-kDa subunits, Homodimeric)	Antiproliferative activity toward tumor cell lines and mitogenic activity toward splenocytes (Wang et al., 2003)
Psathyrella velutina	Lectin (40-kDa, Monomeric having a regular seven-bladed β-propeller fold)	Used in detection of glycosylation abnormality in rheumatoid IgG (Cioci et al., 2006)
Polyporus squamosus	Lectin (28-kDa subunits, Homodimeric)	Can be a valuable tool for glycobiological studies in biomedical and cancer research (Mo et al., 2000)
Russula lepida	Lectin (16-kDa subunits, Homodimeric)	Antiproliferative activity towards hepatoma Hep G2 cells and human breast cancer MCF-7 cells (Zhang et al., 2010b)

Table 2 contd. ...

Table 2 contd. ...

Mushroom	Structural property	Biological property
Russula delica	Lectin (60-kDa, Homodimeric)	Potent inhibitor for proliferation of HepG2 hepatoma and MCF 7 breast cancer cells, also inhibits HIV-1 reverse transcriptase activity (Zhao et al., 2010)
Stropharia rugosoannulata	Lectin (38-kDa, Homodimeric)	Exhibits anti-proliferative activity toward both hepatoma Hep G2, cells and leukemia L1210 cells, along with anti HIV-1 reverse transcriptase activity (Zhang et al., 2014)
Schizophyllum commune	Lectin (64-kDa, Homodimeric)	Potent mitogenic activity toward mouse splenocytes, antiproliferative activity toward tumor cell lines, and inhibitory activity toward HIV-1 reverse transcriptase (Han et al., 2005)
Tricholoma mongolicum	Lectin (37-kDa, Homodimeric, non-glycoprotein in nature)	Exhibits antiproliferative activities against mouse monocyte-macrophage PU5-1.8 cells and mouse mastocytoma P815 cells *in vitro*. Stimulates production of nitrite ions by macrophages in normal and tumor-bearing mice (Wang et al., 1995; Wang et al., 1997)
Volvariella volvacea	Lectin (32-kDa, Homodimeric, Non-glycoprotein)	Potent stimulatory activity towards murine splenic lymphocytes showing immuno-modulatory activity. Also found to enhance transcriptional expression of interleukin-2 and interferon-γ (She et al., 1998)
Xerocomus chrysenteron	Lectin (15-kDa)	It possesses a high insecticidal activity against the dipteran *Drosophila melanogaster* and the hemipteran, Acyrthosiphon pisum (Trigueros et al., 2003)
Xerocomus spadiceus	Lectin (32.2-kDa [16-kDa subunits], Dimeric)	Capable of eliciting an approximately four-fold stimulation of mitogenic response in murine splenocytes (Liu et al., 2004)
Xylaria hypoxylon	Lectin (28.8-kDa, Homodimeric)	Potent hemagglutinating activity. Antiproliferative activity towards tumor cell lines, and anti-mitogenic activity on mouse splenocytes (Liu et al., 2006)
Hypsizigus marmoreus	Ribosome-inactivating protein, marmorin (9 kDa)	Antitumor (Wong et al., 2008)
Pleurotus eryngii	Laccase	Antiviral (Wang and Ng, 2006)
Pleurotus ostreatus	Laccase	Antiviral (El Fakharany et al., 2010)
Clitocybe maxima	Laccase	Antitumor (Zhang et al., 2010a)

Table 2 contd. ...

Table 2 contd. ...

Mushroom	Structural property	Biological property
Ganoderma lucidum	Fungal immunomodulatory proteins (FIPs) ling zhi-8 (LZ-8)	Immunomodulatory (Kino et al., 1989)
Flammulina velutipes	Fip-fve	Immunomodulatory (Lin et al., 1997)
Ganoderma tsugae	Fip-gts	Immunomodulatory (Ko et al., 1995)
Volvariella volvacea	Fip-vvo	Immunomodulatory (Hsu et al., 1997)

Bioactive Terpenes from Basidiomycetes

Terpenes, the largest group of compounds in mushrooms (Sánchez, 2017) includes the monoterpenoids, sesquiterpenoids, diterpenoids and triterpenoids. Several terpenes have been isolated from mushrooms with various biological and pharmacological activities such as anticancer (Wang et al., 2012; Arpha et al., 2012; Wang et al., 2013b), anti-inflammatory (Kamo et al., 2004), anticholinesterase (Lee et al., 2011) antimalarial (Isaka et al., 2013), antiviral (Mothana et al., 2003) and antibacterial activities. Detailed information on some terpenes isolated from the basidiomycetes has been provided in Table 3.

Table 3. Bioactive terpenes from basidiomycetes.

Mushroom	Terpene	Biological property
Sesquiterpenoid		
Stereum hirsutum	Hirsutenol A, B and C	Antimicrobial against Escherichia coli (Yun et al., 2002)
Inonotus rickii	Inonotic acid A, 3-O-formyl inonotic acid A, Inonotic acid A and 3α,6α-Hydroxycinnamolide	Cytotoxic (Chen et al., 2014)
Pleurotus cornucopiae	Pleurospiroketal A, B and C	Inhibition of NO production and Cytotoxic (Wang et al., 2013a)
Anthracophyllum sp. BCC18695	Anthracophyllic acid and Anthracophyllone	Antimalarial, antibacterial and cytotoxic (Intaraudom et al., 2013)
Flammulina velutipes	Enokipodin F, G and I	Antifungal (Wang et al., 2012)
Flammulina velutipes	Enokipodin J, 2,5-Cuparadiene-1,4-dione, Enokipodin B and D	Cytotoxic, antioxidant and antibacterial (Wang et al., 2012)
Flammulina velutipes	Flammulinolide A, B, C, D, E, F and G	Cytotoxic and antibacterial (Wang et al., 2012)
Flammulina velutipes	Enokipodin A and C	Antimicrobial (Ishikawa et al., 2001)

Table 3 contd. ...

... Table 3 contd.

Mushroom	Terpene	Biological property
Sesquiterpenoid		
Agrocybe salicacola	Agrocybin H, Agrocybin I and Illudosin	Cytotoxic (Liu et al., 2012)
Neonothopanus nambi	Nambinone A and C, and 1-epi-nambinone B	Antimalarial, Antitubercular and Cytotoxic (Kanokmedhakul et al., 2012)
Neonothopanus nambi	Nambinone B and D	Antitubercular and cytotoxic (Kanokmedhakul et al., 2012)
Russula lepida and *R. amarissima*	Rulepidadiol B and C	Cytotoxic (Clericuzio et al., 2012)
Strobilurus ohshimae	Strobilol A	Brine shrimp toxicity and antimicrobial (Hiramatsu et al., 2007)
Strobilurus ohshimae	Strobilol B, C and D	Antimicrobial (Hiramatsu et al., 2007)
Diterpenoid		
Cyathus africanus	Cyathin D, E, F, G and H	Inhibition of NO production and cytotoxic (Han et al., 2013)
Cyathus africanus	Neosarcodonin O	Cytotoxic (Han et al., 2013)
Cyathus africanus	Cyathatriol and 11-O-acetylcyathatriol	NO production inhibitory and Cytotoxic (Han et al., 2013)
Hericium erinaceum	Erinacine A	Antibacterial (Kawagishi et al., 2006)
Pleurotus eryngii	Eryngiolide A	Cytotoxic (Wang et al., 2012)
Sarcodon scabrosus	Scabronine M, Sarcodonin I, Scabronine K and L	Cytotoxic (Shi et al., 2011)
Sarcodon scabrosus	Sarcodonin G and A	Antibacterial and Cytotoxic (Dong et al., 2009)
Sarcodon scabrosus	Scabronine H and 19-O-acetylsarcodonin G	Cytotoxic (Shi et al., 2011)
Sarcodon scabrosus	Sarcodonin M and L	Antibacterial (Shibata et al., 1998)
Sarcodon scabrosus	19-O-linoleoyl sarcodonin A, 19-O-Oleoyl sarcodonin A, 19-O-Steroyl sarcodonin A, Allocyathin B2, Neosarcodonin A, 19-O-Octanoyl sarcodonin A, 19-O-Butryl sarcodonin A, 19-O-Acetyl sarcodonin A, 19-O-Benzoyl sarcodonin A and 19-O-Pivaloyl sarcodonin A	Anti-inflammatory (Kamo et al., 2004)
Tricholoma sp.	Tricholomalide A, B and C	Cytotoxic (Tsukamoto et al., 2003)

Table 3 contd. ...

... Table 3 contd.

Mushroom	Terpene	Biological property
Triterpenoid		
Ganoderma boninense	Ganoboninketal A, B and C	Antiplasmodial, NO inhibition and Cytotoxic (Ma et al., 2014)
Ganoderma orbiforme	Ganorbiformin A, D, E, F and G. The C-3 epimer of ganoderic acid T, Ganoderic acid T and S	Antitubercular, Antimalarial and Cytotoxic (Isaka et al., 2013)
Ganoderma lucidum	Methyl ganoderate A acetonide, n-Butyl ganoderate H, Methyl ganoderate A, Ganoderic acid B, E and Y, Ganolucidic acid A, Ganodermadiol, Ganoderiol F, Lucidumol B, Ganodermanondiol, Ganodermanontriol and Lucidadiol	Anticholinesterase (Lee et al., 2011)
Ganoderma lucidum	Lucidenic acid A and N	Anticholinesterase (Lee et al., 2011) and anti-invasive (Weng et al., 2007)
Ganoderma lucidum and *Ganoderma pfeifferi*	Methyl lucidenate E2	Anticholinesterase (Lee et al., 2011) and antiviral (Mothana et al., 2003)
Ganoderma lucidum	Butyl ganoderate A and B, Ganoderic acid A, Methyl ganoderate B, D, E and H, Methyl lucidenate A, F and P, n-Butyl lucidenate A and N	Effect of on adipocyte differentiation in 3T3L1 cells (Lee et al., 2010).
Elfvingia applanata	Elfvingic acid A, B, C, D, E, F, G and H	Cytotoxic (Yoshikawa et al., 2002)
Tricholoma saponaceum	Saponaceol A, B and C	Cytotoxic (Yoshikawa, et al., 2004)
Spongiporus leucomallellus	Spongiporic acid A	Antibacterial (Ziegenbein et al., 2006)
Leucopaxillus gentianeus	Cucurbitacin B, Cucurbitacin B esters, Leucopaxillone A and B, Cucurbitacin D, and 16 -Deoxycucurbitacin B	Antiproliferative (Clericuzio et al., 2006)
Hebeloma versipelle	24(E)-3β-Hydroxylanosta-8,24-dien-26-al-21-oic acid	Cytotoxic (Shao et al., 2005)
Astraeus hygrometricus	Astrakurkurol	Anticandidal (Lai et al., 2012)
Astraeus pteridis	Astrakurkurone and 3-epi-astrahygrol	Antitubercular and Cytotoxic (Stanikunaite et al., 2008)
Antrodia camphorata	Astrahygrone, astrapteridone, astrapteridiol, 3-epi-astrapteridiol, Antcin K and C, Zhankuic acid A and C and Dehydroeburicoic acid	Cytotoxic (Du et al., 2012)
Russulalepida and *Russula amarissima*	3,4-Secocucurbita-4,24E-diene-3,26-dioic acid and Cucurbitane hydroxyacid	Cytotocic (Clericuzio et al., 2012)

Table 3 contd. ...

... Table 3 contd.

Mushroom	Terpene	Biological property
Triterpenoid		
Spongiporus leucomallellus	Spongiporic acid A	Antibacterial (Ziegenbein et al., 2006)
Ganoderma lucidum	Lucidenic acid C, 7-Oxo-ganoderic acid Z and 15-Hydroxy-ganoderic acid S	Inhibitory activity against HMG-CoA reductase and acyl CoA acyltransferase (Li et al., 2006)
Ganoderma lucidum	Ganoderic acid C1 and Lucidenic acid B	Anti-invasive (Weng et al., 2007)
Ganoderma lucidum	Butyl ganoderate A and B, Ganoderic acid A, Methyl ganoderate B, D, E and H, Methyl lucidenate A, P and F	Effect of on adipocyte differentiation in 3T3L1 cells (Lee et al., 2010)
Ganoderma lucidum	Lucialdehyde A	Cytotoxic (Gao et al., 2002)
Ganoderma amboinenese	Lucialdehyde B and C, Ganodermanonol, Ganodermacetal and Methyl ganoderate C	Toxic activity against brine shrimp larvae (Yang et al., 2012)
Ganoderma amboinenese	Ganoderic acid X	Cytotoxic (El Dine et al., 2008)
Ganoderma colossum	Colossolactone V	Anti-HIV-1 Protease (El Dine et al., 2008)
Astraeus odoratus	Astraodorol, Astraodoric acid A, B and C	Antitubercular and cytotoxic (Arpha et al., 2012)
Ganoderma pfeifferi	Colossolactone VI, VII, VIII and A, E, G, Schisanlactone A, Lucialdehyde D and Ganoderone A	Antiviral (Niedermeyer et al., 2005)
Ganoderma pfeifferi	Ganoderone C, Lucialdehyde B, Ganoderol A and Applanoxidic acid G	Antiviral (Mothana et al., 2003)
Ganoderma fornicatum	Fornicatin A and B	The inhibitory effects on platelet aggregation (Niu et al., 2004)
Ganoderma concinna	5α-Lanosta-7,9(11),24-triene-3β-hydroxy-26-al, 5α-Lanosta-7,9(11),24-triene-15α-26-dihydroxy-3one and 8α,9α-Epoxy-4,4,14α-trimethyl-3,7,11,15,20-pentaoxo-5α-pregrane	Cytotoxic (Gonzalez et al., 2002)
Naematoloma fasciculare	Fusciculol C, G, J, K, L and M	Cytotoxic Kim et al., 2013)
Ganoderma tsuage	Tsugarioside C, Tsugaric acid A and B, Tsugarioside A, 3β-Hydroxy-5α-lanosta-8, 24-dien-21-oic acid and 3-Oxo-5α-lanosta-8, 24-dien-21-oic acid	Cytotoxic (Su et al., 2000)

Other Bioactive Metabolites from Basidiomycetes

Several other compounds, including polysaccharide–protein complexes, alkaloids, infractine and psilocybin with several biological activities, have been identified in the basidiomycetes (Table 4).

Table 4. Other Bioactive Metabolites from Basidiomycetes.

Mushroom	Polysaccharide	Biological property
Trametes versicolor	Polysaccharide-K (krestin), protein bind with b(1,6) side chain, and b(1,3)-branched b(1,4) main chain glucan (94–100 kDa)	Antimetastatic activity (Wasser, 2002)
Trametes versicolor	Coriolan, a b-glucanprotein	Hypoglycemic effects and ameliorate the symptoms of diabetes (Rathee et al., 2012)
Calvatia gigantea	Calvacin	Antitumor activity (Chatterjee et al., 2011)
Phellinus linteus	Proteoglycan	Anti-inflammatory properties (Kim et al., 2004)
Agaricus macrosporus and *Grifola frondosa*	Agaricoglycerides (chlorinated 4-hydroxy benzoic acid and glycerol)	Anti-inflammatory activity (Han and Cui, 2012)
Hericium erinaceum	Dilinoleoylphosphatidylethanolamine	Anti-neurodegenerative activity (Nagai et al., 2006)
Termitomyces albuminosus	Termitomycesphins A, B, C, D, G and H (cerebrosides) and termitomycamide A, B, C, D and E (fatty acid amides)	Anti-neurodegenerative activity (Choi et al., 2010)
Dictyophora indusiata	Dictyophorine A and B	Improve the amount of nerve growth factor (Kawagishi et al., 1997)
Dictyophora indusiata	Dictyoquinazol A, B and C	Neuroprotective properties (Lee et al., 2002a)
Mycoleptodonodes aitchisonii	3(hydroxymethyl)-4-methylfuran-2(5H)-one, (3R, 4S, 1' R)-3-(1'-hydroxyethyl)4methyldihydrofuran-2(3H)-one, 5-hydroxy-4-(1-hydroxyethyl)-3-methylfuran-2 (5H)-one, and 5-phenylpentane-1,3,4-triol	Anti-neurodegenerative activity (Choi et al., 2014)
Cortinarius infractus	6-hydroxyinfractine and infractopicrine (alkaloids infractine)	AChE-inhibiting activity with non-detectable cytotoxicity (Geissler et al., 2010)
Daldinia concentrica	Caruilignan C and 1-(3,4,5-trimethoxyphenyl) ethanol	Neuroprotective activity (Lee et al., 2002b)
Psilocybe spp.	Psilocybin	Antidepressant (Petri et al., 2014)

Table 4 contd. ...

...*Table 4 contd.*

Mushroom	Polysaccharide	Biological property
Boletus spp.	Antioxidant metabolites, 2, 4, 6-trimethylacetophenoneimine, glutamyltryptophan, azatadine and lithocholic acid glycine conjugate	Antioxidant (Yuswan et al., 2015)
Lactarius necator	Alkaloids necatorin and necotoron	Antibacterial and antifungal (Badalyan, 2012)
Amauroderma rude	Ergosterol peroxide	Antitumour (Li et al., 2015)
Lactarius deliciosus, Lactarius sanguifluus, Lactarius semisanguifluus, Russula delica and *Suillus bellinii*	Ergosta-5,7-dienol, ergosta-7,22-dienol, ergosta-7-enol, lanosterol, lanosta-8,24-dienol, and 4α-methylzymosterol	Antioxidant (Kalogeropouls et al., 2013)

Conclusions

There is substantial evidence that the Basidiomycetous mushrooms have nutritional as well as medicinal properties. Basidiomycetes have been generally recognized as a valuable source of biologically important compounds, which mankind can exploit as biological and pharmacological agents formulated to serve mankind.

References

Aksel, B. 2010. A brief review on bioactive compounds in plants. pp. 11–17. *In*: Bernhoft, A. (ed.). Bioactive Compounds in Plants—Benefits and Risks for Man and Animals. Oslo. The Norwegian Acad. Sci. Lett.

Alborés, S., Mora, P., Bustamante, M.J. et al. 2014. Purification and applications of a lectin from the mushroom *Gymnopilus spectabilis*. Appl. Biochem. Biotechnol., 172: 2081–2090.

Alexopoulos, C.J., Mims, C.W. and Blackwell, M. 1996. Introductory Mycology, 4th Edition. John Wiley and Sons, New York.

Arpha, K., Phosri, C., Suwannasai, N. et al. 2012. Astraodoric acids A-D: New lanostane triterpenes from edible mushroom *Astraeus odoratus* and their anti-Mycobacterium tuberculosis H37Ra and cytotoxic activity, J. Agric. Food Chem., 60: 9834–9841.

Badalyan, S.M. 2012. Edible Ectomycorrhizal Mushrooms. pp. 317–334. *In*: Zambonelli, A. and Bonito, G. (eds.). Edible Ectomycorrhizal Mushrooms, Soil Biology series, Volume 34. Springer Verlag.

Barros, L., Venturini, B.A., Baptista, P. et al. 2008. Chemical composition and biological properties of Portuguese wild mushrooms: A Comprehensive study. J. Agri. Food Chem., 56: 3856–3862.

Barros, L., Duenas, M., Ferreira, I.C.F.R. et al. 2009. Phenolic acids determination by HPLCDAD-ESI/MS in sixteen different Portuguese wild mushrooms species. Food Chem. Toxicol., 47: 1076–1079.

Bhanja, S.K., Nandan, C.K., Mandal, S. et al. 2012. Isolation and characterization of the immunostimulating β-glucans of an edible mushroom *Termitomyces robustus* var. Carbohyd. Res., 357: 83–89.

Bhunia, S. K., Dey, B., Maity, K. K. et al. 2010. Structural characterization of an immunoenhancing heteroglycan isolated from an aqueous extract of an edible mushroom, *Lentinus squarrosulus* (Mont.) Singer Carbohyd. Res., 345: 2542–2549.

Bhunia, S.K., Dey, B., Maity, K.K. et al. 2011. Isolation and characterization of an immunoenhancing glucan from alkaline extract of an edible mushroom, *Lentinus squarrosulus* (Mont.) Singer Carbohyd. Res., 346: 2039–2044.

Bhunia, S.K., Dey, B. and Maity, K.K. 2012. Heteroglycan from an alkaline extract of a somatic hybrid mushroom (PfloVv1aFB) of *Pleurotus florida* and *Volvariella volvacea*: Structural characterization and study of immunoenhancing properties. Carbohyd. Res., 354: 110–115.

Bovi, M., Cenci, L., Perduca, M. et al. 2013. BEL β-trefoil: A novel lectin with antineoplastic properties in king bolete (*Boletus edulis*) mushrooms. Glycobio., 23: 578–592.

Carocho, M. and Ferreira, I.C.F.R. 2013. The role of phenolic compounds in the fight against cancer a review. Anti-Cancer Agents in Med. Chem., 13: 1236–1258.

Chang, H.H., Hsieh, K.Y., Yeh, C.H. et al. 2010. Oral administration of an Enoki mushroom protein FVE activates innate and adaptive immunity and induces anti-tumor activity against murine hepatocellular carcinoma. Int. Immunopharmacol., 20: 239–246.

Chatterjee S., Biswas, G. and Basu, S.K. 2011. Antineoplastic effect of mushrooms: A review. Aust. J. Crop Sci., 5: 904–911.

Chen, H.P., Dong, W.B., Feng, T. et al. 2014. Four new sesquiterpenoids from fruiting bodies of the fungus *Inonotus rickii*, J. Asian Nat. Prod. Res., 16: 581–586.

Chihara, G., Hamuro, J., Maeda, Y.Y. et al. 1970. Fractionation and purification of the polysaccharides with marked antitumour activity, especially lentinan from *Lentinan edodes* (Berk) Sing (an edible mushroom) Cancer Res., 30: 2776–2781.

Choi, J.H., Maeda, K., Nagai, K. et al. 2010. Termitomycamides A to E, fatty acid amides isolated from the mushroom *Termitomyces titanicus*, suppress endoplasmic reticulum stress. Org. Lett., 12: 5012–5015.

Choi, J.H., Suzuki, T., Okumura, H. et al. 2014. Endoplasmic reticulum stress suppressive compounds from the edible mushroom *Mycoleptodonoides aitchisonii*. J. Nat. Prod., 77: 1729–1733.

Cioci, G., Mitchell, E.P., Chazalet, V. et al. 2006. A β-propeller crystal structure of *Psathyrella velutina* lectin: An integrin-like fungal protein interacting with monosaccharides and calcium. J. Mol. Biol., 357: 1575–1591.

Clericuzio, M., Tabasso, S., Bianco, M.A. et al. 2006. Cucurbitane triterpenoids from *Leucopaxillus gentianeus*. J. Nat. Prod., 69: 1796–1799.

Clericuzio, M., Cassino, C., Corana, F. et al. 2012. Terpenoids from *Russula lepida* and *R. amarissima* (Basidiomycota, Russulaceae), Phytochem., 84: 154–159.

Cordara, G., Winter, H.C., Goldstein, I.J. et al. 2014. The fungal chimerolectin MOA inhibits protein and DNA synthesis in NIH/3T3 cells and may induce BAX-mediated apoptosis. Biochem. Biophys. Res. Commun., 447: 586–589.

Dey, B., Bhunia, S.K., Maity, K.K. et al. 2010. Chemical analysis of an immunoenhancing water-soluble polysaccharide of an edible mushroom, *Pleurotus florida* blue variant Carbohyd. Res., 345: 2736–2741.

Dey, B., Bhunia, S.K., Maity, K.K. et al. 2012. Glucans of *Pleurotus florida* blue variant: Isolation, purification, characterization and immunological studies. Int. J. Biol. Macromol., 50: 591–597.

Ding, Y., Seow, S.V., Huang, C.H. et al. 2009. Coadministration of the fungal immunomodulatory protein FIP-Fve and a tumour-associated antigen enhanced antitumour immunity. Immunol., 128: 881–894.

Diyabalanage, T., Mulabagal, V., Mills, G. et al. 2008. Health beneficial qualities of the edible mushroom, *Agrocybe aegerita*. Food Chem., 108: 97–102.

Dong, M., Chen, S.P., Kita, K. et al. 2009. Anti-proliferative and apoptosis- inducible activity of sarcodonin G from *Sarcodon scabrosus* in Hela cells. Int. J. Oncol., 34: 201–207.

Du, Y.C., Chang, F.R., Wu, T.Y. et al. 2012. Antileukemia component, dehydroeburicoic acid from *Antrodia camphorate* induces DNA damage and apoptosis *in vitro* and *in vivo* models. Phytomed., 19: 788–796.

Dudareva, N. and Pichersky, E. 2000. Biochemical and molecular genetic aspects of floral scent. Plt. Physiol., 122: 627–633.

El Dine, R.S., El Halawany, A.M., Ma C.M. et al. 2008. Anti-HIV-1 protease activity of lanostane triterpenes from the Vietnamese mushroom *Ganoderma colossum*. J. Nat. Prod., 71: 1022–1026.

El Fakharany, E.M., Haroun, B.M., Ng, T.B. et al. 2010. Oyster mushroom laccase inhibits hepatitis C virus entry into peripheral blood cells and hepatoma cells. Protein Pept. Lett., 17: 1031–1039.

Feng, K., Liu, Q.H., Ng, T.B. et al. 2006. Isolation and characterization of a novel lectin from the mushroom *Armillaria luteo-virens*. Biochem. Biophys. Res. Commun., 345: 1573–1578.

Ferreira, I.C., Heleno, S.A., Reis, F.S. et al. 2015. Chemical features of *Ganoderma* polysaccharides with antioxidant, antitumor and antimicrobial activities. Phytochem., 114: 38–55.

Fraga, G.C. 2010. Plant phenolics and human health: Biochemistry, nutrition and pharmacology. John Whiley & Sons: New Jersey.

Gao, J.J., Min, B.S., Ahn, E.M. et al. 2002. New triterpene aldehydes, lucialdehydes A-C, from *Ganoderma lucidum* and their cytotoxicity against murine and human tumor cells. Chem. Pharm. Bull., 50: 837–840.

Gao, W., Sun, Y., Chen, S. et al. 2013. Mushroom lectin enhanced immunogenicity of HBV DNA vaccine in C57BL/6 and HBsAg-transgenic mice. Vaccine, 31: 2273–2280.

Geissler, T., Brandt, W., Porzel, A. et al. 2010. Acetylcholinesterase inhibitors from the toadstool *Cortinarius infractus*. Bioorg. Med. Chem., 18: 2173–2177.

Goldstein, I.J., Winter, H.C., Aurandt, J. et al. 2007. A new alpha-galactosyl-binding protein from the mushroom *Lyophyllum decastes*. Arch. Biochem. Biophys., 467: 268–274.

Gonzalez, A.G., Leon, F., Rivera, A. et al. 2002. New lanostanoids from the fungus *Ganoderma concinna*. J. Nat. Prod., 65: 417–421.

Han, C. and Cui, B. 2012. Pharmacological and pharmacokinetic studies with agaricoglycerides, extracted from *Grifola frondosa*, in animal models of pain and inflammation. Inflammat., 35: 1269–1275.

Han, C.H., Liu, Q.H., Ng, T.B. et al. 2005. A novel homodimeric lactose-binding lectin from the edible split gill medicinal mushroom *Schizophyllum commune*. Biochem. Biophys. Res. Commun., 336: 252–257.

Han, J.J., Chen, Y.H., Bao, L. et al. 2013. Anti-inflammatory and cytotoxic cyathane diterpenoids from the medicinal fungus *Cyathus africanus*, Fitoter., 84: 22–31.

Hara, C., Kiho, T., Tanaka, T. et al. 1982. Anti-inflammatory activity and conformational behavior of a branched (1 leads to 3)-beta-D-glucan from an alkaline extract of *Dictyophora indusiata* Fisch. Carbohyd. Res., 110: 77–87.

He, X.R., Wang, X.X., Fang, J.C. et al. 2017. Polysaccharides in *Grifola frondosa* mushroom and their health promoting properties: A review. Int. J. Biol. Macromol., 101: 910–921.

Heleno, S.A., Barros, L., Martins, A. et al. 2012. Fruiting body spores and *in vitro* produced mycelium of *Ganoderma lucidum* from Northeast Portugal: a comparative study of the antioxidant potential of phenolic and polysaccharidic extracts. Food Res. Int., 46: 135–140.

Heleno, S.A., Barros, L., Sousa, M.J. et al. 2011. Targeted metabolites analysis in wild *Boletus* species. LWT Food Sci. Technol., 44: 1343–1348.

Hiramatsu, F., Murayama, T., Koseki T. et al. 2007. Strobilols A-D: Four cadinane-type sesquiterpenes from the edible mushroom *Strobilurus ohshimae*. Phytochem., 68: 1267–1271.

Hsu, H.C., Hsu, C.I., Lin R.H. et al. 1997. Fip-vvo, a new fungal immunomodulatory protein isolated from *Vovariella volvacea*. Biochem. J., 323: 557–565.

Intaraudom, C., Boonyuen, N., Supothina, S. et al. 2013. Novel spiro-sesquiterpene from the mushroom *Anthracophyllum* sp. BCC18695. Phytochem. Lett., 6: 345–349.

Isaka, M., Chinthanom, P., Kongthong, S. et al. 2013. Lanostane triterpenes from cultures of the basidiomycete *Ganoderma orbiforme* BCC 22324. Phytochem., 87: 133–139.

Ishikawa, N.K., Fukushi, Y., Yamaji, K. et al. 2001. Antimicrobial cuparene- type sesquiterpenes, enokipodins C and D, from a mycelial culture of *Flammulina velutipes*, J. Nat. Prod., 64: 932–934.

Jayakumar, T., Thomas, P.A. and Geraldine, P. 2009. *In vitro* antioxidant activities of an ethanolic extract of the oyster mushroom, *Pleurotus ostreatus*. Inn. Food Sci. Emerg. Tech., 10: 228–234.

Jedinak, A., Dudhgaonkar, S., Wu, Q.L. et al. 2011. Anti-inflammatory activity of edible oyster mushroom is mediated through the inhibition of NF-κB and AP-1 signaling. Nutr. J., 10: 52.

Jung, E.C., Kim, K.D., Bae, C.H. et al. 2007. A mushroom lectin from ascomycete *Cordyceps militaris*. Biochim. Biophys. Acta., 1770: 833–838.

Kalogeropoulos, N., Yanni, A.E., Koutrotsios, G. et al. 2013. Bioactive microconstituents and antioxidant properties of wild edible mushrooms from the island of Lesbos, Greece. Food Chem. Toxicol., 55: 378–385.

Kamo, T., Imura, Y., Hagio, T. et al. 2004. Anti-inflammatory cyathane diterpenoids from *Sarcodon scabrosus*. Biosci. Biotech. Biochem., 68: 1362–1365.

Kanokmedhakul, S., Lekphroma, R., Kanokmedhakul, K. et al. 2012. Cytotoxic sesquiterpenes from luminescent mushroom *Neonothopanus nambi*. Tetrahedr., 68: 8261–8266.

Kawagishi, H., Nomura, A., Mizuno, T. et al. 1990. Isolation and characterization of a lectin from *Grifola frondosa* fruiting bodies. Biochim. Biophys. Acta, 1034: 247–252.

Kawagishi, H., Mori, H., Unoa, A. et al. 1994. A sialic acid-binding lectin from the mushroom *Hericium erinaceum*. FEBS Lett., 340: 56–58.

Kawagishi, H., Ishiyama, D., Mori, H. et al. 1997. Dictyophorines A and B, two stimulators of NGF-synthesis from the mushroom *Dictyophora indusiata*. Phytochem., 45: 1203–1205.

Kawagishi, H., Masui, A., Tukuyama, S. et al. 2006. Erinacines J and K from the mycelia of *Hericium erinaceum*. Tetrahedr., 62: 8463–8466.

Kim, K.H., Moon, E., Choi, S.U. et al. 2013. Lanostane triterpenoids from the mushroom *Naematoloma fasciculare*. J. Nat. Prod., 76: 845–851.

Kim, M.Y., Seguin, P., Ahn, J.K. et al. 2008. Phenolic compound concentration and antioxidant activities of edible and medicinal mushrooms from Korea. J. Agric. Food Chem., 56: 7265–7270.

Kim, S.H., Song, Y.S., Kim, S.K. et al. 2004. Anti-inflammatory and related pharmacological activities of the n-BuOH subfraction of mushroom *Phellinus linteus*. J. Ethnopharmacol., 93: 141–146.

Kino, K., Yamashita, A., Yamaoka, K. et al. 1989. Isolation and characterization of a new immunomodulatory protein, Ling Zhi-8 (LZ-8), from *Ganoderma lucidum*. J. Biol. Chem., 264: 472–478.

Ko, J.L., Hsu, C.T., Lin, R.H. et al. 1995. A new fungal immunomodulatory protein, FIP-fve isolated from the edible mushroom, *Flammulina velutipes* and its complete amino acid sequence. Eur. J. Biochem., 228: 244–249.

Kobayashi, Y., Kobayashi, K., Umehara, K. et al. 2004. Purification, characterization, and sugar binding specificity of an N-Glycolylneuraminic acid-specific lectin from the mushroom *Chlorophyllum molybdites*. J. Biol. Chem., 279: 53048–53055.

Kochibe, N. and Furukawa, K. 1980. Purification and properties of a novel fucosespecific hemagglutinin of *Aleuria aurantia*. Biochem., 19: 2841–2846.

Kojima, H., Akaki, J., Nakajima, S. et al. 2010. Structural analysis of glycogen-like polysaccharides having macrophage activating activity in extracts of *Lentinula edodes* mycelia J. Nat. Med., 64: 16–23.

Koyama, Y., Katsuno, Y., Miyoshi, N. et al. 2002. Apoptosis induction by lectin isolated from the mushroom *Boletopsis leucomelas* in U937 cells. Biosci. Biotechnol. Biochem., 66: 784–789.

Kozarski, M., Klaus, A., Niksic, M. et al. 2011. Antioxidative and immunomodulating activities of polysaccharide extracts of the medicinal mushrooms *Agaricus bisporus*, *Agaricus brasiliensis*, *Ganoderma lucidum* and *Phellinus linteus*. Food Chem., 129: 1667–1675.

Lai, T.K., Biswas, G., Chatterjee, S. et al. 2012. Leishmanicidal and anticandidal activity of constituents of Indian edible mushroom *Astraeus hygrometricus*. Chem. Biodivers., 9: 1517–1524.

Lee, I.K., Yun, B., Kim, Y. et al. 2002a. Two neuroprotective compounds from mushroom *Daldinia concentrica*. J. Microbiol. Biotechnol., 12: 692–694.

Lee, I.K., Yun, B.S., Han, G. et al. 2002b. Dictyoquinazols A, B, and C, new neuroprotective compounds from the mushroom *Dictyophora indusiata*. J. Nat. Prod., 65: 1769–1772.

Lee, I.S., Seo, J.J., Kim, J.P. et al. 2010. Lanostane triterpenes from the fruiting bodies of *Ganoderma lucidum* and their inhibitory effects on adipocyte differentiation in 3T3-L1 cells. J. Nat. Prod., 73: 172–176.

Lee, I.S., Ahn, B.R., Choi, J.S. et al. 2011. Selective cholinesterase inhibition by lanostane triterpenes from fruiting bodies of *Ganoderma lucidum*. Bioorg. Med. Chem. Lett., 21: 6603–6607.

Li, C., Li, Y. and Sun, H.H. 2006. New ganoderic acids, bioactive triterpenoid metabolites from the mushroom *Ganoderma lucidum*. Nat. Prod. Res., 20: 985–991.

Li, X., Wu, Q., Xie, Y. et al. 2015. Ergosterol purified from medicinal mushroom *Amauroderma rude* inhibits cancer growth *in vitro* and *in vivo* by up-regulating multiple tumor suppressors. Oncotarget., 6: 17832–17846.

Li, Y.R., Liu, Q.H., Wang, H.X. et al. 2008. A novel lectin with potent antitumor, mitogenic and HIV-1 reverse transcriptase inhibitory activities from the edible mushroom *Pleurotus citrinopileatus*. Biochim. Biophys. Acta, 1780: 51–57.

Lin, W.H., Huang, C.H., Hsu, C.I. et al. 1997. Dimerization of the N-terminal amphipathic a-helix domain of the fungal immunomodulatory protein from *Ganoderma tsugae* (Fip-gts) defined by a yeast two-hybrid system and site-directed mutagenesis. J. Biol. Chem., 272: 2044–2204.

Lindequist, U., Niedermeyer, T.H.J. and Julich, W.D. 2005. The pharmacological potential of mushrooms. ECAM, 2: 285–299.

Liu, L.Y., Li, Z.H., Dong, Z.J. et al. 2012. Two novel fomannosane-type sesquiterpenoids from the culture of the basidiomycete *Agrocybe salicacola*. Nat. Prod. Bioprospect., 2: 130–132.

Liu, Q., Wang, H. and Ng, T.B. 2004. Isolation and characterization of a novel lectin from the wild mushroom *Xerocomus spadiceus*. Peptid., 25: 7–10.

Liu, Q., Wang, H. and Ng, T.B. 2006. First report of a xylose-specific lectin with potent hemagglutinating, antiproliferative and anti-mitogenic activities from a wild ascomycete mushroom. Biochim. Biophys. Acta, N1760: 1914–1919.

Lyimo, B., Funakuma, N., Minami, Y. et al. 2012. Characterization of a new α-galactosyl-binding lectin from the mushroom *Clavaria purpurea*. Biosci. Biotechnol. Biochem., 76: 336–342.

Ma, K., Ren, J., Han, J. et al. 2014. Ganoboninketals A-C, antiplasmodial 3,4-seco-27-norlanostane triterpenes from *Ganoderma boninense*. Pat., J. Nat. Prod., 77: 1847–1852.

Maiti, S., Bhutia, S.K., Mallick, S.K. et al. 2008. Antiproliferative and immunostimulatory protein fraction from edible mushrooms. Environ. Toxicol. Pharmacol., 26: 187–191.

Maity, K.K., Patra, S., Dey, B. et al. 2011. A heteropolysaccharide from aqueous extract of an edible mushroom, *Pleurotus ostreatus* cultivar: Structural and biological studies Carbohyd. Res., 346: 366–372.

Maity, K.K., Patra, S., Dey, B. et al. 2013. A β-glucan from the alkaline extract of a somatic hybrid (PfloVv5FB) of *Pleurotus florida*and *Volvariella volvacea*: Structural characterization and study of immunoactivation Carbohyd. Res., 370: 13–18.

Maji, P.K., Sen, I.K., Behera, B. et al. 2012. Structural characterization and study of immunoenhancing properties of a glucan isolated from a hybrid mushroom of *Pleurotus florida* and *Lentinula edodes*. Carbohyd. Res., 358: 110–115.

Mancheño, J.M., Tateno, H., Goldstein, I.J. et al. 2005. Structural analysis of the *Laetiporus sulphureus* hemolytic pore-forming lectin in complex with sugars. J. Biol. Chem., 280: 17251–17259.

Mandal, E.K. Maity, K., Maity, S. et al. 2012. Chemical analysis of an immunostimulating (1→4)-, (1→6)-branched glucan from an edible mushroom *Calocybe indica*. Carbohyd. Res., 347: 172–177.

Margulis, L. and Schwartz, K.V. 1988. Five Kingdoms: An Illustrated Guide to the Phyla of Life on Earth. 2nd edition, W.H. Freeman and Company, New York, 25–28.

Mattila, P., Konko, K., Eurola, M. et al. 2001. Contents of vitamins, mineral elements, and some phenolic compounds in cultivated mushrooms. J. Agric. Food Chem., 49: 2343–2348.

Mizuno, M., Morimoto, M., Minato, K. et al. 1998. Polysaccharides from *Agaricus blazei* stimulates lymphocyte T-cell subsets in mice. Biosci. Biotech. Biochem., 62: 434–437.

Mizuno, T., Hagiwara, T., Nakamura, T. et al. 1990. Antitumour activity and some properties of water soluble polysaccharides from "Himematsutake" the fruiting body of *Agaricus blazei* Murril. Agric. Biol. Chem., 54: 2889–2896.

Mizuno, T., Minato, K., Ito, H. et al. 1999. Antitumor polysaccharide from the mycelium of liquid-cultured *Agaricus blazei* Murrill. Biochem. Mol. Biol. Int., 47: 707–714.

Mo, H., Winter, H.C., Goldstein, I.J. 2000. Purification and characterization of a Neu5Acalpha2-6Galbeta1-4Glc/GlcNAc-specific lectin from the fruiting body of the polypore mushroom *Polyporus squamosus*. J. Biol. Chem., 275: 10623–10629.

Mothana, R.A.A., Ali, N.A.A., Jansen, R. et al. 2003. Antiviral lanostanoid triterpenes from the fungus *Ganoderma pfeifferi*. Fitoter., 74: 177–180.

Nagai, K., Chiba, A., Nishino, T. et al. 2006. Dilinoleoyl-phosphatidylethanolamine from *Hericium erinaceum* protects against ER stress-dependent neuro-2a cell death via protein kinase C pathway. J. Nutr. Biochem., 17: 525–530.

Nandi, A.K. Sen, I.K. Samanta, S. et al. 2012. Glucan from hot aqueous extract of an ectomycorrhizal edible mushroom, *Russula albonigra* (Krombh.) Fr.: structural characterization and study of immunoenhancing properties Carbohyd. Res., 363: 43–50.

Ng, T.B., Ngai, P.H., Xia, L. 2006. An agglutinin with mitogenic and antiproliferative activities from the mushroom *Flammulina velutipes*. Mycologia, 98: 167–171.

Ngai, P.H. and Ng, T.B. 2004. A mushroom (*Ganoderma capense*) lectin with spectacular thermostability, potent mitogenic activity on splenocytes, and antiproliferative activity toward tumor cells. Biochem. Biophys. Res. Commun., 314: 988–993.

Niedermeyer, T.H.J., Lindequist, U., Mentel, R. et al. 2005. Antiviral terpenoid constituents of *Ganoderma pfeifferi*. J. Nat. Prod., 68: 1728–1731.

Niu, X., Qiu, M., Li, Z. et al. 2004. Two novel 3,4-seco-trinorlanostane triterpenoids isolated from *Ganoderma fornicatum*. Tetrah. Lett., 45: 2989–2993.

Ojha, K.A., Krishnendu, C., Kaushik, G. et al. 2010. Glucans from the alkaline extract of an edible mushroom, *Pleurotus florida*, cv Assam Florida: Isolation, purification, and characterization. Carbohyd. Res., 345: 2157–2163.

Olausson, J., Tibell, L., Jonsson, B.H. et al. 2008. Detection of a high affinity binding site in recombinant *Aleuria aurantia* lectin. Glycoconj. J., 25: 753–762.

Ooi, D.V.E. and Liu, F. 2000. Immunomodulation and anti-cancer activity of polysaccharide-protein complexes Current Med. Chemist., 7: 715–729.

Pacheco-Sanchez, M., Boutin, Y., Angers, P. et al. 2006. A bioactive (1-3)-,(1-4)-β-D-glucan from *Collybia dryophila* and other mushrooms. Mycolog., 98: 180–185.

Panchak, L.V. and Antoniuk, V.O. 2007. Purification of lectin from fruiting bodies of *Lactarius rufus* (Scop.: Fr.) Fr. and its carbohydrate specificity. Ukr. Biokhim. Zh., 79: 123–128.

Patel, S. and Goya, A. 2012. Recent developments in mushrooms as anti-cancer therapeutics: A review. Biotechnol., 2: 1–15.

Patra, P., Bhanja, S.K., Sen, I.K. et al. 2012. Structural and immunological studies of hetero polysaccharide isolated from the alkaline extract of *Tricholoma crassum* (Berk.) Sacc Carbohyd. Res. 362: 1–7.

Pemberton, R.T. 1994. Agglutinins (lectins) from some British higher fungi. Mycol. Res. 98: 277–290.

Petri, G., Expert, P., Turkheimer, F. et al. 2014. Homological scaffolds of brain functional networks. J. R. Soc. Interface., 11: 1–10.

Pohleven, J., Renko, M., Magister, Š. et al. 2012. Bivalent carbohydrate binding is required for biological activity of *Clitocybe nebularis* lectin (CNL), the N,N'-diacetyllactosediamine (GalNAcβ1–4GlcNAc, LacdiNAc)-specific lectin from basidiomycete *C. nebularis*. J. Biol. Chem., 287: 10602–10612.

Prasad, Y. and Wesely, W.E.G. 2008. Antibacterial activity of the bio-multidrug (*Ganoderma lucidum*) on Multidrug resistant *Staphylococcus aureus* (MRSA). Advanced Biotechnology, 10: 9–16.

Puttaraju, N.G., Venkateshaiah, S.U., Dharmesh, S.M. et al. 2006. Antioxidant activity of indigenous edible mushrooms. J. Agric. Food Chem., 54: 9764–9772.

Rathee, S. Rathee, D. Rathee, D. et al. 2012. Mushrooms as therapeutic agents. Braz. J. Pharmacog., 22: 459–474.

Reis, F.S., Heleno, S.A., Barros, L. et al. 2011. Toward the antioxidant and chemical characterization of mycorrhizal mushrooms from Northeast Portugal. J. Food Sci., 76: 824–830.

Ribeiro, B., Rangel, J., Valentao, P. et al. 2006. Contents of carboxylic acids and two phenolics and antioxidant activity of dried Portuguese wild edible mushrooms. J. Agric. Food Chem., 54: 8530–8537.

Ribeiro, B., Valentao, P., Baptista, P. et al. 2007. Phenolic compounds, organic acids profiles and antioxidative properties of beefsteak fungus (*Fistulina hepatica*). Food Chem. Toxicol., 45: 1805–1813.

Ribeiro, B., de Pinho, P.G., Andrade, P.B. et al. 2011. Do Bioactive Carotenoids Contribute to the Color of Edible Mushrooms? Chem. Biomed. Meth. J., 4: 14–18.

Rout, D., Mondal, S., Chakraborty, I. et al. 2004. Structural characterization of an immunomodulating polysaccharide isolated from aqueous extract of *Pleurotus florida* fruit-bodies. Med. Chem. Res., 13: 509–517.

Roy, S.K., Das, D., Mondal, S. et al. 2009. Structural studies of an immunoenhancing water-soluble glucan isolated from hot water extract of an edible mushroom, Pleurotus florida, cultivar Assam Florida Carbohyd. Res., 344: 2596–2601.

Ruthes, A.C., Smiderle, F.R., Iacomini, M. 2016, Mushroom heteropolysaccharides: A review on their sources, structure and biological effects. Carbohydr. Polym., 136: 358–375.

Samanta, S. Maity, K. Nandi, A.K. et al. 2013. A glucan from an ectomycorrhizal edible mushroom *Tricholoma crassum* (Berk.) Sacc.: Isolation, characterization, and biological studies Carbohyd. Res., 367: 33–40.

Sánchez, C. 2017. Bioactives from Mushroom and their Application. Puri, M. (ed.). Springer International Publishing, Mexico.

Sarkar, R. Nandan, C.K. Bhunia, S.K. et al. 2012. Glucans from alkaline extract of a hybrid mushroom (backcross mating between PfloVv12 and *Volvariella volvacea*): Structural characterization and study of immunoenhancing and antioxidant properties Carbohyd. Res., 347: 107–113.

Sasaki, T. and Takasuka, N. 1976. Further study of the structure of lentinan, an anti-tumor polysaccharide from *Lentinus edodes* Carbohyd. Res., 47: 99–104.

Shao, H.J., Qing, C., Wang, F. et al. 2005. A new cytotoxic lanostane triterpenoid from the basidiomycete *Hebeloma versipelle*. J. Antibiot., 58: 828–831.

She, Q.B., Ng, T.B., Liu, W.K. et al. 1998. A novel lectin with potent immunomodulatory activity isolated from both fruiting bodies and cultured mycelia of the edible mushroom *Volvariella volvacea*. Biochem. Biophys. Res. Commun., 247: 106–111.

Shi, X.W., Liu, L., Gao J.M. et al. 2011. Cyathane diterpenes from chinese mushroom *Sarcodon scabrosus* and their neurite outgrowth-promoting activity. Eur. J. Med. Chem., 46: 3112–3117.

Shibata, H., Irie, A. and Morita, Y. 1998. New antibacterial diterpenoids from the *Sarcodon scabrosus*. Biosci. Biotech. Biochem., 62: 2450–2452.

Singdevsachana, S.K., Auroshreeb, P., Mishrab, J. et al. 2016. Mushroom polysaccharides as potential prebiotics with their antitumor and immunomodulating. Bioact. Carbohyd. Diet. Fibre., 7: 1–14.

Stanikunaite, R., Radwan, M.M., Trappe, J.M. et al. 2008. Lanostane-type triterpenes from the mushroom *Astraeus pteridis* with antituberculosis activity. J. Nat. Prod., 71: 2077– 2079.

Stepanova, L.V., Nikitina, V.E. and Boĭko, A.S. 2007. Isolation and characterization of lectin from the surface of *Grifola frondosa* (Fr.) S.F. Gray mycelium. Mikrobiol., 76: 488–493.

Su, H.J., Fann, Y.F., Chung, M.I. et al. 2000. New lanostanoids of *Ganoderma tsugae*, J. Nat. Prod., 63: 514–516.

Suzuki, T., Sugiyama, K., Hirai, H. et al. 2012. Mannose-specific lectin from the mushroom Hygrophorus russula. Glycobiol., 22: 616–629.

Svajger, U., Pohleven, J., Kos, J. et al. 2011. A ricin B-like lectin from mushroom *Clitocybe nebularis*, induces maturation and activation of dendritic cells via the toll-like receptor 4 pathway. Immunol., 134: 409–418.

Trigueros, V., Lougarre, A., Ali-Ahmed, D. et al. 2003. *Xerocomus chrysenteron* lectin: Identification of a new pesticidal protein. Biochim. Biophys. Acta, 1621: 292–298.

Tsukamoto, S., Macabalang, A.D., Nakatani, K. et al. 2003. Tricholomalides A-C, new neurotrophic diterpenes from the mushroom *Tricholoma* sp. J. Nat. Prod., 66: 1578–1581.

Valentao, P., Andrade, P.B. and Rangel, J. 2005. Effect of the conservation procedure on the contents of phenolic compounds and organic acids in Chanterelle (*Cantharellus cibarius*) mushroom. J. Agric. Food Chem., 53: 4925–4931.

Vaz, J.A., Barros, L., Martins, A. et al. 2011a. Phenolic profile of seventeen Portuguese wild mushrooms. LWT Food Sci. Technol., 44: 343–346.

Vaz, J.A., Barros, L., Martins, A., et al. 2011b. Chemical composition of wild edible mushrooms and antioxidant properties of their water soluble polysaccharidic and ethanolic fractions. Food Chem., 126: 610–616.

Wang, H., Gao, J. and Ng, T.B. 2000. A new lectin with highly potent antihepatoma and antisarcoma activities from the Oyster mushroom *Pleurotus ostreatus*. Biochem. Biophys. Res. Commun., 275: 810–816.

Wang, H., Ng, T.B., Liu, Q. 2002. Isolation of a new heterodimeric lectin with mitogenic activity from fruiting bodies of the mushroom *Agrocybe cylindracea*. Life Sci., 70: 877–885.

Wang, H. and Ng, T.B. 2003. Isolation of a novel N-acetylglucosamine-specific lectin from fresh sclerotia of the edible mushroom *Pleurotus tuber-regium*. Protein Expr. Purif., 29: 156–160.

Wang, H., Ng, T.B., Liu, Q. 2003. A novel lectin from the wild mushroom *Polyporus adusta*. Biochem. Biophys. Res. Commun., 307: 535–539.

Wang, H.X., Ng, T.B., Liu, W.K. et al. 1995. Isolation and characterization of two distinct lectins with antiproliferative activity from the cultured mycelium of the edible mushroom *Tricholoma mongolicum*. Int. J. Pept. Protein Res., 46: 508–513.

Wang, H.X., Ng, T.B., Ooi, V.E. et al. 1997. Actions of lectins from the mushroom *Tricholoma mongolicum* on macrophages, splenocytes and life-span in sarcoma-bearing mice. Anticancer Res., 17: 419–424.

Wang, H.X., Ng, T.B. and Ooi, V.E.C. 1999. Studies on purification of a lectin from fruiting bodies of the edible shiitake mushroom *Lentinus edodes*. Int. J. Biochem. Cell Biol., 31: 595–599.

Wang, H.X. and Ng, T.B. 2006. Purification of a laccase from fruiting bodies of the mushroom *Pleurotus eryngii*. Appl. Microbiol. Biotechnol., 69: 521–525.

Wang, S., Bao, L., Zhao, F. et al. 2013b. Isolation, identification, and bioactivity of monoterpenoids and sesquiterpenoids from the mycelia of edible mushroom *Pleurotus cornucopiae*, J. Agric. Food Chem., 61: 5122–5129.

Wang, S.J., Bao, L., Han, J.J. et al. 2013a. Pleurospiroketals A-E, perhydrobenzannulated 5,5-spiroketal sesquiterpenes from the edible mushroom *Pleurotus cornucopiae*. J. Nat. Prod., 76: 45–50.

Wang, Y., Bao, L., Liu, D. et al. 2012. Two new sesquiterpenes and six norsesquiterpenes from the solid culture of the edible mushroom *Flammulina velutipes*, Tetrahedr., 68: 3012–3018.

Wasser, S.P. 2002. Medical mushrooms as a source of antitumor and immunomodulating polysaccharides. Appl. Microbiol. Biotechnol., 60: 258–274.

Weng, C.J., Chau, C.F., Chen, K.D. et al. 2007. The anti-invasive effect of lucidenic acids isolated from a new *Ganoderma lucidum* strain. Mol. Nutr. Food Res., 51: 1472–1477.

Wong, J.H., Wang, H.X. and Ng, T.B. 2008. Marmorin, a new ribosome inactivating protein with antiproliferative and HIV-1 reverse transcriptase inhibitory activities from the mushroom *Hypsizigus marmoreus*. Appl. Microbiol. Biotechnol., 81: 669–674.

Wu, Y., Wang, H. and Ng, T.B. 2011. Purification and characterization of a lectin with antiproliferative activity toward cancer cells from the dried fruit bodies of *Lactarius flavidulus*. Carbohydr. Res., 346: 2576–2581.

Xu, C.J., Wang, Y.X., Niu, B.N. et al. 2014. Isolation and characterization of a novel lectin with mitogenic activity from *Pleurotus ferulae*. Pak. J. Pharm. Sci., 27: 983–989.

Yagi, F. and Tadera, K. 1988. Purification and characterization of lectin from *Auricularia polytricha*. Agric. Biol. Chem., 52: 2077–2079.

Yaltirak, T., Aslim, B., Ozturk, S. et al. 2009. Antimicrobial and antioxidant activities of *Russula delica* Fr. Food Chem. Toxicol., 47: 2052–2056.

Yan, J.K., Wang, W.Q. and Wu, J.Y. 2014. Recent advances in *Cordyceps sinensis* polysaccharides: Mycelial fermentation, isolation, structure, and bioactivities: A review. J. Funct. Foods, 6: 33–47.

Yang, N., Tong, X., Xiang, Y. et al. 2005a. Molecular character of the recombinant antitumor lectin from the edible mushroom *Agrocybe aegerita*. J. Biochem., 138: 145–150.

Yang, N., Tong, X., Xiang, Y. et al. 2005b. Crystallization and preliminary crystallographic studies of the recombinant antitumour lectin from the edible mushroom *Agrocybe aegerita*. Biochim. Biophys. Acta, 1751: 209–212.

Yang, S.X., Yu, Z.C., Lu, Q.Q. et al. 2012. Toxic lanostane triterpenes from the basidiomycete *Ganoderma amboinense*. Phytochem. Lett., 5: 576–580.

Yatohgo, T., Nakata, M., Tsumuraya, Y. et al. 1988. Purification and properties of a lectin from the fruitbodies of *Flammulina velutipes*. Agric. Biol. Chem., 52: 1485–1493.

Yoshikawa, K., Kuroboshi, M., Ahagon, S. et al. 2004. Three novel crustulinol esters, saponaceols A-C from *Tricholoma saponaceum*. Chem. Pharm. Bull., 52: 886–888.

Yoshikawa, K., Nishimura, N., Bando, S. et al. 2002. New lanostanoids, elfvingic Acids A-H, from the fruit body of *Elfvingia applanata*. J. Nat. Prod., 65: 548–552.

Yu, Z., Ming, G., Kaiping, W. et al. 2010. Structure, chain conformation and antitumor activity of a novel polysaccharide from *Lentinus edodes* Fitoterapia, 81: 1163–1170.

Yun, B.S., Lee, I.K., Cho, Y. et al. 2002. New tricyclic sesquiterpenes from the fermentation broth of *Stereum hirsutum*, J. Nat. Prod., 65: 786−788.

Yuswan, M.H.M.Y, Al-Obaidi, J.R., Rahayu, A. 2015. New bioactive molecules with potential antioxidant activity from various extracts of wild edible Gelam mushroom (*Boletus* spp.). Adv. Biosci. Biotechnol., 6: 320−329.

Zhang, G., Sun, J., Wang, H. et al. 2010b. First isolation and characterization of a novel lectin with potent antitumor activity from a Russula mushroom. Phytomed., 17: 775−781.

Zhang, G.Q., Sun, J., Wang, H.X. et al. 2009. A novel lectin with antiproliferative activity from the medicinal mushroom *Pholiota adiposa*. Acta Biochim. Pol., 56: 415−421.

Zhang, G.Q., Wang, Y.F., Zhang, X.Q. et al. 2010a. Purification and characterization of a novel laccase from the edible mushroom *Clitocybe maxima*. Process Biochem., 45: 627−633.

Zhang, W., Tian, G., Geng, X. et al. 2014. Isolation and Characterization of a Novel Lectin from the Edible Mushroom *Stropharia rugosoannulata*. Molecules, 19: 19880−19891.

Zhao, S., Zhao, Y. and Li, S. 2010. A novel lectin with highly potent antiproliferative and HIV-1 reverse transcriptase inhibitory activities from the edible wild mushroom *Russula delica*. Glycoconj. J., 27: 259−265.

Zheng, R., Jie, S., Hanchuan, D. et al. 2005. Characterization and immunomodulating activities of polysaccharide from *Lentinus edodes* Int. Immunopharmacol., 5: 811−820.

Zheng, S., Li, C., Ng, T.B. et al. 2007. A lectin with mitogenic activity from the edible wild mushroom *Boletus edulis*. Process Biochem., 42: 1620−1624.

Ziegenbein, F.C., Hanssen, H.P. and Konig, W.A. 2006. Secondary metabolites from *Ganoderma lucidum* and *Spongiporus leucomallellus*. Phytochem., 67: 202−211.

Zong, A., Cao, H. and Wang, F. 2012. Anticancer polysaccharides from natural resources: A review of recent research. Carbohyd. Polym., 90: 1395−1410.

Žurga, S., Pohleven, J., Renko, M. et al. 2014. Novel β-trefoil lectin from the parasol mushroom (*Macrolepiota procera*) is nematotoxic. FEBS J., 281: 3489−3506.

11

Terpenoids of *Russula* (Basidiomycota), with Emphasis on Cucurbitane Triterpenes

Marco Clericuzio[1,*] and *Alfredo Vizzini*[2,*]

INTRODUCTION

Fungi are eukaryotic organisms known to colonize and inhabit almost all the ecological niches of the Earth, being able to utilize a noteworthy variety of growing substrates (Taylor et al., 2014; Tedersoo et al., 2014). They play key roles in ecosystems as mutualists, pathogens, and decomposers. Fungi comprise about 100,000 described species but the analysis by Taylor et al. (2014) suggested that < 2% of fungal species have been described, implying that the Fungi are equalled only by the Insecta with respect to Eukaryote diversity. A characteristic feature of most fungi is the ability to produce a wide array of secondary metabolites, probably even more numerous and diverse than those of plants and animals. These are specialized molecules, different from the so-called primary metabolites (as proteins, nucleic acids and sugars), that serve the more generalized purpose of interacting with the environment. Secondary metabolites include defensive tools against predators and parasites (Anke and Sterner, 1991), aggressive weapons (e.g., those utilised by nematode predating fungi (Anke and Sterner, 1991), pigments (Gill and Steglich, 1987; Velíšek and Cejpek, 2011), molecular signalling in mycorrhizal interactions and others (Schmidt-Dannert, 2015;

---

[1] Dipartimento di Scienze ed Innovazione Tecnologica – Università del Piemonte Orientale - Via T. Michel 11, 15121 Alessandria, Italy.

[2] Dipartimento di Scienze della Vita e Biologia dei Sistemi – Università di Torino, Viale Mattioli 25, 10125 Torino, Italy.

* Corresponding authors: marco.clericuzio@uniupo.it; alfredo.vizzini@unito.it

Chen and Liu, 2017). The key evolutionary role of many such compounds has started to be understood only recently (Fox and Howlett, 2008; Pusztahelyi et al., 2015; Rohlfs and Churchill, 2011; Pang et al., 2016; Karpyn Esqueda et al., 2017; Kellogg and Raja, 2017; Singh et al., 2017), but surely much is still to be discovered. Many secondary metabolites exhibit useful biological activities and are of interest to the pharmaceutical, food, and agrochemical industries (Smedsgaard and Nielsen, 2004; Frisvad et al., 1998, 2007, 2008). Several mycotoxins are responsible for severe human poisonings (mycetismus), sometimes with fatal outcomes (Rumack and Spoerke, 1994; Saviuc and Flesch, 2003; Diaz, 2005; Smith and Davis, 2016).

Fungal metabolites have been used by men since prehistory for nutritive, pharmaceutical and other practical aims (Grienke et al., 2014). These uses have been increasing with times, and nowadays the vast range of products derived from fungi, sold on the Internet web market, is a testament to this enduring relationship between men and fungi.

Fungal Chemotaxonomy

In contrast to primary metabolites such as proteins, DNA, RNA, polysaccharides, and lipids, which occur universally, secondary metabolites are small organic compounds usually found restricted to a particular species, genus or family. While plant chemotaxonomy is a well-established discipline (see for instance Singh, 2016, and references therein), in fungi, a taxonomic application of metabolite profiling is still in its infantry. More recently, some useful contributions of metabolite markers to fungal classification, has led to interesting results (Allen et al., 2003; Larsen et al., 2005; Nielsen and Oliver, 2005; Aliferis et al., 2013). Indeed, more chemotaxonomic studies can be found in the literature for Ascomycota (Frisvad et al., 2008), but only a few for Basidiomycota (Gluchoff and Lebreton, 1970; Gill and Steglich, 1987; Frisvad et al., 1998, 2008; Lorenzen and Anke, 1998; Gill, 1996, 2003).

What is often uneasy to understand in this field, is whether the presence of similar metabolites is due to true phylogenetic relationships, or due to simple evolutionary convergence. Thus, the toxic cyclic peptides α-amanitines, isolated for the first time from *Amanita* species (Wieland, 1983; Faulstich and Zilker, 1994), were later found also in species of the unrelated genera *Lepiota*, *Galerina* and *Conocybe* (Enjalbert et al., 2004); pulvinic acids have been detected in many Boletales (Marumoto et al., 1986; Davoli and Weber, 2002; Gill, 2003; Winner et al., 2004), but also in unrelated *Omphalotus* species (Kirchmair et al., 2002), and even in lichens (Müller, 2001); styrylpyrones in *Gymnopilus*, *Cortinarius*, *Pholiota*, *Hypholoma*, but also in the polyporoid *Phaeolus* and *Phellinus* (Benedict, 1970; Gill and Steglich, 1987; Høiland, 1990); anthraquinones are found in some *Cortinarius* groupings, but also in the unrelated genus *Tricholoma* (Vizzini, 2003); marasmane sesquiterpenes are typical of Russulales (*Russula, Lactarius, Heterobasidion, Lentinellus, Auriscalpium, Bondarzewia, Vararia, Dichostereum, Peniophora, Artomyces*) (Daniewski and Vidari, 1999), but they occur also in some Agaricales (Arnone et al., 1997). In such cases, it is evident that their presence is of limited taxonomic significance.

Better results can be achieved if the comparison is made within a lower, more restricted taxonomic group. Høiland (1983), for example, used anthraquinoid compounds as systematic markers in the subgenus *Dermocybe* of *Cortinarius*; these metabolites can be very informative at specific level, but not at higher rank, as we have previously seen. Within the *Calochroi-Fulvi* grouping of *Cortinarius*, the two sections can be differentiated by observing that *Fulvi* synthesize anthraquinones, while *Calochroi* do not. Instead, at least some species among *Calochroi,* synthesize the triterpenoid sodagnitin (Sontag et al., 1999; Høiland and Holst-Jensen, 2000).

In addition, chemotaxonomic analyses have been used with success in Boletales (Besl et al., 1986; Besl and Bresinsky, 1997; Andary et al., 1992; Binder and Bresinsky, 2002; Zeng et al., 2014), *Rhizoctonia* (Cantharellales) (Aliferis et al., 2013) and among Agaricales, *Amanita* (Beutler and der Marderosian, 1981), *Cortinarius* (Høiland, 1983; Keller and Ammirati, 1983; Arnold et al., 1987; Sontag et al., 1999), and Hygrophoraceae (Lodge et al., 2014).

Now-a-days, chemical markers may be useful for taxonomic inferences mainly in conjunction with DNA sequence analyses. In fact, incorporation of molecular data has profoundly changed the systematics of fungal phyla and lower ranks (Hibbett et al., 2007, 2013, 2014; Gherbawy and Voigt, 2010; Hibbett and Taylor, 2013; Hibbett, 2014; Jayasiri et al., 2015; Ekanayaka et al., 2017; Zhao et al., 2017). In particular, ribosomal RNA genes (rDNA) sequences, recently supplemented by protein-coding genes (e.g., tef1α, rpb1 and rpb2; Matheny et al., 2007; Zhao et al., 2017), have shed light on understanding the relationships within the Basidiomycota, and have pointed out that some morphological features which have been given importance in high-level classification, such as the form of basidia and basidiomata, are subjected to homoplasy (Swann and Taylor, 1993, 1995; Swann et al., 1999; Hibbett et al., 2014). As a consequence of the phylogenetic data, we now have a better picture of the taxonomic meaning of secondary metabolites, in particular, whether they can be considered relevant to systematics, or whether they are the outcome of evolutionary convergence. We shall discuss these topics, as concerns *Russula* terpenes, in the outgoing of this paper.

The Genus *Russula*

Russula Pers. is the major genus in Russulaceae Lotsy, a family of Agaricomycetes Doweld belonging to order Russulales Kreisel ex P.M. Kirk, P.F. Cannon & J.C. David. Russulales currently includes more than 1,700 described species (Kirk et al., 2008), showing an amazing diversity of basidiome morphologies and trophic strategies (Larsson and Larsson, 2003; Miller et al., 2006; Hibbett et al., 2014). The shape of the basidioma varies from effuse-resupinate to reflexed and bracket-like, coralloid, pileate-stipitate and sequestrate. The hymenophore is often smooth (Stereaceae, Peniophoraceae), or sometimes hydnoid (e.g., *Auriscalpium, Hericium*); a poroid hymenophore is less common (e.g., *Albatrellus, Bondarzewia, Heterobasidion*), while a lamellate hymenophore is only known from Auriscalpiaceae (*Lentinellus*) and Russulaceae. Most species in Russulales share basidiospores with an amyloid reaction of the spore wall, and for most of them, the amyloidity is associated with an ornamented pattern of the wall. The presence of specialized hyphae, i.e., "gloeocystidia" or "gloeoplerous hyphae" is characteristic of most Russulales. These elements are filled with lipidic substances that are not but the sesquiterpene fatty acid esters discussed later in this chapter.

Within Russulales, the family Russulaceae (consisting of six genera) is the richest in species due to the high diversity detected in the agaricoid genera *Lactarius* and *Russula*. These genera are characterized by a classical mushroom-like basidiome, that is, stipitate-pileate (provided with pileus and stipe), with a lamellate hymenium. The monophyletic genus *Russula* (Buyck et al., 2008) has a truly worldwide distribution, being equally represented in the temperate regions of both hemispheres, as well as in the tropical regions, up to polar ecosystems (Miller and Buyck, 2002; Geml et al., 2010; Geml and Taylor, 2013; Bazzicalupo et al., 2017; Lee et al., 2017). *Russula* is one of the most highly diverse groups of Agaricomycetes: Some 2800 *Russula* names are reported in Index Fungorum (http://www.indexfungorum.org/), corresponding to no less than 800 specific binomials; there are few doubts that many species shall still be described, mainly in extra-European regions. Inside the genus, the various infra-generic taxonomic units show a rather marked differentiation between the boreal-temperate regions and the tropics (Buyck and Mitchell, 2003). The basal, most archaic groups seem to be exclusive of tropical biomes. The totality of the species in *Russula* are obliged ecto-mycorrhizal, with a large variety of plant hosts, both woody and herbaceous (Singer, 1986); they are also an important nutrient source for insects (Yamashita and Hijii, 2007) and mammals (Fogel and Trappe, 1978), including humans (Guo, 1992; Hu and Zeng, 1992).

The Terpenes of *Russula*

Fungi are known to produce a wide spectrum of terpenoid compounds, including antibiotics, antifeedant and antitumor molecules, mycotoxins and phytohormones (Schmidt-Dannert, 2015; Chen and Liu, 2017).

Here we will review the *Russula* terpenoids: except for the pigmented metabolites which impart colour to the cap cuticle, i.e., the russupteridines (Eugster and Iten, 1975), the *Russula* secondary metabolites are constituted by terpenoids for the great majority. The prenylated phenols found in *R. ochroleuca* and *R. viscida* (Sontag et al., 2006), will not be treated, as they are only in small part of terpenic origin.

Sesquiterpenes

Sesquiterpenes, the C-15 molecules derived from farnesyl-pyrophosphate, are dominant in Russulaceae, and up to a few years ago even believed to be the only terpene type present in the family, i.e., in *Russula* and *Lactarius*. The Russulaceae sesquiterpenes have been the object of various reviews (Vidari and Vita-Finzi, 1995; Daniewski and Vidari, 1999; Clericuzio et al., 2008) but none of them has dealt with *Russula* triterpenes.

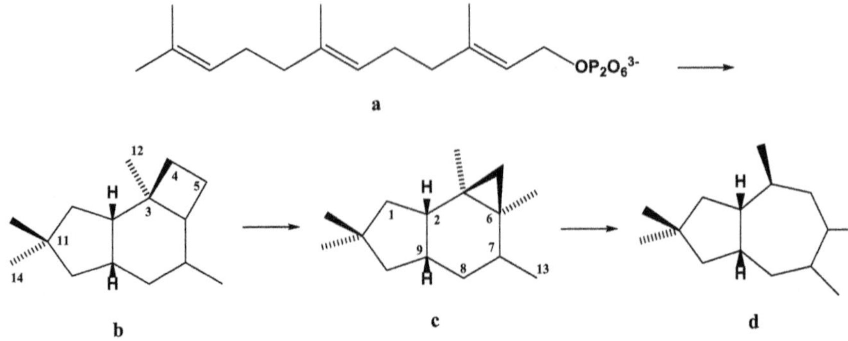

Fig. 1. The biosynthetic path leading from the protoilludane skeleton (b), to the marasmane skeleton (c), and then to the lactarane skeleton (d). In turn, all these structures derive from farnesyl pirofosfate (a).

The sesquiterpenes skeletons found to date in *Russula* belong to far less different types than in *Lactarius*; to be more precise, they mostly belong to a single sesquiterpene "family", i.e., to the proto-illudane family. For sesquiterpene family, we mean a group of structures biosynthetically derived from a single precursor skeleton (Vidari and Vita-Finzi, 1995). Within this sesquiterpene family, only protoilludane, marasmane and lactarane structures have been isolated from *Russula*; the latter two sesquiterpene types are derived from the former one via a known biosynthetic path (Fig. 1) (De Bernardi et al., 1982).

Marasmane, Lactarane and Allied Sesquiterpenes

In Russulaceae, a major role is played by the penta-cyclic marasmane sesquiterpene velutinal (**1a**), a hemi-acetal always found esterified by different fatty acids (velutinal esters (**1b**), but for convenience here referred to as "stearoyl-velutinal") (Favre-Bonvin et al., 1982). Stearoyl-velutinal is mainly stored in cystidia, both in hymenial cystidia and in dermatocystidia, two types of gloeocystidia; it is the lipidic component that reacts by turning black with sulfo-aldehyde reagents, such as sulfo-vanilline (SV) or sulfo-anisaldehyde (SA) (Gluchoff-Fiasson and Kühner, 1982). This is a micro-chemical reaction relevant to taxonomic purposes (Romagnesi, 1985). In *Lactarius*, where it is also commonly found, **1b** is stored in laticiferous hyphae (Camazine and Lupo, 1984).

Stearoyl-velutinal (**1b**) is a labile molecule, which can be isolated from intact basidiomes if careful operations are performed, as the use of proper solvents, and the absence of even weak acidic media (Sterner et al., 1983; 1985a,c); it is a mild-tasting molecule, and is not provided with any particular bioactivity (Sterner et al., 1985b). When brought in contact with specific enzymes, such as lipases (typically after rupture of the basidiome), it is more or less rapidly hydrolysed with release of the free fatty acid. The hemiacetal moiety, velutinal (**1a**), is never found free in basidiomes (apart from minuscule amounts visible if the transformation occurs at 4°C, Sterner et al., 1985b). What happens is that the oxirane ring is immediately cleaved under cationic conditions, leading to different structures that may still bear the original marasmane skeleton, or else rearrange to different skeletons, such as the lactarane one (Hansson and Sterner, 1991). Among the molecules formed after hydrolysis and transformation of **1b**, noteworthy are the α,β unsaturated dialdehydes iso-velleral **2** (marasmane skeleton), velleral **3** and piperdial **4** (lactarane skeleton), the substances responsible for the acrid-pungent taste of many *Russula* and *Lactarius* species (Sterner, 1989). Interestingly enough, such a bio-transformation from **1b** to **3** can also be performed by human saliva, due to the enzymes there contained (Hashimoto et al., 1994).

Observing the products formed in *L. vellereus*, at different time intervals after grinding fresh mushrooms, Sterner et al. (1985b) have demonstrated that dialdehydes **2-3** are transformed into the corresponding alcohol-aldehydes **5-6**, after reduction of the carbonyl at C-13. Compounds **5-6** (as well as precursor **1b**) showed, in a toxicity assay against *Escerichia coli*, *Micrococcus luteus*, and *Candida utilis*, one order of magnitude lower values than those of the parent dialdehydes **2-3**. Actually, the list

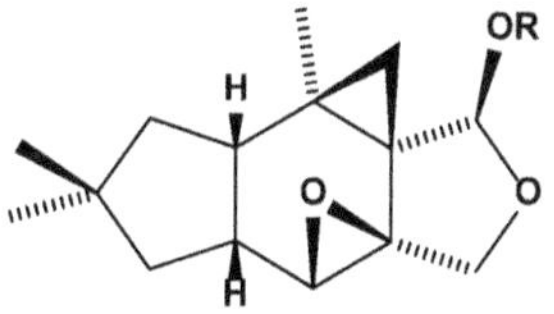

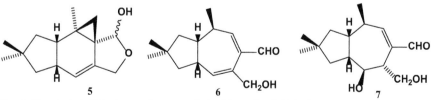

1a-b

Fig. 2. 1a velutinal (R = H); **1b** velutinal esters (R = stearoyl, oleoyl, linoleoyl, palmitoyl; also 6-keto-and 6-*S*-hydroxy-stearoyl in *Lactarius*). In the present Chapter, velutinal esters **1b** will be collectively named stearoyl-velutinal.

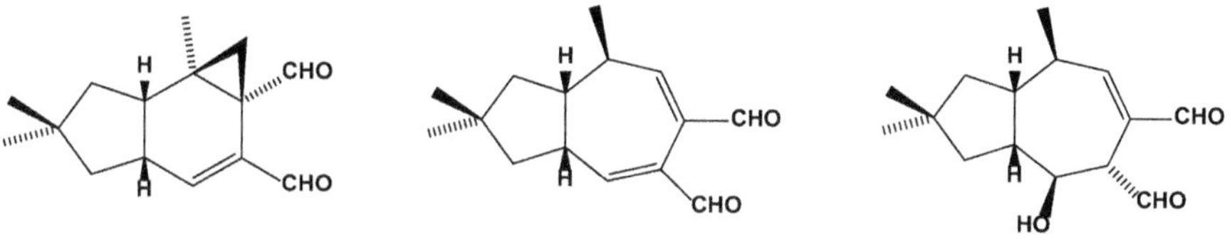

2 **3** **4**

Fig. 3. The structures of the most common hot pungent dialdehydes found in Russulaceae: Isovelleral **2**; velleral **3**; piperalol **4**.

5 **6** **7**

Fig. 4. The alcohol-aldehyde sesquiterpenes, derived from the dialdehydes **2-4**, after reduction of the aldehyde function at C-13: Isovellerol (**5**) (drawn in the hemiacetal form); vellerol (**6**); piperalol (**7**).

of bioactivities allied to Russulaceae dialdehydes, in particular to isovelleral **2**, is impressive: Cytotoxic-antiproliferative (Anke and Sterner, 1991), anti-fungal (Sterner et al., 1985b), antibacterial (Sterner et al., 1985b) and anti-feedant (Daniewski et al., 1993, 1995). Finally, these dialdehydes are in various extents harmful to men, after ingestion of the relative basidiomes: Poisoning with acrid Russulaceae frequently leads to hospitalization, even if serious effects have been scarcely reported (Rumack and Spoerke, 1994).

1b has been found in all *Russula* species provided with Dcy so far investigated: In strongly acrid species, such as *R. foetens, R. cuprea, R. fellea*, in fugitively acrid ones, such as several *Griseinae* or *Tenellae*, and in completely mild ones, such as *R. paludosa* or the *Xerampelinae*. The occurrence of **1b** in little acrid species may be explained by rapid enzymatic reduction of the pungent dialdehydes to harmless aldehyde-alcohols: For example, relatively large amounts of piperalol (**7**) were found in *R. versatilis, R. exalbicans* and *R. paludosa*, all slightly to not pungent species (Clericuzio et al., 2008). Anyway, a direct path from **1b** to mild terpenoids, i.e., not passing through the pungent dialdehydes, cannot be excluded. The fugitive or not-acrid taste may also be the consequence of the very small amount of **1b** initially present. This has been verified in *R. xerampelina*, a species whose Dcy (dermatocystidia or pileocystidia) are known to give a very weak reaction with sulfo-aldehydic reagents (Clericuzio et al., 2008). In *L. mitissimus*, a mild species, we observed that the rate of enzymatic conversion of **1b** to free sesquiterpenes was extremely slow, and still incomplete 12 h after mushroom grinding (unpublished data). The amount of precursor **1b**; the rate of its conversion to free sesquiterpenes; the formation or not of pungent dialdehydes; the chemical structure of the dialdehydes formed; the rate of reduction of dialdehydes to alcohols; the formation of furans and lactones (*vide infra*), have all been found to be diverse in the various Russulaceae species, very likely due to different enzymatic patterns (Vidari and Vita-Finzi, 1995). It seems that during natural evolution, several mechanisms have been selected concerning the presence and the enzymatic transformation of **1b**, contributing, at least partly, to the observed biodiversity of Russulaceae. A list of *Russula* and *Lactarius* species containing stearoyl-velutinal **1b** can be found in Clericuzio et al. (2008).

Recently Described Marasmane and Lactarane Sesquiterpenes

After the publication of our latest review (Clericuzio et al., 2008), a few more sesquiterpenes belonging to marasmanes were reported. From basidiomes of *R. foetens*, the new lactone russulfoen, endowed with moderate cytotoxicity, was isolated (Kim et al., 2010). Three novel marasmane lactones, viz. russulanigrins A-C, together with the nor-marasmane russulanigrin D, have been isolated from *R. nigricans* (Isaka et al., 2017a). This latter compound shows an unusual ketone function at C-7. The marasmanes from *R. nigricans* were all inactive in a series of bioassays.

Compounds **8-11** are all lactarane lactones, a numerous class of sesquiterpenes, particularly abundant in *Lactarius* (see for instance, the lactarorufins: Daniewski and Vidari, 1999), but also found in *Russula*. They have been often considered artifacts derived from air oxidation of the corresponding furans; at their turn, furanolactaranes

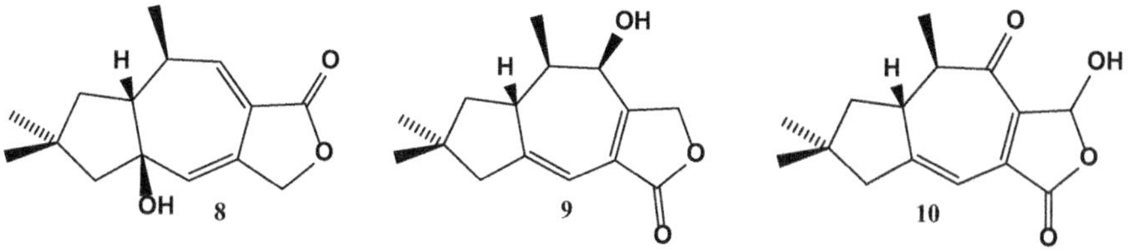

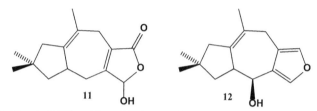

Fig. 5. Molecular structures of lactarane lactones recently isolated from *R. nobilis*: Russulanobilins A (**8**), B (**9**), and C (**10**).

Fig. 6. Blennin B (**11**), and furanol (**12**), two widespread sesquiterpenes in Russulaceae.

(as furanol 12) have been demonstrated to be, at least partly, products formed by silicagel degradation of stearoylvelutinal **1b** (Sterner et al., 1985a). Actually, after biotransformation experiments carried out in strictly controlled conditions, it is now very likely that there are enzymatic mechanisms implied in the transformation of **1b** to furans and lactones (Gilardoni et al., 2014), so that some of them should not be viewed as artefacts anymore. This is for example, the case of russulanobilin A (**8**), B (**9**), and C (**10**), recently isolated in our lab, from basidiomes of *R. nobilis* (Malagòn et al., 2014). We propose that lactones **8-10** indeed derive from oxidation of velleral **3**. Strictly speaking, an oxidative path from furans cannot be excluded, but no trace of furans was found in the *R. nobilis* extract. Conversely, the similar hydroxy-lactone blennin B(**11**) (Vidari et al., 1976) has been isolated, together with furanol **12**, a widespread furanolactarane. Therefore, in this case, the oxidation of the furan moiety to lactone seems the most likely biogenetic pathway to **11**. Interestingly enough, both lactarane sesquiterpenes **11-12** have been recently reported from Chilean basidiomes of *R. austrodelica* (Alarcón et al., 2013), one of the few reports of *Russula* terpenoids from the American continent.

Protoilludane Sesquiterpenes

In contrast with widespread marasmane derivatives, the rarer protoilludane sesquiterpenes have been isolated only from two *Russula* species to date, i.e., *R. japonica* Hongo, and its closely allied *R. pseudodelica* Lange (some protoilludanes have been found in *Lactarius* too, see Clericuzio et al., 2008). These two species belong to sect. *Pallidosporinae*, and possess only rudimentary Dcy, little reacting to SV and SA. Both investigated species were collected in Eastern Asia, either Japan or China.

Compounds **13-15** were found in *R. pseudodelica* (Clericuzio et al., 1997a), erroneously reported as *R. delica* in the original paper. In intact basidiomes, only

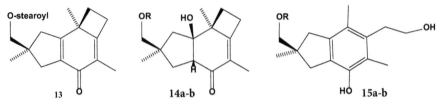

Fig. 7. The structures of some selected protoilludane and illudalane sesquiterpenes isolated from *R. pseudodelica*. Stearoyl delicone (**13**), stearoyl plorantinone (**14a**, R = stearoyl), and the illudalane ester **15a** (R = stearoyl) were found in intact basidiomes. Plorantinone B (**14b**, R = H), and illudalane alcohol **15b** (R = H) instead, were found in injured basidiomes. The aromatic illudalane structures (**15a,b**) are all artefacts obtained after rearrangement of the protoilludane skeleton.

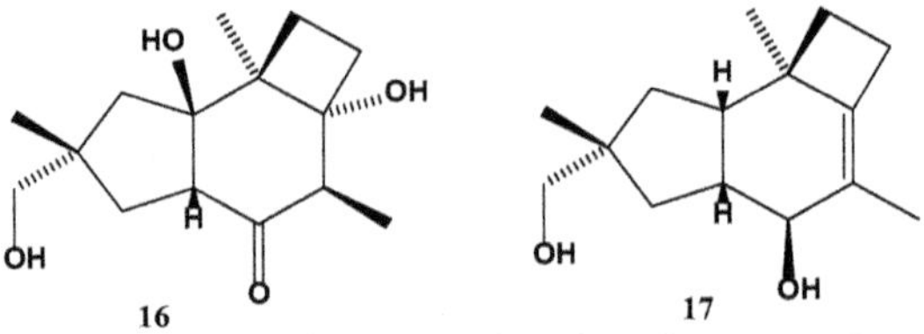

Fig. 8. Two selected protoilludane sesquiterpenes isolated from *R. japonica*: Russujaponol A (**16**), and Russujaponol C (**17**).

fatty acid esters of sesquiterpene alcohols **13** (stearoyl delicone, a very labile molecule) and **14a** (stearoyl plorantinone) were found. When the basidiome is broken ("injured"), a mechanism analogous to that seen for stearoylvelutinal **1b** happens: Esters **13-14a** are hydrolysed, probably by lipase enzymes, and the terpene part is released, even if only alcohol **14b** could be isolated, together with more oxidized protoilludane sesquiterpenes (Clericuzio et al., 1997a,b). In the course of our work on *R. pseudodelica*, illudalane ester **15a** and the relative alcohol **15b** were also encountered; however, we could demonstrate that the rearrangement from the protoilludane structure to the illudalane one, was due to traces of protonic acids present in the extraction medium, or to the use of silicagel during liquid chromatography (Clericuzio et al., 1997b).

Protoilludane sesquiterpenes, closely related to plorantinones, namely russujaponols A-D, were isolated from *R. japonica* (Yoshikawa et al., 2006). In contrast with the preceding case, no fatty acid ester was reported. We wonder whether they were really absent in intact basidiomes, or else it depended on the incorrect extraction procedure followed, which featured extraction for 4 weeks with aqueous ethanol.

Compounds such as **16** and **17** differ from plorantinones **13-14** mainly in the inverted stereochemistry at C-11. Also, in *R. japonica* illudalane sesquiterpenes were isolated (russujaponols E-F), but, as we have seen in the case of *R. pseudodelica*, these may be formed after a proton-initiated rearrangement of the protoilludane skeleton.

Our protoilludane terpenoids, as well as those isolated by the Japanese authors, were inactive in cytotoxic assays against a few selected human tumour cells; however, russujaponol A (**16**) inhibited migration of human fibrosarcoma cells (Yoshikawa et al., 2006).

The occurrence of protoilludane sesquiterpenes may represent a chemotaxonomic marker of sect. *Pallidosporinae*. Although in the past they were generally considered part of *Compactae*, a section to which the well-known *R. delica* belongs, *Pallidosporinae* have been later placed in a different clade by molecular analysis, in particular in a more archaic one (Miller and Buyck, 2002). The metabolite profile well confirms this difference: The *Compactae* so far chemically investigated, viz. *R. delica*, *R. austrodelica*, and *R. brevipes*, all contain marasmane and lactarane sesquiterpenes, and not protoilludane ones, such as the *Pallidosporinae*.

Aristolane Sesquiterpenes and Derivatives

A sesquiterpene class completely unrelated to the protoilludane family (from a biosynthetic point of view), i.e., aristolane sesquiterpenes, has been isolated from *R. lepida*. In the basidiomes collected in Italy, only aristolane sesquiterpenes, and nardosinanes derived from them, were found in an initial work (Vidari et al., 1998). It was carefully investigated if any fatty acid ester was present in intact basidiomes, but this was not the case. This experimental datum, which has been confirmed also in successive experiments, marks a profound difference with almost all other Russulaceae species so far investigated. Actually, *R. lepida* does not show true Dcy. In the cuticle cap there are scattered cystidia-like cells which, however, do not react with sulfo aldehydes (Romagnesi, 1985). From a systematic point of view, it has been traditionally placed in sect. *Lepidinae*, an isolated small group of species that do not show a hot taste, but rather have a slightly astringent or bitter taste.

Aristolanes **18-19** and nardosinane **20** belong to a rare family of sesquiterpenes isolated from higher plants and soft corals, but in the fungal kingdom only found in *R. lepida*. No specific bioactivity could be observed for the above sesquiterpenes.

A paper dealing with the terpenoids found in the basidiomes of *R. lepida*, collected in China, appeared in 2000 (Tan et al., 2000). This mushroom is frequently sold on Chinese street markets, being used both as food and as ethnopharmacologic medicine (Yuan and Sun, 1995). Moreover, an ethanol extract of this fungus has shown unspecified anti-tumour activity (Tan et al., 2000). The authors could isolate a series of aristolanes identical, or structurally closely related, to those found in the European species. A detailed comparison between the two sets of sesquiterpenes can be found in our second investigation of the same fungus (Clericuzio et al., 2012).

Actually, it should be carefully checked if the Chinese fungus is identical to the European one, or if a single species rather than more than one, is collected and commercialized under the same name. Given the similarity in secondary metabolites, we can safely accept that also the Chinese *Russula* belongs to *Lepidinae*.

A paper on the terpenoids of *R. lepida* has recently appeared from a third scientific team (Lee et al., 2016), dealing with basidiomes collected in Japan. Two novel hydroxy-aristolones **21-22** were isolated in this work.

Triterpenes

In Basidiomycota, the two more diffused triterpene skeleton types are the lanostane (Fig. 11, left), and the cucurbitane ones (Fig. 11, right). Cucurbitanes are biogenetically

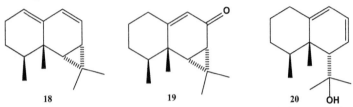

Fig. 9. Structures of two aristolane sesquiterpenes (**18-19**), and one nardosinane sesquiterpene (**20**) isolated from *R. lepida*.

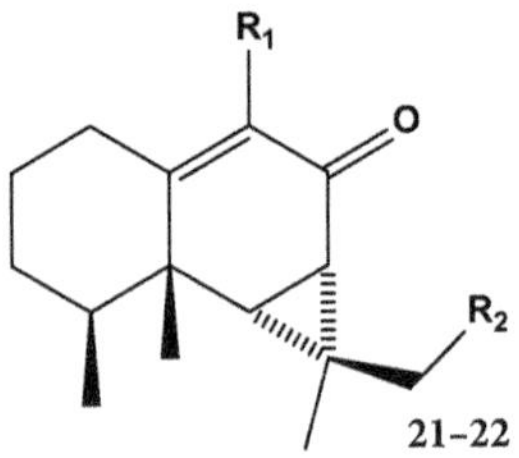

Fig. 10. Recently isolated aristolanes from Japanese specimens of *R. lepida*. **21** R$_1$ = –OH; R$_2$ = –H; **22** R$_1$ = –H, R$_2$ = –OH.

Fig. 11. Molecular structure, atom numbering and ring identification of the lanostane triterpene skeleton (left), vs. the cucurbitane skeleton (right).

derived from lanostanes (more exactly from the lanosteryl cation) after 1,2 shift of methyl 19, and successive hydride shift, leading to *trans-cis* isomerization of the B-C ring closure (Xu et al., 2004). Lanostane triterpenoids are widespread in fungi, having been found in many polyporoid species, first of all in *Ganoderma* (Xia et al., 2014; Baby et al., 2015; Chen et al., 2017), but also in *Poria, Pycnoporus, Inonotus, Ceriporia, Antrodia,* among the others (Zjawiony, 2004); in some agaricoid genera, as for example, *Hebeloma, Pholiota, Stropharia* (Shiono et al., 2007); and in species of order Boletales (*Astraeus*) (Isaka et al., 2017b). Cucurbitane triterpenes, instead, are much rarer in higher fungi (see next section).

Russula Triterpenes: Cucurbitane and Seco-cucurbitane Terpenoids

Together with the previously seen aristolane sesquiterpenes, Tan et al. (2000, 2001) also isolated triterpene metabolites in the basidiomes of *R. lepida* collected in China. Compounds **23-24** possess a cucurbitane skeleton, a kind of triterpene structure much less common than the lanostane one in fungi. In fact, a recent review on the subject (Liu, 2014), reports a total of three Agaricomycetes species so far known to possess

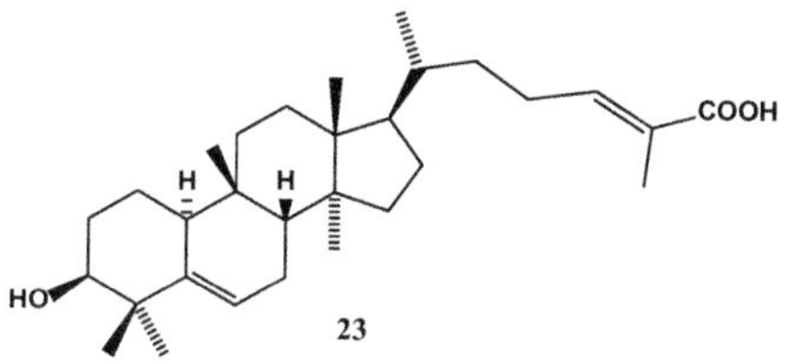

Fig. 12. Structures of the epimeric (mono)-carboxylic acids **23-24** isolated from both Chinese and European specimens of *R. lepida*.

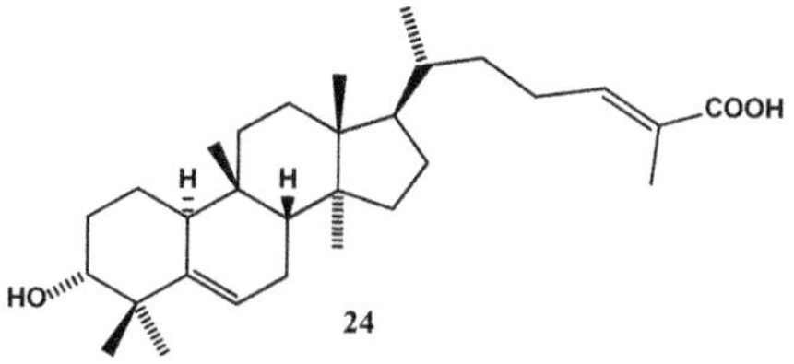

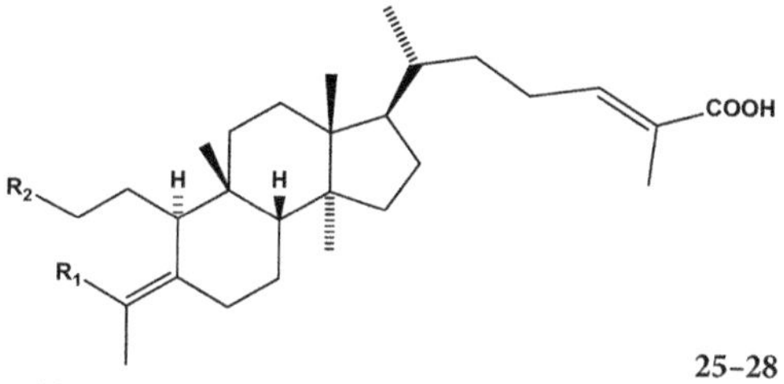

Fig. 13. Some seco-cucurbitanes isolated from *R. lepida*. **25** R_1 = H; R_2 = COOH; **26** R_1 = H, R_2= CH$_2$OH; **27** R_1 = R_2 = COOH; **28** R_1 = COOH, R_2 = CH$_2$OH.

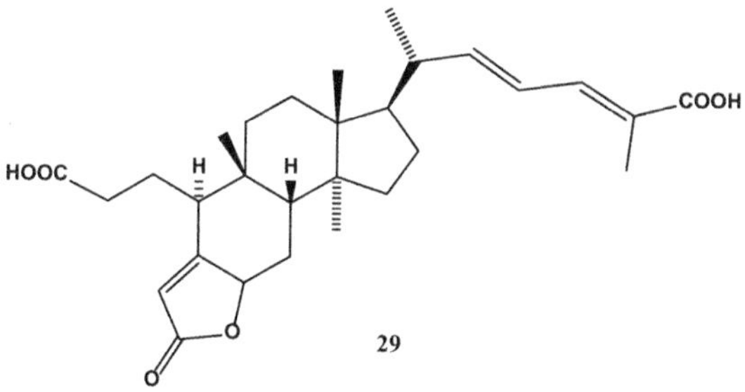

Fig. 14. Lepidolide (**29**), a seco-cucurbitane triterpene found in Chinese specimens of *R. lepida*.

cucurbitane triterpenes, that is, *Hebeloma vinosophyllum* (Fujimoto et al., 1986, 1987, 1991), *Leucopaxillus gentianeus* (Clericuzio et al., 2004, 2006) and *Russula lepida* (Tan et al., 2000), to which *R. aurora*, discussed below, should be added. Given the great phylogenetic distance among these fungi (they belong to different orders), it is very likely that the simple occurrence of cucurbitane triterpenes represents a case of evolutionary convergence.

On the other hand, cucurbitane triterpenes are not rare in higher plants (mainly in the Cucurbitaceae family), and many of them, such as the cucurbitacins, are known to be extremely bitter and cytotoxic. Their medicinal application has been limited, however, owing to high unspecific toxicity (Chen et al., 2005), even if currently there is a renewed interest in the anti-proliferative effect of some cucurbitacins (Ren et al., 2012).

It should be noted that almost all of *Russula* cucurbitanes bear a carboxylic acid functionality at C-27, which can be considered a marker of *Russula* cucurbitane triterpenes.

Fig. 15. Basidiomes of *Russula aurora*.

Noteworthy are compounds **25-28**, also isolated by Tan et al. (2000) from *R. lepida*: they are 3,4-*seco*-cucurbitane triterpenes, i.e., molecules where the A-ring of the triterpene structure has been cleaved, in correspondence of the 3-4 C-C bond. While A-*seco* triterpenes bearing a lanostane skeleton are fairly diffused in fungi (see for example the cytotoxic poricoic acids isolated form *Poria cocos* (Ukiya et al., 2002), A-*seco* triterpenes belonging to the cucurbitane series are much rarer, and those isolated from *R. lepida* represent the first known case, to our knowledge.

An interesting biosynthetic mechanism has been suggested for **25** and related *seco*-triterpenes (Xiong et al., 2006): The 3-4 bond cleavage is generated by Grob fragmentation of a bicyclic triterpene structure, where only rings A and B have been formed from the squalene epoxide precursor. This means that bond cleavage does not occur after the tetracyclic triterpene skeleton is formed, but instead at an intermediate stage: First, a bicyclic A-B ring structure is synthesized, and then it is cleaved at the A ring. Final cyclization to the tricyclic structure of these terpenoids is driven by a specific cyclase, whose codifying gene has been isolated and cloned by the authors. It is likely that this biosynthetic mechanism is at the basis of all A-*seco* triterpenes known, as well as of triterpenes cleaved in other rings of the structure (Shibuya et al., 2007).

In the second paper on *R. lepida* by our research team (Clericuzio et al., 2012), we could isolate both aristolane sesquiterpenes, and cucurbitane triterpenes, as found by the Chinese authors. In the former paper by Vidari et al. (1998), the carboxylic acid triterpenes had been lost owing to an incorrect chromatographic procedure, which featured the use of neutral alumina, a stationary phase that strongly retains acidic molecules.

In the already mentioned paper by Lee et al. (2016), the new alcohol-diacid **28** was reported from Japanese specimens of *R. lepida*. Apart from small differences, the pattern of terpenoids isolated from the Italian, the Chinese, and the Japanese basidiomes of *R. lepida* is fundamentally the same. For this reason, the finding of lactone **29** (lepidolide) by Tan et al. (2002), is puzzling, since it was not found both by the Japanese authors and by ourselves. We suspect that **29** was actually isolated from a *Roseinae* species (close to *R. aurora*, see next section), incorrectly determined as *R. lepida*.

Another representative of sect. *Lepidinae* was investigated by us: *R. amarissima*, a taxon very closely related to *R. lepida*. Phytochemical investigation of the crude extract of this fungus showed a complete identity of terpene metabolites as in *R. lepida* (Clericuzio et al., 2012).

The bioactivities of the various cucurbitane triterpenoids of *R. lepida* have been rather disappointing to date: In particular, concerning cytotoxicity against human cancer cells, only a weak activity or no activity at all, was observed. Conversely, Lee et al. (2016) have reported that compounds **27-28** were very effective in the inhibition of protein tyrosine phosphate 1B (PTP1B) activity, and may, therefore, be useful in treating diabetes and obesity.

Section Roseinae: The Chemical Base of the Pink Reaction with Sulpho-vanilline

Section *Roseinae* is a small group of *Russula*, composed of a handful species in Europe, the two most diffused being *R. aurora* and *R. minutula*. The section comprises species completely devoid of dermatocystidia (Dcy), but instead provided with a different kind of specialized cap cuticle hyphae, known as "primordial hyphae" (HP).

The members of this section all share a peculiar macrochemical reaction: A drop of sulpho-vanilline stains the flesh of the basidiome bright coral pink (the colour change is especially evident on recently dried specimens, see Fig. 17a). In order to yield a chemical rationale to this observation, we collected and extracted intact basidiomes of *R. aurora* from Italy. An acetone raw extract, loaded on a silicagel TLC plate, gave the result shown in Fig. 17b, after spraying with SV at rt; this colorimetric assay is noteworthy, since most coloured reactions obtained by SV on TLC are visible only after heating at > 100°C. The TLC plate of Fig. 17 also showed that more than one compound was involved in the pink reaction.

Column liquid chromatography on silicagel allowed us to isolate four compounds **30-33**, all giving the above reaction with SV; they turned out to be cucurbitane triterpenes having the A-ring cleaved in the 3-4 position, i.e., *seco*-cucurbitane triterpenes (Clericuzio et al., 2014). A remarkable feature of **30-33** is the presence of a furan ring fused to carbons 5-6 of the B-ring: it is very likely that the presence of this functionality is responsible for the SV pink reaction. It has already been reported that furan-bearing terpenes (as the lactarane furanol **12**) give a pink reaction with SV on TLC, even if usually heating is required (Clericuzio et al., 2014).

Dicarboxylic acid **30** is the most abundant of the four molecules; we named it roseic acid A.[1] It shows a triple conjugated double-bond sequence on the side chain, viz. two C=C bonds conjugated with the C=O bond of a carboxylic acid. The configuration of the olefin bonds was found to be *Z,Z* by analysis of the ^{1}H NMR coupling constants. Roseic acid A is found together with the correspondent 22-23 dihydro derivative **31**, here named by us roseic acid B, bearing only two

---

[1] In our original paper (Clericuzio et al., 2014), compound **23** was simply named roseic acid. However, once the related diacid **24** was isolated and characterized, the label A was here added for the sake of clarity.

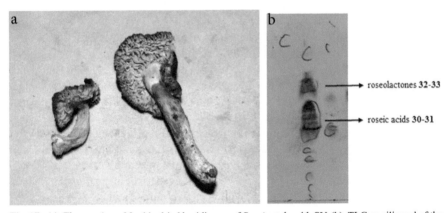

Fig. 16. The four furan-carboxylic acids isolated from *R. aurora* and *R. minutula*.

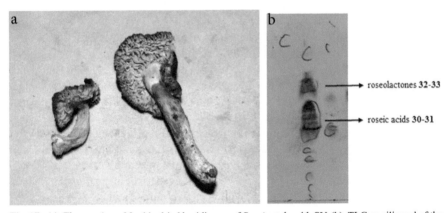

Fig. 17. (a). The reaction of freshly dried basidiomes of *R. minutula* with SV. (b). TLC on silicagel of the *R. aurora* crude extract. Eluent: 70% toluene; 28% ethyl acetate; 2% methanol. The spots absorbing at 254 nm are marked by a pencil. Coloured with SV at rt.

conjugated double bonds; it approximately amounts 40% of roseic acid A, based on ^{1}H NMR integral comparison of the crude extract. The two compounds are difficult to separate, and in particular, diacid **31** could never be isolated in a pure form, if not in tiny amounts.

The two less polar compounds **32-33** bear an unusual γ-lactone moiety in the side chain, hence they were named roseolactone A and B, respectively. The relative configurations at C-23 were determined by an accurate comparison between experimental NMR data and *ab-initio* theoretical calculations (Clericuzio et al., 2014). Diastereomeric lactones **32-33** are formed by ring closure of the carboxylic function at C-27 with a secondary alcoholic function at C-23, at its turn derived from non-stereoselective hydration of the 22-23 double bond of diacid **30**. The two isomers are very difficult to separate from each other, and were usually obtained as mixtures. Compounds **32-33** are rather unstable, and once isolated from the raw extract, undergo unspecific degradation in a few months, even if stored at –20°C.

Our furan-cucurbitanes should be compared with lepidolide **29**, isolated by Tan et al. (2002) from basidiomes of *R. lepida* collected in China. It appears evident that **29** is very close to roseic acid A **30**. The main difference lies in the presence of a γ-lactone in **29**, rather than a furan ring in **30**. In nature, γ-lactones are often formed by oxidation of furan rings (Daniewski and Vidari, 1999), so diacid **29** may be a likely oxidation product of **30**. As concerns the configuration at C-C double bonds 22-23 and 24-25, it was reported as *E,E* in **29**, while we inferred a *Z,Z* configuration in **30**, from the analysis of ^{1}H NMR coupling constants and NOE effects. As discussed in Clericuzio et al. (2014), the *trans, trans* stereochemistry of **29** should be re-investigated, at the light of the NMR coupling constants reported by the authors, which were in better agreement with a *cis* geometry.

Metabolites **30-33** were tested for selected bioactivities. In a MTT anti-proliferative assay on some human cancer cell lines, only a mixture of lactones **32-33** was found to be weakly active (Clericuzio et al., 2014). In an anti-fungal assay on five different species of *Candida,* neither the *R. aurora* crude extract, nor any of the single metabolites, was found to be active. The same negative result was obtained when they were tested against a few selected bacteria (*Escherichia coli, Pseudomonas aeruginosa,* and *Staphylococcus aureus*) (unpublished data).

Insect-Fungi Relationships: The Anti-Feedant Properties of Russulaceae Terpenes

The basidiomes of *R. aurora, R. minutula* and *R. lepida* are little affected by maggots, while several other *Russula* are often infested by them. This empirical observation may be related to the presence of anti-feedant compounds. Daniewski et al. (1993) carried out a series of anti-feedant experiments using some sesquiterpenes isolated from *Lactarius*, together with a few synthetic analogues. These compounds were tested on two adults and two larvae of Coleoptera: Several natural products, in particular lactarorufine A (a widespread lactarane lactone), showed potent deterrence against Coleoptera feeding.

Actually, one of the main interactions faced by fungi in the environment is that with fungivorous insects (Nakamori and Suzuki, 2007). It is likely that many fungal secondary metabolites have evolved specifically to fight this predation, while at their turn, arthropods have counter-evolved to resist these weapons. The arthropod-fungi co-evolution is nicely described in the review by Rohlfs and Churchill (2011). Antifeedant tests on *Russula* triterpenes are underway in our lab.

Enzymatic Biotransformation of *R. aurora* Triterpenes

Another important issue to be considered is the biotransformation of the metabolites initially present in intact basidiomes. When placed in contact with the enzymatic system of the fungus, several chemical reactions can take place, as thoroughly studied in the case of stearoylvelutinal (**1b**). Not rarely, the metabolites formed by such enzymatic transformations may be more bioactive than those initially present.

Early experiments in our lab demonstrate that, if the basidiomes of *R. aurora* are minced and left at RT for a few hours (without adding solvent), all of the metabolites **30-33** transform in time, at different rates. In particular, roseolactones **32-33** react more rapidly, and after 3h are completely gone, while roseic acids react more slowly (data not shown). From TLC and NMR experiments, it appears that each of the four metabolites **30-33** is transformed in more than one product; this makes the final mixtures very complex and difficult to separate. Some of the transformed products still reacted pink with SV, while others did not. We believe than in the former case the furan ring remained untouched, while in the latter case it was chemically transformed in some way. From the ^{13}C NMR spectra of the reacted mixtures at different times, signals corresponding to oxygenated sp^3 carbons (in the range 60–80 ppm) appear, meaning that oxidations have occurred.

The Taxonomic Meaning of the Cucurbitane Triterpenes found in *Russula*

Since section *Roseinae* of *Russula* was defined as "basidiomes reacting pink with SV" (Romagnesi, 1985), we can safely assume that *seco*-cucurbitane triterpenoids are a chemotaxonomic marker of the section. On the other hand, the chemotaxonomy of sect. *Lepidinae* is less clear. The presence of sesquiterpenes characterized by a molecular skeleton with no equals in Agaricomycetes, together with cucurbitane triterpenes similar to those of *Roseinae*, leaves this intriguing puzzle still unresolved.

It has been proposed that *Russula* species totally devoid of Dcy, and instead provided with HP, should all be assigned to a subgenus named *Incrustatula* (Romagnesi, 1987). We have extracted the basidiomes of *R. lilacea* (sect. *Lilaceinae*), *R. risigallina* (sect. *Chamaeleontinae*), *R. olivacea* and *R. vinosobrunnea* (sect. *Olivaceinae*), all groups that should be part of *Incrustatula*, to check if any similarity with the chemical contents of *Roseinae* and *Lepidinae* could be found. Even if our experimental data are still incomplete, we can say that to date none of them contains cucurbitane triterpenes. *R. olivacea* and *R. vinosobrunnea* synthesize tiny amounts of ergostane triterpenoids, which, however, should really be included in the class of sterols (primary metabolites). As a consequence, the cucurbitane triterpenoids of *Roseinae* and *Lepidinae* probably represent an isolated class of compounds, even if, on a worldwide perspective, they will be likely found in other species.

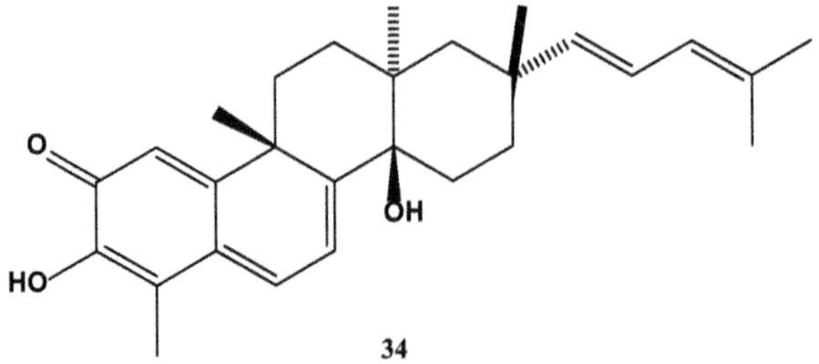

Fig. 18. Russulaflavidin (**34**), a shionane triterpenoid isolated from basidiomes of *R. flavida*.

Shionane Triterpenoids

For the sake of completeness, a mention should be made of two chromogenic triterpenoids found in *R. flavida*, a species growing in Eastern Asia and in Eastern North America. From this toadstool, a C28 *nor*-triterpenoid, named russulaflavidin (**34**), and its dihydro-derivative, were isolated (Fröde et al., 1995). They bear an unusual shionane skeleton, never found before in fungi.

Conclusions

The terpene secondary metabolites of Russulaceae are an intriguing class of secondary metabolites, often endowed with potent bioactivities. The present review is mainly focussed on the recently discovered cucurbitane triterpenes in members of sect. *Lepidinae* and *Roseinae*: This is a particularly rare class of metabolites, which may be regarded as a chemotaxonomic marker of this restricted group of *Russula*.

From an applicative point of view, the bioactivities of *Russula* triterpenes is still an open field for future work. Even if no particular cytotoxic activity has been found to date, the discovery of diabetes-fighting properties of compounds **27-28** indicates that, by widening the bioactivity assays, surely more important results will be found.

References

Alarcón, J., Villalobos, N., Lamilla, C. and Céspedes, C.L. 2013. Ceramides and terpenoids from *Russula austrodelica* Singer. Bol. Latinoam. Caribe, 12: 493–498.

Aliferis, K.A., Cubeta, M.A. and Jabaji, S. 2013. Chemotaxonomy of fungi in the *Rhizoctonia solani* species complex performing GC/MS metabolite profiling. Metabolomics, 9: S159–S169.

Allen, J., Davey, H.M., Broadhurst, D., Heald, J.K., Rowland, J.J., Oliver, S.G. and Kell, D.B. 2003. High throughput classification of yeast mutants for functional genomics using metabolic foot printing. Nat. Biotechnol., 21: 692–696.

Andary, C., Cosson, L., Bourrier, M.J., Wylde, R. and Heitz, A. 1992. Chimiotaxonomie des bolets de la section *Luridi*. Cryptogamie Mycol., 13: 103–114.

Anke, H. and Sterner, O. 1991. Comparison of the antimicrobial and cytotoxic activities of twenty unsaturated sesquiterpene dialdehydes from Plants and Mushrooms. Planta Med., 57: 344–346.

Arnold, N., Besl, H., Bresinsky, A. and Kemmer, H. 1987. Notizen zur Chemotaxomomie der Gattung *Dermocybe* (Agaricales) und zu ihrem Vorkommen in Bayern. Z. Mykol., 53: 187–194.

Arnone, A., Nasini, G. and Vajna de Pava, O. 1997. Marasmane sesquiterpenes from the basidiomycete *Clitocybe hydrogramma*. Phytochemistry, 46: 1099–1101.

Baby, S., Johnson, A.J. and Govindan, B. 2015. Secondary metabolites from *Ganoderma*. Phytochemistry, 114: 66–101.

Bazzicalupo, A.L., Buyck, B., Saar, I., Vauras, J., Carmean, D. and Berbee, M.L. 2017. Troubles with mycorrhizal mushroom identification where morphological differentiation lags behind barcode sequence divergence. Taxon, 66: 791–810.

Benedict, R.G. 1970. Chemotaxonomic relationships among the Basidiomycetes. Adv. Appl. Microbiol., 13: 1–23.

Besl, H., Bresinsky, A. and Kämmerer, A. 1986. Chemosystematik der Coniophoraceae. Z. Mykol., 52: 277–286.

Besl, H. and Bresinsky, A. 1997. Chemosystematics of Suillaceae and Gomphidiaceae (sub-order Suillineae). Plant Syst.Evol., 206: 223–242.

Beutler, J.A. and der Marderosian, A.H. 1981. Chemical variation in *Amanita*. J. Nat. Prod., 44: 422–431.

Binder, M. and Bresinsky, A. 2002. *Retiboletus*, a new genus for a species-complex in the Boletaceae producing retipolides. Feddes Report., 113: 30–40.

Buyck, B. and Mitchell, D. 2003. *Russula lentiginosa* spec. nov. from West Virginia, USA: A probable link between tropical and temperate *Russula*-groups. Cryptogamie Mycol., 24: 317–325.

Buyck, B., Hofstetter, V., Eberhardt, U., Verbeken, A. and Kauff, F. 2008. Walking the thin line between *Russula* and *Lactarius*: The dilemma of *Russula* sect. *Ochricompactae*. Fungal Divers., 28: 15–40.

Camazine, S. and Lupo, A.T.J. 1984. Labile toxic compounds of the Lactarii: The role of the lacticiferous hyphae as a storage depot for precursors of pungent dialdehydes. Mycologia, 76: 355–358.

Chen, H.-P. and Liu, J.-K. 2017. Secondary metabolites from Higher Fungi. pp. 1–201. *In*: Kinghorn, A.D., Falk, H., Gibbons, S. and Kobayashi, J. (eds.). Progress in the Chemistry of Organic Natural Products, Vol. 106, Springer International Publishing, Cham, Switzerland.

Chen, J.C., Chiu, M.H., Nie, R.L., Cordell, G.A. and Qiu, S.X. 2005. Cucurbitacins and cucurbitane glycosides: Structures and biological activities. Nat. Prod. Rep., 22: 386–399.

Chen, S., Li, X., Yong, T., Wang, Z., Su, J., Jiao, C., Xie, Y. and Yang, B.B. 2017. Cytotoxic lanostane-type triterpenoids from the fruiting bodies of *Ganoderma lucidum* and their structure–activity relationships. Oncotarget, 8: 10071–10084.

Clericuzio, M., Fu, J., Pan, F., Pang, Z. and Sterner, O. 1997a. Structure and absolute configuration of protoilludane sesquiterpenes from *Russula delica*. Tetrahedron, 28: 9735–9740.

Clericuzio, M., Pan, F., Han, F., Pang, Z. and Sterner, O. 1997b. Stearoyldelicone, an unstable protoilludane sesquiterpenoid from intact fruit bodies of *Russula delica*. Tetrahedron Lett., 38: 8237–8240.

Clericuzio, M., Han, F., Pan, F., Pang, Z. and Sterner, O. 1998. The sesquiterpenoid contents of fruit bodies of *Russula delica*. Acta Chem. Scand., 52: 1333–1337.

Clericuzio, M., Mella, M., Vita-Finzi, P., Zema, M. and Vidari, G. 2004. Cucurbitane triterpenoids from *Leucopaxillus gentianeus*.J. Nat. Prod., 67: 1823–1828.

Clericuzio, M., Tabasso, S., Bianco, M.A., Pratesi, G., Beretta, G., Tinelli, S., Zunino, F. and Vidari, G. 2006. Cucurbitane triterpenes from the fruiting bodies and cultivated mycelia of *Leucopaxillus gentianeus*. J. Nat. Prod., 69: 1796–1799.

Clericuzio, M., Gilardoni, G., Malagòn, O., Vidari, G. and Vita-Finzi, P. 2008. Sesquiterpenes of *Lactarius* and *Russula* (Mushrooms): An update. Nat. Prod. Commun., 3: 951–974.

Clericuzio, M., Cassino, C., Corana, F. and Vidari, G. 2012. Terpenoids from *Russula lepida* and *R. amarissima* (Basidiomycota, Russulaceae). Phytochemistry, 84: 154–159.

Clericuzio, M., Vidari, G., Cassino, C., Legnani, L. and Toma, L. 2014. Roseic acid and Roseolactones A and B, furan-cucurbitane triterpenes from *Russula aurora* and *R. minutula* (Basidiomycota). Eur. J. Org. Chem., 25: 5462–5468.

Daniewski, W.M., Gumulka, M., Ptaszyńska, K., Skibicki, P., Błoszyk, E., Drożdż, B., Stromberg, S., Norin, T. and Holub, M. 1993. Antifeedant activity of some sesquiterpenoids of the genus *Lactarius* (Agaricales: Russulaceae). Eur. J. Entomol., 90: 65–70.

Daniewski, W.M., Gumułka, M., Przesmycka, D., Ptaszyńska, K., Błoszyk, E. and Drożdż, B. 1995. Sesquiterpenes of *Lactarius* origin, antifeedant structure-activity relationships. Phytochemistry, 38: 1161–1168.

Daniewski, W.M. and Vidari, G. 1999. Constituents of *Lactarius* (Mushrooms). pp. 69–171. *In*: Herz, W., Falk, H., Kirby, G.W., Moore, R.E. and Tamm Ch. (eds.). Fortschritte der Chemie organischer Naturstoffe. Vol. 77. Springer, New York.

Davoli, P. and Weber, R.W.S. 2002. Simple method for reversed-phase high-performance liquid chromatographic analysis of fungal pigments in fruit bodies of Boletales (Fungi). J. Chromatogr A, 964: 129–135.

De Bernardi, M., Vidari, G. and Vita-Finzi, P. 1982. Biogenesis-like conversion of marasmane to lactarane and seco-lactarane skeleton. Tetrahedron Lett., 23: 4623–4624.

Diaz, J.H. 2005. Evolving global epidemiology, syndromic classification, general management, and prevention of unknown mushroom poisonings. Crit. Care Med., 33: 419–426.

Ekanayaka, A.H., Ariyawansa, H.A., Hyde, K.D., Jones, E.B.G., Daranagama, D.A., Phillips, A.J.L., Hongsanan, S., Jayasiri, S.C. and Zhao, Qi. 2017. DISCOMYCETES: the apothecial representatives of the phylum Ascomycota. Fungal Divers., 87: 237–298.

Enjalbert, F., Cassanas, G., Rapior, S., Renault, C. and Chaumont, J.-P. 2004. Amatoxins in wood-rotting *Galerinamarginata*. Mycologia, 96: 720–729.

Eugster, C.H. and Iten, P.X. 1975. Russupteridines. pp. 881–917. *In*: Chemistry and Biology of Pteridines, Proc. Int. Symp., 5th. De Gruyter, Berlin-New York.

Faulstich, H. and Zilker, T. 1994. Amatoxins. pp. 233–248. *In*: Spoerke, D.G. and Rumack, B.H. (eds.). Handbook of Mushroom Poisoning, Diagnosis and Treatment. CRC Press Inc., Boca Raton.

Favre-Bonvin, J., Gluchoff-Fiasson, K. and Bernillon, I. 1982. Structure du stearyl-velutinal, sesquiterpenoids naturel de *Lactarius velutinus* Bert. Tetrahedron Lett., 23 : 1907–1908.

Fogel, R. and Trappe, J.M. 1978. Fungus consumption (mycophagy) by small animals. Northwest Sci., 52: 1–31.

Fox, E.M. and Howlett, B.J. 2008. Secondary metabolism: Regulation and role in fungal biology. Curr. Opin. Microbiol., 11: 481–487.

Frisvad, J.C., Thrane, U. and Filtenborg, O. 1998. Role and use of secondary metabolites in fungal taxonomy. pp. 289–319. *In*: Frisvad, J.C., Bridge, P.D. and Arora, D.K. (eds.). Chemical Fungal Taxonomy. Marcel Dekker, New York.

Frisvad, J.C., Larsen, T.O., de Vries, O., Meijer M., Houbraken, J., Cabañes, F.J., Ehrlich, K. and Samson, R.A. 2007. Secondary metabolite profiling, growth profiles and other tolls for species recognition and important *Aspergillus* mycotoxins. Stud. Mycol., 59: 31–37.

Frisvad, J.C., Andersen, B. and Thrane, U. 2008. The use of secondary metabolite profiling in chemotaxonomy of filamentous fungi. Mycol. Res., 112: 231–240.

Fröde, R., Brockelmann, M., Steffan, B., Steglich, W. and Marumoto, R. 1995. A novel type of triterpenoid quinone methide pigment from the toadstool Russula flavida (Agaricales). Tetrahedron, 51: 2553–2560.

Fujimoto, H., Hagiwara, H., Suzuki, K. and Yamazaki, M. 1986. New toxic metabolites from a mushroom, *Hebeloma vinosophyllum*. I. Isolation and structures of hebevinosides I, II, III, IV, X, and V. Chem Pharm. Bull., 34: 88–99.

Fujimoto, H., Hagiwara, H., Suzuki, K. and Yamazaki, M. 1987. New toxic metabolites from a mushroom, *Hebeloma vinosophyllum*. II. Isolation and structures of hebevinosides VI, VII, VIII, IX, X, and XI. Chem. Pharm. Bull., 35: 2254–2260.

Fujimoto, H., Maeda, K. and Yamazaki, M. 1991. New toxic metabolites from a mushroom, *Hebeloma vinosophyllum*. III. Isolation and structures of hebevinosides XII, XIII, and XIV, and productivity of hebevinosides at three growth stages of the mushroom. Chem. Pharm. Bull., 39: 1958–1961.

Geml, J., Laursen, G.A., Herriott, I.C., McFarland, J.M., Booth, M.G., Lennon, N., Chad Nusbaum, H. and Lee Taylor, D. 2010. Phylogenetic and ecological analyses of soil and sporocarp DNA sequences reveal high diversity and strong habitat partitioning in the boreal ectomycorrhizal genus *Russula* (Russulales; Basidiomycota). New Phytol., 187: 494–507.

Geml, J. and Taylor, D.L. 2013. Biodiversity and molecular ecology of *Russula* and *Lactarius* in Alaska based on soil and sporocarp DNA sequences. Scr. Bot. Belg., 51: 132–145.

Gherbawy, Y. and Voigt, K. (eds.). 2010. Molecular identification of fungi. Springer, Berlin.

Gilardoni, G., Malagòn, O., Tosi, S., Clericuzio, M. and Vidari, G. 2014. Lactarane sesquiterpenes from the European mushrooms *Lactarius aurantiacus*, *L. subdulcis*, and *Russula sanguinaria*. Nat. Prod. Commun., 9: 319–322.

Gill, M. and Steglich, W. 1987. Pigments of fungi (Macromycetes). pp. 1–317. *In*: Herz, W., Grisebach, H., Kirby, G.W. and Tamm, C. (eds.). Progress in the Chemistry of Organic Natural Products, 51. Springer Verlag, Berlin.

Gill, M. 1996. Pigments of fungi (macromycetes). Nat. Prod. Rep., 13: 513–528.

Gill, M. 2003. Pigments of fungi (macromycetes). Nat. Prod. Rep., 20: 615–639.

Gluchoff, K. and Lebreton, P. 1970. Fungal chemotaxonomy. Preliminary data on properties and structure of *Russules* (Basidiomycetes). C. R. Acad. Sci. Hebd. Seances Acad. Sci. D, 270: 213–216.

Gluchoff-Fiasson, K. and Kühner, R. 1982. Le principe responsable du bleuissement au réactif sulfovanillique des cystides et lacticifères de divers Homobasidiomycètes: Intérèt systématique. C. R. Seances Acad. Sci. III, 294: 1067–1071.

Grienke, U., Zoll, M., Peintner, U. and Rollinger, J.M. 2014. European medicinal polypores—A modern view on traditional uses. J. Ethnopharmacol., 154: 564–583.

Guo, W. 1992. Resources of wild edible fungi in Tibet, China. Zhongguo Shiyongjun, 11: 33–34.

Hansson, T. and Sterner, O. 1991. Studies of the conversions of sesquiterpenes in injured fruit bodies of *Lactarius vellereus*. A biomimetic transformation of stearoylvelutinal to isovelleral. Tetrahedron Lett., 32: 2541–2544.

Hashimoto, T., Tanaka, H. and Asakawa, Y. 1994. Stereostructure of plagiochiline A and conversion of plagiochiline A and stearoylvelutinal into hot-tasting compounds by human saliva. Chem. Pharm. Bull., 42: 1542–1544.

Hibbett, D.S., Binder, M., Bischoff, J.F., Blackwell, M., Cannon, P.F., Eriksson, O.E., Huhndorf, S., James, T., Kirk, P.M., Lücking, R., Lumbsch, H.T., Lutzoni, F., Mathenya, P.B., McLaughlin, D.J., Powell, M.J., Redhead, S., Schoch, C.L., Spatafora, J.W., Stalpers, J.A., Vilgalys, R., Aime, M.C., Aptroot, A., Bauer, R., Begerow, D., Benny, G.L., Castlebury, L.A., Crous, P.W., Dair, Y.C., Gams, W. and Geisers, D.M. 2007. A higher-level phylogenetic classification of the fungi. Mycol. Res., 111: 509–547.

Hibbett, D.S. and Taylor, J.W. 2013. Fungal systematics: Is a new age of enlightenment at hand? Nat. Rev. Microbiol., 11: 129–133.

Hibbett, D.S., Stajich, J.E. and Spatafora, J.W. 2013. Towards genome-enabled mycology. Mycologia, 105: 1339–1349.

Hibbett, D.S. 2014. Major events in the evolution of the Fungi. pp. 152–158. *In*: Losos, J. (ed.). Princeton guide to evolution. Princeton University Press, Princeton.

Hibbett, D.S., Bauer, R., Binder, M., Giachini, A.J., Hosaka, K., Justo, A., Larsson, E., Larsson, K.H., Lawrey, J.D., Miettinen, O., Nagy, L.G., Nilsson, R.H., Weiss, M. and Thorn, R.G. 2014. Agaricomycetes. pp. 373–429. *In*: McLaughlin, D.J. and Spatafora, J.W. (eds.). The Mycota, vol. VII, part A. Systematics and evolution, 2nd edn. Springer, Berlin.

Høiland, K. 1983. *Cortinarius* subgenus *Dermocybe*. Opera Botanica, 71: 1–113.

Høiland, K. 1990. The genus *Gymnopilus* in Norway. Mycotaxon, 39: 257–279.

Høiland, K. and Holst-Jensen, A. 2000. *Cortinarius* phylogeny and possible taxonomic implications of ITS rDNA sequences. Mycologia, 92: 694–710.

Hu, L. and Zeng, L. 1992. Investigation on wild edible mushroom resources in Wanxian Country, Sichuan Province. Zhongguo Shiyongjun, 11: 35–37.

Isaka, M., Yangchum, A., Wongkanoun, S. and Kongthong, S. 2017a. Marasmane and normarasumane sesquiterpenenoids from the edible mushroom *Russula nigricans*. (2017). Phytochem. Lett., 21: 174–178.

Isaka, M., Palasarn, S., Sommai, S., Veeranondha, S., Srichomthong, K., Kongsaeree, P. and Prabpai, S. 2017b. Lanostane triterpenoids from the edible mushroom *Astraeus asiaticus*. Tetrahedron, 73: 1561–1567.

Jayasiri, S.C., Hyde, K.D., Ariyawansa, H.A., Bhat, J., Buyck, B., Cai, L., Dai, Y.-C., Abd-Elsalam, K.A., Ertz, D., Hidayat, I., Jeewon, R., Jones, E.B.G., Bahkali, A.H., Karunarathna, S.C., Liu, J.-K., Luangsa-ard, J.J., Lumbsch, H.T., Maharachchikumbura, S.S.N., McKenzie, E.H.C., Moncalvo, J.-M., Ghobad-Nejhad, M., Nilsson, H., Pang, K.-L., Pereira, O.L., Phillips, A.J.L., Raspé, O., Rollins, A.W., Romero, A.I., Etayo, J., Selçuk, F., Stephenson, S.L., Suetrong, S., Taylor, J.E., Tsui, C.K.M., Vizzini, A., Abdel-Wahab, M.A., Wen, T.-C., Boonmee, S., Dai, D.Q., Daranagama, D.A., Dissanayake, A.J., Ekanayaka, A.H., Fryar, S.C., Hongsanan, S., Jayawardena, R.S., Li, W.-J., Perera, R.H., Phookamsak, R., de Silva, N.I., Thambugala, K.M., Tian, Q., Wijayawardene, N.N., Zhao, R.-L., Zhao, Qi., Kang, J.-C. and Promputtha, I. 2015. The faces of fungi database: Fungal names linked with morphology, phylogeny and human impacts. Fungal Divers., 74: 3–18.

Karpyn Esqueda, M., Yen, A.L., Rochfort, S., Guthridge, K.M., Powell, K.S., Edwards, J. and Spangenberg, G.C. 2017. A review of perennial ryegrass endophytes and their potential use in the management of African Black Beetle in perennial grazing systems in Australia. Front. Plant Sci., 8: 3.

Keller, G. and Ammirati, J.F. 1983. Chemotaxonomic significance of anthraquinone derivatives in North American species of *Dermocybe*, section *Dermocybe*. Mycotaxon, 18: 357–377.

Kellogg, J.J. and Raja, H.A. 2017. Endolichenic fungi: A new source of rich bioactive secondary metabolites on the horizon. Phytochem. Rev., 16: 271–293.

Kim, K.H., Noh, H.J., Choi, S.U., Park, K.M., Seok, S.-J. and Lee, K.R. 2010. Russulfoen, a new cytotoxic marasmane sesquiterpene from *Russula foetens*. J. Antibiot. (Tokyo), 63: 575–577.

Kirchmair, M., Pöder, R., Huber, C.G. and Miller, Jr., O.K. 2002. Chemotaxonomical and morphological observations in the genus *Omphalotus* (Omphalotaceae). Persoonia, 17: 583–600.

Kirk, P.M., Cannon, P.F., Minter, D.W. and Stalpers, J.A. 2008. Dictionary of the Fungi, 10th ed. CABI, Wallingford.

Larsen, T.O., Smedsgaard, J., Nielsen, K.F., Hansen, M.E. and Frisvad, J. 2005. Phenotypic taxonomy and metabolite profiling in microbial drug discovery. Nat. Prod. Rep., 22: 672–695.

Larsson, E. and Larsson, K.-H. 2003. Phylogenetic relationships of russuloid basidiomycetes with emphasis on aphyllophoralean taxa. Mycologia, 95: 1037–1065.

Lee, H., Park, M.S., Jung, P.E., Eimes, J.A., Seok, S.J. and Lim, Y.W. 2017. Re-evaluation of the taxonomy and diversity of *Russula* section *Foetentinae* (Russulales, Basidiomycota) in Korea. Mycoscience, 58: 351–360.

Lee, J-S., Maarisit, W., Abdjul, D.B., Yamazaki, H., Takahashi, O., Kirikoshi, R., Kanno, S. and Namikoshi, M. 2016. Structures and biological activities of triterpenes and sesquiterpenes obtained from *Russula lepida*. Phytochemistry, 127: 63–68.

Liu, D.Z. 2014. A review of ergostane and cucurbitane triterpenoids of mushroom origin. Nat. Prod. Res., 28: 1099–1105.

Lodge, D.J., Padamsee, M., Matheny, P.B., Aime, M.C., Cantrell, S.A., Boertmann, D., Kovalenko, A., Vizzini, A., Dentinger, B.T.M., Kirk, P.M., Ainsworth, A.M., Moncalvo, J.-M., Vilgalys, R., Larsson, E., Lücking, R., Griffith, G.W., Smith, M.E., Norvell, L.L., Desjardin, D.E., Redhead, S.A., Ovrebo, C.L., Lickey, E.B., Ercole, E., Hughes, K.W., Courtecuisse, R., Young, A., Binder, M., Minnis, A.M., Lindner, D.L., Ortiz-Santana, B., Haight, J., Læssøe, T., Baroni, T.J., Geml, J. and Hattori, T. 2014. Molecular phylogeny, morphology, pigment chemistry and ecology in Hygrophoraceae (Agaricales). Fungal Divers., 64: 1–99.

Lorenzen, K. and Anke, T. 1998. Basidiomycetes as a source of new bioactive natural products. Curr. Org. Chem., 2: 329–364.

Malagòn, O., Porta, A., Clericuzio, M., Gilardoni, G., Gozzini, D. and Vidari, G. 2014. Structures and biological significance of lactarane sesquiterpenes from the European mushroom *Russula nobilis*. Phytochemistry, 107: 126–134.

Marumoto, R., Kilpert, C. and Steglich, W. 1986. Fungal dyes. 51. New pulvinic acid derivatives from *Pulveroboletus* types (Boletales). Z. Naturforsch. C, 41: 363–365.

Matheny, P.B., Wang, Z., Binder, M., Curtis, J.M., Lim, Y.W., Nilsson, R.H., Hughes, K.W., Hofstetter, V., Ammirati, J.F., Schoch, C.L., Langer, E., Langer, G., McLaughlin, D.J., Wilson, A.W., Frøslev, T., Ge, Z.W., Kerrigan, R.W., Slot, J.C., Yang, Z.L., Baroni, T.J., Fischer, M., Hosaka, K., Matsuura, K., Seidl, M.T., Vauras, J. and Hibbett, D.S. 2007. Contributions of rpb2 and tef1 to the phylogeny of mushrooms and allies (Basidiomycota, Fungi). Mol. Phylogenet. Evol., 43: 430–451.

Miller, S.L. and Buyck, B. 2002. Molecular phylogeny of the genus *Russula* in Europe with a comparison of modern infrageneric classifications. Mycol. Res., 106: 259–276.

Miller, S.L., Larsson, E., Larsson, K.-H., Verbeken, A. and Nuytinck, J. 2006. Perspectives in the new Russulales. Mycologia, 98: 960–970.

Müller, K. 2001. Pharmaceutically relevant metabolites from lichens. Appl. Microbiol. Biotechnol., 56: 9–16.

Nakamori, T. and Suzuki, A. 2007. Defensive role of cystidia against *Collembola* in the basidiomycetes *Russula bella* and *Strobilurus ohshimae*. Mycol. Res., 111: 1345–1351.

Nielsen, J. and Oliver, S. 2005. The next wave in metabolome analysis. Trends Biotechnol., 23: 544–546.

Pang, K.L., Overy, D.P., Jones, E.B.G., Calado, M.D.L., Burgaud, G., Walker, A.K., Johnson, J.A., Kerr, R.G., Cha,H.J. and Bills, G.F. 2016. 'Marine fungi' and 'marine-derived fungi' in natural product chemistry research: Toward a new consensual definition. Fungal Biol. Rev., 30: 163–175.

Pusztahelyi, T., Holb, I.J. and Pócsi, I. 2015. Secondary metabolites in fungus-plant interactions. Front. Plant Sci., 6: 573.

Ren, S., Ouyang, D.Y., Saltis, M., Xu, L.H., Zha, Q.B., Cai, J.Y. and He, X.H. 2012. Anti-proliferative effect of 23,24-dihydrocucurbitacin F on human prostate cancer cells through induction of actin aggregation and cofilin-actin rod formation. Cancer Chemother. Pharmacol., 70: 415–424.

Rohlfs, M. and Churchill, A.C.L. 2011. Fungal secondary metabolites as modulators of interactions with insects and other arthropods. Fungal Genet. Biol., 48: 23–34.

Romagnesi, H. 1985. Les Russules d'Europe et d'Afrique du Nord. J. Cramer, Vaduz.

Romagnesi, H. 1987. Novitates—Statuts et noms nouveaux pour les taxa infrageneriques dans le genre *Russula*. Doc. Mycol., 18: 39–40.

Rumack, B.H. and Spoerke, D.G. 1994. Handbook of mushroom poisoning: Diagnosis and treatment. CRC press, Boca Raton.

Saviuc, P. and Flesch, F. 2003. Acute higher fungi mushroom poisoning and its treatment. Presse Med., 32: 1427–1435.

Schmidt-Dannert, C. 2015. Biosynthesis of terpenoid natural products in Fungi. Adv. Biochem. Eng. Biotechnol., 148: 19–61.

Shibuya, M., Xiang, T., Katsube, Y., Otsuka, M., Zhang, H. and Ebizuka, Y. 2007. Origin of structural diversity in natural triteprenes: Direct synthesis of seco-triterpene skeletons by oxidosqualene cyclase. J. Am. Chem. Soc., 129: 1450–1455.

Shiono, Y., Sugawara, H., Nazarova, M., Murayama, T., Takahashi, K. and Ikeda, M. 2007. Three lanostane triterpenoids, aeruginosols A, B and C, from the fruiting bodies of *Stropharia aeruginosa*. J. Asian Nat. Prod. Res., 9: 531–535.

Singer, R. 1986. The Agaricales in modern taxonomy. Koeltz Scientific, Koenigstein.

Singh, R. 2016. Chemotaxonomy: A tool for plant classification. J. Medic. Pl. Studies, 4: 90–93.

Singh, B.N., Upreti, D.K., Gupta, V.K., Dai, X.F. and Jiang, Y.M. 2017. Endolichenic Fungi: A hidden reservoir of next generation biopharmaceuticals. Trends Biotechnol., 35: 808–813.

Smedsgaard, J. and Nielsen, J. 2004. Metabolite profiling of fungi and yeast: From phenotype to metabolome by MS and informatics. J. Exp. Bot., 56: 273–286.

Smith, M.R. and Davis, R.L. 2016. Mycetismus: A review. Gastroenterol. Rep., 4: 107–112.

Sontag, B., Frode, R., Bross, M. and Steglich, W. 1999. Chromogenic triterpenoids from *Cortinarius fulvoincarnatus*, *C. sodagnitus* and related toadstools (Agaricales). Eur. J.Org. Chem., 1: 255–260.

Sontag, B., Rüth, M., Spiteller, P., Arnold, N., Steglich, W., Reichert, M. and Bringmann, G. 2006. Chromogenic meroterpenoids from the mushrooms *Russula ochroleuca* and *R. viscida*. Eur. J. Org. Chem., 4: 1023–1033.

Sterner, O., Bergman, R., Kesler, E., Nilsson, L., Oluwadiya, J. and Wickberg, B. 1983. Velutinal esters of *Lactarius vellereus* and *L. necator*. The preparation of free velutinal. Tetrahedron Lett., 24: 1415–1418.

Sterner, O., Bergman, R., Kihlberg, J., Oluwadiya, J., Wickberg, B., Vidari, G., De Bernardi, M., De Marchi, F., Fronza, G. and Vita-Finzi, P. 1985a. Basidiomycete sesquiterpenes: The silica gel induced degradation of velutinal derivatives. J.Org. Chem., 50: 950–953.

Sterner, O., Bergman, R., Kihlberg, J. and Wickberg, B. 1985b. The sesquiterpenes of *Lactarius vellereus* and their role in a proposed chemical defence system. J. Nat. Prod., 48: 279–288.

Sterner, O., Bergman, R., Franzén, C. and Wickberg, B. 1985c. New Sesquiterpenes in a proposed Russulaceae chemical defense system. Tetrahedron Lett., 26: 3163–3166.

Sterner, O. 1989. The co-formation of sesquiterpene aldehydes and lactones in injured fruit bodies of *Lactarius necator* and *L. circellatus*. The isolation of epi-piperalol. Acta Chem. Scand., 43: 694–697.

Swann, E.C. and Taylor, J.W. 1993. Higher taxa of basidiomycetes: An 18S rRNA gene perspective. Mycologia, 85: 923–936.

Swann, E.C. and Taylor, J.W. 1995. Phylogenetic perspectives on basidiomycete systematics: Evidence from the 18S rRNA gene. Can. J. Bot., 73: S862–S868.

Swann, E.C., Frieders, E.M. and McLaughlin, D.J. 1999. *Microbotryum*, *Kriegeria* and the changing paradigm in basidiomycete classification. Mycologia, 91: 51–66.

Tan, J.W., Dong, Z.J. and Liu, J.K. 2000. New Terpenoids from Basidiomycetes *Russula lepida*. Helv. Chim. Acta., 83: 3191–3197.

Tan, J.W., Dong, Z.J. and Liu, J.K. 2001. A new sesquiterpenoid from *Russula lepida*. Acta Bot. Sin., 43: 329–330.

Tan, J.W., Dong, Z.J., Ding, Z.H. and Liu, J.K. 2002. Lepidolide, a novel seco-ring-A cucurbitane triterpenoid from *Russula lepida* (Basidiomycetes). Z. Naturforsch. C, 57: 963–965.

Taylor, D.L., Hollingsworth, T.N., McFarland, J.W., Lennon, N.J., Nusbaum, C. and Ruess, R.W. 2014. A first comprehensive census of fungi in soil reveals both hyperdiversity and fine-scale niche partitioning. Ecol. Monogr., 84: 3–20.

Tedersoo, L., Bahram, M., Polme, S., Koljalg, U., Yorou, N.S., Wijesundera, R., Ruiz, L.V., Vasco-Palacios, A.M., Thu, P.Q., Suija, A., Smith, M.E., Sharp, C., Saluveer, E., Saitta, A., Rosas, M., Riit, T., Ratkowsky, D., Pritsch, K., Poldmaa, K., Piepenbring, M., Phosri, C., Peterson, M., Parts, K., Partel, K., Otsing, E., Nouhra, E., Njouonkou, A.L., Nilsson, R.H., Morgado, L.N., Mayor, J., May, T.W., Majuakim, L., Lodge, D.J., Lee, S.S., Larsson, K.-H., Kohout, P., Hosaka, K., Hiiesalu, I., Henkel, T.W., Harend, H., Guo, L.-d., Greslebin, A., Grelet, G., Geml, J., Gates, G., Dunstan, W., Dunk, C., Drenkhan, R., Dearnaley, J., De Kesel, A., Dang, T., Chen, X., Buegger, F., Brearley, F.Q., Bonito, G., Anslan, S., Abell, S. and Abarenkov, K. 2014. Global diversity and geography of soil fungi. Science, 346: 1256688.

Ukiya, M., Akihisa, T., Tokuda, H., Hirano, M., Oshikubo, M., Nobokuni, Y., Kimura, Y., Tai, T., Kondo, S. and Nishino H. 2002. Inhibition of tumor-promoting effects by Poricoic Acids G and H and other Lanostane-type triterpenes and cytotoxic activity of Poricoic Acids A and G from *Poria cocos*. J. Nat Prod., 65: 462–465.

Velíšek J. and Cejpek, K. 2011. Pigments of higher fungi – a review. Czech J. Food Sci., 29: 87–102.

Vidari, G., De Bernardi, M., Vita-Finzi, P. and Fronza, G. 1976. Sesquiterpenes from *Lactarius blennius*. Phytochemistry, 15: 1953–1955.

Vidari, G. and Vita-Finzi, P. 1995. Sesquiterpenes and other secondary metabolites of genus *Lactarius* (Basidiomycetes): Chemistry and biological activity. pp. 153–206. *In*: Atta-ur-Rahman (ed.). Studies in Natural Products Chemistry. Structure and Chemistry (Part D). Vol. 17. Elsevier, London.

Vidari, G., Che, Z. and Garlaschelli, L. 1998. New nardosinane and aristolane sesquiterpenes from the fruiting bodies of *Russula lepida*. Tetrahedron Lett., 39: 6073–6076.

Vizzini, A. 2003. I metaboliti secondari delle specie di *Cortinarius* s.l.: Aspetti tossico-farmacologici (tossine, pigmenti, cromogeni bioattivi). pp. 29–64. *In*: Consiglio, G. and Antonini, D. & M. Il genere *Cortinarius* in Italia, Tomo 1. A.M.B., Fondazione Centro Studi Micologici, Vicenza.

Wieland, T. 1983. The toxic peptides from *Amanita* mushrooms. Int. J. Pept. Protein Res., 22: 257–276.

Winner, M., Gimenez, A., Schmidt, H., Sontag, B., Steffan, B. and Steglich, W. 2004. Unusual pulvinic acid dimers from the common fungi *Scleroderma citrinum* (Common Earthball) and *Chalciporus piperatus* (Peppery Bolete). Angew Chem. Int. Ed. Engl., 43: 1883–1886.

Xia, Q., Zhang, H., Sun, X., Zhao, H., Wu, L., Zhu, D., Yang, G., Shao, Y., Zhang, X., Mao, X., Zhang, L. and She, G. 2014. A comprehensive review of the structure elucidation and biological activity of Triterpenoids from *Ganoderma* spp. Molecules, 19: 17478–17535.

Xiong, Q., Wilson, W.K. and Matsuda, S.P.T. 2006. An *Arabidopsis* oxidosqualene cyclase catalyzes iridal skeleton formation by Grob fragmentation. Angew Chem. Int. Ed. Engl., 45: 1285–1288.

Xu, R., Fazio, G.C. and Matsuda, S.P.T. 2004. On the origins of triterpenoid skeletal diversity. Phytochemistry, 65: 261–291.

Yamashita, S. and Hijii, N. 2007. The role of fungal taxa and developmental stage of mushrooms in determining the composition of the mycophagous insect community in a Japanese forest. Eur. J. Entomol., 104: 225–233.

Yoshikawa, K., Kaneko, A., Matsumoto, Y., Hama, H. and Arihara, S. 2006. Russujaponols A-F, Illudoid Sesquiterpenes from the fruiting body of *Russula japonica*. J. Nat. Prod., 69: 1267–1270.

Yuan, M.S. and Sun, P.Q. 1995. Sichuan Mushrooms. Sichuan Academic Press, Chengdu.

Zeng, N.K., Wu, G., Li, Y.C., Liang, Z.Q. and Yang, Z.L. 2014. *Crocinoboletus*, a new genus of Boletaceae (Boletales) with unusual boletocrocin polyene pigments. Phytotaxa, 175: 133–140.

Zhao, R.L., Li, G.J., Sánchez-Ramírez, S., Stata, M., Yang, Z.-L., Wu, G., Dai, Y.-C., He, S.-H., Cui, B.-K., Zhou, J.-L., Wu, F., He, M.-Q., Moncalvo, J.-M. and Hyde, K.D. 2017. A six-gene phylogenetic overview of Basidiomycota and allied phyla with estimated divergence times of higher taxa and a phyloproteomics perspective. Fungal Divers., 84: 43–74.

Zjawiony, J.K. 2004. Biologically active compounds from Aphyllophorales (Polypore) fungi. J. Nat. Prod., 67: 300–310.

12

The Genus *Phellinus*
A Rich Source of Diverse Bioactive Metabolites

Sunil K. Deshmukh,[1],* Ved Prakash[2] and Manish K. Gupta[1]

INTRODUCTION

Phellinus is a member of family Hymenochaetaceae, Aphyllophorales, Hymenomycetes, Basidiomycota. The members of Hymenochaetaceae form a cosmopolitan group of wood inhabiting fungi, capable of utilizing structural polymers of wood. Overall 427 records comprising of 310 species, 69 forma species, 42 varieties and 6 sub-species of *Phellinus* have been reported till date (Ranadive et al., 2012).

Medicinal mushroom *Phellinus linteus* ('meshimakobu' in Japanese, 'Sanghuang' in Chinese) is a year-round growing fungus, belonging to *Phellinus* Quel. Traditional Chinese medicine (TCM) consists of the fruit body of *P. linteus* in dehydrated form. The *P. linteus* named by parasitizing on the trunk of *Morus alba* (Moraceae) are primarily found in Japan, China, Korea, Mongolia, other Asian countries, Africa and America. The fungus also inhabits and parasitizes other plants, such as *Betula*, *Salix babylonica* and *Populus*.

In the Indian state of Kerala, *Phellinus* is known as 'Pilamangal' (pila=jack fruit tree, mangal=turmeric) and in Maharashtra 'Phansomba" (phanas=jack fruit tree, alombe=mushroom) (Vaidhya et al., 2005). Uses of *Phellinus* (in Kerala and Karnataka region): (1) For pittha-pittha gulika (tablet), *Phellinus* is used along with "kantham" (a mineral taken from the earth) and "annupedi" (an ayurvedic medicine).

[1] TERI-Deakin Nano Biotechnology Centre, The Energy and Resources Institute, Darbari Seth Block, IHC Complex, Lodhi Road, New Delhi 110003, India.
[2] Department of Biotechnology, Motilal Nehru National Institute of Technology, Allahabad, India.
* Corresponding author: sunil.deshmukh1958@gmail.com

These three are ground together for seven days in lemon juice and then taken to treat ailments that come out of pittha; (2) *Phellinus* is a constituent of the "Marmanithailam", an ayurvedic oil used for curing major and minor fractures; (3) *Phellinus* is also a constituent of the "Narayanithailam", an ayurvedic oil used for various massages. This is also used to reduce burning sensations and severe joint pains; (4) *Phellinus* is used to cure oral cancers by an ayurvedic decoction called "kashayam"; (5) *Phellinus* is used to cure brain tumours as well as rectal and intestinal cancers. In combination with other medicines, *Phellinus* is used against severe headaches by sniffing as "nasya" through the nostrils. For all the above uses, *Phellinus* is collected from jack fruit tree and *Strychnos nux-vomica*. For medicinal use, it is first soaked in rice gruel to release all the toxic chemicals and then ground to prepare various formulations. These specific details were obtained based on personal discussion with many local traditional vaidyas (healers). In Karnataka, it is commonly used against throat cancer, mumps and for children having ailments for apothae and excessive salivation (Vaidya et al., 2005). The *P. linteus* possesses various pharmaceutical applications in the regulation of blood glucose levels, enhancing circulation of blood, hepatoprotective activity and enhancement of the immune system. Besides, *P. linteus* is also known for its anti-cytotoxic, anti-diabetic, immunomodulation, anti-oxidant, anti-inflammatory and anti-hyperlipidemic properties (Song et al., 2003; Kim et al., 2007; Kim et al., 2004; Nakamura et al., 2004; Inagaki et al., 2005, Chen et al., 2016). Some literature reveals that polysaccharides present in the fruit body of this genus account for diverse biological properties (Wasser and Didukh, 2005; Yan et al., 2017) and those are not covered in this review. This review focuses on giving an insight into the different strains of *Phellinus* identified for their roles in eliciting bioactive compounds. Major emphasis is given to the pharmacological activities of various metabolites of *Phellinus*. The details of bioactive compounds isolated along with their biological properties are given in Table 1.

Styrylpyrone

Representative therapeutic fungi, *Phellinus linteus*, *P. igniarius*, *P. ribis*, *P. pini*, *P. baumii*, *Inonotus obliquus* and *I. xeranticus* species, have been identified as producing a wide array of styrylpyrone-type polyphenol pigments that have been shown to possess various pharmacological attributes, such as anti-platelet aggregation, anti-inflammatory, anti-oxidative, anti-dementia, anti-diabetic, anti-viral and cytotoxic properties. The unique arrangement of the carbon backbone of fused styrylpyrone makes it an interesting candidate for pharmacological usage and an attractive molecule scaffold (Lee at al., 2011).

Hispidin (**1**), a styrylpyrone occurring in nature, was isolated from *Inonotus hispidus* (formerly *Polyporus hispidus*) for the first time in 1889. From the genus *Phellinus* and *Inonotus*, a range of metabolites of other styrylpyrones have been discovered. Styrylpyrone pigments form the basic composition of fungi, specifically of Hymenochaetaceae family, comprising *Phellinus* and *Inonotus* (Fiasson, 1982). Styrylpyrones are considered as phenylalanine (Phe)-derived fungal metabolites involved in defense, as signaling molecules and pigmentation.

Table 1. Occurrence of Bioactive compounds in the genus *Phillinus.*

	Compounds	Producing fungi	Biological activity	References
1	Hispidin (1)	*Phellinus igniarius, P. pini, P. ribis, P. linteus, P. baumii*	Anti-oxidant, cytotoxic, anti-inflammatory, anti-viral, anti-dementia	Gill and Steglich (1987); Gonindard et al. (1997); Awadh Ali et al. (2003); Park et al. (2004a, 2004b); Jung et al. (2008); Kemami Wangun et al. (2006); Singh et al. (2003)
2	Phelligridin A (2)	*P. igniarius*	Cytotoxic	Mo et al. (2004)
3	Phelligridin B (3)	*P. igniarius*	Cytotoxic	Mo et al. (2004)
4	Phelligridin C (4)	*P. igniarius, P. linteus*	Cytotoxic	Mo et al. (2004); Nagatsu et al. (2004)
5	Phelligridin D (5)	*P. igniarius, P. linteus, P. baumii*	Cytotoxic	Mo et al. (2004); Nagatsu et al. (2004); Lee et al. (2007)
6	Phelligridin E (6)	*P. igniarius*	Cytotoxic	Mo et al. (2004); Lee et al. (2007)
7	Phelligridin F (7)	*P. igniarius*	Cytotoxic	Mo et al. (2004)
8	Phelligridin G (8)	*P. igniarius*	Anti-oxidant, cytotoxic	Wang et al. (2005); Lee et al. (2007)
9	Phelligridin H (9)	*P. igniarius*	Anti-oxidant, cytotoxic	Wang et al. (2007)
10	Phelligridin I (10)	*P. igniarius*	Anti-oxidant, cytotoxic	Wang et al. (2007); Lee et al. (2007)
11	Phelligridin J (11)	*P. igniarius*	Anti-oxidant, cytotoxic	Wang et al. (2007)
12	Phellifuropyranone A (12)	*P. linteus*	Anti-oxidant, cytotoxic	Kojima et al. (2008); Lee et al. (2007)
13	Hypholomine B (13)	*P. linteus, P. ribis*	Anti-oxidant, anti-diabetes, anti-inflammatory	Fiasson (1982); Gill and Steglich (1987); Fiasson et al. (1977); Jung et al. (2008); Kemami Wangun et al. (2006); Lee et al. (2008b,c)
14	Squarrosidine (14)	*P. pini, P. linteus*	Anti-inflammatory	Kemami Wangun and Hertweck (2007); Kemami Wangun et al. (2006)
15	1,1-Distyrylpyrylethan (15)	*P. linteus, P. pini*	Anti-oxidant, anti-inflammatory	Kemami Wangun and Hertweck, (2007); Jung et al. (2008); Kemami Wangun et al. (2006)

Table 1 contd. ...

...Table 1 contd.

	Compounds	Producing fungi	Biological activity	References
16	Phellinusfuran A (16)	*P. linteus*	Anti-inflammatory	Min et al. (2006)
17	Phellinusfuran B (17)	*P. linteus*	Anti-inflammatory	Min et al. (2006)
18	Davallialactone (18)	*P. igniarius, P. baumi*	Anti-oxidant, antidiabetes, anti-inflammatory, anti-platelet aggregation	Lee and Yun (2006); Cui et al. (1990); Lee et al. (2008a,b,c); Kim et al. (2008)
19	Interfungin A (19)	*P. linteus, P. baumii*	Anti-oxidant, anti-diabetes	Lee and Yun (2007); Lee et al. (2008c,d)
20	Inoscavin A (20)	*P. igniarius, P. linteus*	Anti-oxidant, antidiabetes	Kim et al. (1999); Lee et al. (2008c,d)
21	Phelligridimer A (21)	*P. igniarius*	Anti-oxidant	Wang, et al. (2005b)
22	3,14'-Bihispidinyl (22)	*P. linteus, P. ignarius*	Anti-oxidant	Fiasson (1982); Gill and Steglich (1987); Klaar and Steglich (1977); Jung et al. (2008); Lee and Yun (2008)
23	Inoscavin C (23)	*P. linteus*	Prolylendopeptidase inhibitory activity	Yoon et al. (2013)
24	Phellinstatin (24)	*P. linteus*	Antimicrobial Activity	Cho et al. (2011)
25	Hispolon (25)	*P. linteus*	Cytotoxic, Anti-inflammatory activity, antiviral	Mo et al. (2004); Chen et al. (2006); Hsiao et al. (2013); Yang et al. (2014); Arcella et al. (2017); Wu et al. (2017)
26	4-(3,4-Dihydro-xyphenyl)-3-buten-2-one (26)	*P. vaniniiwere*	Antitumor activity	Cheng et al. (2011)
27	8-Methyl-13-phenyltri-deca-4,6,8,10,12-pentaene-3-one (27)	*P. pini*	Antifungal activity	Ayer et al. (1996)
28	4-Vinylresorcinol (28)	*P. pini*	Antifungal activity	Ayer et al. (1996)
29	Phellinsin A (29)	*Phellinus* sp. PL3.	Antifungal activity	Hwang et al. (2000)
30	Igniarens A (30)	*P. igniarius*	Anti-inflammatory	Wang et al. (2009)
31	Igniarens B(31)	*P. igniarius*	Anti-inflammatory	Wang et al. (2009)
32	Igniarens C (32)	*P. igniarius*	Anti-inflammatory	Wang et al. (2009)
33	Igniarens D (33)	*P. igniarius*	Anti-inflammatory	Wang et al. (2009)
34	Ergostanesgilvsin C (34)	*P. igniarius*	Anti-inflammatory	Wang et al. (2009)
35	Gilvsin A (35)	*P. igniarius*	Anti-inflammatory	Wang et al. (2009)
36	5α, 8α-Epidioxy-22E-ergosta-6,22-dien-3β-ol (36)	*P. igniarius*	Anti-inflammatory	Wang et al. (2009)

Table 1 contd. ...

...Table 1 contd.

	Compounds	Producing fungi	Biological activity	References
37	5α-Ergosta-7,22-dien-3-one **(37)**	*P. igniarius*	Anti-inflammatory	Wang et al. (2009)
38	Ergosta-7,24(28)-dien-3-ol **(38)**	*P. pini*	Anti-inflammatory	Jang et al. (2007)
39	Ergosterol peroxide **(39)**	*P. pini*	Anti-inflammatory	Jang et al. (2007)
40	Inotilone**(40)**	*P. linteus*	Anti-inflammatory	Huang et al. (2012)
41	Phelligrin A **(41)**	*P. baumii*	Anti-inflammatory	Wu et al. (2011)
42	Methylphelligrins A **(42)**	*P. baumii*	Anti-inflammatory	Wu et al. (2011)
43	Methylphelligrins B **(43)**	*P. baumii*	Anti-inflammatory	Wu et al. (2011)
44	5α,6α-Epoxy-ergosta-8(14),22-dien-3β,7α-diol **(44)**	*P. linteus*	HNE-inhibitory activity	Lee et al. (2012)
45	5α,6α-Epoxyergosta-8(9),22-dien-7-on-3β-ol **(45)**	*P. linteus*	HNE-inhibitory activity	Lee et al. (2012)
46	5α,6α;8α,9α-Diepoxy-ergost-22-en-3β,7α-diol **(46)**	*P. linteus*	HNE-inhibitory activity	Lee et al. (2012)
47	14α-Hydroxy-ergosta-4,7,9(11),22-tetraen-3,6-dione **(47)**	*P. linteus*	HNE-inhibitory activity	Lee et al. (2012)
48	Ergosta-4,7,22-trien-3,6-dione **(48)**	*P. linteus*	HNE-inhibitory activity	Lee et al. (2012)
49	3β,5α-Dihydroxy-6β-methoxyergosta-7,22-diene **(49)**	*P. linteus*	HNE-inhibitory activity	Lee et al. (2012)
50	Ergosta-7,22-dien-3β,5α,6β,9α-tetraol **(50)**	*P. linteus*	HNE-inhibitory activity	Lee et al. (2012)
51	Ergosta-4,6,8(14),22-tetraen-3-one **(51)**	*P. pini*	Anti-inflammatory	Hong et al. (2012)
52	5,8-Epidioxyergosta-6,22-dien-3-ol **(52)**	*P. pini*	Anti-inflammatory	Hong et al. (2012)
53	Atractylenolide I **(53)**	*P. linteus*	Cytotoxic activity	Jeon et al. (2013)
54	Phelliribsin A **(54)**	*P. ribis*	Cytotoxic activity	Kubo et al. (2014)
55	Ergosta-7,22-dien-3β-ylpentadecanoate **(55)**	*P. baumii*	Cytotoxic activity	Feng et al. (2015)
56	Ergosta-7,22-dien-3β-ol **(56)**	*P. baumii*	Cytotoxic activity	Feng et al. (2015)
57	Ganoderiol B **(57)**	*P. baumii*	Cytotoxic activity	Feng et al. (2015)
58	Ergosta-6,2-dien-3β,5α,8α-triol **(58)**	*P. baumii*	Cytotoxic activity	Feng et al. (2015)
59	Ganoderic acid DM **(59)**	*P. baumii*	Cytotoxic activity	Feng et al. (2015)
60	3,4-Dihydroxybenzaldehyde **(60)**	*P. igniarius*	Prevent acrolein toxicity	Suabjakyong et al. (2015)
61	4-(3,4-Dihydroxyphenyl)-3-buten-2-one **(61)**	*P. igniarius*	Prevent acrolein toxicity	Suabjakyong et al. (2015)

Table 1 contd. ...

...Table 1 contd.

	Compounds	Producing fungi	Biological activity	References
62	Inonoblin C **(62)**	*P. igniarius*	Prevent acrolein toxicity	Suabjakyong et al. (2015)
63	Inoscavin C **(63)**	*P. igniarius*	Prevent acrolein toxicity	Suabjakyong et al. (2015)
64	Interfungin B **(64)**	*P. igniarius*	Prevent acrolein toxicity	Suabjakyong et al. (2015)
65	Phellibarins B **(65)**	*P. rhabarbarinus*	Anti-inflammatory	Feng et al. (2016a)
66	Phellibarins C **(66)**	*P. rhabarbarinus*	Anti-inflammatory	Feng et al. (2016a)
67	Igniaren C **(41)**	*P. rhabarbarinus*	Anti-inflammatory	Feng et al. (2016a)
68	Igniaren D **(42)**	*P. rhabarbarinus*	Anti-inflammatory	Feng et al. (2016a)
69	Gilvsin A **(35)**	*P. rhabarbarinus*	Anti-inflammatory	Feng et al. (2016a)
70	Gilvsin C **(34)**	*P. rhabarbarinus*	Anti-inflammatory	Feng et al. (2016a)
71	Gilvsin D **(67)**	*P. rhabarbarinus*	Anti-inflammatory	Feng et al. (2016a)
72	Phellibarin D **(68)**	*P. rhabarbarinus*	Cytotoxic activity	Feng et al. (2016b)
73	(-)-Trans-γ-monocyclofarnesol **(69)**	*P. linteus*	Antibacterial	Shirahata et al. (2017)
74	(+)-γ-Ionylideneacetic acid **(70)**	*P. linteus*	Antibacterial	Shirahata et al. (2017)
75	Phellidene E **(71)**	*P. linteus*	Antibacterial	Shirahata et al. (2017)
76	3,4-Dihydroxy benzaldehyde **(60)**	*P. baumii*	Cytotoxic activity	Zhang et al. (2017)
77	Inoscavin A **(20)**	*P. baumii*	Cytotoxic activity	Zhang et al. (2017)
78	Baicalein **(72)**	*P. baumii*	Cytotoxic activity	Zhang et al. (2017)
79	3,4-Dihydroxybenzaldehyde **(60)**	*P. baumii*	Anti-inflammatory	Lee et al. (2017)
80	4-(4-Hydroxyphenyl)-3-buten-2-one **(73)**	*P. baumii*	Anti-inflammatory	Lee et al. (2017)
81	4-(3,4-Dihydroxyphenyl)-3-buten-2-one **(61)**	*P. baumii*	Anti-inflammatory	Lee et al. (2017)
82	9,11-Dehydroergosterol peroxide **(74)**	*P. baumii*	Anti-inflammatory	Lee et al. (2017)
83	Ethyl linoleate **(75)**	*P. baumii*	Anti-inflammatory	Lee et al. (2017)
84	8,14-Labdadien-13-ol **(76)**	*P. pini*	Anti-inflammatory	Jang and Yang (2011); Zhu et al. (2017)
85	Dehydroabietic acid **(77)**	*P. pini*	Anti-inflammatory	Jang and Yang (2011); Zhu et al. (2017)
86	3,4-Dihydroxybenzalacetone **(78)**	*P. linteus*	Anti-inflammatory	Chao et al. (2017)

Cytotoxicity

Hispidin (**1**) is reported to inhibit isoform-b of protein kinase C (IC_{50} 2 mM) and exhibited cytotoxic activity against the pancreatic duct and keratinocyte cancerous cells compared to normal cells (fibroblast) (Gonindard et al., 1997). Phelligridins (**A-J**) (**2-11**), obtained from *P. igniarius*, and phellifuropyranone A (27) (**12**), from *P. linteus*, exhibited anti proliferative activity *in vitro* against many cancer cell lines of humans (Mo et al., 2004; Wang et al., 2005a, 2007; Kojima et al., 2008).

It is reported that hispidin (**1**) leads to an onset of extrinsic and intrinsic apoptotic paths in colon cancer cells lead by ROS (Lim et al., 2014). Hispidin (**1**) plays a protective role in H9c2 cells of cardiomyoblast when exposed to hydrogen peroxide by regulating the ROS generation, apoptotic proteins and Akt/GSK-3β activation (Kim et al., 2014).

Hispidin is (**1**) reported to reduce effects of cytotoxicity mediated by peroxynitrite, hydroxyl radical formation and DNA damage (Chen et al., 2012). It is reported that

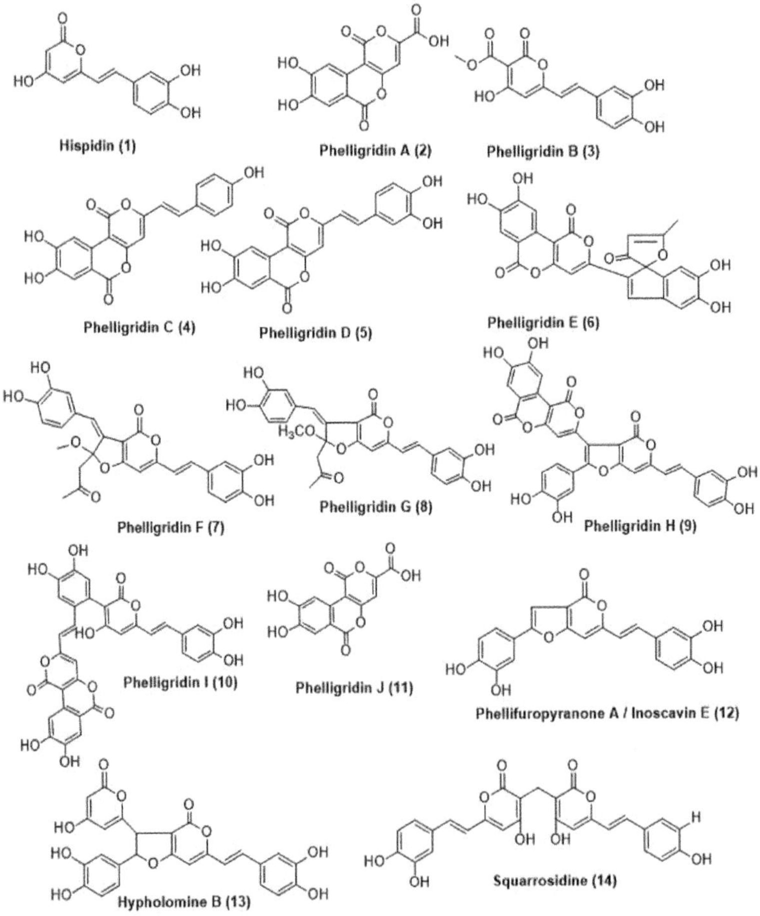

Fig. 1. Structure of compounds 1–14.

hispidin in RAW 264.7 cells of mouse macrophage showed an anti-inflammatory response by reducing ROS induced NF-κB pathway (Shao et al., 2015).

Anti-Inflammatory Activity

Hispidin (1), when compared to selected inhibitors like bimesulide and meloxicam, was found to block cyclooxygenase-2 at almost equal concentrations to that of standard inhibitors. Hispidin (1) also exhibited significant 3-hydroxysteroid dehydrogenase and xanthine oxidase inhibitory properties (Wangun et al., 2006). The tautomeric hispidin a-pyrone appeared to be more active than g-pyrone. Compared with allopurinol (IC_{50} 4.4 mM), hypholomine B (13), squarrosidine (14), and 1,1-distyrylpyrylethan (15), showed to be effective xanthine oxidase inhibitors with IC_{50} values of 6.7, 8.1 and 5.8 mM, respectively (Jung et al., 2008).

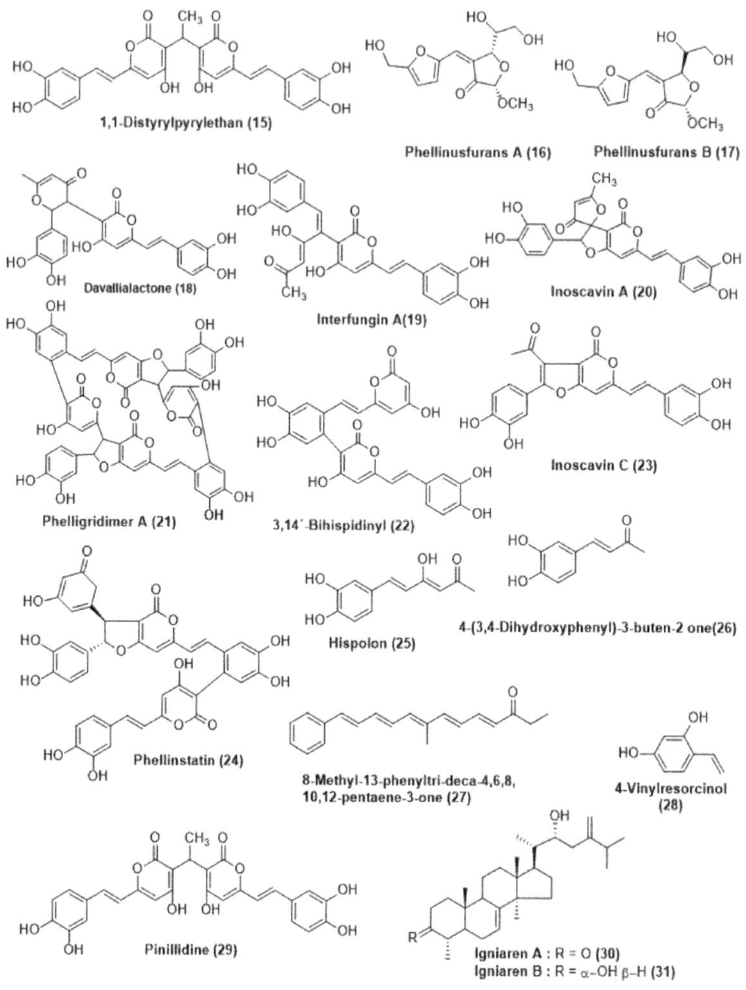

Fig. 2. Structure of compounds 15–31.

The humoral immunity consists of a complement system which controls the inflammation through different pathways. Phellinus furans A **(16)**, and B **(17)**, showed inhibitory potential in the classical pathway of the complement system, with IC_{50} values of 33.6 and 33.7 mM, respectively, with respect to rosmarinic acid as positive control with IC_{50} value of 180 μM (Min et al., 2006).

Davallialactone **(18)**, when tested over RAW264.7 cells, was shown to down regulate LPS-induced inflammation as well as cause a decrease in production of NO, release of E2 prostaglandin, cytokines and expression level of co-stimulatory molecules of cell surface. The cell morphology and viability remained unaltered on treatment with davallialactone **(18)**. The compound **(18)** exhibited anti-inflammatory activity by blocking a series of signaling molecules that activates nuclear factor kB via P13K, Akt and IKK but not protein kinases (Lee et al., 2008a). Compound **(18)** also blocked LPS-induced phosphorylation and kinase activity of Src, inferring Src as a potent target of davallialactone **(18)** (Lee et al., 2008b).

Anti-Diabetic Activity

Phellinus linteus fruiting bodies led to identification of compounds davallialactone **(18)**, hypholomine B **(13)**, interfungin A **(19)** and inoscavin A **(20)**, that showed inhibitory activity against human recombinant aldose reductases and rat lens with IC_{50} values of 0.33, 0.82,1.03, 1.06 μM and 0.56, 1.28, 1.82, 1.40 μM, respectively. Interfungin A **(19)** was found to inhibit glycation of protein and further studies showed that it stopped proteins cross-linking, which was better than amino guanidine, a product well known for inhibition of glycation end products (Lee et al., 2008c,d).

Apoptosis induced by hydrogen peroxide and enhanced release in hydrogen peroxide treated cells was observed to be inhibited on treatment with hispidin. The observation showed the potential of hispidin as an antidiabetic, this basically prevents β-cells from the adverse effects of ROS in diabetes (Jang et al., 2010).

Antioxidant Activity

Styrylpyrones have been shown to exhibit potent anti-oxidant properties. Davallialactone **(18)** and inoscavin A **(20)** were isolated from the fruiting body of *P. linteus*. Davallialactone **(18)** exhibited noteworthy ABTS radical scavenging activity, with IC_{50} values of 0.8 μM (vitamin E, 5.7 μM). Davallialactone **(18)** exhibited DPPH radical scavenging activity, with IC_{50} values of 3.4 μM (vitamin E, 12.3 μM), and superoxide radical quenching activity, with IC_{50} values of 2.3 μM (vitamin E, 4100 μM; caffeic acid, 2.9 μM). Inoscavin A **(20)** inhibited rat liver microsomal lipid peroxidation, with an IC_{50} value of 0.3 mg/ml, which is five times the activity of vitamin E (1.5 mg/ml) (Lee et al., 2008c). Phelligridimer A **(21)**,and Phelligridins G-J **(8-11)** (Gill, 2001; Perrin and Towers, 1973a,b; Hatfied and Brady, 1973) from the fruiting body of *P. igniarius* inhibited microsomal lipid peroxidation in rat liver, with IC_{50} values of 10.2, 3.8, 4.8, 3.7 and 6.5 μM, respectively (Wang et al., 2005a,b, 2007). Other findings showed that hispidin dimers, such as hispidin with hispidin or hispolon, had better radical-scavenging capacity than hispidin alone (Park et al., 2004a; Jeon et al., 2009).

Phellinus linteus mycelial culture broth led to identification of hispidin and evaluate for its antioxidant property. The result showed that 1.0 μM of hispidin exhibited antioxidant activities on DPPH radical, hydroxyl radical and super oxide anion radical at 85.5%, 95.3% and 56.8%, respectively, which was comparable to positive control α-tocopherol, but hydrogen peroxide radical remained unaltered with hispidin (Lee and Yun, 2007; Jung et al., 2008).

Hispidin (1), along with its dimers, 3,14'-bihispidinyl (22), hypholomine B (13), and 1,1-distyrylpyrylethan (15), in a concentration-dependent manner showed significant radical scavenging activity (Jung et al., 2008). Whereas, dimeric hispidins consisting of two catechol moieties, 3,14'-bihispidinyl (22), hypholomine B (13) and 1,1-distyrylpyrylethan (15), showed more activity than hispidin for DPPH and ABTS radical scavenging activity. Therefore, the anti-oxidative effects of hispidin, 3,14'-bihispidinyl (22), hypholomine B (13), and 1,1-distyrylpyrylethan (15), against DPPH and ABTS radicals may be as a result of catechol moiety. Their efficacy is around 2–3 times higher than of trolox, a commercial anti-oxidant (Jung et al., 2008).

Hispidin from *P. linteus* displayed reducing effects against hydrogen peroxide, DPPH radicals, and superoxide radicals, in a concentration-dependent manner. At 30 μM concentration, hispidin exhibited around 55% intracellular reactive oxygen activity (Jang et al., 2010).

Two styrylpyranones inoscavin C (23), and E (12) were identified from dichloromethane fraction of *P. linteus* and their structures were elucidated NMR spectral data. Compounds displayed prolylendopeptidase inhibitory activity with IC_{50} values of 4.08 and 4.26 μM and Ki values of 1.43and 1.50 μM, respectively. It also displayed ABTS radical system with EC_{50} values of 7.64 and 6.47 μM, respectively (Yoon et al., 2013).

Antimicrobial Activity

From culture broth of *P. linteus* led to identification of a novel trimeric hispidin derivative, phellinstatin (24). Compound (24) potently checked *Staphylococcus aureus* enoyl-ACP reductase with an IC_{50} of 6 μM and it also exhibited strong antibacterial activity against *S. aureus* and MRSA (Cho et al., 2011).

Anti-Platelet Aggregation Activity

Davallialactone (18) checked platelet aggregation induced by thrombin or collagen and led by ADP in dose-dependent manner. Additionally, it blocked p38 mitogen-activated protein kinase (MAPK), phosphorylation of ERK-2and intracellular calcium concentration level, in a dose-dependent way. Phosphorylation of tyrosine 60 and 85kDa proteins were differentially blocked by the compound (18) on activation by collagen (Nagatsu et al., 2004). Thus, compound (41) is expected to show anti-platelet aggregation effect via suppression of intracellular downstream signaling pathways (Kim et al., 2008).

Anti-Viral Activity

Hispidin (1) showed significant anti-viral activity against influenza viruses type A (H1N1 and H3N2) and B using the allantois on the shell-test system (Awadh Ali et al., 2003). HIV-1 integrase, along with protease and reverse transcriptase, plays a prominent role in viral replication. The HIV-1 protease and reverse transcriptase inhibitors help in control of HIV infection. For retroviral therapy, integrase inhibition can be targeted to control the HIV-1, although new drug resistant viruses and drug naïve patients have also emerged. In coupled assay system, hispidin (1) exhibited HIV-1 integrase inhibitory activity, with an IC_{50} value of 2 μM. On capping of phenolic group with methyl ester the HIV-1integrase inhibitory activity was lost, indicating that the acidic phenolic functional group is a must for integrase inhibitory activity, a well-established feature of the enzyme (Singh et al., 2003).

Hispidin and hypholomine B (13) were identified from the culture broth of *P. linteus*. These compounds were characterized based on spectral data. Hispidin (1) and hypholomine B (13) inhibited neuraminidase, with IC_{50} values of 13.1 and 0.03 μM, respectively (Yeom et al., 2012).

Phelligridins E (6) and G (8) were isolated from the fruiting bodies of *Phellinus igniarius*. The compounds were identified using ^{1}H NMR and electrospray ionization mass measurements. Compounds (6) and (8) blocked neuraminidases from recombinant rvH1N1, H3N2, and H5N1 influenza viruses, with IC_{50} values in the range of 0.7~8.1 μM (Kim et al., 2016a).

Anti-Dementia Activity

Alzheimer's disease is a progressive neurodegenerative disorder; amyloid protein deposits in the neocortex, hippocampus and parenchyma of the amygdale is a distinguishing feature of the disease. Amyloid plaque is mostly made of β-amyloid peptide, formed by α-, β-, and γ-secretase, formed by the breakdown of the amyloid precursor protein, which forms a prominent part of amyloid plaques. For the inhibition of amyloid formation, BACE1 (β-site amyloid precursor protein leaving enzyme) is a promising target among all secretases. Thus, inhibitors of BACE1 should be helpful in the curing of disease. Hispidin (1) isolated from the mycelial culture of *P. linteus* non-competitively inhibited BACE1 in a dose-dependent manner, exhibiting an IC_{50} value of 4.9 μM. In addition, hispidin (1) showed as specific inhibitor of BACE1 and prolylendopeptidase, and it exhibited no activity against TACE (tumor necrosis factor alpha-converting enzyme) and other serine proteases, such as elastase trypsin and chymo trypsin (Park et al., 2004b).

Hispolon

Hispolon (25) is a natural bioactive polyphenol found in several medicinal mushrooms. It was isolated initially from *Inonotus hispidus* (Ali et al., 1996) and hence the name, hispolon. Subsequently, hispolon was isolated from *P. linteus, Phellinus igniarius* and other species. Hispolon was first synthesized in the year 2002 (Venkateswarlu et al., 2002). A modified synthesis and a number of its derivatives were reported (Ravindran et al., 2010; Balaji et al., 2015, 2017). Hispolon (25), has been used in the treatment of various pathological conditions, namely gastroenteric disorders, inflammation, lymphatic diseases and various cancer sub-types. Hispolon studies reveal its potential for use in the treatment of various cancer cells, such as gastric cancer cells, hepato carcinoma, melanoma, leukemia and bladder cancer cells.

Anticancer Activity

Phelligridins C-D (**3-4**), together with hispolon (**25**), were isolated from *P. igniarius*. Their structures were deciphered by spectroscopic methods, including IR, MS, and 1D and 2D NMR expts. Phelligridins C (**3**) and D (**4**) showed selectivity against A549 and Bel7402, with IC_{50} values of 0.012, 0.016, 0.010, and 0.008 μM, respectively. Hispolon (**25**) was found to be more sensitive to MCF-7 and Bel7402, with IC_{50} values of 0.025 and 0.038 μM, respectively (Mo et al., 2004).

In a study on human epidermoid KB cells, Hispolon (**25**) exhibited inhibition of KB cell division in a dose-dependent manner, with IC_{50} of 4.62 ± 0.16 μg/ml. It was shown that hispolon induces the apoptosis in epidermoid KB cells. This process was supported by the collapse of mitochondrial membrane potential, the activation of caspase-3 and release of cytochrome c. The study suggested that hispolon induces the mitochondria mediated apoptotic pathway which led to the death of KB cells (Chen et al., 2006).

Hispolon (**25**) exhibited cytotoxicity against MCF7, MDA-MB-231, T24 and J82 cell lines, with IC_{50} values of 20, 10, 10 and 40 μg/ml, respectively. Hispolon down regulates MDM2 expression through ERK1/2-mediated MDM2 ubiquitination. Hispolon-induced activated ERK1/2 directly binds to MDM2, thereby mediating MDM2 ubiquitination. Cell lines with higher ERK1/2 activity are more sensitive to hispolon, indicating that the sensitivity of cell lines to hispolon is largely related to their ERK1/2 activity. Inhibiting ERK1/2 activity also attenuates hispolon-induced caspase-7 cleavage. The involvement of ERK1/2-mediated MDM2 ubiquitination provides a strategy for cancer therapy in cells with constitutive ERK1/2 activity and over expression of MDM2. In addition, hispolon inhibits cell growth independent of p53 status. The elevated p21WAF1 may be regulated by MDM2, which may contribute to the effects of hispolon on cell apoptosis and cell cycle arrest (Lu et al., 2009).

Huang et al. (2010) reported that hispolon (**25**) inhibits the metastasis of SK-Hep1 cells by reducing expression of MMP-2, MMP-9, and uPA through the suppression of the FAK signaling pathway and of the activity of PI3K/Akt and Ras homolog gene family, member A (RhoA).

Hispolon (**25**) had an anti-proliferative effect on human hepatocellular carcinoma Hep3B cells. Hispolon caused cell cycle arrest at S phase in a time-dependent and dose-dependent manner, leading to cellular growth with down-regulation in expressions of cyclins A and E and cyclin-dependent kinases 2, with concomitant induction of p21waf1/Cip1 and p27Kip1 (Huang et al., 2011).

The antitumor activity of sakuranetin, 4-(3,4-dihydro-xyphenyl)-3-buten-2-one (**26**) and hispolon (**25**) from the fruiting body of *Phellinus vaninii* was studied by MTT method *in vitro*. The inhibitory rates of compound (**26**) and hispolon on SMMC-7721 cells were 34.83% and 48.09%, respectively. The inhibitory rates of compound (**26**) and hispolon on MCF-7 cells were 71.09% and 74.57%, respectively, at 100 μg/ml. The IC_{50} values were 69.48, 61.57, 30.22 and 24.68 μg/ml, respectively. Compound **(26)** and hispolon had good inhibitory effects on the proliferation of SMMC-7721 cells and MCF-7 cells (Chang et al., 2011).

To check the anti-proliferative activity of hispolon NB4, cell lines of human hepatocellular carcinoma were used and analysed using DAPI (4, 6-diamidino-2-phenylindole dihydrochloride) staining, MTT assay, DNA fragmentation studies and flow cytometric studies. Studies based on flow cytometry and apoptotic cell death using DNA laddering method showed that hispolon checked cell division in a dose-dependent manner by cell cycle arrest at the G0/G1 stage in NB4 cells. Treatment of NB4 cells with hispolon lead to expression of proteins related to apoptosis, namely the cleavage form of caspase 9, caspase 8, caspase 3, poly (ADP ribose) polymerase and the Bax protein. Furthermore, it was found that intrinsic and extrinsic protein levels and Bax/Bcl-2 ration were also increased on treatment of NB4 cells with hispolon. Down regulation in expression level of p53, cyclins D1 and cyclins E, and cyclin-dependent kinases (CDKs) 2 and 4, with concomitant induction of p21waf1/Cip1 and p27Kip1 was inferred for hispolon-induced G0/G1-phase arrest. It was concluded that extrinsic and intrinsic apoptotic pathways in NB4 human leukemia cells was induced by hispolon *in vitro* (Chen et al., 2013).

It was reported that hispolon induces apoptosis through JNK1/2-mediated activation of a caspase-8, -9, and -3-dependent pathway in acute myeloid leukemia (AML) cells and inhibits AML xenograft tumor growth *in vivo* (Hsiao et al., 2013). Hispolon significantly inhibited cell proliferation of HONE-1 and NP-039 cell lines. Hispolon could induce the phosphorylation of ERK1/2, JNK1/2 and p38 MAPK, and stimulate the activation of caspase-3, -8, and -9, which eventually results in the cleavage of PARP, inhibition of proliferation and apoptosis induction of HONE-1 and NPC-039 cells (Hsieh et al., 2014).

At lower concentrations (< 2 μM), hispolon in α-MSH stimulated B16-F10 cells represses the expression of tyrosinase and the microphthalmia-associated transcription factor in order to decrease the production of melanin. In contrast, at higher concentrations (> 10 μM), hispolon can induce activity of caspase-3, -8 and -9 and trigger apoptosis of B16-F10 cells but not of Detroit 551 normal fibroblast cells. It suggests that treatment of hyper pigmentation and melanoma could be done using hispolon in the near future (Chen et al., 2014).

Wang et al. (2014) reported the estrogenic and anti-estrogenic double directional adjusting effects of hispolon, which might behave as a partial agonist of the estrogen receptors. The estrogenic agonist activity of hispolon was investigated at low concentration or lack of endogenous estrogen, and the estrogenic antagonistic effect was stimulated at high concentrations or excess endogenous estrogen. Hispolon had significant binding capacity for ERs, and the ER β-binding capacity was larger than ERα in the lower hispolon concentration. Hispolon might be safer than E_2 when used as estrogen replacement drug, which might be used in the treatment of the estrogen deficiency-related diseases with the benefit of non-toxic to normal cells, good antitumor effects and estrogenic activity.

Hispolon reduces cell viability in a concentration-dependent way. The aggregation of cells in G0/G1 phase was enhanced by hispolon treatment. Hispolon down regulated expression of G1-S transition-related proteins, namely Cyclin D1, Cyclin E, CDK2, CDK4 and CDK6, but upregulated CDK inhibitor p21CIP1 and p27KIP1. Moreover, hispolon regulated mitochondrial pathway to cause apoptosis of cell. Additionally, hispolon enhanced the expression of p53, specific silencing of which almost completely reversed hispolon-mediated antitumor activity. Moreover, hispolon treatment was more effective on H661 cells than on A549 cells in inhibiting cell viability and inducing cell apoptosis. The study showed cell viability is inhibited by hispolon as it induces G0/G1 cell cycle arrest and apoptosis in lung cancer cells and p53 plays a critical role in hispolon-mediated antitumor activity (Wu et al., 2014).

Hispolon against breast cancer growth showed reduced expression of ERα at protein and mRNA levels in MCF7 and T47D cells. Luciferase reporter assay deciphered decreased transcriptional activity of ERα on treatment with hispolon. It also inhibits cell growth through modulation of ERα in estrogen-positive breast cancer cells and has potential for further chemotherapy for breast cancer treatment (Jang et al., 2015).

Hispolon shown to activate caspase-3, caspase-8 and caspase-9, while reducing the expression of cFLIP, Bcl-2 and Bcl-xL and upregulating the expression of Bax. It was further found that hispolon induces death receptors in a non-cell type-specific manner. Overall, hispolon was found to potentiate the apoptotic effects of TNF-related apoptosis-inducing ligand (TRAIL) through down regulation of anti-apoptotic proteins and upregulation of death receptors linked with CHOP and pERK elevation (Kim et al., 2016b).

Hong et al. (2017) investigated the effects of hispolon on the epithelial-mesenchymal transition in human epithelial cancer cells and found that transforming growth factor β induced increased EMT-associated phenotypic changes and cell migration and invasion. Hispolon down regulated Snail and Twist, an effect that was enhanced by TGF-β. These studies provide novel proofs that hispolon reduces metastasis and invasion by inhibiting EMT (Hong et al., 2017).

It was demonstrated that treatment with hispolon inhibited cell metastasis in two cervical cancer cell lines. In addition, the down regulation of the lysosomal protease Cathepsin S (CTSS) was critical for hispolon-mediated suppression of tumor cell metastasis in both *in vitro* and *in vivo* models. Moreover, hispolon induced autophagy, which increased LC3 conversion and acidic vesicular organelle formation. Mechanistically, hispolon inhibited the cell motility of cervical cells through the extracellular signal-regulated kinase (ERK) pathway and blocking of the ERK pathway reversed autophagy-mediated cell motility and CTSS inhibition. These results indicate that autophagy is essential for decreasing CTSS activity to inhibit tumor metastasis by hispolon treatment in cervical cancer. (Hsin et al., 2017).

Hispolon decreased human glioblastoma cells U87MG viability in a dose and time-dependent manner. The cell cycle distribution showed that hispolon enhanced the accumulation of the cells in G2/M phase. Hispolon decreased the expression of G1–S transition-related protein cyclin D4 but increased the expression of CDK inhibitor p21. Additionally, hispolon enhanced the expression of p53. Moreover, hispolon treatment was effective on U87MG cells in inhibiting cell viability and inducing cell apoptosis. It is reported that hispolon inhibits the cell viability, induces

G2/M cell cycle arrest and apoptosis in glioblastoma U87MG cells, and p53 should play a role in hispolon-mediated antitumor activity (Arcella et al., 2017).

Anti-Inflammatory Activity

Administration of hispolon (10 and 20 mg/kg) to male ICR mice considerably inhibited the numbers of writhing response induced by acetic acid. The result also revealed that hispolon (20 mg/kg) inhibited the formalin-induced pain. The λ-carrageenin (Carr) induced inflammation was reduced by hispolon (20 mg/kg) which reduced the paw edema at the 4th and 5th hr., and superoxide dismutase, glutathione peroxidase and glutathione reductase was increased in the liver tissue. At the 5th hour of Carr injection hispolon (10 and 20 mg/kg) the nitric oxide (NO) levels were decreased on both the edema paw and serum level and diminished the serum TNF-α. It is expected that hispolon might be associated with the decrease in the level of MDA in the edema paw by enhancing the activities of SOD, GPx and GRx in the liver. Suppression of TNF-α and NO is probably the cause of anti-inflammatory activity (Chang et al., 2011).

Hispolon showed inhibition of LPS, LTA, and PGN-induced iNOS protein expressions and production of NO by RAW264.7 macrophages in a dose-dependent manner. Accordingly, hispolon preserved RAW264.7 cells from LPS, LTA, and PGN induced apoptosis. A study to identify the mechanism indicated that activator protein (AP)-1 and nuclear factor (NF)-κB activation and inhibition of c-Jun N-terminal kinase (JNK) protein phosphorylation were involved in the anti-inflammatory activities of HIS in macrophages (Yang et al., 2014).

Hispolon inhibited production of NO from LPS-activated macrophages by 72.1% at 10 μg/ml concentration. The effect of hispolon on the proliferation of murine macrophage (RAW264.7) cells was assessed by MTT assay. The results indicate that hispolon may play an important role in the suppression of iNOS expression and production of NO in macrophages. In addition, hispolon, dramatically inhibited the luciferase activity by 61.3% at 10 mg/ml, compared to LPS treatment alone. These results indicate that hispolon inhibited the LPS-induced inflammatory responses through attenuation of NF-kB activity followed by suppression of NO production and TNF-a secretion in murine macrophages (Lin et al., 2014).

Further, it is reported that hispolon possesses inhibitory activity against LPS- or LTA-induced inflammatory responses including iNOS/NO production and apoptosis in BV-2 microglial cells and that the mechanisms involve upregulation of the HO-1 protein and down regulation of JNK/NF-κB activation (Wu et al., 2017).

Miscellaneous Activities

A differential effect of hispolon on the production of antigen induced Th1 and Th2 cytokines, in which the suppression of IFN- was associated with the diminishment of intracellular glutathione in antigen-primed splenocytes. These findings provide new insights to the direct effect of hispolon on the antigen-induced T cell immune responses and its underlying mechanisms (Wang et al., 2015). Hispolon (**25**) showed significant anti-viral activity against influenza viruses type A (H1N1 and H3N2) and B, using the allantois on the shell-test system (Awadh Ali et al., 2003).

Other Bioactive Compounds of the Genus *Phellinus*

8-methyl-13-phenyltri-deca-4,6,8,10,12-pentaene-3-one **(27)**, and 4-vinylresorcinol **(28)** were isolated from *Phellinus pini*, a fungus pathogenic to conifer trees and tested for antifungal activity against several tree pathogens. Compound **(28)** inhibited *Ophiostoma crassivaginatum, O. piliferum* and *P. tremulla* completely at 100 ppm concentration and compound **(27)** inhibited completely the growth of *O. crassivaginatum O. piliferum* at 100 and 1000 ppm concentration, respectively (Ayer et al., 1996).

Phellinsin A **(29)**, a novel chitin synthase inhibitor, was identified from the cultured broth of *Phellinus* sp. PL3. The structure of phellinsin A **(29)** was assigned as a phenolic compd. on the basis of various spectroscopic analyses. Phellinsin A **(29)** selectively inhibited chitin synthase I and II of *Saccharomyces cerevisiae* with an IC_{50} value of 76 and 28 µg/ml, respectively, in cell free assay system. It showed antifungal activity against *Colletotrichum lagenarium, Pyricularia oryzae, Rhizoctonia solani, Aspergillus fumigatus*, and *Trichophyton mentagrophytes* with MIC value of 12.5,50.0,50.0,50.0 and 50.0 µg/ml, respectively (Hwang et al., 2000).

Compounds 3pyrano[4,3-c]isochromen-4-one derivatives, phelligridins H **(9)**, I **(10)** and J **(11)** together with the known compounds davallialactone **(18)** were identified from *P. igniarius*. Compounds **(9-11** and **18)** exhibited antioxidant activity inhibiting rat liver microsomal lipid peroxidation with IC_{50} values of 4.8, 3.7, 6.5, and 8.2 µM, respectively. Both **(9)** and **(10)** inhibited protein tyrosine phosphatase 1B (PTP1B), with IC_{50} values of 3.1 and 3.0 µM, respectively, but phelligridin J **(11)** was inactive (IC_{50} > 10 µM). Compound **(11)** exhibited cytotoxicity against A2708, A549, Bel-7402, HCT-8, cell line with IC_{50} value of 7.2, 4.2, 9.2, 8.4 µM while topotecan exhibited cytotoxicity against A2708, A549, Bel-7402, HCT-8, cell line with IC_{50} value of 1.2, 3.1, 1.4, 1.6 µM respectively (Wang et al., 2007).

Hypholomine B **(13)**, squarrosidine **(14)** and pinillidine **(29)** were isolated from a culture of *P. pini*. Compared to the standard allopurinol (IC_{50} 4.4 µM) **(13), (14)** and **(29)** proved to be potent inhibitors of xanthine oxidase with IC_{50} values of 6.7, 8.1, and 5.8 µM, respectively (Kemami Wangun et al., 2007).

Four new lanostanol-type triterpenoids, igniarens A - D **(30-33)**, were isolated from the fruit body of *P. igniarius* together with two known triterpenoids and two known ergostanes gilvsin C **(34)**, gilvsin A **(35)**, 5α, 8α-epidioxy-22E-ergosta-6,22-dien-3β-ol **(36)** and 5α-ergosta-7,22-dien-3-one **(37)**. Their effects on production of NO in lipopolysaccharide (LPS)-activated macrophages were assessed. Compounds **(30-37)** inhibited LPS-induced production of NO in activated RAW 264.7 cells with IC_{50} of > 100, 47.89, 48.28, 91.74, 60.02, > 100, 37.57 and >100 µM respectively. The most potent compound **(36)** significantly inhibited LPS-induced production of NO in a concentration-dependent manner without affecting the cell viability, with an IC_{50} of 37.57 µM (Wang et al., 2009).

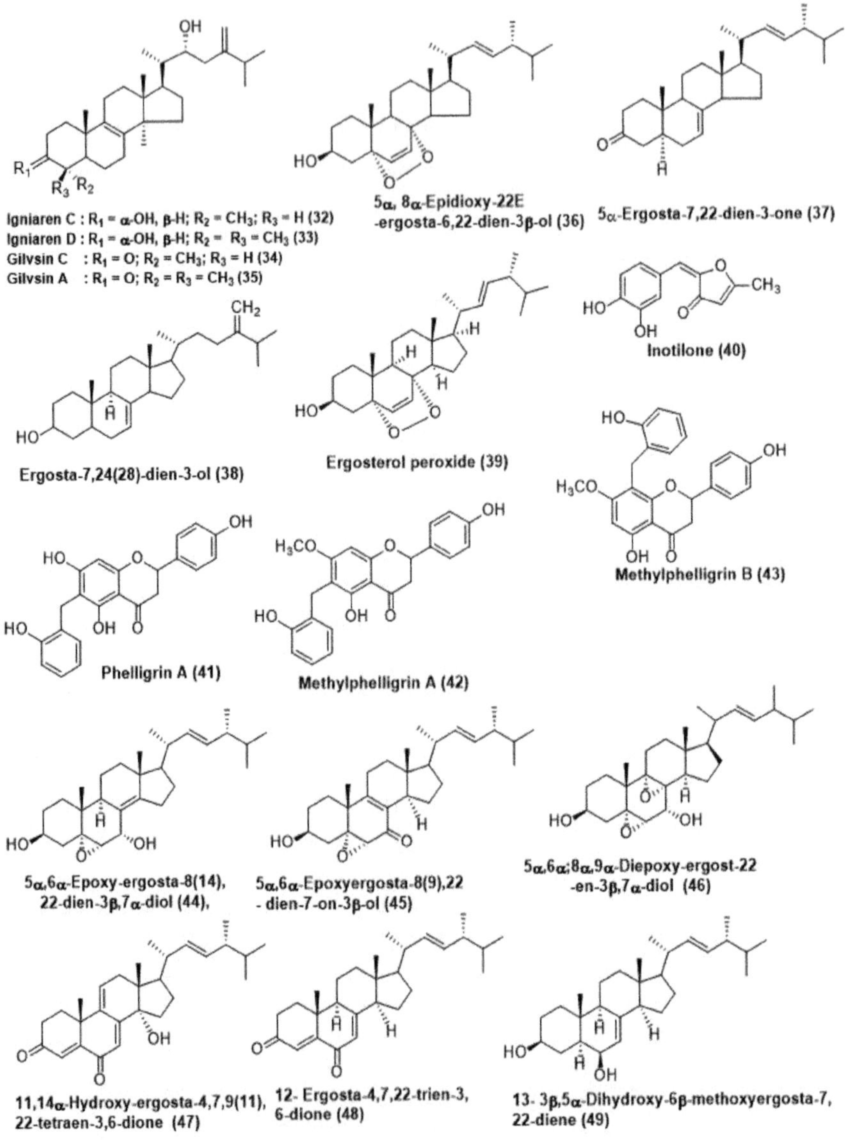

Igniaren C : R₁ = α-OH, β-H; R₂ = CH₃; R₃ = H (32)
Igniaren D : R₁ = α-OH, β-H; R₂ = R₃ = CH₃ (33)
Gilvsin C   : R₁ = O; R₂ = CH₃; R₃ = H (34)
Gilvsin A   : R₁ = O; R₂ = R₃ = CH₃ (35)

5α, 8α-Epidioxy-22E
-ergosta-6,22-dien-3β-ol (36)

5α-Ergosta-7,22-dien-3-one (37)

Ergosta-7,24(28)-dien-3-ol (38)

Ergosterol peroxide (39)

Inotilone (40)

Phelligrin A (41)

Methylphelligrin A (42)

Methylphelligrin B (43)

5α,6α-Epoxy-ergosta-8(14),
22-dien-3β,7α-diol (44),

5α,6α-Epoxyergosta-8(9),22
- dien-7-on-3β-ol (45)

5α,6α;8α,9α-Diepoxy-ergost-22
-en-3β,7α-diol (46)

11,14α-Hydroxy-ergosta-4,7,9(11),
22-tetraen-3,6-dione (47)

12- Ergosta-4,7,22-trien-3,
6-dione (48)

13- 3β,5α-Dihydroxy-6β-methoxyergosta-7,
22-diene (49)

Fig. 3. Structure of compounds 32–49.

Ergosta-7,24(28)-dien-3-ol **(38)** and ergosterol peroxide **(39)** were isolated fruit body of *P. pini*. Ergosterol derivatives showed inhibitory activity of NO production in lipopolysaccharide (LPS) activated RAW 264.7 cells, IC_{50} of **(38)** was 18.9 μM and 20.4 uM for **(39)** (Jang et al., 2007). Phellinsin A **(29)**, a potent antifungal agent was isolated from the cultured broth of *Phellinus* sp. PL3.5. Phellinsin A **(29)** exhibited inhibition of chitin synthase II with an IC_{50} value of 28 μg/ml and showed 2.5 times stronger inhibitory activity than polyoxin D (Lee et al., 2009).

Huang et al. (2012) reported inotilone **(40)** from *P. linteus* (0, 6.25, 12.5, and 25 µM) which inhibited production of NO and iNOS protein expression in dose-dependent manner. Further inotilone also showed to inhibit expressions of iNOS, NF-κB and MMP-9 as well as ERK, JNK and p38 phosphorylation in LPS-stimulated RAW264.7 macrophages. Inotilone **(40)** increased the activities of catalase (CAT), superoxide dismutase (SOD), and glutathione peroxidase (GPx) (Huang et al., 2012). However, intra-peritoneal injection of inotilone **(40)** could result in the decrease of neutrophil infiltration into sites of inflammation. Several phenolic compounds were isolated from *Phellinus baumii* and their anti-inflammatory activities were assessed. It was found that pretreatment with phelligrin A **(41)**, methylphelligrins A **(42)** and B **(43)** substantially inhibited LPS-induced NF-κB activation in human prostate cancer cells with IC_{50} values of 54.50, 36.44 and 22.46 µM, respectively (Wu et al., 2011).

Compounds 5α,6α-epoxy-ergosta-8(14),22-dien-3β,7α-diol **(44)**, 5α,6α-epoxyergosta-8(9),22-dien-7-on-3β-ol **(45)**, 5α,6α;8α,9α-diepoxy-ergost-22-en-3β,7α-diol **(46)**, 14α-hydroxy-ergosta-4,7,9(11),22-tetraen-3,6-dione **(47)**, ergosta-4,7,22-trien-3,6-dione**(48)**, 3β,5α-dihydroxy-6β-methoxyergosta-7,22-diene **(49)**, Ergosta-7,22-dien-3β,5α,6β,9α-tetraol **(50)**, were isolated from the mycelium of *P. linteus*. The inhibitory activity of the compounds **(44-50)** on HNE *in vitro* was evaluated. Epigallocatechin gallate was used as a positive control. Of the compounds tested, 3β,5α-dihydroxy-6β-methoxyergosta-7,22-diene **(49)** exhibited the strongest HNE-inhibitory activity with an IC_{50} value of 14.6 µM, comparable to that of EGCG (IC50 = 12.5 µM). Other compounds **(44-48 and 50)** exhibited considerable HNE inhibition, with IC_{50} values of 28.2, 75.1, 35.2, 20.5, 55.2, 77.5µM respectively. Kinetic study indicated that **(49)** inhibited HNE in a competitive manner (Lee et al., 2012).

Ergosta-4,6,8(14),22-tetraen-3-one **(51)**, ergosta-7,24(28)-dien-3-ol **(38)**, and 5,8-epidioxyergosta-6,22-dien-3-ol **(52)** were identified from the fruit body of *P. pini*. Their structures were deciphered using spectroscopic methods including IR, MS, and NMR (1D and 2D). These compounds were checked for their potential to inhibit production of NO in LPS-activated RAW 264.7 cells. Compounds **(51, 38 and 52)** reduced production of NO in the assay with IC_{50} values of 29.7 µM **(51)**, 15.1 µM **(38)**, and 18.4 µM **(52)**, respectively. They also suppressed the expression of protein and m-RNA of iNOS and COX-2 in a dose-dependent manner when analyzed by western blot analysis and RT-PCR experiment in LPS-activated microglial cells (Hong et al., 2012).

An antitumor compound atractylenolide I **(53)** a eudesmane-type sesquiterpene lactone, was isolated from ethyl acetate extract of *P. linteus* grown on germinated brown rice. Atractylenolide I was then used to assess the inhibitory effect on HT-29 cells. Atractylenolide I dose-dependently inhibited the growth of HT-29 human colon cancer cells (Jeon et al., 2013).

A new spiroindene pigment, phelliribsin A **(54)** was isolated from *Phellinus ribis*, and its structure was determined by 2-D NMR method. Compound **(54)** is an unprecedented spiroindene compound and was found to possess cytotoxic activity against PC12 cells at 30 µM (Kubo et al., 2014).

Six compounds were isolated from fruiting bodies of *P. baumii*, these include ergosta-7,22-dien-3β-ylpentadecanoate **(55)**, ergosta-7,22-dien-3β-ol **(56)**, ganoderiol B (**57**), ergosta-6,2-dien-3β,5α,8α-triol (**58**), ganoderic acid DM (**59**) and inoscavin A (**20**). In *in vitro* cytotoxicity screening, compounds (**55-58 and 20**) were found to inhibit the proliferation of K562 tumor cells with IC_{50} of 63.5, 10.3, 70.6, 35.9 and 3.5 µg/ml, respectively. Compound (**59**) showed strong inhibition to the proliferations of HepG2 with IC_{50} of 50.3 µg/ml (Feng et al., 2015).

Suabjakyong et al. (2015) reported that the complete inhibition of FM3A cell growth by 5 µM acrolein could be prevented by crude ethanol extract of *P. igniarius* at 0.5 µg/ml. Seven polyphenol compounds named 3,4-dihydroxybenzaldehyde, (**60**) 4-(3,4-dihydroxyphenyl)-3-buten-2one, (**61**) inonoblin C, (**62**) phelligridin D (**5**), inoscavin C (**63**), phelligridin C (**4**) and interfungin B (**64**) were identified from this ethanolic extract by LCMS and ¹H NMR. Polyphenol-containing extracts of *P. igniarius* were then used to prevent acrolein toxicity in a mouse neuroblastoma (Neuro-2a) cell line. It suggested that Neuro-2a cells were protected from acrolein toxicity at 2 and 5 µM by this polyphenol extract at 0.5 and 2 µg/ml, respectively. Furthermore, in mice with experimentally induced stroke, intra-peritoneal treatment with *P. igniarius* polyphenol extract at 20 µg/kg caused a reduction of the infarction volume by 62.2% compared to untreated mice (Suabjakyong et al., 2015).

Systemic investigation on the chemical constituents of fruiting bodies of *Phellinus rhabarbarinus* resulted in the isolation of lanostane triterpenoids Phellibarins B-C (**65,66**), igniaren D (**42**), igniaren C (**41**), gilvsin A (**35**), gilvsin C (**34**), gilvsin D (**67**), Compounds (**65, 66, 35 and 34**), showed inhibitory activities against production of NO in LPS-activated RAW264.7 macrophages, whereas compounds (**65, 66, 42, 41, 35 and 67**), exhibited cytotoxicity against human cancer cell lines (Feng et al., 2016a).

Phellibarin D (**68**), the first B/C ring-rearranged lanostane triterpenoid possessing a 6/5/7/5 ring framework, was isolated from *P. rhabarbarinus*. Phellibarin D (**68**) exhibited cytotoxicity against five human cancer cell lines and inhibitory activity against production of NO in LPS-activated RAW264.7 macrophages. Phellibarin D (**68**) revealed cytotoxicity against HL-60, SMMC-7721, A-549, MCF-7, SW480 cell line with IC_{50} value of 14.5, 19.1, 18.4, 8.8, and 17.8 µM respectively while cisplatin showed cytotoxicity against HL-60, SMMC-7721, A-549, MCF-7, SW480 cell line with IC_{50} value of 1.8, 14.2, 7.2, 14.7 and 11.0 µM respectively. In addition, (**68**) also demonstrated inhibitory activity against nitric oxide production in LPS-activated RAW264.7 macrophages, with an IC_{50} value of 22.3 µM (Feng et al., 2016b).

Two known γ-ionylidenesesquiterpenoids, (-)-trans-γ-monocyclofarnesol (**69**) and (+)-γ-ionylideneacetic acid (**70**), as well as a new compound phellidene E (**71**), were isolated from *P. linteus*. Compounds (**69-71**) exhibited MICs against *Porphyromonas gingivalis* at 5.9, 34.1, and 155 µg/ml, respectively (Shirahata et al., 2017).

Compounds 3,4-dihydroxy benzaldehyde (**60**), InoscavinA (**20**) and baicalein (**72**) were isolated from the solid-state fermentation of *P. baumii* mycelia inoculated in rice medium *in vitro*. Compounds (**60, 20 and 72**) had potent inhibition effects on the proliferation of a series of tumor cell lines, including K562, L1210, SW620, HepG2, LNCaP and MCF-7cells. The inhibition rate of (**20**) at a concentration of 100 µg/ml was more than 80% against K562, SW620 and LNCaP cell lines and the

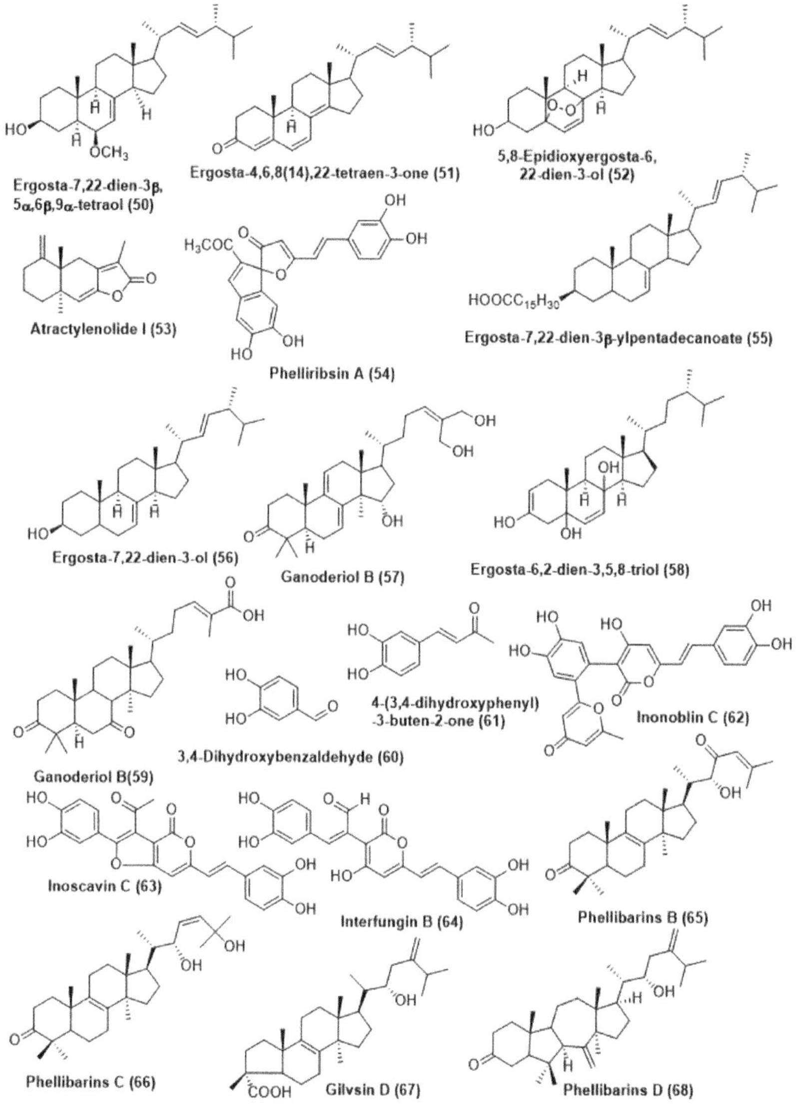

Fig. 4. Structure of compounds 50–68.

inhibition rate of **(60)** at concentrations of 50 and 100 μg/ml was more than 60% in L1210 and LNCaP cell lines. Compound **(72)** exhibited cytotoxic activity against HepG2, K562, L1210, LNCaP, MCF-7, SW620, with IC_{50} value of 60, 22, 39.61, 23.62 , 19.515 , 37.73 , 33.17 (μg/ml), respectively (Zhang et al., 2017).

Compounds 3,4-dihydroxybenzaldehyde **(60)**, 4-(4-hydroxyphenyl)-3-buten-2-one **(73)**, 4-(3,4-dihydroxyphenyl)-3-buten-2-one **(61)**, 9,11-dehydroergosterol peroxide **(74)** and ethyl linoleate **(75)** were isolated from fruiting bodies of *P. baumii.*

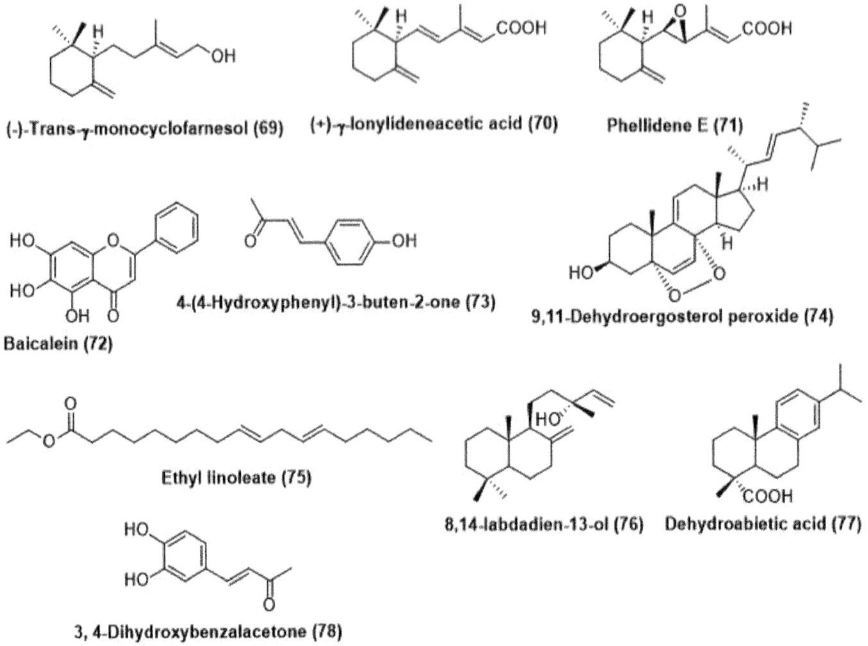

(-)-Trans-γ-monocyclofarnesol (69)   (+)-γ-lonylideneacetic acid (70)   Phellidene E (71)

Baicalein (72)   4-(4-Hydroxyphenyl)-3-buten-2-one (73)   9,11-Dehydroergosterol peroxide (74)

Ethyl linoleate (75)   8,14-labdadien-13-ol (76)   Dehydroabietic acid (77)

3, 4-Dihydroxybenzalacetone (78)

Fig. 5. Structure of compounds 69–78.

These compounds were subjected to anti-inflammatory effects in lipopolysaccharide (LPS)-stimulated RAW264.7 cells. Compounds **(60)**, **(73)**, **(61)**, **(74)** and **(75)** inhibited LPS-stimulated NO production, and compounds **(60)**, **(73)** and **(61)** significantly inhibited production of NO in LPS-activated RAW264.7 macrophages, with IC_{50} values of 9.1, 0.8, and 0.7 µM, respectively. Treatment of LPS-stimulated RAW264.7 cells with compounds **(60)**, **(73)**, **(61)**, inhibited phosphorylation of IKKα and IκBα. In addition, treatment with compounds **(60)**, **(73)** and **(61)**, reduced LPS-induced increases of nuclear factor-kappa B (NF-κB) p65, iNOS and COX-2 protein expressions. Collectively, compounds **(60)**, **(73)**, **(61)**, inhibited NF-κB-dependent inflammation in RAW264.7 cells (Lee et al., 2017).

Two compounds, 8,14-labdadien-13-ol **(76)** and dehydroabietic acid **(77)**, were isolated from the fruit body of *P. pini* and were screened for their ability to inhibit NO production in LPS-activated RAW 264.7 cells. Compounds **(76)** and **(77)** at 30 µM and 50 µM, respectively, inhibited production of NO in activated macrophages (Jang and Yan, 2011; Zhu et al., 2017).

Compounds 3,4-Dihydroxybenzalacetone **(78)** a constituent of *P. linteus*, demonstrated the protective effect on lipopolysaccharide (LPS)-induced acute lung injuries in mice. Pretreatment with DBL significantly improved LPS-induced histological alterations in lung tissues. In addition, DBL markedly reduced the total cell count, the leukocytes, the protein concentrations, and decreased the release of nitrite, tumor necrosis factor (TNF)-α, interleukin (IL)-1β, IL-6 and the activities of matrix metalloproteinase (MMP)-2 and -9 in the bronchoalveolar lavage fluid. Western blot Analysis indicated that DBL efficiently blocked the protein expressions of inducible nitric oxide synthase, cyclooxygenase-2, MMP-2, MMP-9, and the phosphorylation of MAPK, phosphoinositide-3-kinase (PI3K), AKT, Toll-like receptor 4 and nuclear factor (NF)-κB. Moreover, DBL enhanced the expression of anti-oxidant proteins, such as superoxide dismutase, catalase and glutathione peroxidase (Chao et al., 2017).

Concluding Remarks

Phellinus has been used in traditional medicines for the treatment of various ailments and diseases. The crude extract of *Phellinus* and its powder form were used for curing bone fractures, joint pains, headache, burning sensations, regulation of blood glucose level, improving blood circulation, liver protection, enhancement of immunologic function and fighting various forms of cancers. A wide range of bioactive metabolites from *Phellinus* have been isolated and characterized through modern instrumentation techniques. These include hispolon, hispidin, phelligridins, davallialactone, interfungins, inoscavins, phellinusfurans, phellinsin and sterols. These metabolites possess diverse pharmacological properties such as antitumor, anti-inflammatory, antioxidant, antihyperlipidemic, antiviral, antifungal and antidiabetic activities. Hispolon and hispidin are the most active metabolites of *Phellinus*. Hispolon possesses remarkable anticancer activity which is attributed to its action on cell apoptosis, inhibition of metastasis and cell cycle arrest. Hispolon can also suppress the activity of various inflammatory mediators such as TNF-α and NO and showed excellent anti-inflammatory activity. Hispidin and its derivatives exhibited antiviral, antioxidant, antidementia and immunomodulatory activity. The target specific activities of other metabolites include inhibition of human neutrophil elastase **(44-50)**, inhibition of xanthine oxidase **(13, 14, 15** and **19)**, fungal chitin synthase II **(29)** and BACE1 (hispidin).

The diverse scaffolds (metabolites) from *Phellinus* associated with various biological activities have opened the door for researchers to develop new compounds with drug-like properties. The novel compounds can be assessed for their pharmacological activities for the treatment of various diseases including cancer. Apart from this, combination of existing drugs with metabolites of *Phellinus* can be investigated for their synergistic effect in the treatment of specific diseases.

Acknowledgement

The authors are thankful to Dr. Alok Adholeya, Senior Director, Sustainable Agriculture Division, The Energy and Resources Institute, New Delhi (India) for continuous support.

References

Ali, N.A., Lüdtke, J., Pilgrim, H. and Lindequist, U. 1996. Inhibition of chemiluminescence response of human mononuclear cells and suppression of mitogen-induced proliferation of spleen lymphocytes of mice by hispolon and hispidin. Pharmazie., 51: 667–670.

Arcella, A., Oliva, M.A., Sanchez, M., Staffieri, S., Esposito, V., Giangaspero, F. and Cantore, G. 2017. Effects of hispolon on glioblastoma cell growth. Environ. Toxicol., 32: 2113–2123.

Awadh Ali, N.A., Mothana, R.A., Lesnau, A., Pilgrim, H. and Lindequist, U. 2003. Antiviral activity of *Inonotus hispidus*. Fitoterapia, 74: 483–485.

Ayer, W.A., Muir, D.J. and Chakravarty, P. 1996. Phenolic and other metabolites of *Phellinus pini*, a fungus pathogenic to pine. Phytochemistry, 42: 1321–1324.

Balaji, N.V., Ramani, M.V., Viana, A.G., Sanglard, L.P., White, J., Mulabagal, V. and Tiwari, A.K. 2015. Design, synthesis and *in vitro* cell-based evaluation of the anti-cancer activities of hispolon analogs. Bioorg Med Chem., 23: 2148–2158.

Balaji, N.V., Hari Babu, B., Subbaraju, G.V., Purna Nagasree, K., Murali and Krishna Kumar, M. 2017. Synthesis, screening and docking analysis of hispolon analogs as potential antitubercular agents. Bioorg. Med. Chem. Lett. 27: 11–15.

Chang, H.Y., Sheu, M.J., Yang, C.H., Lu, T.C., Chang, Y.S., Peng, W.H., Huang, S.S. and Huang, G.J. 2011. Analgesic effects and the mechanisms of anti-inflammation of hispolon in mice. Evid. Based Complement. Alternat. Med. eCAM 2011478246.

Chao, W., Deng, J.S., Huang, S.S., Li, P.Y., Liang, Y.C. and Huang, G.J. 2017. 3,4-dihydroxybenzalacetone attenuates lipopolysaccharide-induced inflammation in acute lung injury via down-regulation of MMP-2 and MMP-9 activities through suppressing ROS-mediated MAPK and PI3K/AKT signaling pathways. Int. Immunopharmacol., 50: 77–86.

Chen, W., He, F.Y. and Li, Y.Q. 2006. The apoptosis effect of hispolon from *Phellinus linteus* (Berkeley & Curtis) Teng on human epidermoid KB cells. J. Ethnopharmacol., 105: 280–285.

Chen, W., Feng, L., Huang, Z. and Su, H. 2012. Hispidin produced from *Phellinus linteus* protects against peroxynitrite-mediated DNA damage and hydroxyl radical generation. Chem.-Biol. Interact, 199: 137–142.

Chen, Y.C., Chang, H.Y., Deng, J.S., Chen, J.J., Huang, S.S., Lin, I.H., Kuo, W.L., Chao, W. and Huang, G.J. 2013. Hispolon from *Phellinus linteus* Induces G0/G1 Cell Cycle Arrest and Apoptosis in NB4 Human Leukaemia Cells. Am. J. Chin. Med., 41: 1439–1457.

Chen, Y.S., Lee, S.M., Lin, C.C. and Liu, C.Y. 2014. Hispolon decreases melanin production and induces apoptosis in melanoma cells through the down regulation of tyrosinase and microphthalmia-associated transcription factor (MITF) expressions and the activation of caspase-3, -8 and -9. Int J Mol Sci., 15: 1201–1215.

Chen, H., Tian, T., Miao, H. and Zhao, Y.Y. 2016. Traditional uses, fermentation, phytochemistry and pharmacology of *Phellinus linteus*: A review Fitoterapia, 113: 6–26.

Cheng, X., Bao, H., Ding, Y. and Young, H.K. 2011. Antitumor activity of *Phellinus vaninii in vitro* Junwu Yanjiu, 9: 176–179.

Cho, J.Y., Kwon, Y.J., Sohn, M.J., Seok, S.J. and Kim, W.G. 2011. Phellinstatin, a new inhibitor of enoyl-ACP reductase produced by the medicinal fungus *Phellinus linteus*. Bioorganic Med. Chem. Lett., 21: 1716–1718.

Cui, C.B., Tezuka, Y., Kikuchi, T., Nakano, H., Tamaoki, T. and Park, J.H. 1990. Constituents of a fern, davallia mariesii moore. I. isolation and structures of davallialactone and a new flavanone glucuronide. Chem. Pharm. Bull. 38: 3218–3225.

Feng, N., Wu, N., Yang, Y., Zhang, J.S., Tang, Q.J. and Shao, Q. 2015. Compounds from fruiting bodies of and their inhibition to tumor cell proliferation. Junwu Xuebao, 34: 124–130.

Feng, T., Cai, J.L., Li, X.M., Zhou, Z.Y., Li, Z.H. and Liu, J.K. 2016a. Chemical constituents and their bioactivities of mushroom *Phellinus rhabarbarinus*. J. Agric. Food Chem., 64: 1945–1949.

Feng, T., Cai, J.L., Li, X.-M., Zhou, Z.Y., Huang, R., Zheng, Y.S., Li, Z.H. and Liu, J.K. 2016b. Phellibarin D with an unprecedented triterpenoid skeleton isolated from the mushroom *Phellinus rhabarbarinus*, Tetrahedron Lett., 57: 3544–3546.

Fiasson, J.L., Gluchoff-Fiasson, K. and Steglich, W. 1977. über die Farb-und Fluoreszenstoffe des Grünblattrieng Schwefelkopfes (Hypholomafasiculare, Agaricales). Chem. Ber., 110: 1047–1057.

Fiasson, J.L. 1982. Distribution of styrylpyrones in the basidiocarps of various Hymenochaetaceae. Biochem. Syst. Ecol., 10: 289–296.

Gill, M. and Steglich, W. 1987. Progress in the chemistry of organic natural products. Fortschr. Chem. Org. Naturst., 51: 88–99.

Gill, M. 2001. The biosynthesis of pigments in basidiomycetes. Aust. J. Chem., 54: 721–734.

Gonindard, C., Bergonzi, C., Denier, C., Sergheraert, C., Klaebe, A., Chavant, L. and Hollande, E. 1997. Synthetic hispidin, a PKC inhibitor, is more cytotoxic toward cancer cells than normal cells *in vitro*. Cell Biol. Toxicol., 13: 141–153.

Hatfied, G.M. and Brady, L.R. 1973. Biosynthesis of hispidin in cultures of *Polyporus schweinitzii*. Lloydia, 36: 59–65.

Hong, D., Park, M.J., Jang, E.H., Jung, B., Kim, N.J. and Kim, J.H. 2017. Hispolon as an inhibitor of TGF-β-induced epithelial-mesenchymal transition in human epithelial cancer cells by co-regulation of TGF-β-Snail/Twist axis. Oncology letters, 14: 4866–4872.

Hong, Y.J., Jang, A.R., Jang, H.J. and Yang, K.S. 2012. Inhibition of nitric oxide production, iNOS and COX-2 expression of ergosterol derivatives from *Phellinus pini*. Natural Product Sciences, 18: 147–152.

Hsiao, P.C., Hsieh, Y.H., Chow, J.M., Yang, S.F., Hsiao, M., Hua, K.T., Lin, C.H., Chen, H.Y. and Chien, M.H. 2013. Hispolon Induces Apoptosis through JNK1/2-Mediated Activation of a Caspase-8, -9, and -3-Dependent Pathway in Acute Myeloid Leukemia (AML) Cells and Inhibits AML Xenograft Tumor Growth *in vivo*. J. Agric. Food Chem., 61: 10063–10073.

Hsieh, M.J., Chien, S.Y., Chou, Y.E., Chen, C.J., Chen, J., Chen, M.K. 2014. Hispolon from *Phellinus linteus* possesses mediate caspases activation and induces human nasopharyngeal carcinomas cells apoptosis through ERK1/2, JNK1/2 and p38 MAPK pathway. Phytomedicine, 21: 1746–1752.

Hsin, M.C., Wang, P.H., Ko, J.L., Yang, S.F., Hsieh, Y.H., Wang, P.H., Hsin, I.L. and Yang S.F. 2017. Hispolon Suppresses Metastasis via Autophagic Degradation of Cathepsin S in Cervical Cancer Cells Cell Death & Disease, 8: e3089.

Huang, G.J., Yang, C.M., Chang, Y.S., Amagaya, S., Wang, H.C., Hou, W.C., Huang, S.S. and Hu, M.L. 2010. Hispolon Suppresses SK-Hep1 Human Hepatoma Cell Metastasis by Inhibiting Matrix Metalloproteinase-2/9 and Urokinase-Plasminogen Activator through the PI3K/Akt and ERK Signaling Pathways. J. Agric. Food Chem., 58: 9468–9475.

Huang, G.J., Deng, J.S., Huang, S.S. and Hu, M.L. 2011. Hispolon induces apoptosis and cell cycle arrest of human hepatocellular carcinoma Hep3B cells by modulating ERK phosphorylation, J. Agric. Food Chem., 59: 7104–7113.

Huang, G.J., Huang, S.S. and Deng, J.S. 2012. Anti-inflammatory activities of inotilone from *Phellinus linteus* through the inhibition of MMP-9, NF-κB and MAPK activation *in vitro* and *in vivo*. PLoS One, 7: e35922.

Hwang, E.I., Yun, B.S., Kim, Y.K., Kwon, B.M., Kim, H.G., Lee, H.B., Jeong, W.J. and Kim, S.U. 2000. Phellinsin A, a novel chitin synthase inhibitor produced by *Phellinus* sp. PL3. J. Antibiot., 53: 903–911.

Inagaki, N., Shibata, T., Itoh, T., Suzuki, T., Tanaka, H., Nakamura, T., Akiyama, Y., Kawagishi, H. and Nagai, H. 2005. Inhibition of igE-dependent mouse triphasic cutaneous reaction by a boiling water fraction separated from mycelium of *Phellinus linteus*, Evid. Based Complement. Alternat. Med., 2: 369–374.

Jang, E.H., Jang, S.Y., Cho, I.H., Hong, D., Jung, B., Park, M.J. and Kim, J.H. 2015. Hispolon inhibits the growth of estrogen receptor positive human breast cancer cells through modulation of estrogen receptor alpha. Biochem. Biophys. Res. Commun., 463: 917–922.

Jang, H.J., Kim, A.K., Pyo, M.Y. and Yang, K.S. 2007. Inhibitors of nitric oxide synthesis from *Phellinus pini* in murine macrophages. Yakhak Hoechi, 51: 430–434.

Jang, H.J. and Yang, K.S. 2011. Inhibition of Nitric Oxide Production in RAW 264.7 Macrophages by Diterpenoids from *Phellinus pini*. Arch. Pharm. Res., 34: 913–917.

Jang, J.S., Lee, J.S., Lee, J.H., Kwon, D.S., Lee, K.E., Lee, S.Y. and Hong, E.K. 2010. Hispidin produced from *Phellinus linteus* protects pancreatic β-cells from damage by hydrogen peroxide. Arch. Pharm. Res., 33: 853–861.

Jeon, T.I., Jung, C.H., Cho, J.Y., Park, D.K. and Moon, J.H. 2013. Identification of an anticancer compound against HT-29 cells from *Phellinus linteus* grown on germinated brown rice. Asian Pacific Journal of Tropical Biomedicine, 3: 785–789.

Jeon, Y.E., Lee, Y.S., Lim, S.S., Kim, S.J., Jung, S.H., Bae, Y.S.,Yi, J.S. and Kang, I.J. 2009. Evaluation of the antioxidant activity of the fruiting body of *Phellinus linteus* using the on-line HPLC-DPPH method, J. Korean Soc. Appl. Bi., 52: 472–479.

Jung, J.Y., Lee, I.K., Seok, S.J., Lee, H.J., Kim, Y.H. and Yun, B.S. 2008. Antioxidant polyphenols from the mycelial culture of the medicinal fungi *Inonotus xeranticus* and *Phellinus linteus*. J. Appl. Microbiol., 104: 1824–1832.

Kemami Wangun, H.V., Härtl, A., Kiet, T.T. and Hertweck, C. 2006. Inotilone and related phenyl propanoid polyketides from *Inonotus* sp. and their identification as potent COX and XO inhibitors. Org. Biomol. Chem., 4: 2545–2548.

Kemami Wangun, H.V. and Hertweck, C. 2007. Squarrosidine and Pinillidine: 3,3'-Fused Bis (styrylpyrones) from *Pholiota squarrosa* and *Phellinus pini*. Eur. J. Org. Chem. 2007: 3292–3295.

Kim, D.E., Kim, B., Shin, H.S., Kwon, H.J. and Park, E.S. 2014. The protective effect of hispidin against hydrogen peroxide-induced apoptosis in H9c2 cardio myoblast cells through Akt/GSK-3β and ERK1/2 signaling pathway. Experimental Cell Research, 327: 264–275.

Kim, H.G., Yoon, D.H., Lee, W.H., Han, S.K., Shrestha, B., Kim, C.H. Lim, M.H., Chang, W., Lim, S., Choi, S., Song, W.O., Sung, J.M., Hwang, K.C. and Kim, T.W. 2007. *Phellinus linteus* inhibits inflammatory mediators by suppressing redox-based NF-κBand MAPKs activation in lipopolysaccharide-induced RAW 264.7 macrophage, J. Ethnopharmacol., 114: 307–315.

Kim, J.Y., Kim, D.W., Hwang, B.S., Woo, E.E., Lee, Y.J., Jeong, K.W., Lee, I.K. and Yun, B.S. 2016a. Neuraminidase inhibitors from the fruiting body of *Phellinus igniarius*. Mycobiology, 44: 117–120,

Kim, J.H., Kim, Y.C. and Park, B. 2016b. Hispolon from *Phellinus linteus* induces apoptosis and sensitizes human cancer cells to the tumor necrosis factor-related apoptosis-inducing ligand through upregulation of death receptors Oncology Reports, 35: 1020–1026.

Kim, S.D., Lee, I.K., Lee, W.M., Cho, J.Y., Park, H.J., Oh J.W., Park, S.C., Kim, S.K., Kwak, Y.S., Yun, B.S. and Rhee, M.H. 2008. The mechanism of anti-platelet activity of davallialactone: Involvement of intracellular calcium ions, extracellular signal-regulated kinase 2 and p38 mitogen activated protein kinase. Eur. J. Pharmacol., 584: 361–367.

Klaar, M. and Steglich, W. 1977. Isolierung von Hispidin und 3,14-Bihispidinyl aus *Phellinus pomaceus* (Poriales). Chem. Ber., 110: 1058–1062.

Kojima, K., Ohno, T., Inoue, M., Mizukami, H. and Nagatsu, A. 2008. Phellifuropyranone A: A new furopyranone compound isolated from fruit bodies of wild *Phellinus linteus*. Chem. Pharm. Bull., 56: 173–175.

Kubo, M., Liu, Y., Ishida, M., Harada, K. and Fukuyama, Y. 2014. A new spiroindene pigment from the medicinal fungus *Phellinus ribis*, Chem. Pharm. Bull., 62: 122–124.

Lee, I.K. and Yun, B.S. 2006. Hispidin analogs from the mushroom *Inonotus xeranticus* and their free radical scavenging activity. Bioorg. Med. Chem. Lett., 16: 2376–2379.

Lee, I.K. and Yun, B.S. 2007. Highly oxygenated and unsaturated metabolites providing a diversity of hispidin class antioxidants in the medicinal mushrooms *Inonotus* and *Phellinus*, Bioorg. Med. Chem. Lett., 15: 3309–3314.

Lee, I.K. and Yun, B.S. 2008a. Peroxidase-mediated formation of the fungal polyphenol 3,14'-bihispidinyl. J. Microbiol. Biotechnol., 18: 107–109.

Lee, I.K. and Yun, B.S. 2011. Styrylpyrone-class compounds from medicinal fungi *Phellinus* and *Inonotus* spp., and their medicinal importance J. Antibiot., 64: 349–359.

Lee, I.S., Bae, K.H., Yoo, J.K., Ryoo, I.J., Kim, B.Y., Ahn, J.S. and Yoo, I.D. 2012. Inhibition of human neutrophil elastase by ergosterol derivatives from the mycelium of *Phellinus linteus*. J. Antibiot., 65: 437–440.

Lee, S., Kim, J.N., Kim, E., Kim, M.S. and Lee, H.K. 2009. Biological evaluation of dilactonelignan analogs of phellinsin A as chitin synthase II inhibitors. Bull. Korean Chem. Soc., 30: 3092–3094.

Lee, S., Lee, D., Jang, T.S., Kang, K.S., Nam, J.W., Lee, H.J. and Kim, K.H. 2017. Anti-inflammatory phenolic metabolites from the edible fungus *Phellinus baumii* in LPS-stimulated RAW264.7 cells. Molecules, 22: 1583/1-1583/10.

Lee, Y.G., Lee, W.M., Kim, J.Y., Lee, J.Y., Lee, I.-K., Yun, B.-S., Rhee, M.H and Cho, J.Y. 2008b. Src kinase-targeted anti-inflammatory activity of davallialactone from *Inonotus xeranticus* in lipopolysaccharide-activated RAW2647 cells. Br. J. Pharmacol., 154: 852–863.

Lee, Y.S., Kang, Y.H., Jung, J.Y., Kang, I.J., Han, S.N., Chung, J.S., Shin, H.K. and Lim, S.S. 2008c. Inhibitory constituents of aldose reductase in the fruiting body of *Phellinus linteus*. Biol. Pharm. Bull., 31: 765–768.

Lee, Y.S., Kang, Y.H., Jung, J.Y., Lee, S., Ohuchi, K., Shin, K.H., Kang, I.J., Park, J.H., Shin, H.K. and Lim, S.S. 2008d. Protein glycation inhibitors from the fruiting body of *Phellinus linteus*. Biol. Pharm. Bull., 31: 1968–1972.

Lim, J.H., Lee, Y.M., Park, S.R., Kim, D.H. and Lim, B.O. 2014. Anticancer activity of hispidin via reactive oxygen species-mediated apoptosis in colon cancer cells. Anticancer Res., 34: 4087–4093.

Lu, T.L., Huang, G.J., Lu, T.J., Wu, J.B., Wu, C.H., Yang, T.C., Iizuka, A. and Chen, Y.F. 2009. Hispolon from *Phellinus linteus* has antiproliferative effects via MDM2-recruited ERK1/2 activity in breast and bladder cancer cells. Food Chem. Toxicol., 47: 2013–2021.

Lin, C.J., Lien, H.M., Chang, H.Y., Huang, C.L., Liu, J.J., Chang, Y.C., Chen, C.C. and Lai, C.H. 2014. Biological evaluation of *Phellinus linteus*-fermented broths as anti-inflammatory agents. J. Biosci. Bioeng., 118: 88–93.

Mo, S., Wang, S., Zhou, G., Yang, Y., Li, Y., Chen, X. and Shi, J. 2004. Phelligridins C-F: Cytotoxic pyrano[4,3-c][2]benzopyran-1,6-dione and furo[3,2-c] pyran-4-one derivatives from the fungus *Phellinus igniarius*. J. Nat. Prod., 67: 823–828.

Min, B.S., Yun, B.S., Lee, H.K., Jung, H.J., Jung, H.A. and Choi, J.S. 2006. Two novel furan derivatives from *Phellinus linteus* with anti-complementactivity. Bioorg. Med. Chem. Lett., 16: 3255–3257.

Nagatsu, A., Shizueltoh, Tanaka, R., Kato, S., Haruna, M., Kishimoto, K., Hirayama, H., Goda, Y., Mizukami, H. and Ogihara, Y. 2004. Identification of novel substituted fused aromatic compounds, meshimakobnol A and B, from natural *Phellinus linteus* fruit body. Tetrahedron Lett., 45: 5931–5933.

Nakamura, T., Matsugo, S., Uzuka, Y., Matsuo, S. and Kawagishi, H. 2004. Fractionation andanti–tumor activity of the mycelia of liquid–cultured *Phellinus linteus*, Biosci. Biotechnol. Biochem., 68: 868–872.

Park, I.H., Chung, S.K., Lee, K.B., Yoo, Y.C., Kim, S.K., Kim, G.S. and Song, K.S. 2004a. An antioxidant hispidin from the mycelial cultures of *Phellinus linteus*, Arch. Pharm. Res., 27: 615–618.

Park, I.H., Jeon, S.Y., Lee, H.J., Kim, S.I. and Song, K.S. 2004b. A β-secretase (BACE1) Inhibitor hispidin from the mycelial cultures of *Phellinus linteus*. Planta Med., 70: 143–146.

Perrin, P.W. and Towers, G.H.N. 1973a. Metabolism of aromatic acids by *Polyporus hispidus*. Phytochemistry, 12: 583–588.

Perrin, P.W. and Towers, G.H.N. 1973b. Hispidin biosynthesis in cultures of *Polyporus hispidus*. Phytochemistry, 12: 589–592.

Ranadive K., Jagtap, N. and Vaidya J.G. 2012. Host diversity of genus *Phellinus* from world. Elixir Appl. Botany, 52: 11402–11408.

Ravindran, J., Subbaraju, G.V., Ramani, M.V., Sung, B. and Aggarwal, B.B. 2010. Bisdemethylcurcumin and structurally related hispolon analogues of curcumin exhibit enhanced prooxidant, anti-proliferative and anti-inflammatory activities *in vitro*. Biochem Pharmacol., 9: 1658–1666.

Shao, H.J., Jeong, J.B., Kim, K.J. and Lee, S.H. 2015. Anti-inflammatory activity of mushroom-derived hispidin through blocking of NF-κB activation. J. Sci. Food Agric., 95: 2482–2486.

Shirahata, T., Ino, C., Mizuno, F., Asada, Y., Hirotani, M., Petersson, G.A., Omura, S., Yoshikawa, T. and Kobayashi, Y. 2017. γ-Ionylidene-typesesquiterpenoids possessing antimicrobial activity against *Porphyromonas gingivalis* from *Phellinus linteus* and their absolute structure determination. J. Antibiot., 70: 695–698.

Singh, S.B., Jayasuriya, H., Dewey, R., Polishook, J.D., Dombrowski, A.W., Zink, D.L., Guan, Z., Collado, J., Platas, G., Pelaez, F., Felock, P.J. and Hazuda, D.J. 2003. Isolation, structure, and HIV-1-integrase inhibitory activity of structurally diverse fungal metabolites. J. Ind. Microbiol. Biotechnol., 30: 721–731.

Song, Y.S., Kim, S.H., Sa, J.H., Jin, C., Lim, C.J. and Park, E.H. 2003. Anti-angiogenic, antioxidant and xanthine oxidase inhibition activities of the mushroom *Phellinus linteus*, J. Ethnopharmacol., 88: 113–116.

Suabjakyong, P., Saiki, R., Van Griensven, Leo, J.L., D., Higashi, K., Nishimura, K., Igarashi, K. and Toida, T. 2015. Polyphenol extract from *Phellinus igniarius* protects against acrolein toxicity *in vitro* and provides protection in a mouse stroke model. PLoS One, 10: e0122733/1–e0122733/14.

Vaidya, J.G., Lamrood, P.Y. and Bhosle, S.R. 2005. Phansomba, a folk medicinal mushroom from the western ghats of Maharashtra: Future perspective. pp. 349–364. *In* Rai, M.K. and Deshmukh, S.K. (eds.). Fungi: Diversity and Biotechnology. Jodhpur, India: Scientific Publishers.

Venkateswarlu, S., Ramachandra, M.S., Sethuramu, K. and Subbaraju, G.V. 2002. Synthesis and antioxidant activity of hispolon, a yellow pigment from *Inonotus hispidus*. Indian J. Chem., 41B: 875–877.

Wang, G.J., Tsai, T.H., Chang, T.T., Chou, C.J. and Lin, L.C. 2009. Lanostanes from *Phellinus igniarius* and their iNOS inhibitory activities. Planta Medica, 75: 1602–1607.

Wang, J., Hu, F., Luo, Y., Luo, H., Huang, N., Cheng, F., Deng, Z., Deng, W. and Zou, K. 2014. Estrogenic and anti-estrogenic activities of hispolon from *Phellinus lonicerinus* (Bond.) Bond. et sing. Fitoterapia, 95: 93–101.

Wang, P.Y., Wu, H.Y., Cheng, C.H., Hou, W.C. and Jan, T.R. 2015. Hispolon differentially modulated the production of antigen-induced t cell cytokines via the regulation of cellular glutathione. Taiwan Veterinary Journal, 41: 59–65.

Wang, Y., Mo, S.Y., Wang, S.J., Li, S., Yang, Y.C. and Shi, J.G. 2005a. A unique highly oxygenated pyrano[4,3-c][2]benzopyran-1,6-dionederivative with antioxidant and cytotoxic activities from the fungus *Phellinus igniarius*.Org. Lett., 7: 1675–1678.

Wang, Y., Wang, S.J., Mo, S.Y., Li, S., Yang, Y.C. and Shi, J.G. 2005b. Phelligridimer A, a highly oxygenated and unsaturated 26-memberedmacrocyclic metabolite with antioxidant activity from the fungus *Phellinus igniarius*. Org. Lett., 7: 4733–4736.

Wang, Y., Shang, X.Y., Wang, S.J., Mo, S.Y., Li, S., Yang, Y.C., Ye, F., Shi, J.G. and He, L. 2007. Structures, biogenesis, and biological activities of pyrano[4,3-c]isochromen-4-one derivatives from the fungus *Phellinus igniarius*. J. Nat. Prod., 70: 296–299.

Wasser, S.P. and Didukh, M.Y. 2005. Mushroom polysaccharides in human health care. pp. 289–328. *In*: Deshmukh, S.K. and Rai, M.K. (eds.). Biodiversity of Fungi—Their Role in Human Life. Enfield, N.H. USA: Science Publishers.

Wu, C.S., Lin, Z.M., Wang, L.N., Guo, D.X., Wang, S.Q., Liu, Y.Q. Yuan, H.Q. and Lou, H.X. 2011. Phenolic compounds with NF-κB inhibitory effects from the fungus *Phellinus baumii*, Bioorg. Med. Chem. Lett., 21: 3261–3267.

Wu, M.S., Chien, C.C., Cheng, K.T., Subbaraju, G.V. and Chen, Y.C. 2017. Hispolon suppresses LPS- or LTA-Induced iNOS/NO Production and Apoptosis in BV-2 Microglial Cells. Am. J. Chin. Med., 45: 1649–1666.

Wu, Q., Kang, Y., Zhang, H., Wang, H., Liu, Y. and Wang, J. 2014. The anticancer effects of hispolon on lung cancer cells. Biochem. Biophys. Res. Commun., 453: 385–391.

Wu, X., Lin, S., Zhu, C., Zhao, F., Yu, Y., Yue, Z., Liu, B., Yang, Y., Dai, J. and Shi, J. 2011. Studies on chemical constituents of cultures of fungus *Phellinus igniarius*. Zhongguo Zhongyao Zazhi, 36: 874–880.

Yan, J.K., Pei, J.J., Ma, H.L., Wang, Z.B. and Liu, Y.S. 2017. Advances in antitumor polysaccharides from *Phellinus sensu lato*: Production, isolation, structure, antitumor activity, and mechanisms. Crit. Rev. Food Sci. Nutr., 57: 1256–1269.

Yang, L.Y., Shen, S.C., Cheng, K.T., Subbaraju, G.V., Chien, C.C. and Chen, Y.C. 2014. Hispolon inhibition of inflammatory apoptosis through reduction of iNOS/NO production via HO-1 induction in macrophages. J. Ethnopharmacol., 156: 61–72.

Yeom, J.H., Lee, I.K., Ki, D.W., Lee, M.S., Seok, S.J. and Yun, B.S. 2012. Neuraminidase inhibitors from the culture broth of *Phellinus linteus*. Mycobiology, 40: 142–144.

Yoon, H.R., Han, A.R. and Paik, Y.S. 2013. Prolyl endopeptidase inhibitory activity of two styrylpyranones from *Phellinus linteus*, J. Appl. Biol. Chem., 56: 183–185.

Zhang, H., Shao, Q., Wang, W., Zhang, J., Zhang, Z., Liu, Y. and Yang, Y. 2017. Characterization of compounds with tumor-cell proliferation inhibition activity from mushroom (*Phellinus baumii*) mycelia produced by solid-state fermentation. Molecules (Basel, Switzerland), 22: E698.

Zhu, F., Lu, W., Feng, W., Song, Z., Wang, C. and Chen, X. 2017. Preliminary investigation on the chemical constituents and antioxidant activities of two *Phellinus* mushrooms collected in foshan. Int. J. Org. Chem., 7: 25–33.

13

Commercial Inoculation of *Pseudotsuga* with an Ectomycorrhizal Fungus and its Consequences

Ian R. Hall,[1,]* Phil De La Mare,[2] Gina Bosselmann,[3] Chris Perley[4] and Yun Wang[5]

INTRODUCTION

Prior to the arrival of man in the mid-1300s, more than 80% of New Zealand was covered in forest, from the coast to the tree line at 1050 to 1500 metres (Cieraad and McGlone, 2013). At lower altitudes podocarp forest was the norm, whilst New Zealand beeches dominated up to the tree line and in some areas down to sea level (Fig. 1) (Te Ara, 2010; McGlone et al., 1995; Ogden et al., 1998). Elsewhere stands of *Leptospermum scoparium* and *Kunzea robusta* (previously *Kunzea ericoides*) would not have been uncommon almost up to treeline (Burrell, 1965; Stephens et al., 2005). Within 200 years of Māori settlement, up to 40% of the forest was burnt (McWethy et al., 2009). Europeans were the next to arrive (about 1800) and continued deforestation (Fig. 2) so by 2005 only a quarter of New Zealand was still forested (Warne and Gasteiger, 2017).

---

[1] Truffles & Mushrooms (Consulting) Ltd, P.O. Box 268, Dunedin 9054, New Zealand.
[2] Ernslaw One Ltd, PO Box 36 Tapanui, West Otago 9542, New Zealand.
[3] University of Otago, P.O. Box 268, Dunedin, New Zealand.
[4] Thoughtscapes, 900 Dufferin Street, Ākina, Hastings, Hawke's Bay, New Zealand.
[5] Plant & Food Research, Canterbury Agriculture & Science Centre, Lincoln, Private Bag 4704, Christchurch 8140, New Zealand.
* Corresponding author: truffle@trufflesandmushrooms.co.nz

Fig. 1. Prior to human settlement around 80% of New Zealand was covered in forest. (https://teara.govt.nz/en/map/23596/forest-cover-before-human-habitation Courtesy of Te Ara and Bateman Ltd).

Fig. 2. By 1840 up to 40% of New Zealand's forest had been burnt. https://teara.govt.nz/en/map/23597/forest-cover-around-1840 Courtesy of Te Ara and Bateman Ltd).

Deforestation was recognised as undesirable as early as the mid-1800s, but it was not until the 1920s and 1930s that government took a strong hand in the management of forests. In 1960 the government proposed to double exotic forests to 800,000 hectares by 2000 and plant a further 400,000 hectares by 2025. The aim was to increase income from exotic forests to 25% of export earnings by 2000 (Poole, 1969; Roach, 1990). Export income from forests currently stands at 4% of export earnings (New Zealand Institute for Economic Research, 2017) quite possibly because New Zealand's forestry industry has been myopic and has increasingly centred on growing relatively low value *Pinus radiata*, which has to be chemically treated before use, and *Eucalyptus* spp. for pulp (Hall and Perley, 2007).

By 1993 there were 1.34 million hectares of exotic forest in New Zealand (Ministry of Forestry, 1993). Radiata pine (*Pinus radiata*) made up 90% of these forests and Douglas-fir 5% (*Pseudotsuga menziesii*), although in Otago and Southland Douglas-fir accounted for 10% of the total 137,000 ha (Ministry of Forestry, 1993). The rotation time for Douglas fir is longer than radiata pine but it produces higher quality timber that does not need to be chemically treated to ensure durability and protection against wood borer. In the 1990s, plans were made by Ernslaw One Ltd, a New Zealand forest production company, to extend Douglas fir plantings on poor quality upland pastures in Otago and Southland (Floate and Cossens, 1992) (Fig. 3). It was also planted on similar land in Canterbury (Burdon and Miller, 1995; New Zealand Forest Research Institute, 1995).

Douglas-fir can become stunted and chlorotic after outplanting, which can lead to uneven growth of stands and, in severe cases, may result in the need to replant (Fig. 4–7). The cause of chlorosis in Australia, Germany, Canada, Poland, New Zealand, Switzerland, UK and USA had been attributed to nutritional problems,

Fig. 3. An overgrazed deforested site at Gowan Hill, near Ohai, Southland. This is typical of areas being used for exotic forests in New Zealand. Note the remnants of native forest in the gulley.

Fig. 4. Nurseries in New Zealand producing containerised plants frequently had problems with chlorosis in Douglas fir.

Fig. 5. After outplanting some Douglas fir can become stunted and chlorotic, sometimes adjacent to healthy green ones.

in particular N and P (Idczak, 1975; Belton and Davis, 1986; Radwan and Brix, 1986; van den Driessche, 1989; Weetman et al., 1992; Walker et al., 1995); lime-

Fig. 6. Another site in New Zealand where a patch of chlorotic Douglas firs grows adjacent to a stand of green trees.

Fig. 7. A seriously chlorotic Douglas fir that is unlikely to survive.

induced chlorosis (Motschalow, 1988; Dumroese et al., 1990) linked to iron and magnesium deficiency (Foerst, 1981); Mn-induced Fe deficiency (Schone, 1987);

pathogens (Singh and Bhure, 1974; Hood and van der Pas, 1979); herbicide damage (Cole and Newton, 1989) or a lack of effective mycorrhizal fungi (Gilmore, 1958; Hall and Garden, 1984; Davis, 1989; Walker, 1992; Marschner and Colinas et al., 1994; Ledgard, 2002; Parke et al.; Teste et al., 2003; Davis and Smaill, 2009). The last of these is particularly likely in New Zealand because only a handful of native plants form ectomycorrhizas—the native beeches *Fuscospora* spp. and *Lophozonia menziesii*, and tea trees (*Kunzea robusta* and *Leptospermum scoparium*). Consequently, large areas of New Zealand, such as the tussock grasslands as well as improved pasture, are devoid of ectomycorrhizal fungi, although the lack of even a mention that mycorrhizas might be important is characteristic of many papers written in the 1980s and 1990s.

To cater for the increased demand for Douglas fir seedlings by Ernslaw One, a new nursery, Forestart Nurseries, later to become Oregon Nurseries, was built near Oamaru. This was capable of producing up to 10 million containerised Douglas firs annually. Plants were successfully produced in the first season where duff from an existing Douglas fir plantation was incorporated into the potting mix as recommended by Gilmore (1958). However, the following year, this practice was discontinued for fear that pests and diseases would be introduced. In December 1994, all the seedlings were consistently yellow and deemed unsuitable for outplanting, a problem that was not unique at that time (Fig. 4). This chapter is a retrospective look at the confidential research that was carried out by Ian Hall's research group at Invermay Agricultural Centre near Dunedin, the staff of Forestart Nurseries, and Gina Bosselmann, who was studying for a Master's degree at the University of Otago (Bosselmann, 2002). The information that came from this research is directly applicable to other parts of the world where ectomycorrhizal trees new to an area are being introduced.

Field Survey of Douglas Fir

In an attempt to determine potential causes of chlorosis and stunting, leaf samples were taken in late spring and early summer from thirty healthy stands of Douglas fir aged between 5 and 50-years-old in Southland, Otago and Canterbury. At each site and at least 50 m from access roads, representative samples of the end 250 mm sections of branches were collected and chilled. Later these were dried, and the leaves were removed from the flush of new growth on the tips and the previous year's leaves. The procedure was repeated in stands of chlorotic trees but as these had ceased to grow the leaves represented growth in previous years. To give a comparison, random samples of leaves and roots were also taken from the 6 month old chlorotic Douglas firs in Forestart Nurseries. Samples of the surface feeder roots were also collected from all sites as well as samples of the litter layer and the top 10 cm of soil under the litter layer. The shoots were dried and analysed for nutrients and the roots observed for mycorrhizal formation. The litter layer and soil samples were analysed using standard AgResearch techniques (Cornforth and Sinclair, 1982). Estimates of mycorrhizal formation were made by scanning each root system and assessing on a 0–5 scale where 0 = no infection, 1 = up to 5%, 2 = 5–20%, 3 = 20–40%, 4 = 40–80%, 5 = 80–100%.

Observations

While the soil under the chlorotic trees had a lower nitrogen concentration, this difference was not significant (Table 1). All but one of the chlorotic trees in our survey were non-mycorrhizal and had significantly lower N and P concentrations in their foliage (Table 2). The exception was at Glen Dhu, where although the trees were mycorrhizal, they were mycorrhized by fungi very different from those on the adjacent green, healthy trees. It was not possible to identify these fungi from the morphology of their mycorrhizas (Agerer, 2001). There were no mycorrhizas on root samples taken from the nursery and the nutrient concentrations in their leaves were much higher than anything observed in the field (Table 2).

Greenhouse Experiment

The lack of mycorrhizas on the plants in the nursery and the chlorotic and stunted plants in the field, as well as previous research, strongly suggested that a lack of mycorrhizas and chlorosis were linked (Gilmore, 1958; Hall and Garden, 1984; Chu Chou and Grace, 1987). It also suggested that the non-mycorrhizal plants in the nursery were unlikely to suit the outplant sites where the soils were likely to be low in nutrients (Table 1) and devoid of ectomycorrhizal fungi. What was required was a quick way of inoculating the 7 million plants already in the greenhouse and the additional 90 million that were needed over the following decade. Speed was of the essence and a long research programme which determined the most suitable mycorrhizal fungus and preferably one with edible mushrooms, was out of the question.

Chu-Chou and Grace (1981a,b, 1983) had shown that *Rhizopogon parksii* was the most common ectomycorrhizal fungus associated with Douglas fir in New Zealand and had demonstrated that inoculating with it could eliminate chlorosis in Douglas fir in Edendale Nursery, Southland (Chu-Chou and Grace, 1987). It was therefore decided to inoculate the plants in Forestart Nurseries with *R. parksii* even though no information was available on whether *R. parksii* was the best species to use in upland areas of New Zealand (Castellano et al., 1985; Chu-Chou and Grace, 1985; Le Tacon et al., 1992; Walker, 1992; Amaranthus and Perry, 1994).

A multifactorial experiment was conducted in Forestart Nurseries' main greenhouse (Fig. 8). This investigated where on the asymptotic growth response curve to fertiliser (Pete Lite Special high N) the plants were being grown at in the nursery, and the best way of storing *R. parksii* inocula from year to year. The fungicide Terrazole 35WP (Nufarm, 2017) was routinely incorporated into potting mixes by the nursery to prevent and control damping-off by *Pythium* and *Phytophthora*. Consequently, an additional treatment was included in order to investigate what effect Terrazole had on mycorrhizal formation.

The nursery fertilised its plants with a nutrient solution that was introduced via the overhead irrigation system. A stock solution was first made by dissolving 45 kg Pete Lite special Hi N in 200 litres of water, which was assumed to make 245 litres of stock solution. During irrigation, 29 litres of this stock solution was injected per hour into the irrigation water and each hour 9720 litres of nutrient plus water was

Table 1. Mean nutrient concentrations in the soils beneath stands of stunted and chlorotic, and healthy Douglas fir in New Zealand ((P, S, B and Cu are in ppm, and total N and C are percentages. Ca, K and Mg are in standard AgResearch test units (to convert to μg/ml, multiply by 125, 20, and 5 respectively). SEMs are beneath each mean).

pH	Total N	Organic C	C/N	P	S	Mg	Ca	K	Cu	B
Litter layer										
Healthy trees										
5.5	0.62	10.1	17.3	21.4	16	39	8	8.9	2.8	1.8
0.1	0.06	0.8	0.95	4.4	2.2	3.7	0.9	1.1	0.4	0.3
Apparently healthy trees adjacent to chlorotic and stunted ones										
4.8	0.23	4.2	18.5	4.6	2	9	2	4.1	0.9	0.5
0.2	0.05	0.8	1.2	1.04	8.1	2.3	0.6	0.8	0.3	0.2
Stunted and chlorotic trees										
4.9	0.34	5.6	16.3	5.3	15	11	2	5.4	1.9	0.7
0.2	0.14	2.2	2.81	1.5	9	1.9	0.6	0.8	0.5	0.2
10 cm soil beneath the litter layer										
Healthy trees										
5.3	0.24	3.2	14.4	19.2	20	29	6	6.9	2.2	1.0
0.6	0.02	0.3	0.81	0.45	3	4.1	1.0	0.9	0.6	0.3
Apparently healthy trees adjacent to chlorotic and stunted ones										
4.9	0.19	0.6	13.3	3.3	33	5	1	2.7	1.0	0.4
0.4	0.03	0.6	1.65	0.87	18.6	1.6	0.3	0.6	0.2	0.1
Stunted and chlorotic trees										
5.0	0.13	1.9	13.6	4.3	44	4	1	2.1	1.0	0.3
0.1	0.05	0.6	3.33	1.2	29.6	1.8	0.3	0.5	0.1	0.2

Table 2. Nutrient standards for Douglas fir in the USA (Jones et al. 1991), mean nutrient concentrations in Douglas fir needles from healthy green trees, green and chlorotic trees adjacent to each other, and seedlings from Forestart Nurseries (N, P, S, Mg, Ca, and K are percentages. Mn, Zn, Cu, Fe and B are in ppm. *Significant differences at $P < 5\%$).

	N	P	S	Mg	Ca	K	Mn	Zn	Cu	Fe	B
USA standards											
Needles from the topmost lateral branch of 2-5-year-old trees	1.50-2.30	0.18-2.30	0.15-0.25	0.09-0.15	0.30-0.50	0.75-1.10	200-600	25-45	3-12	70-200	4-15
Needles from upper crown of mature trees	0.98-1.34	0.19-0.29	0.14-0.29	0.09-0.18	0.25-0.82	0.48-0.70	452-1045	21-35	2.4-5.3	52-68	6.7-10.0
Mean of green trees	**1.29***	**0.15***	**0.08**	**0.10**	**0.29**	**0.51**	**282**	**12**	**3.3**	**86**	**10**
SEM	0.052	0.011	0.005	0.011	0.047	0.044	106	0.93	0.42	19.2	0.42
Mean of chlorotic trees adjacent to the green trees	**0.89***	**0.13***	**0.08**	**0.10**	**0.24**	**0.57**	**352**	**11**	**2.6**	**115**	**14**
SEM	0.11	0.011	0.012	0.007	0.047	0.044	71	0.95	0.18	27.4	2
Needles from plants in Forestart Nurseries	**3.28**	**0.55**	**0.36**	**0.18**	**0.41**	**1.26**	**120**	**100**	**6**	**71**	**23**
SEM	4.40	0.62	0.34	0.34	0.44	1.45	247	61	4	52	34

applied. From this, the nursery calculated that 1.63 litres of irrigation water (plus nutrient) was applied per m² during each pass of the irrigation booms. There were 12 passes per irrigation so that 19.56 litres of nutrient irrigation water was applied per m² per irrigation. There were 546 cells per m² and, hence, each cell received 0.03582 litres per plant. The weight of nutrient applied per plant per irrigation was therefore:

0.03582 x 29/9720 x 45/245 x 1000 = 0.0196 g per cell per irrigation.

The nursery irrigated on average 2.35 times per week, so the amount of nutrient applied per week per plant was 0.0461 g. We identified this as the x1 rate of nutrient application for our experiment.

The experimental treatments were.

1. Pete Lite Special high N applied each time the nursery applied nutrients in the greenhouse
   a. 0.125 normal dose per plant in 1 mL/application = 0.003 g/L
   b. 0.25   "   = 0.006 g/L
   c. 0.5   "   = 0.012 g/L
   d. 1.0   "   = 0.024 g/L

2. Inocula (applied only once) consisting of 10⁷ *R. parksii* spores in 1 mL of water made from fruiting bodies that had been either:
   a. Blended and then stored frozen
   b. Stored in moist sand in the refrigerator
   c. Stored dry
   d. Blended and stored in a refrigerator
   e. Stored in moist sand at 4°C and jellified with Crystal rain (Alter-Natives 2016)

3. Fungicide
   a. No Terrazole
   b. Terrazole applied at the same rate as that applied by the nursery

4. There were 3 replicates

The experiment was established in September (spring in New Zealand) in the main greenhouse at Forestart Nurseries where there was a 25°C, 12-hour day, and 20°C night. Forestart's standard potting mix was used and the 81 cell BCC side slit seedling trays, those normally used by the nursery (BCC, 2017), were filled with the nursery's automatic tray filling and seeding machinery. At the start of the experiment the seedlings were about 3 cm high. The trays were first watered to excess, and the treatments applied to the centre 25 cells in each tray with the outer 16 plants acting as guard rows, i.e., there were 9 experimental plants per tray. The remaining 56 plants in each tray were left untreated. The nutrients and fungicide treatments were applied using long-nosed 60 ml syringes without a needle (Fig. 8). The trays were misted regularly at the normal rate using the greenhouse's overhead irrigation system but without nutrients.

Fig. 8. Applying a fertiliser treatment in the greenhouse experiment.

The inocula treatments were applied 12 weeks after sowing. Fresh *R. parksii* fruiting bodies had been collected the previous autumn from under 90-year-old Douglas firs near the gondola in Queenstown. Within a few days of collection these were either air dried on the laboratory bench, frozen, or blended (100 g in 500 ml water) and then stored in Schott screw-topped bottles either in a refrigerator at 4°C or at –20°C in a deep freeze. Prior to the application of the inocula, dried *R. parksii* fruiting body was rehydrated by soaking in water for 2 hours. One gram of this was placed in a mortar and ground with a pestle until smooth (without abrasive) and then diluted until there were 10^7 spores/mL of suspension. This required the addition of about 300 ml of water per g of dried fruiting body. Frozen fruiting body was treated similarly. The chilled and frozen suspensions were also diluted until there were 10^7 spores/mL of suspension. All the inocula treatments were applied using a bottle top dispenser with the contents agitated continuously in order to ensure the spores remained in suspension.

Observations

The results of our experiment are presented in Figs. 9 and 10 and can be summarised as follows.

- The level of nutrients normally applied in the nursery was antagonistic to mycorrhizal formation.
- Of the 72 plants sampled that had received less than the full nutrient solution and that had been inoculated with the dried or refrigerated inocula, only one plant remained completely non-mycorrhizal, i.e., 98% were mycorrhizal. The mean mycorrhizal formation on these plants was approximately 35%.

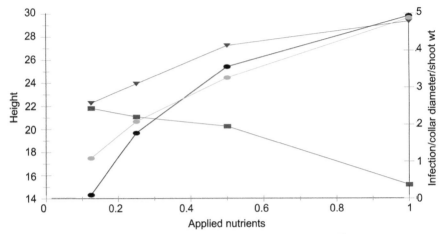

Fig. 9. Effect of rates of applied nutrients on mycorrhizal formation ■, height ■, collar diameter ■ and dry weight ■ of greenhouse grown containerised Douglas fir.

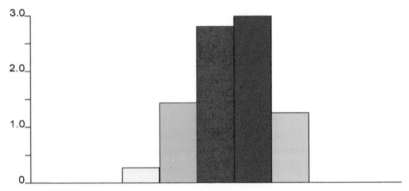

Fig. 10. Effect of inocula on the establishment of mycorrhizal infection (blended and frozen, ■ stored in moist sand, ■ dried, ■ blended and stored in a refrigerator, blended, stored in a refrigerator and then mixed with a gelling agent). (0 = 0%, 1 = trace, 2 = 1–25%, 3 = 25–50%, 4 = 50–75%, 5 = 75–100%).

- A single application of 10^7 *R. parksii* spores was sufficient to produce good mycorrhizal formation.

- Mycorrhizas plus half the normal level of applied nutrients produced plants almost equal to the size to those produced in the nursery—a mean height of 25 cm and a mean collar diameter of 4 mm. These were considered acceptable for outplanting.

- Dried or blended then refrigerated *R. parksii* inocula produced the heaviest infections provided that less than the maximum rate of nutrients was applied.

- As expected of a fungicide that is virtually specific to the control of Oomycota (Radzuhn and Lyr 1984), Terrazole had no ill effects on mycorrhizal formation. It also had no significant effect on shoot dry weight.

Application of the Technology

When left in a plastic bag *R. parksii* quickly turns into a slimy mess, so Lance Freer, an Ernslaw One afforestation planner, collected about 50 kg of *R. parksii* and left these to rot down. These were then blended with a kitchen wizz, mixed with adequate water and then injected into the irrigation system once. This one application was found to be sufficient to establish mycorrhizas on 10 million trees and became the standard practice in the nursery for the following decade. Rotten *R. parksii* is not an easy product to store and ship around the country, so Lance Freer developed a method of producing a dried product that could be stored and distributed more easily (De La Mare pers. comm.).

A plantation established with our trees is shown 20 years later in Fig. 11. There is no suggestion of chlorosis and the stand is even, an important feature when the forest is eventually harvested. In 2015 the Otago-Southland plantation forests covered 207,000 ha with 25% Douglas fir, up from 8.8% in 1994. This is about half of New Zealand's Douglas fir plantings (Southern Wood Council, 2015).

Fig. 11. A plantation established with Forestart trees at Gowan Hill (see Fig. 3) 20 years after planting. There is no suggestion of chlorosis and the stand is even, an important feature when the forest is eventually harvested.

Discussion

We chose to inoculate Douglas fir with a mycorrhizal fungus that had probably been present in New Zealand at least since Douglas fir had been planted at Walter Peak, Queenstown, in 1880. While we would have preferred to have inoculated with an ectomycorrhizal fungus with edible fruiting bodies, such as *Suillus lakei* (painted bolete), the technology for doing this was not available and would not become available for another 7 years (Wang et al., 2002; Hall, pers. comm.). Such a delay was simply out of the question and Forestart Nurseries needed a rapid resolution to their problem just to stay in business. The method devised was quick, easy and cost a fraction of a cent to inoculate each plant. We now know that had we chosen to inoculate the plants with cultures of *S. lakei*, massive quantities of pure mycorrhizal inocula would have had to have been grown in laboratories and the cost of this plus the labour required to inoculate each plant would have been prohibitively expensive.

There is no doubt that wilding pines are a significant problem in New Zealand (Paul, 2015; Department of Conservation, 2017) and it could be argued that the use of *R. parksii* as an inoculant for Douglas fir and *Rhizopogon roseolus* (= *R. rubescens*) for radiata pine (*Pinus radiata*) (De La Mare, pers comm.) has contributed to their spread in tussock grasslands. Furthermore, our work has developed ways of ensuring that these fungi spread faster now than when our predecessors accidentally introduced them in the 1800s. However, many areas of the "iconic landscapes" (Department of Conservation, 2017) that we have learned to love in New Zealand, are not stable and without fire they would eventually return to forest that covered these areas prior to the arrival of man (see the Introduction). This would be populated by New Zealand natives if the seed source was available or pine forest if seeds of these species got there first.

In an ideal world our predecessors would have planted stone pines with large, heavy, edible nuts that are not dispersed by wind, and mycorrhizal fungi that produce edible mushrooms. Similarly, our predecessors would not have brought with them stoats, cats, deer, dogs, pigs, possums, rats and sheep because they had the potential of decimating our native flora or fauna. They would also have been careful not to have allowed the spread of a *Phytophthora* that now threatens our iconic kauri forests (Landcare Research, 2017). Clearly these things are irreversible, but we can re-establish forest on areas where forest once existed. If we do, do we want native trees that may pay a dividend in hundreds of years or do we want the alternative tussock grassland parkland much of which was produced by man, and then pay the cost of maintaining it by weeding out wilding pines? Another alternative is to plant exotic trees that suit the landscape, pay a dividend within a lifetime, produce timbers that are something more than a commodity and perhaps produce crops of mycorrhizal mushrooms with a value that might outstrip the value of the timber (Hall et al., 2011; Hall and Zambonelli, 2012a,b). Another is to burn the tussock grasslands as our forebears did in order to help maintain the artificial landscape we are so used to, or fertilise the land and oversow with clovers and grasses so it can carry more stock. Of course, none of these options will suit everyone!

Mycorrhizas form an integral part of the way the majority of plants obtain the nutrients they require, in particular P and N (Smith and Read, 2010), and a lack of these nutrients can enhance the beneficial effects mycorrhizas confer on their plant hosts. But in forest nurseries, the aim is for plants to have an acceptable height and collar diameter and to get them out of the nursery in time for planting the following year's crop. The consequences of this in Forestart Nurseries were nutrient concentrations in plants that received the "normal" rate of fertilisers, that were multiples, or in the case of Zn an order of magnitude higher than what were found in healthy plantation forests (Table 2). The consequence of this in our study and others (e.g., Diaz et al., 2010) was a reduction in mycorrhizal formation. Other consequences might have been a shift in the balance between mycorrhizal fungi on the roots (Lilleskov and Bruns, 2001) or the occupation of space on the roots by apparently ineffective mycorrhizas, as on the chlorotic plants at the Glen Dhu site. There is plenty of scope for this to happen with Douglas fir because 2000 fungi are suspected as being mycorrhizal candidates (Trappe, 1977). Of course, not all will necessarily produce the maximum stimulation of plant growth nor produce effective mycorrhizas even though they might produce mycorrhizas in pure culture experiments (Parladé et al., 1996). There is also likely to be some degree of "ecological specificity" (Harley and Smith, 1983) or "mycorrhizal specificity" (Molina et al., 1992; Molina and Horton, 2015). Clearly our guess that *R. parksii* would be well suited to Douglas fir over a wide range of edaphic and climatic conditions in New Zealand was indeed fortunate and did not expose just how little we knew about exploiting mycorrhizal macro-mushrooms in forestry.

Acknowledgements

We wish to thank Ernslaw One for funding and its generosity in allowing us to publish the work described here. We are also extremely grateful for technical assistance that was provided by the staff at Invermay in particular Dr. Rachael Byars, Julie Hamilton and Lynley Hall. Also, to Ray Hannagan and the staff of the Soil and Plant Analysis Group for the countless analyses they carried out for us. Thanks also go to Te Ara for the use of Figs. 1 and 2 from Te Ara's website and to Bateman Ltd that produced the initial maps. Our thanks also go to the late Malcolm Douglas for his careful reading of the manuscript and some very useful additions.

References

Agerer, R. 2001. Exploration types of ectomycorrhizae. A proposal to classify ectomycorrhizal mycelial systems according to their patterns of differentiation and putative ecological importance. Mycorrhiza, 11: 107–114.

Alter-Natives. 2016. Crystal rain. http.//www.alter-natives.co.nz/html/ http.//www.alter-natives.co.nz/html/documents/PotsorbBrochureJune2012.pdf.

Amaranthus, M. and Perry, D.A. 1994. The functioning of ectomycorrhizal fungi in the field. linkages in space and time. Plant Soil, 159: 133–140.

BCC. 2017. BCC Side Slit 81 seedling trays. http://www.bccab.com/Produkt-Growing-Systems-eng.php?sd_esp_1_pid=44.

Belton, M.C. and Davis, M.R. 1986. Growth decline and phosphorus response by Douglas fir on a degraded high-country yellow-brown earth. NZ. J. For. Sci., 16: 55–68.

Bosselmann, G.M. 2002. Ectomycorrhizal fungi of Douglas fir. Master of Science thesis, University of Otago.

Burdon, R.D. and Miller, J.T. 1995. Alternative species revisited. Categorisation and issues for strategy and research. NZ. J. For. Sci., 40: 4–9.

Burrell, J. 1965. Ecology of *Leptospermum* in Otago, N.Z. J. Bot., 3: 3–16.

Castellano, M., Trappe, J.M. and Molina, R. 1985. Inoculation of container-grown Douglas-fir seedlings with basidiospores of *Rhizopogon vinicolor* and *R. colossus*. Effects of fertility and spore application rate. Can. J. For. Res., 15: 10–13.

Chu-Chou, M. and Grace, L.J. 1981a. Mycorrhizal fungi of *Pseudotsuga menziesii* in the North Island of New Zealand. Soil Biol. Biochem., 13: 247–249.

Chu-Chou, M. and Grace, L.J. 1981b. *Tuber* sp. as a mycorrhizal fungus of Douglas fir in New Zealand. Trans. Br. Mycol. Soc., 77: 652–654.

Chu-Chou, M. and Grace, L.J. 1983. Characterisation and identification of mycorrhizas of Douglas fir in New Zealand. Eur. J. Forest Pathol., 13: 251–260.

Chu-Chou, M. and Grace, L.J. 1985. Comparative efficiency of the mycorrhizal fungi *Laccaria laccata, Hebeloma crustuliniforme* and *Rhizopogon* species on growth of radiata pine seedlings. N.Z. J. Bot., 23: 417–424.

Chu-Chou, M. and Grace, L. 1987. Mycorrhizal fungi of *Pseudotsuga menziesii* in the South Island of New Zealand. Soil Biol. Biochem., 19: 243–246.

Cieraad, E. and McGlone, M.S. 2013. Thermal environment of New Zealand's gradual and abrupt treeline ecotones. NZ. J. Ecol., 38: 12–25.

Colinas, C., Perry, D., Molina, R. and Amaranthus, M. 1994. Survival and growth of *Pseudotsuga menziesii* seedlings inoculated with biocide-treated soils at planting in a degraded clearcut. Can. J. For. Res., 24: 1741–1749.

Cole, E.C. and Newton, M. 1989. Height growth response in Christmas trees to sulfometuron and other herbicides. Proceedings of the Western Society of Weed Science, 42: 129–135.

Cornforth, I.S. and Sinclair, A.G. 1982. Fertilisers and lime recommendations for pastures and crops in New Zealand. Agricultural Research and Advisory Service Divisions, New Zealand Ministry of Agriculture and Fisheries, Wellington.

Davis, M.R. 1989. Establishment of conifer plantations in the South Island high country by direct drilling. NZ. J. For. Sci., 34: 21–24.

Davis, M. and Smaill, S. 2009. Mycorrhizal colonisation of exotic conifers in kānuka and mānuka shrublands. NZ. J. Ecol., 33: 147–155.

Department of Conservation. 2017. Wilding pines control work nears million hectare mark. http.//www.doc.govt.nz/news/media-releases/2017/wilding-pines-control-work-nears-million-hectare-mark/.

Diaz, G., Carrillo, C. and Honrubia, M. 2010. Mycorrhization, growth and nutrition of *Pinus halepensis* seedlings fertilized with different doses and sources of nitrogen. Ann. For. Sci., 67: 1–9.

Dumroese, R.K., Thompson, G. and Wenny, D.L. 1990. Lime-amended growing medium causes seedling growth distortions. Tree Planters' Notes, 41: 12–17.

Fitter, A.H. and Garbaye, J. 1994. Interactions between mycorrhizal fungi and other soil organisms. pp. 123–132. *In*: Robson, A.D., Abbott, L.K. and Malajczuk, N. (eds.). Management of Mycorrhizas in Agriculture, Horticulture and Forestry. Kluwer, Dordrecht.

Floate, M. and Cossens, G. 1992. Land resources. pp. 23–37. *In*: Floate, M. (ed.). Guide to Tussock Grassland Farming, New Zealand Pastoral Research Institute, Mosgiel.

Foerst, K. 1981. Die Ertragsleistung der Douglasie in Bayern. Allgemeine forstzeitschrift, 33: 842–844.

Gilmore, J.W. 1958. Chlorosis of Douglas fir. NZ. J. For. Sci., 7: 84–106.

Hall, I.R., Fearn, K., De La Mare, P., Perley, C. and Douglas, M. 2011. The cultivation of edible mycorrhizal mushrooms in plantation forests. A Power Point presentation and supporting text from the 6th Internal Workshop on Edible Mycorrhizal Mushrooms, Rabat, Morocco, 6–10 April 2011. https.//www.researchgate.net/publication/312032386_The_cultivation_of_edible_mycorrhizal_mushrooms_in_plantation_forests.

Hall, I.R. and Garden, E. 1984. Effect of fertilisers and ectomycorrhizal inoculum on stunted Douglas firs. Proceedings of the 6th North American conference on mycorrhizae (June 25–29, 1984, Bend, Oregon). 224.

Hall, I.R. and Perley, C. 2007. Removing a significant constraint limiting the diversification of the forestry estate: Research on the exploitation of mycorrhizas in forestry practice. Sustainable Farming Fund, New Zealand Ministry of Agriculture and Fisheries, http.//maxa.maf.govt.nz/sff/about-projects/search/05-142/05-142-final-report.pdf and http.//maxa.maf.govt.nz/sff/about-projects/search/05-142/05-142-appendices.pdf.

Hall, I.R. and Zambonelli, A. 2012a. Laying the foundations. Chapter 1. pp. 3–16. *In*: Zambonelli, A. and Bonito, G. (eds.). Edible Mycorrhizal Mushrooms, Soil Biology # 34. Dordrecht, Springer.

Hall, I.R. and Zambonelli, A. 2012b. The cultivation of mycorrhizal mushrooms - still the next frontier! pp. 16–27. *In*: Zhang, J., Wang, H. and Chen, M. (eds.). Mushroom Science XVIII. Beijing, China Agricultural Press.

Harley, J.L. and Smith, S.E. 1983. Mycorrhizal symbiosis. Academic Press, London.

Hood, I.A. and van der Pas, J.B. 1979. Fungicidal control of *Phaeocryptopus gaeumannii* infection in a 19-year-old Douglas fir stand. NZ. J. For. Sci., 9: 272–283.

Idczak, R. 1975. Silviculture—Research activity 74. Forests Commission. Melbourne, Victoria, 5–15.

Jones, J.B., Wolf, B. and Mills, H.A. 1991. Plant analysis handbook. Micro-Macro Publishing, Athens, Georgia.

Landcare Research. 2017. Kauri dieback. Kia toitu he kauri. https.//www.landcareresearch.co.nz/ publications/newsletters/biological-control-of-weeds/issue-67/kauri-dieback.

Ledgard, N. 2002. The spread of Douglas-fir into native forests. NZ. J. For. Sci., 36–38.

Le Tacon, F., Alvarez I.F., Bouchard, D., Henrion, B., Jackson, R.M., Luff, S., Parlade, J.I., Pera, J., Stenstrom, E., Villeneuve, N. and Walker, C. 1992. Variations in field response of forest trees to nursery ectomycorrhizal inoculation in Europe. pp. 119–134. *In*: Read, D.J., Lewis, D.H., Fitter, A.H., Alexander, I.J. (eds.). Mycorrhizas in Ecosystems, CAB, Wallingford.

Lilleskov, E.A. and Bruns, T.D. 2001. Nitrogen and ectomycorrhizal fungal communities: What we know, what we need to know. New Phytol., 149: 156–158.

McGlone, M.S., Mark, A.F. and Bell, D. 1995. Late Pleistocene and Holocene vegetation history, Central Otago, South Island. J. R. Soc. N.Z., 25: 1–22.

McWethy, D.B., Whitlock, C., Wilmshurst, J.M., McGlone, M.S. and Li, X. 2009. Rapid deforestation of South Island, New Zealand, by early Polynesian fires. The Holocene, 19: 883–897.

Marschner, H. and Dell, B. 1994. Nutrient uptake in mycorrhizal symbiosis. pp. 89–102. *In*: Robson, A.D., Abbott, L.K. and Malajczuk, N. (eds.). Management of Mycorrhizas in Agriculture, Horticulture and Forestry. Kluwer, Dordrecht.

Ministry of Forestry. 1993. A National Exotic Forest Description. Edit. 10. Ministry of Forestry.

Molina, R. and Horton, T.R. 2015. Mycorrhiza specificity. Its role in the development and function of common mycelial networks. pp. 1–15. *In*: Horton, T. (ed.). Mycorrhizal Networks: Ecological studies (Analysis and Synthesis) 224. Springer, Dordrecht.

Molina, R., Massicotte, H. and Trappe, J.M. 1992. Specificity phenomena in mycorrhizal symbioses. Community-ecological consequences and practical implications. pp. 357–423. *In*: Allen, M.F. (ed.). Mycorrhizal Functioning. Chapman and Hall, New York.

Motschalow, S. 1988. Zur Jugendentwicklung der Douglasie auf kalkboden. Schweizerische Zeitschrift für Forstwesen, 139: 675–689.

New Zealand Forestry Research Institute 1995. Douglas fir research. *In*: New Zealand FRI Research Directions # 7. New Zealand Forest Research Institute, Rotorua.

New Zealand Institute for Economic Research. 2017. Plantation forestry statistics. https.//nzier.org.nz/ static/media/filer_public/c6/a5/c6a55bbf-8f36-484e-82a0-91bb59211880/plantation_forestry_ statistics.pdf.

Nufarm. 2017. Terrazole 35 WP. http://www.nufarm.com/assets/18031/1/Terrazole_35WP_2Kg_label. pdf.

Ogden, J., Basher, L. and Mcglone, M. 1998. Fire, forest regeneration and links with early human habitation: Evidence from New Zealand.

Parke, J.L., Linderman, R.G. and Black, C.H. 1983. The role of ectomycorrhizae in drought tolerance of Douglas fir seedlings. New Phytol., 95. 83–95.

Parladé, J., Álvarez, I.F. and Pera, J. 1996. Ability of native ectomycorrhizal fungi from northern Spain to colonize Douglas-fir and other introduced conifers. Mycorrhiza, 6: 51–55.

Paul, T.S.H. 2015. Guidelines for the use of the Decision Support System "Calculating Wilding Spread Risk From New Plantings". New Zealand Forest Research Institute Limited (Trading As Scion), Report S0019. http://www.wildingconifers.org.nz/images/wilding/articles/Strategy/Guidelines_for_ using_the_DSS_for_new_forest_plantings.pdf.

Poole, A.L. 1969. Forestry in New Zealand the Shaping of Policy. London, Hodder and Stoughton.

Radwan, M.A. and Brix, H. 1986. Nutrition of Douglas-fir. 112 p. *In*: Oliver, Chadwick Dearing, Hanley, Donald P. and Johnson, Jay A. (eds.). Douglas-fir: Stand Management for the Future. Proceedings of a symposium, 18–20 June 1985, Seattle, WA. Seattle College of Forest Resources, University of Washington.

Radzuhn, B. and Lyr, H. 1984. On the mode of action of the fungicide etridiazole. Pestic. Biochem. Physiol., 22: 14–23.

Roche, M. 1990. History of forestry. Wellington, New Zealand Forestry Corporation and GP Publications.

Schöne, D. 1987. Eine Mangan-induzierte Eisenchlorose bei Douglasie. Allg. Forst. Z., 42: 1154–1157.

Singh, P., Bhure, N.D. 1974. Influence of Armillaria root rot on the foliar nutrients and growth of some coniferous species. Eur. J. Forest Pathol., 4: 20–26.

Smith, S.E. and Read, D.J. 2010. Mycorrhizal Symbiosis, 3rd Edition. Academic Press, London.

Southern Wood Council. 2015. Otago–Southland Forestry Profile. https://www.southernwoodcouncil. co.nz/wp-content/uploads/2015/11/Otago-Southland-Forestry-Profile-2015-03.pdf.

Stephens, J.M.C., Molan, P.C. and Clarkson, B.D. 2005. A review of *Leptospermum scoparium* (Myrtaceae) in New Zealand, N. Z. J. Bot., 43: 431–449. DOI: 10.1080/0028825X.2005.9512966.

Te Ara. 2010. Forest cover before human habitation and Forest cover around 1840. https://teara.govt. nz/en/map/23596/forest-cover-before-human-habitation and https://teara.govt.nz/en/map/23597/ forest-cover-around-1840.

Teste, F.P., Schmidt, M.G., Berch, S.M., Bulmer, C. and Egger, K.N. 2004. Effects of ectomycorrhizal inoculants on survival and growth of interior Douglas-fir seedlings on reforestation sites and partially rehabilitated landings. Can. J. For. Res., 34: 2074–2088. doi: 10.1139/X04-083.

Trappe, J.M. 1977. Selection of fungi for ectomycorrhizal inoculation in nurseries. Ann. Rev. Phytopathol., 15: 203–222.

van den Driessche, R. 1989. Nutrient deficiency symptoms in container-grown Douglas-fir and white spruce seedlings. B.C. Ministry of Forests/Canada/CC Economic and Regional Development Agreement.

Walker, C. 1992. Inoculating Douglas-fir seedlings with mycorrhizal fungi. Research information note 222. United Kingdom Forestry Authority.

Walker, R.B., Gessel, S.P. and Miller, R.E. 1995. Greenhouse and laboratory evaluation of two soils derived from volcanic ash. Northwest Sci. 68: 250–258.

Warne, K. and Gasteiger, A. The future of our forests: How the bush was lost (click on sidebar). https:// www.nzgeo.com/stories/the-future-of-our-forests/.

Weetman, G.F., McWilliams, G. and Thompson, W.A. 1992. Nutrient Management of Coastal Douglas-fir and Western Hemlock Stands: The Issues. 17–27. *In*: Chappell, H.N., Weetman, G.F. and Miller, R.E. (eds.). Improving Nutrition and Growth of Western Forests. College of Forest Resources, University of Washington, Seattle, Washington, USA.

Zas, R. 2003. Foliar nutrient status and tree growth response of young *Pseudotsuga menziesii* Mirb. (Franco) to nitrogen, phosphorus and potassium fertilization in Galicia (Northwest Spain). For. Syst., 12: 75–85.

14

Barcoding DNA
A Tool for Molecular Identification of Macrofungi

Arun Kumar Dutta and *Krishnendu Acharya**

INTRODUCTION

The term 'macrofungi' refers to the group of fungi that have epigeous or hypogeous fructifications and are clearly visible to the unaided eye (Chang and Miles, 1992). These groups of fungi are commonly known by several names, i.e. conks, mushrooms, puffballs, toadstools, etc. (Redhead, 1997). According to recent estimations, the existence of 140,000 species is predicted to be called macrofungi due to the stature of their fruit bodies (Hawksworth, 2012). Among these predicted numbers of macrofungi, we currently have detailed descriptions of around 14,000 species. This accounts for only 10% of the total predicted number (Chang and Miles, 2004).

Macrofungi play an important role in the evolution of macroscopic life forms and continue to drive many of the ecological processes, such as bioremediation, nutrient cycling, litter decomposition, biogeochemical cycle and soil formation. Wild edible mushrooms have long been considered as a part of the human diet in Asian and European countries because of their medicinal and nutritional properties. They are rich in biologically active compounds and have antibacterial, antifungal, antioxidant, immunomodulation, hepatoprotective, and apoptogenic properties (Chatterjee et al., 2011; Mallick et al., 2014; Pattanayak et al., 2015; Mallick et al., 2016; Chatterjee and Acharya, 2016; Khatua et al., 2013, 2017a, b; Acharya et al., 2017).

Molecular and Applied Mycology and Plant Pathology Laboratory, Department of Botany, University of Calcutta, Kolkata, West Bengal 700019, India.
* Corresponding author: krish_paper@yahoo.com

Precise identification of a macrofungal species is a crucial step across many fields for the sustainable use of macrofungal biodiversity, following conservation strategies, ecological monitoring, supressing fungal pathogens, developing quarantine steps of exotic species and human health. Among the described species, almost half of them still are still devoid of barcode information (Xu, 2016). According to the estimation of Manoharachary et al. (2005), India is enriched with around one-third of the global fungal diversity, and most of the tropical fungal species show high variability in their genetic sequences.

In the era of modern scientific progress, one of the most notable constraints of global concern is the decreasing number of taxonomists (Wägele et al., 2011). Other than this, the extent of civilization and natural calamities are the major concerns in the enhanced extinction of macrofungi compared to the natural process (Pimm et al., 2014). These overwhelming circumstances have substantiated the DNA barcoding as an inevitable and appropriate solution for rapid species discovery and for opening new insights of conservation. Prior to 2005 (first international DNA barcoding meeting in London), researchers around the world had been engaged in providing the DNA barcoding data of fungi across several groups, starting from indoor air samples to lichen-forming fungi, or human fungal pathogens to edible mushrooms (Adamowicz, 2015).

The process of DNA barcoding preferably uses a short and standardized genetic marker for an error-free, feasible and quick species identification (Hebert and Gregory, 2005). Earlier, the choice of a suitable region of DNA for barcode was primarily made based on some criteria, such as its presence in all the studied organisms, PCR amplification success, comparative ease in sequencing, its supposed evolution rate and the absence of pseudogenes, paralogs or orthologs that could cause complicacy in amplification and analysis (Robert et al., 2011). Later, in 2012, the nuclear ribosomal internal transcribed spacer (nrITS) region was accepted by the International Fungal Barcoding Consortium as a formal fungal DNA barcode region (Schoch et al., 2012). The entire ITS region ranges around 600–700 bp long stretches in fungi and it is composed of two variable spacers (ITS-1 and ITS-2) spanning across a highly conserved 5.8S rRNA gene (White et al., 1990; Das and Deb, 2015). The 18S rRNA gene resides at the 5' end of the ITS-1 spacer, while the 3' end of the ITS-2 spacer ends in 28S rRNA gene (Fig. 1). Because of the presence of conserved regions, such as 18S, 5.8S, and 28S rRNA genes, several universal primers have been developed for the successful amplification of the entire ITS region that is applicable to the vast majority of macrofungal species across various families (Table 1). The amplification success of the region is comparatively higher, owing to the presence of its high copy number in the genome (Schoch et al., 2012). The present chapter reviews the history of DNA barcoding for rapid identification of macrofungi, details of possible regions that fit for quick identification across various macrofungal samples, and the possible steps involved in using an accepted fungal barcode region for phylogenetic and systematic studies.

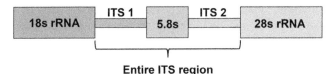

Entire ITS region

Fig. 1. Diagrammatic representation of the entire ITS region, showing positions of two intermediate spacers (ITS1 and ITS2).

Table 1. List of universal primers commonly used for amplification of the nrDNA ITS region, along with their sequence and reference.

Primer name	Primer sequence (5'→3')	Reference
ITS1 (Forward)	TCCGTAGGTGAACCTGCGG	White et al., 1990
ITS1-F (Forward)	CTTGGTCATTTAGAGGAAGTAA	Gardes and Bruns, 1993
ITS5 (Forward)	GGAAGTAAAAGTCGTAACAAGG	White et al., 1990
ITS1-F_KYO1 (Forward)	CTHGGTCATTTAGAGGAASTAA	Toju et al., 2012
ITS1F_KYO2 (Forward)	TAGAGGAAGTAAAAGTCGTAA	Toju et al., 2012
ITS4 (Reverse)	TCCTCCGCTTATTGATATGC	White et al., 1990
ITS4-B (Reverse)	CAGGAGACTTGTACACGGTCCAG	Gardes and Bruns, 1993
LR3 (Reverse)	CCGTGTTTCAAGACGGG	Vilgalys and Hester, 1990
ITS4_KYO1 (Reverse)	TCCTCCGCTTWTTGWTWTGC	Toju et al., 2012
ITS4_KYO2 (Reverse)	RBTTTCTTTTCCTCCGCT	Toju et al., 2012
ITS4_KYO3 (Reverse)	CTBTTVCCKCTTCACTCG	Toju et al., 2012

History of DNA Barcoding

The use of nucleotide variations in investigating the evolutionary relationship across various biological samples has been known to the scientific community since long ago. Carl Woese used the sequence variations of ribosomal RNA gene that acted well as a molecular marker for archaera and became the key source for conducting evolutionary analyses (Bisen, 2014). The use of DNA barcoding was initially begun by an insect taxonomist that later received international collaboration in the establishment of a DNA barcode database for all organisms. In the early 1970s, the utilization of the DNA barcode region for fungal evolutionary analyses emerged soon after the discovery of molecular markers (Dutta and Ojha, 1972). The journey of studying a diverse group of fungi using a sequence of 5S rRNA gene was initiated by Walker and Doolittle (1982). Later, in 2003, Paul D.N. Hebert proposed the compilation of DNA barcodes accessible to public libraries and their linking to known specimens (Bisen, 2014).

The onset of the incorporation of molecular sequence data for fungal studies accelerated with the development of DNA sequencing techniques and designing of the fungal specific primers by White et al. (1990). In his authoritative work, he reported two universal primers (ITS1 and ITS4) that successfully amplified the whole ITS region of the fungal ribosomal operon. Prior to the discovery of ITS1 and ITS4 primers for amplification of the entire nrDNA ITS region encompassing two spacers (ITS1 and ITS2 region) along with an intervening 5.8s region, most studies were solely based on the amplification of short sequences (either ITS1 or ITS2 region alone) due to the length limitations of manual sequencing protocols. Two subsequent studies by Fell et al. (2000) and Scorzetti et al. (2002) established the data set of two variable domains (D1/D2) of the nLSU rRNA of basidiomycetous yeasts. Their study reported that the ITS region is very much applicable for successfully discriminating closely related species in comparison to the D1/D2 domain. Though the variable D1/D2 domain sequences of nLSU rRNA are commonly required for describing any novel taxa of yeasts, more DNA sequence information is necessary for describing macrofungal taxa belonging to other groups (Xu, 2016). Before the announcement of ITS as a consensus barcode region for fungi (Schoch et al., 2012), several earlier studies used the sequence variation of the ITS region for the identification of various novel taxa (Kerrigan et al., 1999; Wang et al., 2007).

The Candidate Regions Across Fungal Samples

Until the unique work of Schoch et al. (2012), researchers around the world were focused on using various candidate loci for the molecular identification of fungal species. The region of the mitochondrial gene encrypting CO1 (cytochrome c oxidase subunit 1) was proposed to be the candidate barcode region for animals (Hebert et al., 2003a,b) and also for fungi (Schindel and Miller, 2005). Seifert et al. (2007) found that the CO1 region is more effective in comparison to the ITS region in a study related to *Penicillium*; meanwhile, Geiser et al. (2007) and Gilmore et al. (2009) demonstrated low divergence between CO1 homologs in species of *Aspergillus* and *Fusarium*. For identification of the species of *Leohumicola*, Nguyen and Seifert (2008) found equal importance in both the regions viz. CO1 and ITS. Vialle et al. (2009) showed the superiority of the CO1 coding region over ITS and LSU regions. Gonzalaz et al. (1998) reported the presence of four introns in the CO1 coding region of *Agrocybe aegerita*, while Haridas and Gantt (2010) reported 15 introns in the CO1 region of a wood rotting fungi *Trametes cingulata*. Overall, the acceptance of the CO1 region as a suitable barcode region for species level identification remained a matter of debate until 2010. Dentinger et al. (2011) performed a huge work in generating 167 partial CO1 sequences from 100 species of Agaricomycotina by designing new primer sets. He reported the presence of large introns (ca. 1500 bp in length) at variable locations and low PCR success rate for CO1 locus. However, their study also compared the ITS region with CO1 and found similar performance of these two regions to be served as possible barcode region. For a densely sampled set of close taxa of mushrooms, they proposed that the CO1 region would be less divergent in comparison to ITS. Finally, the selection of the CO1 region was excluded as a candidate locus for fungal barcode because of the common occurrence of intron regions in fungal CO1 (Yan and Xu, 2005; Seifert et al., 2007).

The SSU region (18S nuclear ribosomal small subunit rRNA gene), commonly used for phylogenetic studies, is also preferred for species resolution in bacteria (Stackebrandt and Goebel, 1994) . But for most of the basidiomycetous fungi, SSU region has fewer number of variable domains and thus has limited applications for fungal phylogenetic studies (Schoch et al., 2012). In yeasts, however, the D1/D2 region of the 28S nuclear ribosomal large sub-unit rRNA gene (LSU) is used to separate species (Fell et al., 2000; Scorzetti et al., 2002). This region, in combination with ITS, is often preferred for more accurate species identification of basidiomycetous fungi (Geml et al., 2009). Moncalvo et al. (2000) performed a massive and most remarkable work by amplifying the nrLSU region from 154 taxa distributed over most fungal families of the order Agaricales. Their study served as the backbone for future systematics studies of Agaricales. Toju et al. (2012) designed some new sets of primer pairs (ITS3_ KYO1/ITS4_KYO1, ITS3_KYO2/ITS4_KYO2) for the successful amplification of the region that showed more accuracy towards identifying most of the ascomycetous and the basidiomycetous fungal taxa. For the phylum Ascomycota, Asemaninejad et al. (2016) designed a new set of primers for amplifying the D1 region of nrLSU and their designed primers (LSU200A-F and LSU476A-R) successfully produced 95–98% reads of target specimens, collected from environmental sources, that were compatible with the Illumina MiSeq platform with comparatively lesser data loss. In comparison to the primers, viz. ITS3_KYO2/ITS4_KYO3 (Toju et al., 2012), the primer pairs designed by Asemaninejad et al. (2016) showed the almost similar read depth and were applicable across various fungal groups.

Among protein-coding genes, the largest sub-unit of RNA polymerase II (RPB1; McLaughlin et al., 2009), second largest sub-unit of RNA polymerase II (RPB2; James et al., 2006) and minichromosome maintenance protein (MCM7; Raja et al., 2011), etc., are most the widely used regions for identifying fungal specimens or in the phylogenetic studies (Aguileta et al., 2008; Schmitt et al., 2009). For Ascomycetous fungi, these protein coding genes provide better resolution to resolve the taxonomic relationships (Schoch et al., 2009). However, RPB1 is omnipresent and has a single copy in the genome where the rate of sequence divergence is very slow (Tanabe et al., 2002).

Schoch et al. (2012) examined the suitability of six marker regions across 700 strains belonging to 200 taxa for providing the best candidate locus that could be used as a suitable fungal barcode region. For the fulfillment of their aim, they considered the criteria of barcode gap that is generally used to indicate the level of sequence difference among various strains (Meyer and Paulay, 2005). Of the regions studied, the PCR success rate was comparatively higher in case of ITS as compared to nrLSU, nrSSU, and RPB1 regions. Beside PCR success rates, the probability of accurate identification up to the species level (PCI) was also evaluated for the four gene markers by Schoch et al. (2012). The obtained results showed a high species discrimination power of RPB1 in Ascomycota and Basidiomycota when compared to other fungal lineages. But combining all of the studied taxa, the PCI value of ITS was found to be slightly lower than that of RPB1 (0.73 vs. 0.76). In Dikarya, the ITS region had the superior species discrimination capability over RPB1 in Basidiomycota (PCI of 0.77 vs. 0.67). Although there were several drawbacks of ITS for some specific taxa owing to the absence of a clear barcode gap (Xu et al., 2000), alignment difficulties across diverge taxonomic groups and intragenic sequence heterogeneity. Schoch et al. (2012) proposed ITS as the consensus primary fungal barcode locus based on a combination of features, such as the universality of primers, rate of PCR success and accurate species discriminating capabilities.

Steps Involved Using Barcode Regions

In the 18th century, Linnaeus introduced the binomial nomenclature and was the first person to draw a phylogenetic tree. Later, Charles Darwin added the incidence of two important processes, branching and divergence, to the phylogeny. Early supporters of the molecular phylogeny claimed that true phylogeny relies upon the molecular data that reflects changes in the genetic level. Though in recent times, this concept has been proven to be inappropriate (Patwardhan et al., 2014).

Recently, molecular phylogenetic studies employing nucleic acid data and protein coding genes have drawn the interest of researchers throughout the world. So in these days, as an easy identification tool, scientists usually isolate the genomic DNA from their collected sample which is subsequently followed by PCR, sequencing of the desired region, and finally comparing that amplified sequence with the available reference sequence in various public databases. The final outcome of these processes leads us to conduct phylogenetic analyses that helps to clear the relationship of similar or distantly related taxa for solving their taxonomic boundaries. The results of phylogenetic analyses are represented by a diagram in the form of phylogenetic trees that indicate an overall idea of the given taxa and their relation with other phylogenetically close taxa. For a given dataset comprising of a number of sequences, there are various possible tree topologies available that can be drawn. So, there lies various methodologies for selecting an optimal tree. For successful construction of a phylogenetic tree with the suitable barcode sequence region of macrofungi, there remains some other processes, such as alignment of multiple sequences, choice of the evolutionary model for the given set of alignment, determination of the method for building up the tree, and finally assessing the tree reliability (Xiong, 2006; Patwardhan et al., 2014; Dutta et al., 2015a,b, 2017; Paloi et al., 2015).

On amplification of a suitable barcode sequence region, the steps involved in constructing a phylogenetic tree are as follows:

Choice of Sequences

Choice of sequences of closely related taxa is considered to be the most crucial step for molecular phylogenetic analysis (Gherbawy and Voigt, 2010). As an example for obtaining the ITS sequences of closely related taxa, various available public databases like GenBank (https://www.ncbi.nlm.nih.gov), UNITE (https://unite.ut.ee), etc., are searched. This process generally involves the comparison of the primary nucleotide sequence of the available reference sequences in the public databases involving a BLAST (Basic Local Alignment Search Tool) search (Camacho et al., 2009) by pairwise alignments and statistical analysis. BLAST E-value denotes the incidence of similarity between sequence pairs but never shows their evolutionary relatedness (Queiroz, 1992; Pertsemlidis et al., 2001).

Alignment of Multiple Sequences

Multiple sequence alignment is another crucial procedure, as it creates positional correspondence that helps in concluding the evolutionary trend (Xiong, 2006). Multiple sequence alignments can be accomplished using various available alignment software programs like Clustal X (Thompson et al., 1997), MUSCLE (Edgar, 2004a,b), T-Coffee (Notredame et al., 2000), MAFFT (Katoh et al., 2002), Multalin (Corpet, 1988), etc. After proper alignment of sequences, several other softwares like AQUA (Muller et al., 2010), Gblocks (Sánchez-Gracia and Castresana, 2012) can be used to check the alignment and remove poorly aligned positions.

Selection of the Evolutionary Model

For a given dataset containing properly aligned multiple sequences, the next step lies in the necessity to test the appropriate substitution model. Although, the visible number of substitutions may not always indicate the true evolutionary process that occurs at the locus of concern. In an alignment, identical nucleotides may be the result of parallel mutations and, in such cases, multiple substitutions at a particular point creates ambiguousness for estimating true evolutionary distances between aligned sequences (Patwardhan, 2014). Such a condition is commonly referred to by the term "homoplasy" and could be resolved by selecting correct statistical models. So, proper selection of a substitution model or evolutionary model is crucial in estimating promising differences between sequences based on the available data set (Choudhuri, 2014). There are several programs available for choosing best-fit models of nucleotide substitution, such as the General time-reversible (GTR) model (Tavaré, 1986), Jukes-Cantor model (Jukes and Cantor, 1969), Kimura model (Kimura, 1980), HKY model (Hasegawa et al., 1985) and General Markov model (Barry and Hartigan, 1987).

Building a Phylogenetic Tree

After choosing a proper model, the next step is the construction of the phylogenetic or evolutionary tree. The time duration for making a successful phylogenetic tree generally alters depending upon the number of taxa included in the dataset. A phylogenetic tree can be built depending on the available characters or based on the distance present within the aligned sequences. A tree constructed based on the available character, via methods like maximum likelihood (ML) and maximum parsimony (MP), accounts for mutational events accumulated within the alignment and provides information on the homoplasy and ancestral states (Patwardhan et al., 2014). In contrast, clustering and optimality-based algorithms compute the true evolutionary distance between aligned sequences. Some software programs used for phylogenetic tree building are MEGA (Tamura et al., 2013), RAxML (Stamatakis, 2014), Mr. Bayes (Ronquist and Huelsenbeck, 2003) and PAUP (Swofford, 2002).

Evaluation of Phylogenetic Tree

The authenticity of a phylogenetic tree can be checked by employing statistical tests like Jackknife and bootstrapping, while Bayesian analysis (BA), Kishino-Hasegawa (KH) test and Shimodaira-Hasegawa (SH) tests can be used to confirm whether the created tree is better than any other tree. For a given data set with 'n' number of observations, Jackknife performs analysis for 'n' times, omitting each observation in turn, then the mean and variance of the estimated parameters is compared. In bootstrapping technique, a randomly placed and positioned sequence from a same part of the molecule is randomly sampled 500–1000 times and a new phylogenetic analysis is performed in order to construct the tree. The Bayesian simulation analysis uses Markov chain Monte Carlo (MCMC), that involves thousands of steps of resampling. For maximum parsimony trees, Kishino-Hasegawa test is generally performed for calculating t-value. Whereas Shimodaira-Hasegawa is performed for ML trees and evaluates the fit using $\chi2$ test (Xiong, 2006).

Conclusions

From the overall review, it is evident that DNA barcoding has an essential role to play in inventorying macrofungal diversity, from the identification of species to the formulation of new species hypotheses. The history of DNA barcoding would help to draw a clear picture of all the background information in the reader's mind. The patterns of species distribution could be analyzed based on the available information in the DNA barcode region, followed by comparison of the primary sequence data with the reference sequence databases deposited in the public repositories. Among several candidate barcode regions (CO1, SSU, LSU, RPB1, RPB2, MCM7), in most cases, ITS fits well for discriminating closely related macrofungal taxa. In a few cases, where ITS is less informative, there is a necessity for the development of a secondary barcode locus. In spite of the availability of different public databases and web-based interfaces, the DNA-based macrofungal identification is a reality; still, there is a need for the availability of more properly identified reference sequence data in the public repositories and the development of more softwares for establishing links between different available databases. The steps for constructing phylogenetic trees discussed in this review would undoubtedly help in inferring evolutionary relationships among related or distantly related taxa based upon their closeness or differences in genetic characteristics.

References

Acharya, K., Das, K., Paloi, S., Dutta, A.K., Hembrom, M.E., Khatua, S. and Parihar, A. 2017. Exploring a novel edible mushroom *Ramaria subalpina*: Chemical characterization and Antioxidant activity. Pharmacog. J., 9: 30–34.

Adamowicz, S.J. 2015. Scientific abstracts from the 6th International Barcode of Life Conference/ Résumés scientifiques du 6e congrès international « Barcode of Life ». Genome, 58(5): iii.

Aguileta, G., Marthey, S., Chiapello, H., Lebrun, M.H., Rodolphe, F., Fournier, E., Gendrault-Jacquemard, A. and Giraud, T. 2008. Assessing the performance of single-copy genes for recovering robust phylogenies. Syst. Biol., 57: 613–627.

Asemaninejad, A., Weerasuriya, N., Gloor, G.B., Lindo, Z. and Thorn, R.G. 2016. New Primers for Discovering Fungal Diversity Using Nuclear Large Ribosomal DNA. PLoS ONE, 11: e0159043.

Barry, D. and Hartigan, J.A. 1987. Asynchronous distance between homologous DNA sequences. Biometrics, 43: 261–276.

Bisen, P.S. 2014. Laboratory Protocols in Applied Life Sciences, CRC Press, 1826 pp.

Camacho, C., Coulouris, G., Avagyan, V., Ma, N., Papadopoulos, J., Bealer, K. and Madden, T.L. 2009. BLAST+: Architecture and applications. BMC Bioinformatics, 10: 421.

Chang, S.T. and Miles, P.G. 1992. Mushroom biology—A new discipline. Mycologist, 6: 64–65.

Chang, S.T. and Miles, P.G. 2004. Mushrooms Cultivation, Nutritional Value, Medicinal Effect, and Environmental Impact. CRC Press, United States.

Chatterjee, A. and Acharya, K. 2016. Include mushroom in daily diet—A strategy for better hepatic health. Food Rev. Int., 32: 68–97.

Chatterjee, S., Biswas, G. and Acharya, K. 2011. Antineoplastic effect of mushrooms: A review. Australian Journal of Crop Science, 5: 904–911.

Choudhuri, S. 2014. Bioinformatics for Beginners, Academic Press, 238 pp.

Corpet, F. 1988. Multiple sequence alignment with hierarchical clustering. Nucl. Acids Res., 16: 10881–10890.

Das, S. and Deb, B. 2015. DNA barcoding of fungi using Ribosomal ITS Marker for genetic diversity analysis: A Review. Int. J. Pure Appl. Biosci., 3: 160–167.

Dentinger, B.T.M., Didukh, M.Y. and Moncalvo, J.M. 2011. Comparing COI and ITS as DNA Barcode Markers for Mushrooms and Allies (Agaricomycotina). PLoS ONE, 6: e25081.

Dutta, A.K., Das, K. and Acharya, K. 2015a. A new species of *Marasmius* sect. Globulares from Indian Himalaya with tall basidiomata. Mycosphere, 6: 560–567.

Dutta, A.K., Wilson, A.W., Antonín, V. and Acharya, K. 2015b. Taxonomic and phylogenetic study on gymnopoid fungi from Eastern India. I. Mycol. Progress, 14: 79.

Dutta, A.K., Nandi, S., Tarafder, E., Sikder, R., Roy, A. and Acharya, K. 2017. *Trogia benghalensis* (Marasmiaceae, Basidiomycota), a new species from India. Phytotaxa, 331: 273–280.

Dutta, S.K. and Ojha, M. 1972. Relatedness between major taxonomic groups of fungi based on the measurement of DNA nucleotide sequence homology. Mol. Gen. Genet., 114: 232–240.

Edgar, R.C. 2004a. MUSCLE: Multiple sequence alignment with high accuracy and high throughput. Nucleic Acids Res., 32: 1792–1797.

Edgar, R.C. 2004b. MUSCLE: A multiple sequence alignment method with reduced time and space complexity. BMC Bioinformatics, 5: 113.

Fell, J.W., Boekhout, T., Fonseca, A., Scorzetti, G. and Statzell-Tallman, A. 2000. Biodiversity and systematics of basidiomycetous yeasts as determined by large-subunit rDNA D1/D2 domain sequence analysis. Int. J. Syst. Evol. Microbiol., 50: 1351–1371.

Gardes, M. and Bruns, T.D. 1993. ITS primers with enhanced specifity for Basidiomycetes: Application to identification of mycorrhizae and rusts. Mol. Ecol., 2: 113–118.

Geiser, D.M., Klich, M.A., Frisvad, J.C., Peterson, S.W., Varga, J. and Samson, R.A. 2007. The current status of species recognition and identification in *Aspergillus*. Stud. Mycol., 59: 1–10.

Geml, J., Laursen, G.A., Timling, I., Mcfarland, J.M., Booth, M.G., Lennon, N., Nusbaum, C. and Taylor D.L. 2009. Molecular phylogenetic biodiversity assessment of arctic and boreal ectomycorrhizal *Lactarius* Pers. (Russulales; Basidiomycota) in Alaska, based on soil and sporocarp DNA. Mol. Ecol., 18: 2213–2227.

Gherbawy, Y. and Voigt, K. 2010. Molecular Identification of Fungi. Springer, Berlin, Heidelberg.

Gilmore, S.R., Gräfenhan, T., Louis-Seize, G. and Seifert, K.A. 2009. Multiple copies of cytochrome oxidase 1 in species of the fungal genus *Fusarium*. Mol. Ecol. Res., 9: 90–98.

Gonzalaz, P., Barroso, G. and Labarère, J. 1998. Molecular analysis of the split cox1 gene from the Basidiomycota *Agrocybe aegerita*: relationship of its introns with homologous Ascomycota introns and divergence levels from common ancestral copies. Gene, 220: 45–53.

Haridas, S. and Gantt, J.S. 2010. The mitochondrial genome of the wood-degrading basidiomycete *Trametes cingulata*. FEMS Microbiol. Lett., 308: 29–34.

Hasegawa, M., Kishino, H. and Yano, T. 1985. Dating of human-ape splitting by a molecular clock of mitochondrial DNA. J. Mol. Evol., 22: 160–174.

Hawksworth, D.L. 2012. Global species numbers of fungi: Are tropical studies and molecular approaches contributing to a more robust estimate. Biodivers. Conserv., 21: 2425–2433.

Hebert, P.D.N., Cywinska, A., Ball, S.L. and deWaard, J.R. 2003a. Biological identifications through DNA barcodes. Proc. Biol. Sci., 270: 313–321.

Hebert, P.D.N., Ratnasingham, S. and deWaard, J.R. 2003b. Barcoding animal life: Cytochrome c oxidase subunit 1 divergences among closely related species. Proc. Biol. Sci., 270: S96–S99.

Hebert, P.D.N. and Gregory, T.R. 2005. The promise of DNA barcoding for taxonomy. Systematic Biology, 54(5): 852–859.

James, T.Y., Kauff, F., Schoch, C.L., Matheny, P.B., Hofstetter, V., Cox, C.J., Celio, G., Gueidan, C., Fraker, E., Miadlikowska, J., Lumbsch, H.T., Rauhut, A., Reeb, V., Arnold, A.E., Amtoft, A., Stajich, J.E., Hosaka, K., Sung, G.H., Johnson, D., O'Rourke, B., Crockett, M., Binder, M., Curtis, J.M., Slot, J.C., Wang, Z., Wilson, A.W., Schüssler, A., Longcore, J.E., O'Donnell, K., Mozley-Standridge, S., Porter, D., Letcher, P.M., Powell, M.J., Taylor, J.W., White, M.M., Griffith, G.W., Davies, D.R., Humber, R.A., Morton, J.B., Sugiyama, J., Rossman, A.Y., Rogers, J.D., Pfister, D.H., Hewitt, D., Hansen, K., Hambleton, S., Shoemaker, R.A., Kohlmeyer, J., Volkmann-Kohlmeyer, B., Spotts, R.A., Serdani, M., Crous, P.W., Hughes, K.W., Matsuura, K., Langer, E., Langer, G., Untereiner, W.A., Lücking, R., Büdel, B., Geiser, D.M., Aptroot, A., Diederich, P., Schmitt, I., Schultz, M., Yahr, R., Hibbett, D.S., Lutzoni, F., McLaughlin, D.J., Spatafora, J.W. and Vilgalys, R. 2006. Reconstructing the early evolution of Fungi using a six-gene phylogeny. Nature 443: 818–822.

Jukes, T.H. and Cantor, C.R. 1969. Evolution of protein molecules. pp. 21–132. *In*: Munro, H.N. (eds). Mammalian Protein Metabolism, Academic Press, New York.

Katoh, K., Misawa, K., Kuma, K.I. and Miyat, T. 2002. MAFFT: A novel method for rapid multiple sequence alignment based on fast Fourier transform. Nucleic Acids Res., 30: 3059–3066.

Kerrigan, R.W., Callac, P., Xu, J. and Noble, R. 1999. Population and phylogenetic structure within the *Agaricus subfloccosus* complex. Mycol. Res., 103: 1515–1523.

Khatua, S., Paul, S. and Acharya, K. 2013. Mushroom as the potential source of new generation of antioxidant: A review. Research Journal of Pharmacy and Technology, 6: 496–505.

Khatua, S., Dutta, A.K., Chandra, S., Paloi, S., Das, K. and Acharya, K. 2017a. Introducing a novel mushroom from mycophagy community with emphasis on biomedical potency. PLoS ONE, 12: e0178050.

Khatua, S., Ghosh, S. and Acharya, K. 2017b. *Laetiporus sulphureus* (Bull.: Fr.) Murr. as Food as Medicine. Pharmacognosy Journal, 9: s1–s15.

Kimura, M. 1980. A simple method for estimating evolutionary rates of base substitutions through comparative studies of nucleotide sequences. J. Mol. Evol., 16: 111–120.

Mallick, S., Dutta, A., Dey, S., Ghosh, J., Mukherjee, D., Sultana, S.S., Mandal, S., Paloi, S., Khatua, S., Acharya, K. and Pal, C. 2014. Selective inhibition of *Leishmania donovani* by active extracts of wild mushrooms used by the tribal population of India: An *in vitro* exploration for new leads against parasitic protozoans. Exp. Parasitol., 138: 9–17.

Mallick, S., Dutta, A., Chaudhuri, A., Mukherjee, D., Dey, S., Halder, S., Ghosh, J., Mukherjee, D., Sultana, S.S., Biswas, G., Lai, T.K., Patra, P., Sarkar, I., Chakraborty, S., Saha, B., Acharya, K. and Pal, C. 2016. Successful therapy of murine visceral leishmaniasis with astrakurkurone, a triterpene isolated from the Mushroom *Astraeus hygrometricus*, involves the induction of protective cell-mediated immunity and TLR9. Antimicrob Agents Ch., 60: 2696–2708.

Manoharachary, C., Sridhar, K., Singh, R., Adholeya, A., Suryanarayanan, T.S., Rawat, S. and Johri, B.N. 2005. Fungal Biodiversity: Distribution, Conservation and Prospecting of Fungi from India. Current Science 89(1): 58–71.

McLaughlin, D.J., Hibbett, D.S., Lutzoni, F., Spatafora, J.W. and Vilgalys, R. 2009. The search for the fungal tree of life. Trends Microbiol., 17: 488–497.

Meyer, C.P. and Paulay, G. 2005. DNA barcoding: Error rates based on comprehensive sampling. PLoS Biol., 3: e422.

Moncalvo, J.M., Lutzoni, F.M., Rehner, S.A., Johnson, J. and Vilgalys, R. 2000. Phylogenetic relationships of agaric fungi based on nuclear large-subunit ribosomal DNA sequences. Syst. Biol., 49: 278–305.

Muller, J., Creevey, C.J., Thompson, J.D., Arendt, D. and Bork, P. 2010. AQUA: Automated quality improvement for multiple sequence alignments. Bioinformatics, 26: 263–265.

Nguyen, H.D.T. and Seifert, K.A. 2008. Description and DNA barcoding of three new species of *Leohumicola* from South Africa and the United States. Persoonia, 21: 57–69.

Notredame, C., Higgins, D.G. and Heringa, J. 2000. T-Coffee: A novel method for fast and accurate multiple sequence alignment. J. Mol. Biol., 302: 205–217.

Paloi, S., Dutta, A.K. and Acharya, K. 2015. A new species of *Russula* (Russulales) from Eastern Himalaya, India. Phytotaxa, 234: 255–262.

Pattanayak, M., Samanta, S., Maity, P., Sen, I.K., Nandi, A.K., Manna, D.K., Mitra, P., Acharya, K. and Islam, S.S. 2015. Heteroglycan of an edible mushroom *Termitomyces clypeatus*: Structure elucidation and antioxidant properties, Carbohyd. Res., 413: 30–36.

Patwardhan, A., Ray, S. and Roy, A. 2014. Molecular markers in phylogenetic studies-a review. J. Phylogenet. Evol. Biol., 2: 131.

Pertsemlidis, A. and Fondon, III J.W. 2001. Having a BLAST with bioinformatics (and avoiding BLASTphemy). Genome Biol., 2: 2002–2110.

Pimm, S.L., Jenkins, C.N., Abell, R., Brooks, T.M., Gittleman, J.L., Joppa, L.N., Raven, P.H., Roberts, C.M. and Sexton, J.O. 2014. The biodiversity of species and their rates of extinction, distribution, and protection. Science, 344(6187): 1246752.

Queiroz, K. 1992. Phylogenetic definitions and taxonomic philosophy. Biol. Philos., 7(3): 295–313.

Raja, H.A., Schoch, C.L., Hustad, V.P., Shearer, C.A. and Miller, A.N. 2011. Testing the phylogenetic utility of MCM7 in the Ascomycota. MycoKeys, 1: 63–94.

Redhead, S. 1997. Macrofungi of British Columbia; requirements for inventory. Res. Br., B.C. Min. For., and Wildl. Br., B.C. Min. Environ., Lands and Parks, Victoria, B.C. Working Paper, 28.

Robert, V., Szöke, S., Eberhardt, U., Cardinali, G., Meyer, W., Seifert, K.A., Lévesque, C.A. and Lewis, C.T. 2011. The quest for a general and reliable fungal DNA barcode. Open Appl. Inform. J., 5: 45–61.

Ronquist, F. and Huelsenbeck, J.P. 2003. MRBAYES 3: Bayesian phylogenetic inference under mixed models. Bioinformatics, 19: 1572–1574.

Sánchez-Gracia, A. and Castresana, J. 2012. Impact of deep coalescence on the reliability of species tree inference from different types of DNA markers in mammals. PLoS ONE, 7: e30239.

Schindel, D.E. and Miller, S.E. 2005. DNA barcoding a useful tool for taxonomists. Nature, 435: 17.

Schmitt, I., Crespo, A., Divakar, P.K., Fankhauser, J.D., Herman-Sackett, E., Kalb, K., Lumbsch, H.T. 2009. New primers for promising single-copy genes in fungal phylogenetics and systematics. Persoonia, 23: 35–40.

Schoch, C.L., Sung, G.H., López-Giráldez, F., Townsend, J.P., Miadlikowska, J., Hofstetter, V., Robbertse, B., Matheny, P.B., Kauff, F., Wang, Z., Gueidan, C., Andrie, R.M., Trippe, K., Ciufetti, L.M., Wynns, A., Fraker, E., Hodkinson, B.P., Bonito, G., Groenewald, J.Z., Arzanlou, M., de Hoog, G.S., Crous, P.W., Hewitt, D., Pfister, D.H., Peterson, K., Gryzenhout, M., Wingfield, M.J., Aptroot, A., Suh, S.O., Blackwell, M., Hillis, D.M., Griffith, G.W., Castlebury, L.A., Rossman, A.Y., Lumbsch, H.T., Lücking, R., Büdel, B., Rauhut, A., Diederich, P., Ertz, D., Geiser, D.M., Hosaka, K., Inderbitzin, P., Kohlmeyer, J., Volkmann-Kohlmeyer, B., Mostert, L., O'Donnell, K., Sipman, H., Rogers, J.D., Shoemaker, R.A., Sugiyama, J., Summerbell, R.C., Untereiner, W., Johnston, P.R., Stenroos, S., Zuccaro, A., Dyer, P.S., Crittenden, P.D., Cole, M.S., Hansen, K., Trappe, J.M., Yahr, R., Lutzoni, F. and Spatafora, J.W. 2009. The Ascomycota tree of life: A phylum-wide phylogeny clarifies the origin and evolution of fundamental reproductive and ecological traits. Syst. Biol., 58: 224–239.

Schoch, C.L., Seifert, K.A., Huhndorf, S., Robert, V., Spouge, J.L., Levesque, C.A., Chen, W. and Fungal Barcoding Consortium. 2012. Nuclear ribosomal internal transcribed spacer (ITS) region as a universal DNA barcode marker for Fungi. Proceedings of the National Academy of Sciences of the United States of America, 109(16): 6241–6246.

Scorzetti, G., Fell, J.W., Fonseca, A. and Statzell-Tallman, A. 2002. Systematics of basidiomycetous yeasts: A comparison of large subunit D1/D2 and internal transcribed spacer rDNA regions. FEMS Yeast Res., 2: 495–517.

Seifert, K.A., Samson, R.A., Dewaard, J.R., Houbraken, J., Lévesque, C.A., Moncalvo, J.M., Louis-Seize, G. and Hebert, P.D. 2007. Prospects for fungus identification using CO1 DNA barcodes, with *Penicillium* as a test case. Proc. Natl. Acad. Sci., 104: 3901–3906.

Stackebrandt, E. and Goebel, B.M. 1994. Taxonomic note: A place for DNA-DNA reassociation and 16S rRNA sequence analysis in the present species definition in bacteriology. Int. J. Syst. Evol. Microbiol., 44: 846–849.

Stamatakis, A. 2014. RAxML Version 8: A tool for Phylogenetic Analysis and Post-Analysis of Large Phylogenies. Bioinformatics, 30: 1312–1313.

Swofford, D.L. 2002. PAUP*. Phylogenetic Analysis Using Parsimony (*and Other Methods). Version # 4. Sinauer Associates, Sunderland, Massachusetts.

Tamura, K., Stecher, G., Peterson, D., Filipski, A. and Kumar, S. 2013. MEGA6: Molecular Evolutionary Genetics Analysis version 6.0. Mol. Biol. Evol., 30: 2725–2729.

Tanabe, Y. and Watanabe, S. 2002. Are Microsporidia really related to Fungi? A reappraisal based on additional gene sequences from basal fungi. Mycol. Res., 106: 1380–1391.

Tavaré, S. 1986. Some probabilistic and statistical problems in the analysis of DNA sequences. Lectures on Mathematics in the Life Sciences (American Mathematical Society), 17: 57–86.

Thompson, J.D., Gibson, T.J., Plewniak, F., Jeanmougin, F. and Higgins, D.G. 1997. The CLUSTAL_X Windows interface: Flexible strategies for multiple sequence alignment aided by quality analysis tools. Nucleic Acids Res., 25: 4876–4882.

Toju, H., Tanabe, A.S., Yamamoto, S. and Sato, H. 2012. High-Coverage ITS Primers for the DNA-Based Identification of Ascomycetes and Basidiomycetes in Environmental Samples. PLoS ONE, 7: e40863.

Vialle, A., Feau, N., Allaire, M., Didukh, M., Martin, F., Moncalvo, J.M. and Hamelin, R.C. 2009. Evaluation of mitochondrial genes as DNA barcode for Basidiomycota. Mol. Ecol. Res., 9: 99–113.

Vilgalys, R. and Hester, M. 1990. Rapid genetic identification and mapping of enzymatically amplified ribosomal DNA from several *Cryptococcus* species. J. Bacteriol., 172: 4238–4246.

Wägele, H., Klussmann-Kolb, A., Kuhlmann, M., Haszprunar, G., Lindberg, D., Koch, A. and Wägele, J.W. 2011. The taxonomist - an endangered race. A practical proposal for its survival. Frontiers in Zoology 8: 25.

Walker, W.F. and Doolittle, W.F. 1982. Redividing the basidiomycetes on the basis of 5S rRNA sequences. Nature, 299: 723–724.

Wang, H., Wang, Y., Chen, J., Zhan, Z., Li, Y. and Xu, J. 2007. Oral yeast flora and its ITS sequence diversity among a large cohort of medical students in Hainan, China. Mycopathologia, 164: 65–72.

White, T.J., Bruns, T.D., Lee, S. and Taylor, J. 1990. Amplification and direct sequencing of fungal ribosomal RNA genes for phylogenetics. pp. 315–322. *In*: Innis, M.A., Gelfand, D.H., Sninsky, J.J. and White, T.J. (eds.). PCR Protocols, a Guide to Methods and Applications. Academic Press, San Diego, California.

Xiong, J. 2006. Essential Bioinformatics. Cambridge University Press. pp. 127–168.

Xu, J. 2016. Fungal DNA barcoding. Genome, 59: 913–932.

Xu, J., Vilgalys, R. and Mitchell, T.G. 2000. Multiple gene genealogies reveal recent dispersion and hybridization in the human pathogenic fungus *Cryptococcus neoformans*. Mol. Ecol., 9: 1471–1481.

Yan, Z. and Xu, J. 2005. Fungal mitochondrial inheritance and evolution. pp. 221–252. *In*: Xu, J. (eds.). Evolutionary Genetics of Fungi. Horizon Scientific Press, England.

15

Mushroom as a Greener Agent to Biosynthesize Nanoparticles

Nor Athirah Kamaliah Ahmad Tarmizi,[1,2]
Chia-Wei Phan,[1,3,*] *Yee-Shin Tan*[1,4] *and*
Vikineswary Sabaratnam[1,4,*]

INTRODUCTION

Nanotechnology is a combination of science, engineering, and technology which deals with construction of functional structures on an atomic and molecular scale (Arun et al., 2014; Sudhakar et al., 2014). One of the earliest mentions of nanotechnology was by Professor Norio Taniguchi from Tokyo Science University in 1974, in which he described nanotechnology as "materials manufactured precisely with nanometers tolerances" which was later used in a book entitled "Engines of Creation: The Coming Era of Nanotech" by Drexler in 1986 (Shinde et al., 2012; Parveen et al., 2016). The U.S National Nanotechnology Initiative defined nanotechnology as "any structure with novel properties that are smaller than 100 nanometers". Nanotechnology's main focus is the synthesis of nanoparticles (NPs) of different sizes, shapes, chemical compositions and functions, which have potential benefits to human beings (Narasimha et al., 2011).

Nanoparticles possess unique and better properties compared to those of bulk materials that have the same chemical composition (Hussain et al., 2016; Iravani,

[1] Mushroom Research Centre, University of Malaya, 50603 Kuala Lumpur.
[2] Department of Anatomy, Faculty of Medicine, University of Malaya, 50603 Kuala Lumpur.
    Email: athirah.tarmizi13@gmail.com
[3] Department of Pharmacy, Faculty of Medicine, University of Malaya, 50603 Kuala Lumpur.
[4] Institute of Biological Sciences, Faculty of Science, University of Malaya, 50603 Kuala Lumpur.
    Email: tanyeeshin@um.edu.my
* Corresponding authors: phancw@um.edu.my; viki@um.edu.my

2011; Parveen et al., 2016). Their properties may be due to their small size and large surface area, which allows them to have a higher reactivity and catalytic activity (Anthony et al., 2014) as compared to other bulk materials (Hussain et al., 2016; Iravani, 2011). NPs' size and shape depend on several factors, such as temperature (Phillip, 2009; Sriramulu and Sumathi, 2017), pH, salt concentration and exposure time (Hussain et al., 2016). They have been proven to exhibit numerous potential applications in different areas, such as drug and gene delivery, bio-therapeutics, electronics, pharmaceuticals, fertilizers (Sriramulu and Sumathi, 2017), textiles, chemical industry (Sudhakar et al., 2014), and other biomedical and engineering fields (Anthony et al., 2014).

Nanoparticles have made their way into the medical field mainly due to the advantages they possess, such as biocompatibility, cost effectiveness, high efficacy and safety (Bhat et al., 2011; Mirunalini et al., 2012; Anthony et al., 2014). They are also of a great interest particularly for researchers in the medical field for their good antibacterial properties due to the recent increase in microbial resistance towards antibiotics (Mirunalini et al., 2012). Owing to their small size and large surface area to volume ratio, they possess a greater toxicity towards pathogens (Gurunathan et al., 2013; Anthony et al., 2014).

Characterization of Nanoparticles

According to Shin et al. (2016), NPs are made of three layers: First, the surface layer, which is composed of small molecules, metal ions, surfactants, and polymers. Second, the shell layer, made up of materials that are chemically different from the core structure, and third, the core which acts as the NP's central portion.

NPs can be also sorted into different classes based on their physical and chemical characteristics. Well-known NPs, such as silver nanoparticles (AgNPs) and gold nanoparticles (AuNPs), belong to the metal NPs group that are purely made of the metal precursors. They possess unique properties owing to their surface plasmon resonance (SPR) characteristics. Other classes include carbon-based NPs (fullerenes and carbon nanotubes), ceramic NPs, semiconductor NPs, polymeric NPs and liquid-based NPs (Shin et al., 2016; Khan et al., 2017).

The AgNP is the most common nanoparticle used in the field of nanotechnology due to its unique properties and biological functions (Al-Bahrani et al., 2017). The AgNP is also effective as an antimicrobial agent against pathogenic bacteria, viruses and other eukaryotic microorganisms (Mirunalini et al., 2012; Sudhakar et al., 2014). AuNP is also another common NPs used in the biomedical field and in the emerging nanobiotechnology field (Iravani, 2011). It has been used in immunoassay, protein assay, cancer nanotechnology and as a biological marker for screening diseases (Iravani, 2011).

Synthesized NPs can be characterized by using techniques that are commonly employed, which includes but is not limited to, UV-visible absorption spectroscopy (UV-vis), scanning electron microscopy (SEM), transmission electron microscopy (TEM), X-ray diffraction (XRD) and Fourier-transform infrared spectroscopy (FTIR). UV-vis can be used to determine the size and shape of NPs in aqueous suspension, whereas SEM and TEM are the common techniques that are used to characterize the size and morphology of NPs. As for metallic NPs, XRD is the technique used for size and phase identification of NPs. Identification of a functional group that is involved in the reduction and stabilization of NPs is usually determined by FTIR spectroscopy (Hussain et al., 2016).

Synthesis of Nanoparticles

There are three different ways in which NPs can be synthesized, namely the physical, chemical and biological methods (Sriramulu and Sumathi, 2017), each having their own pros and cons. NPs synthesized via the physical method usually result in low yields (Anthony et al., 2014), whilst chemical methods use various chemical reagents that can be toxic and harmful to the environment and human beings (Anthony et al., 2014; Nagajyothi et al., 2014). Both methods are considered expensive (Bhat et al., 2011). The physical method requires high temperature and high pressure, a large area for machinery, and it also uses costly equipment. Some of the techniques employed in this method include colloidal dispersion method, vapor condensation, amorphous crystallization and physical fragmentation (Agarwal et al., 2017).

As for the chemical method, some of the techniques used for NPs synthesis include the use of chemical micro emulsion, wet chemical, chemical and direct precipitation, spray pyrolysis electrode position and microwave assisted combustion (Agarwal et al., 2017). Another disadvantage for both methods is the need to add capping and stabilizing agents for NPs synthesis (Agarwal et al., 2017). Overall, there is an increasing need to develop a clean, non-toxic, easy and eco-friendly synthesis of NPs for industry purposes (Narasimha et al., 2011).

Biological methods, also known as the "green synthesis" of NPs, come into play as they offer more advantages over the other two methods in terms of economic impact, efficiency, biosafety, eco-friendliness and ease of process (Bhat et al., 2011; Narasimha et al., 2011; Anthony et al., 2014; Maurya et al., 2016). Biological synthesis of NPs is also more compatible for pharmaceuticals and other biomedical applications (Narasimha et al., 2011). According to Anthony et al. (2014), applications of NPs in the biomedical field increased when biological synthesis of NPs was discovered.

Biosynthesis of NPs often requires natural resources, such as plants, bacteria, fungi, algae, yeast and even viruses (Anthony et al., 2014; Parveen et al., 2016; Agarwal et al., 2017). One of the advantages of using biological method for NPs synthesis is that addition of both reducing and stabilizing agents are not needed because the natural resources used in this approach already contain some phytochemicals that can function as both reducing/ stabilizing agents. However, additional reducing and stabilizing agent need to be added for NPs synthesis via physical and chemical method (Hussain et al., 2016; Agarwal et al., 2017). It is also easier to control the size, shape and morphology of synthesized NPs using the biological method (Gurunathan et al., 2013; Owaid et al., 2017). Bio-coating the surface of NPs with bio-constituents allows them to be more compatible in comparison to NPs synthesized by other methods (Sharma et al., 2015). Other than that, biologically synthesizing NPs helps to produce homogenous NPs with better stability and sizes (Owaid et al., 2017).

Mushrooms for the Green Synthesis of Nanoparticles

Out of all the natural resources available for green synthesis, researchers' main focus is on the utilization of fungi for NPs synthesis. Fungi are one of the most highly diverse and versatile organisms that can be found in all kinds of environment, including but not limited to soil, compost, dead organic substances, and they are also able to grow on plants and animals (Arun et al., 2014). Fungi also has the ability to produce enzymes in a large amount which plays a role as a reducing agent in the biosynthesis of NPs (Gudikandula and Maringanti, 2016; Owaid et al., 2017). Fungal mycelium also confers the ability to better withstand extreme conditions of the bioreactors or other chambers used in the synthesizing process, as compared to other organisms (Gudikandula and Maringanti, 2016).

Mushrooms are fungus that are well known not just in the culinary world, but also in the medicinal field. Aside from the rich flavor that they have, mushrooms also possess a variety of medicinal properties that has led to an increase in mushroom consumption in recent years (Mirunalini et al., 2012; Sujatha et al., 2015; Phan et al., 2017b). Some of the widely-known medicinal properties of mushrooms include anti-inflammatory, anti-microbial, antioxidant, anti-cancer and neuroprotective ability (Phillip, 2009; Phan et al., 2017a). Proteins from mushrooms are the bio-constituents responsible for reducing and capping agents in the biosynthesis of NPs (Nagajyothi et al., 2014).

Due to the medicinal properties, as well as the large amounts of enzymes mushrooms produce, they have been largely employed in the biosynthesis of NPs. Some of the mushroom species that have been utilized in the synthesis of NPs include *Volvariella volvacea* (Bull.) Singer, *Pleurotus* spp., *Ganoderma* spp., *Agaricus bisporus* (J.E. Lange) Imbach, *Schizophyllum commune* Fr., *Lentinula edodes* (Berk.) Pegler, *Inonotus obliquus* (Ach. ex Pers.) Pilát, *Tricholoma matsutake* (S. Ito & S. Imai) Singer, and *Hericium erinaceus* (Bull.) Pers. In this chapter, we describe and focus on the green synthesis of NPs from extracts of these mushroom species (Table 1).

Agaricus bisporus

Agaricus bisporus or the white button mushroom was used by Narasimha et al., (2011) in order to synthesize AgNPs. The aqueous silver nitrate solution turned brown when mixed with the mushroom extract, indicating the deposition of AgNPs. The AgNPs formation were further confirmed by UV-vis analysis which showed absorption peak at 420 nm. Spherical AgNPs with a few agglomerated particles (8 to 50 nm) were observed from TEM micrographs. Proteins, carbonyl groups, esters and carboxylic acids were the bio-constituents that aided in synthesizing AgNPs as well as functioning as a stabilizing agent for bioreduced AgNPs. The white button mushroom-mediated AgNPs exhibited inhibitory activity against both pathogenic and non-pathogenic bacteria, namely *Bacillus* spp., *Escherichia coli*, *Pseudomonas* spp., and *Staphylococcus* spp.

A different set of studies was carried out by Sudhakar et al. (2014) for AgNPs synthesis using *A. bisporus* as the bioreductant. The synthesized AgNPs had the potential to act as an antibacterial agent against *E. coli*, *Klebsiella* sp., *Proteus vulgaris*

Table 1. Summary of myco-synthesized nanoparticles from mushroom extracts.

Mushrooms	Nanoparticles	UV-vis	Shape	Size	Activity	Reference
Agaricus bisporus (J.E. Lange) Imbach	AgNP	420 nm	Spherical	8–50 nm	Anti-bacterial	Narasimha et al., 2011
	AgNP	420 nm	Spherical	15–20 nm	Anti-bacterial	Sudhakar et al., 2014
	AgNP	430 nm	Spherical	4–35 nm	Anti-bacterial	Al-Hamadani and Kareem, 2017
	AgNP	420 nm	–	20–44 nm	Anti-bacterial	Haq et al., 2015
	AgNP	414 nm	Sponge-like	–	Anti-bacterial; Anti-inflammatory; Anti-oxidant; Photocatalytic	Sriramulu and Sumathi, 2017
Ganoderma lucidum (M.A. Curtis: Fr.) P. Karst	AgNP	FRT: 434 nm; FHT: 422 nm	FRT: Rod-shape; FHT: Needle-like	–	Anti-bacterial; Anti-inflammatory; Anti-oxidant; Photocatalytic	Sriramulu and Sumathi, 2017
Ganoderma spp.	AuNP	520 nm	Spherical	20 nm	Bio-compatible; Non-cytotoxic	Gurunathan et al., 2014
Ganoderma applanatum (Pers.) Pat.	AgNP	421 nm	Spherical	133.0 +/- 0.361 nm	Anti-bacterial	Mohanta et al., 2016
Ganoderma neo-japonicum Imazeki	AgNP	420 nm	-	5 nm	Cytotoxic against MDA-MB-231 breast cancer cells	Gurunathan et al., 2013
Hericium erinaceus (Bull.) Pers.	AuNP	520–560 nm	Triangular	20–30 nm	Neurite outgrowth stimulatory effects	Raman et al., 2015a
Inonotus obliquus (Ach. ex Pers.) Pilát	AgNP	427 nm	Triangular, hexagonal	14.7–35.2 nm	Anti-bacterial; Anti-proliferative	Nagajyothi et al., 2014
Lentinula edodes (Berk.) Pegler =	AgNP	430 nm	Walnut-shaped	50–100 nm	Anti-bacterial	Lateef and Adeeyo, 2015
Pleurotus citrinopileatus Singer	AgNP	420–450 nm	Round	6–10 nm	Anti-bacterial	Maurya et al., 2016

Table 1 contd. ...

...Table 1 contd.

Mushrooms	Nanoparticles	UV-vis	Shape	Size	Activity	Reference
Pleurotus cornucopiae var. *citrinopileatus* (Singer) Ohira	AgNP	400–500 nm	Spherical	20–30 nm	Anti-microbial	Owaid et al., 2015
	AuNP	540 nm	Spherical	Fresh: 23–100 nm Dried: 16–91 nm	–	Owaid, Al-Saeedi and Abed, 2017
Pleurotus djamor var. *roseus* Corner	AgNP	440 nm	Spherical	5–50 nm	Anti-proliferative	Raman et al., 2015b
Pleurotus florida (Jacq.) P. Kumm.	AgNP	435 nm	Spherical	20 +/–5 nm	Anti-bacterial	Bhat et al., 2011
	AgNP	425 nm	Spherical	2.445 +/–1.08 nm	Anti-bacterial	Sen et al., 2013
Pleurotus ostreatus (Jacq.) P. Kumm.	AgNP	440 nm	Spherical	< 40 nm	Anti-bacterial	Al-Bahrani et al., 2017
Pleurotus platypus Sacc.	AgNP	435 nm	Spherical	0.56–0.71 um	Anti-bacterial	Sujatha et al., 2013
Pleurotus sapidus Quél.	AuNP	540 nm	Spherical, triangular, hexagonal	15–100 nm	–	Sarkar et al., 2013
Ganoderma sp.	CDS-NP	255.5 nm	Spherical	10 nm	Anti-bacterial; Anti-fungal	Raziya et al., 2016
Schizophyllum commune Fr.	AgNP	440 nm	Spherical	Intracellular: 54–99 nm, Extracellular: 51–93 nm	Anti-bacterial; Anti-fungal; Anti-proliferative	Arun et al.,2014
Tricholoma matsutake (S. Ito & S. Imai) Singer	AgNP	430 nm	Spherical	10 +/–5 nm	Anti-bacterial	Anthony et al., 2014
Volvariella volvacea (Bull.) Singer	AuNP	–	Triangular; Nanoparism; Spherical	20–150 nm		Phillip, 2009
	AgNP	–	Spherical	15 nm		Phillip, 2009

and extended spectrum beta-lactamases (ESBL)-producing bacteria. Furthermore, Haq et al. (2015) also showed that the antibacterial activity of AgNPs obtained from *A. bisporus* against methicillin-resistant *Staphylococcus aureus* (MRSA) increased when combined with an antibiotic, such as gentamycin. Most recently, Al-Hamadani and Kareem (2017) also demonstrated that combination of *A. bisporis*-AgNPs with antibiotics exerted maximum antibacterial activity against *E. coli*, MRSA, *Proteus mirabilis*, and *Pseudomonas aeruginosa*.

Ganoderma lucidium

Gurunathan et al. (2014) were able to synthesize AuNPs by using *Ganoderma* mycelial extract. On combining $HAuCl_4$ with the extract for 24 hours, AuNPs were visually observed, based on a colour change from pale yellow to deep purple. Maximum absorbance at 520 nm was recorded via UV-vis analysis which was concluded to be a characteristic of AuNPs, and the crystallinity of AuNPs was determined by XRD analysis. Amide I, II, and III of polypeptides/proteins were the biological constituents functioning as capping and stabilizing agents for the bioreduced AuNPs. The AuNPs, with an average size of 20nm and spherical shape, were recorded via dynamic light scattering (DLS) and TEM techniques.

Raziya et al. (2016) tested the usefulness of *Ganoderma* mushroom extract in synthesizing a different type of nanoparticle, cadmium sulfide-nanoparticles (CDS-NP). It was observed that *Ganoderma* mushroom extract was able to synthesize CDS-NPs, as observed by the colour change from pale yellow to dark yellow. Confirmation was also obtained from the UV-vis absorption peak at 255.5 nm. SEM analysis showed the crystalline structure of CDS-NPs, which were spherical in shape with a size of 10 nm. The synthesized CDS-NPs also showed antibacterial activity against *S. aureus* and anti-fungal activity against *Aspergillus niger.*

Ganoderma applanatum

A wild *Ganoderma* species, *Ganoderma applanatum* (Pers.) Pat., was used to synthesize AgNPs. The resulting AgNPs were tested against several strains of pathogenic bacteria (Mohanta et al., 2016). Within 24 hours of incubation, appearance of brown coloured solution was recorded suggesting the formation of AgNPs. A single absorption band was recorded at 421 nm via UV-vis analysis which was within the range of AgNPs'SPR band (420 to 430 nm). Hydroxyl and carbonyl groups were identified as the reducing and stabilizing agents, as determined by FTIR. From DLS analysis, it was noted that the AgNPs size were 133.0 ± 0.361 nm with charge of -6.01 ± 5.30 mV. The AgNPs were spherical in shape, as observed using SEM. The synthesized AgNPs also showed potential antimicrobial properties against *Bacillus subtilis*, *E. coli*, *Shigella flexneri*, *Staphylococcus epidermidis* and *Vibrio cholerae*.

Ganoderma neo-japonicum

Study done by Gurunathan et al. (2013) focused on synthesizing AgNPs using the purple reishi mushroom, *G. neo-japonicum* Imazeki, mycelial extract in order to assess its potential as a cytotoxic agent against breast cancer cells. After 24 hours,

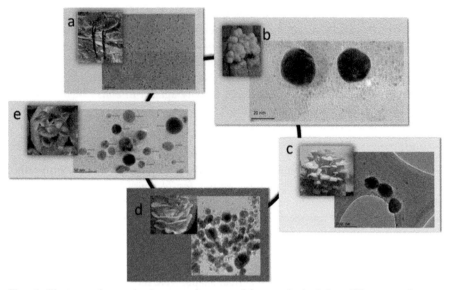

Fig. 1. Electron microscopic images of nanoparticles synthesized by different mushrooms. (a) Representative TEM image of AgNPs produced by *G. neo-japonicum* mycelial extract (Gurunathan et al., 2013), (b) TEM image of AuNPs formed by the reaction of gold (III) chloride with *H. erinaceus* (Raman et al., 2015a), (c) high resolution TEMimage of AgNPs with clear lattice fringes, synthesized by *P. cornucopiae* var. *citrinopileatus* (Owaid et al., 2015), (d) TEM image of AgNPs formed by bioreduction of silver nitrate by *P. djamor* (Raman et al., 2015b), and (e) high resolution TEM image of AgNPs with clear lattice fringes, synthesized by *P. ostreatus* (Al-Bahrani et al., 2017).

the mixture of silver ions and hot aqueous extract of *G. neo-japonicum* mycelial extract changed colour, from pale-yellow to brownish colour. A strong resonance was measured at 420 nm, based on UV-vis spectra analysis, verifying the formation of AgNPs. Dynamic light scattering showed an average AgNPs size of 5nm, which was also observed via TEM (Fig. 1a). Crystalline structure of AgNPs was confirmed by XRD analysis. The synthesized AgNPs also had the potential to become cytotoxic agents against MDA-MB-231 breast cancer cells.

Hericium erinaceus

Another well-known mushroom species especially in the medical field, *Hericium erinaceus* (monkey head/lion's mane mushroom), was also subjected to mycosynthesis of AuNPs (Raman et al., 2015a). Formation of AuNPs were first observed by colour change of the reaction mixture from light yellow to vivid purple. UV-vis analysis also confirmed AuNPs formation as the absorption peak obtained fell between 520 to 560 nm after a 36-hour incubation. The AuNPs formed were mainly spherical, as confirmed by Field emission scanning electron microscopy (FESEM) analysis, with some forming nano-triangular shaped particles, their sizes ranging from 20 to 30 nm as analyzed by TEM (Fig. 1b). The AuNPs crystallinity and purity were confirmed by Energy dispersive X-ray (EDX) analysis which recorded a strong signal for gold atoms. The synthesized AuNPs also exhibited an interesting property that was valuable to the medical field. It was demonstrated that the synthesized AuNPs conferred neurite outgrowth stimulatory effects on rat pheochromocytoma (PC12) cells with a maximum of 13% neurite outgrowth, recorded using 600 ng/ml concentration of the mushroom-mediated AuNPs.

Inonotus obliquus

Another study was done, using *I. obliquus* or Chaga mushroom extract to synthesize AgNPs (Nagajyothi et al., 2014). Within 80 minutes, the reaction mixture changed colour from orange to dark orange due to the formation of AgNPs in the solution. Presence of AgNPs were also confirmed by UV-vis spectra analysis which exhibited absorption peak at 427 nm. The synthesized AgNPs were also very stable for a duration of more than eight months without any signs of NPs aggregation. FTIR analysis was carried out and it was identified that the possible reducing and capping bio-agents were amines, amide I group, and alkynes. The morphology of AgNPs was concluded to be predominantly spherical in shape, as observed using SEM, TEM, and atomic force microscopy (AFM). However, it was also noted that AgNPs of triangular, hexagonal and uneven shape were also discovered. The sizes of AgNPs were measured using SEM, and ranged from 14.7 to 35.2 nm. The AgNPs synthesized in this study demonstrated antibacterial activity against Gram-negative bacteria (*E. coli* and *Proteus mirabilis*), and Gram-positive bacteria, *S. epidermidis*. Finally, this study also showed the anti-proliferative activity of AgNPs against certain cell lines, such as A549 human lung cancer (CCL-185) and MCF-7 human breast cancer cell lines (HTB-22).

Lentinula edodes

The first study of its kind was done by Lateef and Adeeyo (2015), aiming to synthesize AgNPs using extracellular laccase of Shiitake mushroom, *L. edodes*. Early AgNPs formation was noted as the colourless solution turned yellowish brown colour and the intensity increased as the reaction came to completion. UV-vis analysis showed an SPR band reflected at 430 nm, and that the AgNPs were walnut-shaped, with sizes ranging between 50–100 nm. The laccase-mediated mycosynthesis of AgNPs also possessed antibacterial activity against *E. coli*, *K. pneumoniae*, and *P. aeruginosa*.

Pleurotus spp.

Pleurotus citrinopileatus

Maurya et al. (2016) used the "golden" oyster mushroom *P. citrinopileatus* Singer extract as a reducing agent in order to synthesize AgNPs. Mixture of the mushroom extract with aqueous silver nitrate changed colour from colourless to dark brownish, suggesting the formation of AgNPs. Absorption peaks ranging from 420–450 nm were observed through UV-vis analysis which further confirmed the formation of AgNPs. The AgNPs formed were spherical in shape as visualized by XRD and the size range were between 6 to 10 nm, as obtained from TEM analysis. Ethylene groups were the bio-constituent responsible for the capping and stabilization of AgNPs formation, as identified by FTIR. The synthesized AgNPs inhibited the growth of pathogenic bacteria such as *E. coli* and *S. aureus*.

Pleurotus cornucopiae

Pleurotus cornucopiae var. *citrinopileatus* (Singer) Ohiraor, the yellow oyster mushroom, was also used to myco-synthesize AgNPs, subsequently tested against *Candida* species (Owaid et al., 2015). Reaction mixture containing hot water extract of the fresh basidiocarps of *P. cornucopiae* and silver nitrate solution showed a colour change from pale yellow to dark brownish yellow, indicating the formation of AgNPs. UV-vis analysis showed formation of absorption peak between 400–500 nm. FTIR analysis suggested the potential capping and stabilizing agents to be from a functional group of alkenes, carboxylic acids, and heterocyclic compounds such as alkaloids. The synthesized AgNPs were spherical, with an average size between 20 to 30 nm, as analyzed by FESEM and High resolution transmission electron (HRTEM) (Fig. 1c). The crystalline AgNPs were also confirmed by EDX analysis. Finally, the synthesized AgNPs showed moderate inhibitory activity against *Candida* species.

Owaid et al. (2017) also used fresh and dried basidiocarps of *P. cornucopiae* to synthesize AuNPs. Visual observations recorded change in colour from colourless to purple at 24 hours with maximum colour intensity occurred after 48 hours of incubation for both fresh and dried mushroom extract. Two different absorbance peaks were obtained for each sample: 550 nm for fresh mushroom extract, and 540 nm for dried mushroom extract. The AuNPs from both samples were spherical in shape without aggregation, but of different sizes, i.e., 23 to 100 nm, and 16 to 91 nm, for fresh and dried samples, respectively. FTIR analysis also discovered different bio-constituents involved in capping and stabilization of the AuNPs from both samples. For the fresh sample, the bio-constituents were phenol, nitrile, amide and alkyne. On the other hand, for the dried sample, the bio-constituents were nitrile, amide and alkyne. EDX analysis for AuNPs from both samples confirmed both AuNPs' crystallinity.

Pleurotus djamor var. *roseus*

AgNPs synthesized from *P. djamor* var. *roseus* Corner extract were tested for cytotoxic effect on human epithelial teratocarcinoma (P3) cells (Raman et al., 2015b). A brownish yellow coloured reaction mixture was obtained after 12 hours of incubation, suggesting the formation of AgNPs with increased colour intensity at 24 hours. The stability of AgNPs formed was maintained for up to two months. The AgNPs formation was further confirmed by UV-vis analysis which showed formation of SPR bands at 440 nm. The AgNPs' spherical shaped entities were confirmed by TEM analysis, with sizes ranging from 5 to 50 nm (Fig. 1d). Results from XRD analysis confirmed the synthesized AgNPs' crystallinity and purity. The biosynthesized AgNPs also exhibited anti-proliferative activity against PC3 cells that was dose-dependent.

Pleurotus florida (also known as *P. floridanus*)

The edible white oyster mushroom, *P. florida*, was used as bioreductant to synthesize AgNPs via photo-irradiation technique carried out by Bhat et al. (2011). The reaction mixture changed its colour from colourless to reddish brown, claiming the formation of AgNPs which was further confirmed by the absorption peak at 435 nm obtained from UV-vis spectra. Morphology of AgNPs were analyzed using an atomic force microscope (AFM), showing nearly spherical including some irregular shaped AgNPs. The AgNPs size was shown to be in the range of 20 ± 5 nm using TEM analysis, and amide I and amide II groups were identified as the capping and stabilizing agents using FTIR. The AgNPs also conferred antimicrobial activity against *Proteus mirabilis*, *Providencia alcalifaciens*, *S. aureus* and *Salmonella typhi*. It was also noted that the synthesized AgNPs were able to function as anti-microbial, either by mixing with an antibiotic drug or acting as the drug itself.

Sen et al., (2013) used glucan isolated from *P. florida* blue variant to synthesize AgNPs and study the antibacterial activity. Visual observation of AgNPs formation showed the appearance of faint yellow colour after a 2-hour incubation of silver nitrate solution with the glucan solution. The reaction was completed after 10 hours incubation. The maximum absorbance obtained from UV-vis analysis was at 425 nm corresponding to the SPR of AgNPs. TEM analysis showed the average diameter of AgNPs formed to be 2.445 ± 1.08 nm and they are almost spherical in shape. The synthesized AgNPs-glucan conjugates also demonstrated antibacterial activity against multiple antibiotic resistant bacterium, *K. pneumoniae* YSI6A.

Pleurotus ostreatus

A study was carried out by Al-Bahrani et al. (2017) to biosynthesize AgNPs using *P. ostreatus*, which is widely known as tree oyster mushroom or grey oyster mushroom. The AgNPs were first identified by observing the colour change of $AgNO_3$ solution from pale yellow to dark brown, indicating the formation of AgNPs. UV-vis further confirmed the formation of AgNPs, as maximum absorption at 440 nm was obtained and this corresponded to the standard SPR of AgNPs. The synthesized AgNPs conferred spherical shape with size less than 40 nm, as analyzed by FESEM and HRTEM (Fig. 1e). FTIR analysis confirmed that that proteins and carbohydrates of the aqueous extract were the molecules responsible for Ag^+ reduction. It was also observed that the AgNPs exhibited antibacterial activity against pathogenic bacteria such as *B. cereus*, *B. subtilis*, *E. coli*, *P. aeruginosa* and *S. aureus*, even at the minimum concentrations tested (i.e., 100 μg/disc).

Pleurotus platypus

Another variant of *Pleurotus* species, *P. platypus* Sacc. was used by Sujatha et al., (2013) in order to synthesize AgNPs. It was observed that AgNPs were formed after a 72-hour incubation, with colour changing to brown at the end of the reaction. The UV-vis spectra verified that the AgNPs were formed at the maximum absorption of 435 nm. SEM analysis identified that the synthesized AgNPs were spherical in shape with average size of 0.56 to 0.71 μm. A variety of functional groups were detected via FTIR technique, and they were from the amine group, transition metal carbonyls, nitrate, unsaturated nitrogen compounds and halogen compounds. The synthesized AgNPs from *P. platypus* were also found to have a wide antibacterial activity against *E. coli*, *Enterobacter aerogenes*, *K. pneumoniae*, *Pseudomonas putida* and *S. aureus*.

Pleurotus sapidus

Another interesting *Pleurotus* species was tested for its ability to biosynthesize AuNPs. Sarkar et al. (2013) decided to employ *P. sapidus* Quell in reducing chloroaurate ions in order to form AuNPs. A positive outcome was obtained as a colour change from light yellow to pink was noted after one hour of incubation at room temperature, and the reaction was completed after 24 hours as the colour gradually turned to dark red. UV-vis analysis showed SPR bands at 540 nm (note that the SPR band of AuNPs in aqueous solution ranges between 510 to 560 nm). The AuNPs sizes were measured using DLS and were found to be in the range of 15 to 100 nm, with average sizes of 65 ± 5 nm in a variety of shapes (spherical, triangular and hexagonal), as recorded using TEM. The crystallinity of AuNPs was also confirmed by XRD analysis. A protein shell was observed via FTIR analysis, suggesting that the protein was accounted for the stabilizing effects during the biosynthesis of AuNPs.

Schizophyllum commune

The split gill mushroom, *S. commune* was also used to biosynthesize AgNPs (Arun et al., 2014). Extracellular synthesis of AgNPs was confirmed by colour change observed in the reaction mix from pale yellow to brown. It was also shown that *S. commune* had the potential to synthesize AgNPs intracellularly, as indicated by colour change of the mycelium to brown. SEM analysis of the mycelium also discovered the presence of AgNPs associated with hyphal cell wall which were spherical in shape with sizes ranging between 54 to 99 nm. The AgNPs in the aqueous solution were also spherical in shape with sizes ranging from 51 to 93 nm as analyzed by SEM. As determined by UV-vis analysis, the absorbance peak was shown to be at 440 nm. FTIR spectrum analysis suggested the biological function group that acts as capping and stabilizing agents are of proteins and heterocylic compounds. The AgNPs synthesized from *S. commune* also exhibited antibacterial and antifungal activity against *B. subtilis*, *E. coli*, *K. pneumoniae*, *Pseudomonas fluorescens*, *Trichophyton simii*, *T. mentagrophytes* and *T. rubrum*. Apart from that, the synthesized AgNPs also conferred an anti-proliferative effect against HEP-2 carcinoma cell line, with a 65% decrease in cell viability when maximum amount (100 μg/ml) of AgNPs was used.

Tricholoma matsutake

Anthony et al. (2014) used *Tricholoma matsutake*, or pine mushroom extract, for synthesizing AgNPs and also to identify the potential antimicrobial activity of the AgNPs. When mixing the mushroom extract with silver nitrate solution, within 60 minutes of incubation, a colour change from colourless to brown was observed. Absorption peak at 430nm was obtained from UV-vis spectra analysis, which verified the formation of AgNP. As evaluated by TEM analysis, the synthesized AgNPs were spherical with a size range of 10 ± 5 nm. The AgNPs size distribution was within the ranges of 10–20 nm, with an average size of 15 nm. FTIR measurements identified that the possible bio-functional groups that may act as reducing agents were polysaccharide/oligosaccharide present in the culture broth and that the synthesized NPs were stabilized by proteins. The AgNPs synthesized using pine mushroom extract also demonstrated antibacterial activity against *E. coli* and *B. subtilis*, but with better activity against *E. coli* which is Gram-negative.

Volvariella volvacea

In the study done by Phillip (2009), extract of *V. volvacea* was used to synthesize AuNPs, AgNPs and Au-AgNPs extracellularly. Results from UV-vis and TEM analysis showed that the synthesized AuNPs were of different sizes, ranging from 20–150 nm, and also yielded AuNPs of different shapes, i.e., triangular, nanoprismic, spherical and hexagonal. As for the AgNPs synthesized using *V. volvacea* extract, the obtained NPs were spherical in shape with estimated size of ~ 15 nm. The XRD and selected area electron diffraction (SAED) pattern analysis confirmed the crystallinity and photo-luminescent activity. Capping and stabilizing agent for both NPs were identified using FTIR method. It was found out that free amino groups were the responsible capping and stabilizing agent for AuNPs, whereas the carboxylate group of amino acid residues were both agents for AgNPs.

Conclusions

Since the last century, nanotechnology has established its name in the research field. Able to be synthesized in a variety of ways, each NP possesses its own unique characteristics. Current trends in the field of nanotechnology are moving towards a greener way to synthesize NPs to minimize the risk to the environment and human beings. Having to synthesize NPs in a large amount, green synthesis of NPs is definitely a better choice, not only for it being more cost effective, but also for other unique properties it possesses. Mushrooms, which are famous as a food source due to the rich flavor they contain, are also famous due to their nutritional and medicinal benefits. They have now been proven to contain a high amount of proteins, fibers, carbohydrates and vitamins. Due to their high nutritional value and the therapeutic properties they possess, researchers have started to utilize mushroom extracts in order to synthesize NPs. Overall results from the synthesized NPs using mushroom extracts showed formation of NPs with sizes less than 100 nm, in various shapes, possessing a range of beneficial characteristics and properties.

Acknowledgements

We acknowledge the support of this work by the University of Malaya BKP grant (BK011-2017). This work was also supported by the University of Malaya High Impact Research MoE Grants, namely UM.C/625/1/HIR/MoE/SC/02 and UM.C/625/1/HIR/MOHE/ASH/01(H-23001-G000008).

References

Agarwal, H., Kumar, S.V. and Rajeshkumar, S. 2017. A review on green synthesis of zinc oxide nanoparticles—An eco-friendly approach. Resource-Efficient Technologies, 3: 406–413.

Al-Bahrani, R., Raman, J., Lakshmanan, H., Hassan, A.A. and Sabaratnam, V. 2017. Green synthesis of silver nanoparticles using tree oyster mushroom *Pleurotus ostreatus* and its inhibitory activity against pathogenic bacteria. Materials Letters, 186: 21–25.

Al-Hamadani, A. and Kareem, A. 2017. Combination effect of edible mushroom—sliver nanoparticles and antibiotics against selected multidrug biofilm pathogens. Iraq Medical Journal, 1: 68–74.

Anthony, K.J., Murugan, M., Jeyaraj, M., Rathinam, N.K. and Sangiliyandi, G. 2014. Synthesis of silver nanoparticles using pine mushroom extract: A potential antimicrobial agent against *E. coli* and *B. subtilis*. Journal of Industrial and Engineering Chemistry, 20: 2325–2331.

Arun, G., Eyini, M. and Gunasekaran, P. 2014. Green synthesis of silver nanoparticles using the mushroom fungus *Schizophyllum commune* and its biomedical applications. Biotechnology and Bioprocess Engineering, 19: 1083–1090.

Bhat, R., Deshpande, R., Ganachari, S.V., Huh, D.S. and Venkataraman, A. 2011. Photo-irradiated biosynthesis of silver nanoparticles using edible mushroom *Pleurotus florida* and their antibacterial activity studies. Bioinorganic Chemistry and Applications. Doi: 10.1155/2011/650979.

Gudikandula, K. and Maringanti, S.C. 2016. Synthesis of silver nanoparticles by chemical and biological methods and their antimicrobial properties. Journal of Experimental Nanoscience, 11: 714–721.

Gurunathan, S., Sabaratnam, V., Raman, J., Malek, S.N. and John, P. 2013. Green synthesis of silver nanoparticles using *Ganoderma neo-japonicum Imazeki*: A potential cytotoxic agent against breast cancer cells. International Journal of Nanomedicine, 8: 4399–4413.

Gurunathan, S., Han, J., Park, J.H. and Kim, J.-H. 2014. A green chemistry approach for synthesizing biocompatible gold nanoparticles. Nanoscale Research Letters, 9: 248.

Haq, M., Rathod, V., Singh, D., Singh, A.K., Ninganagouda, S. and Hiremath, J. 2015. Dried Mushroom *Agaricus bisporus* mediated synthesis of silver nanoparticles from Bandipora District (Jammu and Kashmir) and their efficacy against Methicillin Resistant *Staphylococcus aureus* (MRSA) strains. Nanoscience and Nanotechnology: An International Journal, 5: 1–8.

Hussain, I., Singh, N.B., Singh, A., Singh, H. and Singh, S.C. 2016. Green synthesis of nanoparticles and its potential application. Biotechnology Letters, 38: 545–560.

Iravani, S. 2011. Green synthesis of metal nanoparticles using plants. Green Chemistry, 13: 2638–2650.

Khan, I., Saeed, K. and Khan, I. 2017. Nanoparticles: Properties, applications and toxicities. Arabian Journal of Chemistry. Doi: 10.1016/j.arabjc.2017.05.011.

Lateef, A. and Adeeyo, A.O. 2015. Green synthesis and antibacterial activities of silver nanoparticles using extracellular laccase of *Lentinus edodes*. Notulae Scientia Biologicae, 7: 405–411.

Maurya, S., Bhardwaj, A.K., Gupta, K.K., Agarwal, S., Kushwaha, A. et al. 2016. Green synthesis of silver nanoparticles using *Pleurotus* and its bactericidal activity. Cellular and Molecular Biology, 62: 131. Doi: 10.4172/1165-158X.1000131.

Mirunalini, S., Arulmozhi, V., Deepalakshmi, K. and Krishnaveni, M. 2012. Intracellular biosynthesis and antibacterial activity of silver nanoparticles using edible mushrooms. Notulae Scientia Biologicae, 4: 55–61.

Mohanta, Y.K., Singdevsachan, S.K., Parida, U.K., Bae, H., Mohanta, T.K. and Panda, S.K. 2016. Green synthesis and antimicrobial activity of silver nanoparticles using wild medicinal mushroom *Ganoderma applanatum* (Pers.) Pat. from Similipal Biosphere Reserve, Odisha, India. IET Nanobiotechnology, 10: 184–189.

Nagajyothi, P.C., Sreekanth, T.V., Lee, J. and Lee, K.D. 2014. Mycosynthesis: Antibacterial, antioxidant and antiproliferative activities of silver nanoparticles synthesized from *Inonotus obliquus* (Chaga mushroom) extract. Journal of Photochemistry and Photobiology B: Biology, 130: 299–304.

Narasimha, G., Praveen, B., Mallikarjuna, K. and Deva Prasad Raju, B. 2011. Mushrooms (*Agaricus bisporus*) mediated biosynthesis of sliver nanoparticles, characterization and their antimicrobial activity. International Journal of Nano Dimension, 2: 29–36.

Owaid, M.N., Raman, J., Lakshmanan, H., Al-Saeedi, S.S., Sabaratnam, V. and Abed, I.A. 2015. Mycosynthesis of silver nanoparticles by *Pleurotus cornucopiae* var. *citrinopileatus* and its inhibitory effects against *Candida sp.* Materials Letters, 153: 186−190.

Owaid, M.N., Al-Saeedi, S.S. and Abed, I.A. 2017. Biosynthesis of gold nanoparticles using yellow oyster mushroom *Pleurotus cornucopiae* var. *citrinopileatus*. Environmental Nanotechnology, Monitoring & Management, 8: 157−162.

Parveen, K., Banse, V. and Ledwani, L. 2016. Green synthesis of nanoparticles: Their advantages and disadvantages. AIP Conference Proceedings, 1724. Doi: 10.1063/1.4945168.

Phan, C.-W., David, P. and Sabaratnam, V. 2017a. Edible and medicinal mushrooms: Emerging brain food for the mitigation of neurodegenerative diseases. Journal of Medicinal Food, 20: 1−10.

Phan, C.-W., Wang, J.-K., Cheah, S.-C., Naidu, M., David, P. and Sabaratnam, V. 2017b. A review on the nucleic acid constituents in mushrooms: Nucleobases, nucleosides and nucleotides. Critical Reviews in Biotechnology. Doi: 10.1080/07388551.2017.1399102.

Philip, D. 2009. Biosynthesis of Au, Ag and Au–Ag nanoparticles using edible mushroom extract. Spectrochimica Acta Part A: Molecular and Biomolecular Spectroscopy, 73: 374−381.

Raman, J., Lakshmanan, H., John, P.A., Zhijian, C., Periasamy, V. et al. 2015a. Neurite outgrowth stimulatory effects of myco-synthesized AuNPs from *Hericium erinaceus* (Bull.: Fr.) Pers. on pheochromocytoma (PC-12) cells. International Journal of Nanomedicine, 10: 5853–5863.

Raman, J., Reddy, G.R., Lakshmanan, H., Selvaraj, V., Gajendran, B., Nanjian, R. et al. 2015b. Mycosynthesis and characterization of silver nanoparticles from *Pleurotusdjamor* var. *roseus* and their *in vitro* cytotoxicity effect on PC3 cells. Process Biochemistry, 50: 140−147.

Raziya, S., Durga, B., Rajamahanthe, S.G., Govindh, B. and Annapurna, N. 2016. Synthesis and characterization of CDS nanoparticles using *Reishi* mushroom. International Journal of Advanced Technology in Engineering and Science, 4: 220−227.

Sarkar, J., Roy, S.K., Laskar, A., Chattopadhyay, D. and Acharya, K. 2013. Bioreduction of chloroaurate ions to gold nanoparticles by culture filtrate of *Pleurotus sapidus Quél*. Materials Letters, 92: 313−316.

Sen, I.K., Mandal, A.K., Chakraborti, S., Dey, B., Chakraborty, R. and Islam, S.S. 2013. Green synthesis of silver nanoparticles using glucan from mushroom and study of antibacterial activity. International Journal of Biological Macromolecules, 62: 439−449.

Sharma, D., Kanchi, S. and Bisetty, K. 2015. Biogenic synthesis of nanoparticles: A review. Arabian Journal of Chemistry. Doi: 10.1016/j.arabjc.2015.11.002.

Shin, W., Cho, J., Kannan, A.G., Lee, Y. and Kim, D. 2016. Cross-linked composite gel polymer electrolyte using mesoporous methacrylate-functionalized sio2 nanoparticles for lithium-ion polymer batteries. Scientific Reports, 6: Doi: 10.1038/srep26332.

Shinde, N.C., Keskar, N.J. and Argade, P.D. 2012. Nanoparticles: Advances in drug delivery systems. International Journal of Advances in Pharmacy, Biology and Chemistry, 1: 132−137.

Sudhakar, T., Nanda, A., Babu, S.G., Janani, S., Evans, M.D. and Markose, T.K. 2014. Synthesis of silver nanoparticles from edible mushroom and its antimicrobial activity against human pathogens. International Journal of PharmTech Research, 6: 1718−1723.

Sujatha, S., Tamilselvi, S., Subha, K. and Panneerselvam, A. 2013. Studies on biosynthesis of silver nanoparticles using mushroom and its antibacterial activities. International Journal of Current Microbiology and Applied Sciences, 2: 605−614.

Sujatha, S., Kanimozhi, G. and Panneerselvam, A. 2015. Synthesis of silver nanoparticles from *Lentinula edodes* and antibacterial activity. International Journal of Pharmaceutical Sciences Review and Research, 33: 189−191.

Sriramulu, M. and Sumathi, S. 2017. Photocatalytic, antioxidant, antibacterial and anti-inflammatory activity of silver nanoparticles synthesized using forest and edible mushroom. Advances in Natural Sciences: Nanoscience and Nanotechnology, 8: Doi: 10.1088/2043-6254/aa92b5.

16

Proteomics of Edible and Medicinal Mushrooms

Shin Yee Fung* and Muhammad Fazril Mohamad Razif

INTRODUCTION

The numerous health benefits of consuming mushrooms have led to the imperative need for investigations into their protein content and potential bioactivities. These studies can provide information on the abundance and type of proteins present, ultimately improving our understanding of their functions within the cell. The potential applications range from biochemistry to drugs discovery, diagnostics and therapeutics. Various studies have uncovered novel proteins and compounds present in mushrooms at different growth stages; these are of interest as their therapeutic effects can be exploited for pharmaceutical development.

The drive towards using complementary technologies, such as (structural and functional) genomics and transcriptomics, to complete proteomic studies is ever growing as these technologies will be able to provide an expansive view of the mushrooms' growth cycles and conditions that may enhance or diminish the synthesis of bioactive enzymes and compounds.

Edible and Medicinal Mushrooms

Over the past decade, there has been an increasing demand for mushrooms due to their vast nutritional and therapeutic content. Traditionally, indigenous communities have used mushrooms in various forms for the treatment of diseases. Numerous

Medicinal Mushroom Research Group (MMRG), Department of Molecular Medicine, Faculty of Medicine, University of Malaya, 50603 Kuala Lumpur, Malaysia.
* Corresponding author: syfung@um.edu.my

studies have uncovered novel proteins and compounds present in different growth stages; these are of interest as their therapeutic effects can be exploited for pharmaceutical development. Mushrooms are macrofungi, defined as an ecologically and phylogenetically heterogeneous organism that produces visible fruiting bodies. For most, a mushroom is identified as the matured mycelium that has produced the fruiting body. The structure consists of a stem (stipe) and a cap (pileus). Some species may exhibit a cup-type (volva) or annulus-type (ring) structure.

In general, there are more than 2,000 edible mushroom varieties worldwide but less than 100 are currently cultivated. The production of edible cultivated mushrooms has increased 30-fold since the late 70s. In 2016, global mushroom production amounted to ~ 10.79 million tonnes, with China being the top producer, 7.80 million tonnes (74% of global production; Food and Agriculture Organization of the United Nations, 2018). The increase in demand and consumption of mushrooms has been boosted by the fact that mushrooms are rich in essential and non-essential amino acids, proteins and minerals, offering higher nutritional content than most vegetables (Wang et al., 2014). In addition, they are known to be low in calories, possess a high content of potassium, phosphorus and selenium, and have low glucose and sodium levels (Wang et al., 2014).

At present, the most cultivated edible mushroom globally is *Lentinus edodes* (shiitake mushroom), accounting for 22% of the world's supply. Other commonly cultivated edible mushrooms include the *Agaricus* species (champignon and button mushroom), the *Pleurotus* species (oyster mushroom), *Auricullaria* species (jelly fungus), *Flammulina* species and the *Volvariella* species. However, not all edible mushrooms are cultivated, many are still sourced from the wild as the conditions for their growth have yet to be understood or developed. They are only seasonally available and may be sold either fresh or dry. These include the *Boletus edulis* (Penny bun mushroom), *Cantharellus cibarius* (chanterelle mushroom), *Grifola frondosa* (maitake mushroom), *Hydnum repandum*, (hedgehog mushroom) and *Lactarius deliciosus* (Saffron milk cap).

Table 1. Types of cultivated and non-cultivated mushrooms.

Type	Nomenclature
Cultivated	*Agaricus bisporus, Auricularia polytricha, Clitocybe nuda, Flammulina velutipes, Fusarium venenatum, Hypsizygus tessulatus, Lentinus edodes, Pleurotus* sp. *(citrinopileatus, cornucopiae, eryngii, ostreatus), Rhizopus oligosporus, Sparassis crispa, Tremella fuciformis, Tuber* sp. *(aestivum, magnatum, melanosporum), Volvariella volvacea*
Non-cultivated	*Amanita caesarea, Armillaria mellea, Agaricus* sp. *(arvensis, silvaticus), Boletus badius, Calvatia* sp. *(gigantea, utriformis), Calocybe gambosa, Clavariaceae* sp., *Coprinus comatus, Cyttaria espinosae, Cortinarius variicolor, Fistulina hepatica, Flammulina velutipes, Hygrophorus chrysodon, Lactarius* sp. *(deterrimus, salmonicolor, subdulcis, volemus), Leccinum* sp. *(aurantiacum, scabrum, versipelle), Macrolepiota procera, Polyporus* sp. *(squamosus, mylittae), Rhizopogon luteolus, Sparassis crispa, Suillus* sp. *(bovinus, granulatus, luteus, tomentosus), Tricholoma terreum*

Fig. 1. *Lignosus rhinocerus* (Tiger milk mushroom).

Mushrooms that are thought to possess medicinal or tonic properties are referred to as medicinal mushrooms. The use of medicinal mushrooms for their therapeutic properties has been found to date back to the Neolithic age. More than 270 documented species of mushrooms are known to have immunotherapeutic properties (Ooi and Liu, 2000). In Asia, specific mushrooms such as the *Lignosus rhinocerus* (Tiger milk mushroom; Fig. 1), *Ganoderma lucidum* (Lingzhi mushroom) and *Trametes versicolor* (Turkey tail mushroom) have been used by locals and Aboriginals for the prevention and treatment of several human diseases. Medicinal mushrooms have been shown to possess anticancer, immunomodulating, antioxidant, antiviral and antidiabetic properties (Abdullah et al., 2012; Alves et al., 2012, 2013; Chang, 2008; Dai et al., 2009; Didukh et al., 2003; Elosta et al., 2012; Ferreira et al., 2010; Gao et al., 2003, 2004; Guillamon et al., 2010; Ichinohe et al., 2010; Lee et al., 2012; Lee et al., 2014; Lee et al., 2018; Lindequist et al., 2005; Liu et al., 2016; Ramkissoon et al., 2013; Rowan et al., 2003; Sullivan et al., 2006; Taofiq et al., 2015; Xu et al., 2011; Yap et al., 2018a,b, Yap et al., 2013; Zhang et al., 2007). There is increasing interest in validating and documenting the medicinal properties of these mushrooms. Researchers are also investigating the presence of various low molecular weight compounds, enzymes and secondary metabolites present throughout the lifespan of mushrooms to determine their biological functions and possible therapeutic uses. Lectins, terpenoids, laccase and superoxide dismutase can be found in certain types of mushroom, all of which can be purified and used for drug discovery and development. Drugs that have been derived from mushroom compounds include illudin (*Omphalotus olearius*; *anti-cancer*) and retapamulin (*Clitopilus scyphoides*; *antibiotic*). At present, medicinal mushrooms are used as dietary supplements, pharmaceuticals, natural bio-control agents in agricultural industries and cosmeceuticals.

Proteomics of Mushrooms

Proteome is defined as the composition of all proteins expressed by the genome of any organism (Westermeier and Naven, 2002). It is the set of proteins expressed in a specific cell under a particular set of conditions. It results from physical laboratory and computational techniques. The term proteomics was coined circa 1994. It had its origin in the electrophoretic separation techniques that was largely popular in the 1970s and 1980s. The simultaneous analysis from this study of proteins can give information on the abundance and type of protein, and improves our understanding on their function within the cell. The application extends from drug discovery to diagnostics, therapy and biochemistry. Pharmaceutical industry and research use proteomics techniques in search of new drug protein targets in transformed cell lines or diseased tissues. Subsequent validation of detected target via *in vitro* or *in vivo* toxicology study and the detection of possible side effects can also be performed using proteomics techniques.

Mushrooms contain appreciable amounts of proteins. Mushroom consumption is associated with better diet quality and improved nutrition. Mushrooms are included in the diets of some for their nutritive and medicinal properties. Due to the many health benefits of consuming mushrooms (of differing kinds), it is therefore imperative that the protein content and combination of amino acid molecules of dietary mushrooms be investigated. This will enable sensible diet combination to fulfill the daily required intake. It will also reveal novel peptides/proteins that are responsible for the medicinal properties of mushrooms. The study into the structure and function of bioactive proteins can also be platform for drugs design.

There have also been various studies done in order to investigate the proteome of mushrooms in different development stages (Chen et al., 2012, 2017; Jia et al., 2017; Liu et al., 2017; Rahmad et al., 2014), effects of differing media (Adav et al., 2012; O'Brien et al., 2014; Xie et al., 2016), effects of different growth conditions (Chang et al., 2011; Fang et al., 2017; Huang et al., 2011; Kim et al., 2017; Lu et al., 2014; Singh et al., 2017; Tang et al., 2016), effects of stimulatory and/or inhibitory agents (Liu et al., 2018; Sathesh-Prabu and Lee, 2016; Zhang et al., 2012) and studies on secretome (Cai et al., 2017; Zorn et al., 2005). Biocontrol agents using certain species of mushrooms have also been investigated with proteomics techniques (Ujor et al., 2012).

Proteomics data integrated with functional analysis of selected proteins of interest may reveal information regarding the role of proteins in metabolic pathways and the interaction of the proteins with other proteins and/or metabolites. The computational (*in silico*) approach for analysis of proteins is adopted as part of the workflow for certain proteomic investigations. This approach predicts the structure and function of unknown proteins and the data obtained can facilitate experimental analysis of proteins of interest so that they can be beneficial for therapeutic purposes.

Techniques used in Mushroom Proteomics

The study of mushroom proteins is challenging due to the dynamic complexity of their proteome. It is also very much dependent on the environmental conditions and period at which the mushrooms are harvested for investigation as they may express different sets of proteins. This hinders the ability of a single technology to be able to detect a mushroom's proteome in its entirety. Difficulties in identifying proteins of low abundance and the lack of efficient purification techniques are also obstacles faced by researchers. Mushrooms contain chitin, a rigid, fractious component that makes up the majority of their cell mass. Good sample preparation, which considers cell lysis and extraction, is needed in order to extract and solubilize proteins of interest so as to increase the likelihood of detecting and identifying them (Fernandez and Novo, 2013). Studies have reported the use of mechanical lysis using beads made from glass (Ebstrup et al., 2005; Grinyer et al., 2007; Melin et al., 2002), cell mill (Bohmer et al., 2007), sonication (Grinyer et al., 2004; 2007; Sulc et al., 2009) and using mortar and pestle on samples in liquid nitrogen (El-Bebany et al., 2010; Horie et al., 2008). Shimizu and Wariishi (2005) generated protoplasts of brown rot fungi (*Tyromyces palustris*) by treating mycelial cells with an enzyme solution containing Novozyme 234, Zymolyase 20 T, mannitol, and maleate; they obtained better 2-DE pattern compared to intact cells. Once the cells have been lysed and the extraction process is completed, good quality protein precipitation methods are required. Several studies have reported using organic solvents such as trichloroacetic acid (TCA), followed by solubilization of the precipitate in an appropriate buffer (Nandakumar et al., 2003). Additional steps to increase protein resolution in-gel can be achieved by adding urea/thiourea, detergents and sodium hydroxide (Nandakumar et al., 2003; Rabilloud, 1996, 1998; Everberg et al., 2008).

Accurate identification and quantitation of proteins has become one of the key goals in mushroom science. Depending on the part of the mushroom and stage of development, different techniques would be needed for investigations, as both yield and quality of mushroom proteins can differ largely. One of the first techniques employed in proteomics is 2D-gel electrophoresis. This technique was subsequently paired with mass spectrometry (MS), and this was regarded as the standard approach to studying mushroom proteins due to its speed, simplicity and robustness. Mass spectrometry is a data-rich technique, capable of identifying several thousand peptides in a single run (Liang et al., 2013). Mushroom protein fractionations using electrophoresis or chromatography techniques prior to identification by MS are

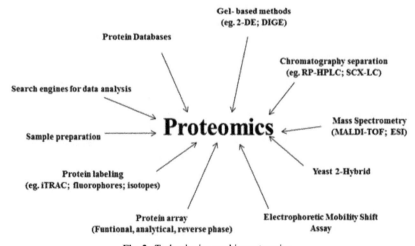

Fig. 2. Technologies used in proteomics.

extremely common as these increase the probability of identifying proteins of low abundance by reducing sample complexity.

The advent of new techniques, such as the isobaric tags for relative and absolute quantitation (iTRAQ) labeling and protein microarrays, coupled to more traditional approaches have given researchers more flexibility to study numerous aspects of mushroom protein expression throughout its developmental process. The former is based on the principle that labeling peptides with isotopes does not affect the chemical properties of proteins. The robustness and sensitivity of the iTRAQ method is exemplified by Liu et al. (2017). They were able to identify and quantify 1198 proteins of *F. velutipes* in three different mycelium growth phases using iTRAQ coupled with 2D two-dimensional liquid chromatography tandem mass spectrometry (2D LC-MS/MS). Chen et al. (2017) was able to use the same technique to identify 160 differentially expressed proteins, out of a total of 1063 proteins detected, in order to better understand the post-harvest maturation progression of *Agaricus bisporus*. Other MS methods that have been employed to study mushroom proteomics include high/ultra-performance liquid chromatography, *matrix-assisted laser desorption/ ionization-time of flight (MALDI-TOF) mass spectrometry* and fourier transform ion cyclotron resonance mass spectrometry. Sugawara et al. (2016) was able to reliably identify and differentiate species of wild mushrooms using MALDI-TOF MS within

Fig. 3. Proteomics Facility, Medical Biotechnology Laboratory, Central Research Facility, Faculty of Medicine, University of Malaya, 50603 Kuala Lumpur, Malaysia. Top Right: Sciex MALDI-TOF/TOF 5800 System; Top Left: UVP800 Bioimaging System; Bottom: Agilent 1260 Infinity Nanoflow Liquid Chromatography System coupled with Agilent 6550 iFunnel Q-TOF Mass Spectrometer.

Table 2. List of selected available protein databases.

Name	Website	Content
DNA Data Bank of Japan (DDBJ)	http://www.ddbj.nig.ac.jp	Sequence
GeneCards	http://bioinfo.weizmann.ac.il/cards/	
InterPro	http://www.ebi.ac.uk/interpro/	
OWL	http://www.bioinf.man.ac.uk/dbbrowser/OWL	
Protein Research Foundation (PRF)	https://www.prf.or.jp/index-e.html	
UniProt	http://www.uniprot.org	
ArchDB	http://sbi.imim.es/cgi-bin/archdb//loops.pl	Structure
Biological Magnetic Resonance Data Bank (BMRB)	http://www.bmrb.wisc.edu	
Enzyme Structures	http://www.biochem.ucl.ac.uk/bsm/enzymes/	
Proteopedia	http://proteopedia.org	
Protein Model Portal (PMP)	http://www.proteinmodelportal.org	
Protein Data Bank (PDB)	http://www.pdb.org	
Clusters of Orthologous Groups (COG)	http://www.ncbi.nlm.nih.gov/COG/	Sequence Family
Pfam	http://pfam.sanger.ac.uk	
PRINTS	http://www.bioinf.man.ac.uk/dbbrowser/PRINTS/	
iProClass	http://pir.georgetown.edu/iproclass/	
ProDom	http://prodom.prabi.fr/	
PROSITE	http://prosite.expasy.org	
Simple Modular Architecture Research Tool (SMART)	http://smart.embl-heidelberg.de	
CATH: Protein Structure Classification Database	http://www.cathdb.info	Structure Family
MODBASE: Comparative Protein Structure Models	http://modbase.compbio.ucsf.edu	
Protein Data Bank (Europe) PDBe	http://www.ebi.ac.uk/pdbe/	
Structural Classification of Proteins (SCOP)	http://scop.mrc-lmb.cam.ac.uk/scop/	
3-Dimensional Interacting Domains (3DID)	http://3did.irbbarcelona.org	Protein interactions
Binding Database (BindingDB)	http://www.bindingdb.org/bind/index.jsp	
Biological General Repository for Interaction Datasets (BioGRID)	http://thebiogrid.org	
Database of Macromolecular Interactions (DOMMINO)	http://dommino.org	
Inferred Biomolecular Interactions Server (IBIS)	http://www.ncbi.nlm.nih.gov/Structure/ibis/ibis.cgi	

30 minutes, using the fruiting body of mushrooms. They found that the technique was more accurate than morphological inspections, with consistency comparable to DNA sequencing.

Despite the abundance of novel proteins that can be found in mushrooms, the 3D structures of many of them have yet to be deciphered. NMR spectroscopy and X-Ray crystallography are two techniques that can be employed in order to determine the 3D structures of these proteins. Zhang et al. (2017) successfully purified and crystalized the Y3 protein, a 130-amino acids glycan-binding protein from the *Corprinus comatus*. *In vitro* studies had previously revealed that the Y3 protein possess potent cytotoxicity towards human T-cell leukemia. Analysis of its crystal structure revealed that the protein has a single-domain αβα-sandwich motif with a large glycan binding pocket on its dimeric interface. They uncovered that Asp26, Asn30, Asp120, Asn122, and Asp123 in the binding pocket were responsible for hydrogen bond interactions between the protein and its target. The major issue often encountered by researchers is the ability to express and purify sufficient protein of interest for crystallography. In many cases, these purified proteins may be difficult to crystallize or may not crystallize at all.

Large amounts of data, regarding proteins, their sequence, structure and functions, are hosted on many different databases. They are often the main reference point when studying a novel protein as information regarding their protein families and relationships across different mushroom species can be obtained. Among all protein sequence databases, UniProt is the most commonly used (UniProt Consortium, 2011). With regards to protein structure, the only international database of protein structures is the Protein Data Bank (Bernstein et al., 1977). Here, structures that have been identified experimentally via NMR, electron microscopy and X-ray crystallography are available. Although protein databases have become crucial resources for studying proteins, the content of these databases may vary significantly. In addition, information regarding mushroom proteins are still lacking on public databases as many researchers tend to host this information on their internal databases.

At present, researchers are progressing towards functional proteomics which includes a combination of protein identification and functional characterization. A high throughput technology that can be employed in order to provide abundant information regarding the biochemical activities of proteins, their interactions and functions are protein microarrays. Albeit costly, protein microarray chips are able to simultaneously analyze protein function and protein-DNA/RNA/protein/ligand interactions in a highthroughput manner (Zhu, 2006; Melton, 2004).

Studies on Mushroom Proteins

Proteomic methodologies have been applied directly on mushroom samples in order to comprehensively probe their protein contents, to reveal possible novel bioactive proteins when referenced against database of known bioactive molecules. In 2008, Horie and colleagues performed proteomic analysis of two edible mushrooms (fruiting bodies) *Sparassis crispa* (cauliflower fungus) and *Hericium erinaceum* (*erinaceus*; lion's mane mushroom) using one- and two-dimensional gel electrophoresis (1-DGE and 2-DGE) coupled with liquid chromatography, mass spectrometry and

N-terminal amino acid sequencing (Horie et al., 2008). Numerous proteins of interest and those with potential use in the food industry were identified including laccase, polygaracturonase, xylose reductase, trehalose phosphorylase, glutamine synthase and some restriction enzymes. Two proteins (14-3-3 proteins and septins) were found to be common in both these mushrooms, suggesting a phylogenetic relationship. It was suggested that protein markers can be used to investigate phylogenetic relationships between mushroom species, rather than relying on morphological properties alone. This study also suggested the use of proteomics databases and 2D gel reference maps in profiling protein changes during the growth of fruiting bodies (mycelia to mature structure) and against diverse environmental factors (stress, artificial culture) in comparative proteomics studies (refer to Section Proteomics of Mushrooms). Most recently, Zeng et al. (2018) discovered proteins involved in the regulation of bioactive metabolites, such as terpenoid, polyketide and sterol, in the same mushroom. It is riveting to note that proteins involved in polyketide biosynthesis were up-regulated in the fruiting body, while some proteins in the mevalonate (MEP) pathway from terpenoid biosynthesis were generally up-regulated in the mycelium. The LC-MS/MS platform was revealed to be a powerful method for investigating putative proteins involved in diverse secondary metabolites biosynthesis. They noted that differential regulation of biosynthesis genes could produce various secondary metabolites with unusual pharmacological effects in lion's mane mushroom.

Matis and colleagues (2005) adopted the isotopic labeling with electrospray ionisation (ESI)-tandem mass spectrometry for *de novo* sequencing in combination with database search, taking advantage of different programs for the identification of proteins from *Pleurotus ostreatus* (oyster mushroom). Yap et al. (2015) used two-dimensional gel electrophoresis, coupled with mass spectrometry analysis, in order to reveal a total of 16 non-redundant, major proteins with high confidence level in *Lignosus rhinocerus* (tiger milk mushroom) sclerotium, based on its genome as custom mapping database.

Proteomic methods have also been adopted in various endeavours in search for bioactive proteins from mushrooms. Lau et al. (2012) used reverse phase high performance liquid chromatography (RP-HPLC) together with electrophoresis and surface-enhanced laser desorption ionization time-of-flight mass spectrometry (SELDI-TOF-MS) to reveal proteins from *Pleurotus cystidiosus* (abalone mushroom) and *Agaricus bisporus* (button mushroom) involved in the inhibition of angiotension-converting enzymes (ACE) that plays a key role in blood pressure homeostasis and salt balance in mammals. Proteins that possess molecular weights ranging from 3–10 kDa were found to be responsible for the activity observed but were not identified. A bioactive protein, PEP, was isolated from an edible mushroom, *Pleurotus eryngii* (king oyster mushroom). Proteomic analysis by MALDI-TOF/MS showed that PEP (molecular weight of 40 kDa) was a novel protein with anti-inflammatory effects inhibiting the overproduction of pro-inflammatory mediators (Yuan et al., 2017).

There have been a number of studies focusing on identifying alterations in protein expression, especially in cancer cell lines that have been treated with mushroom components. Chai and colleagues (2016) worked on the differential expression of proteins in normal HepG2 cells and those treated with polysaccharides that had been isolated from *Phellinus linteus* (black hoof mushroom), *Ganoderma lucidum* (lingzhi mushroom) and *Auricularia auricula* (wood ear mushroom). In their endeavour to discover the mechanisms for the polysaccharide-treated effects of these mushrooms on hepatocellular carcinoma (HCC), they used two-dimensional gel electrophoresis and MALDI-TOF-MS to reveal 59 differentially expressed proteins that were expressed in HepG2 cells treated with the polysaccharides but were absent in the normal HepG2 cells. These proteins can be used as markers for HCC and provide new data to the development of natural anti-tumour foods (Chai et al., 2016). Wang et al. (2004) investigated the differentiation of human myelocytic lymphoma U937 cells into macrophages by the mononuclear cell-conditioned media (MNC-CM) containing protein fractions of *Agrocybe aegerita* (Chestnut mushroom). Treatment of U937 cells with these proteins resulted in a marked inhibition of proliferation and maturation of monocytes/macrophages. Wang et al. (2004) suggested that this may be due to mediators that enable the inhibition of leukemic cell growth, thereby inducing them to be differentiated into mature monocytes/macrophages and functional cells. This same conclusion was also drawn by Chen et al. (2004) and Lieu et al. (1992) who used Poria cocos (Poria, *fu-ling* [Chinese]), tuckahoe, Indian bread, hoelen (Japanese) and Ganoderma lucidum (lingzhi mushroom), respectively. A novel glycoprotein from fermented mycelia *Grifola frondosa* (Hen of the Woods mushroom) has been shown to cause differential expression of 21 proteins in human gastric cancer (SGC-7901) cells. These proteins were associated with cell cycle arrest, apoptosis and stress response. These findings may provide a basis for prevention or treatment of human gastric cancer and provide an understanding of the tumour-inhibitory molecular mechanisms of mushroom glycoproteins (Cui et al., 2016).

Proteomic analysis of the colon and small intestine of mice fed with a heteropolysaccharide from the fruiting body of *Lentinula edodes* (Shiitake mushroom) showed significant changes in abundance of proteins involved in metabolism, binding, structural components and stimuli response. These proteins were suggested to interact and work together in a concerted mechanism in order to regulate the bioactivity observed in mice (Xu et al., 2016). This study was interesting because the proteomic results from the two tissues exhibited spatial specification. The mushroom's heteropolysaccharide was seen to act on the colon and the small intestine separately, but also in concert by regulating sites independently and subsequently a unified, bigger protein network.

5. Future of Proteomics in Mushroom Studies

Future advancements will be focused on decifering the complete proteome of mushrooms, as these organisms possess the ability to modulate their proteome in order to adjust to environmental changes in their saprotrophic life. Environmental and nutritional changes greatly influence the proteins and enzymes synthesized by mushrooms in a species-specific manner. The proteomics of both edible and medicinal mushrooms can offer incredible insights into novel enzymes and their mechanisms of action, which can then be utilized for the development of new drugs and biotechnological applications.

Despite the development of high resolution protein identification and characterization techniques, one major technical challenge that persists is the identification of low abundance proteins. This will require the development of technologies that are sensitive and have a large dynamic range. Identification of low abundance proteins and secondary metabolites are important as they may possess their own distinctive or secondary activities. Zaidman et al. (2005) found that low molecular weight secondary metabolites have the ability to influence processes such as metastasis, apoptosis and cell cycle regulation. They also found that the metabolic source of the mushrooms greatly influences the types of secondary metabolite produced. Reproducibility in protein separation also remains a significant challenge because most of the methods discussed above are not fully-automated and are fundamentally skill-based. At present, there is a drive towards using complementary technologies, such as structural and functional genomics and transcriptomics in order to complete proteomic investigations as these will be able to provide an expansive view of the mushrooms' growth cycles and conditions that may enhance or diminish the synthesis of bioactive enzymes or compounds.

Acknowledgement

Authors have benefited from University of Malaya Programme Research Grant RG034 (A, B, C)-17 AFR in their research endeavours into medicinal mushrooms.

References

Abdullah, N., Ismail, S.M., Aminudin, N., Shuib, A.S. and Lau, B.F. 2012. Evaluation of selected culinary-medicinal mushrooms for antioxidant and ACE inhibitory activities. eCAM 2012: 464238. Doi: http://10.1155/2012/464238.

Adav, S.S., Ravindran, A. and Sze, S.K. 2012. Quantitative proteomic analysis of lignocellulolytic enzymes by Phanerochaete chrysosporium on different lignocellulosic biomass. J. Proteomics, 16; 75(5): 1493–504. Doi: 10.1016/j.jprot.2011.11.020.

Bernstein, F.C., Koetzle, T.F., Williams, G.J., Meyer, E.F. Jr., Brice, M.D., Rodgers, J.R., Kennard, O., Shimanouchi, T. and Tasumi, M. 1977. The protein data bank: A computer-based archival file for macromolecular structures. J. Mol. Biol., 112(3): 535–42.

Bohmer, M., Colby, T., Bohmer, C., Brautigam, A., Schmidt, J. and Bolker, M. 2007. Proteomic analysis of dimorphic transition in the phytopathogenic fungus *Ustilago maydis*. Proteomics, 7: 675–685.

Cai, Y., Gong, Y., Liu, W., Hu, Y., Chen, L., Yan, L., Zhou, Y. and Bian, Y. 2017. Comparative secretomic analysis of lignocellulose degradation by *Lentinula edodes* grown on microcrystalline cellulose, lignosulfonate and glucose. J. Proteomics, 23; 163: 92–101.

Chai, Y., Wang, G., Fan, L. and Zhao, M. 2016. A proteomic analysis of mushroom polysaccharide-treated HepG2 cells. Sci. Rep., 6: 23565. Doi: 10.1038/srep23565.

Chen, Y.Y. and Chang, H.M. 2004. Antiproliferative and differentiating effects of polysaccharide fraction from Fu-Ling (*Poria cocos*) on human leukemic U937 and HL-60 cells. Food Chem. Toxicol. 42: 759–769.

Chen, M., Liao, J., Li, H., Cai, Z., Guo, Z., Wach, M.P. and Wang, Z. 2017. iTRAQ-MS/MS Proteomic analysis reveals differentially expressed proteins during post-harvest maturation of the white button mushroom *Agaricus bisporus*. Curr. Microbiol., 74(5): 641–649. Doi: 10.1007/s00284-017-1225-y.

Chen, L., Zhang, B.B. and Cheung, P.C. 2012. Comparative proteomic analysis of mushroom cell wall proteins among the different developmental stages of *Pleurotus tuber-regium*. J. Agric. Food Chem. 20; 60(24): 6173–82. Doi: 10.1021/jf301198b.

Cui, F., Zan, X., Li, Y., Sun, W., Yang, Y. and Ping, L. 2016. *Grifola frondosa* Glycoprotein GFG-3a arrests S phase, alters proteome and induces apoptosis in human gastric cancer cells. Nutr. Cancer, 68(2): 267–79. Doi: 10.1080/01635581.2016.1134599.

Dai, Y.C.H., Yang, Z.L., Ui, B.K., Yu, Ch.J. and Zhou, L.W. 2009. Species diversity and utilization of medicinal mushrooms and fungi in China (review). Int. J. Med. Mushrooms, 11: 287–302.

Didukh, M.Y., Wasser, S.P. and Nevo, E. 2003. Medicinal value of species of the family Agaricaceae Cohn (higher Basidiomycetes) current stage of knowledge and future perspectives. Int. J. Med. Mushrooms, 5: 133–152.

Ebstrup, T., Saalbach, G. and Egsgaard, H. 2005. A proteomics study of *in vitro* cyst germination and appressoria formation in *Phytophthora infestans*. Proteomics, 5: 2839–2848.

El-Bebany, A.F., Rampitsch, C. and Daayf, F. 2010. Proteomic analysis of the phytopathogenic soil-borne fungus *Verticillium dahliae* reveals differential protein expression in isolates that differ in aggressiveness. Proteomics, 10: 289–303.

Elosta, A., Ghous, T. and Ahmed, N. 2012. Natural products as anti-glycation agents: Possible therapeutic potential for diabetic complications. Curr. Diab. Rev., 8(2): 92–108.

Everberg, H., Gustavasson, N. and Tjerned, F. 2008. Enrichment of membrane proteins by partitioning in detergent/polymer aqueous two-phase systems. Methods Mol. Biol., 424: 403–412.

Fang, D., Yang, W., Deng, Z., An, X., Zhao, L. and Hu, Q. 2017. Proteomic investigation of metabolic changes of mushroom (*Flammulina velutipes*) packaged with nanocomposite material during cold storage. J. Agric. Food Chem. 29; 65(47): 10368–10381. Doi: 10.1021/acs.jafc.7b04393.

Fernandez, R.G. and Novo, J.V.R. 2013. Proteomic protocols for the study of filamentous fungi. *In*: V.K. Gupta et al. (eds.). Laboratory Protocols in Fungal Biology: Current Methods in Fungal Biology, Fungal Biology. Doi: 10.1007/978-1-4614-2356-0_24, Springer Science+Business Media, LLC 2013.

Gao, Y., Zhou, S., Chen, G., Dai, X. and Ye, J.A. 2002. Phase I/II study of a *Ganoderma lucidum* extract (Ganopoly) in patients with advanced cancer. Int. J. Med. Mushrooms, 4: 207–214.

Gao, Y., Zhou, Sh., Huang, M. and Xu, A. 2003. Antibacterial and antiviral value of the genus Species (Aphyllophoromycetideae): A review. Int. J. Med. Mushrooms, 5: 235–246.

Gao, Y., Lan, J., Dai, X., Ye, J. and Zhou, S.H. 2004. A phase I/II study of Ling Zhi mushroom *Ganoderma lucidum* (W.Curt.:Fr.) Lloyd (Aphyllophoromycetideae) extract in patients with type II diabetes mellitus. Int. J. Med. Mushrooms, 6: 96–107.

Grinyer, J., McKay, M., Nevalainen, H. and Herbert, B.R. 2004. Fungal proteomics: Initial mapping of biological control strain *Trichoderma harzianum*. Curr. Genet. 45: 163–169.

Grinyer, J., Hunt, S., McKay, M., Herbert, B.R. and Nevalainen, H. 2005. Proteomic response of the biological control fungus *Trichoderma atroviride* to growth on the cell walls of *Rhizoctonia solani*. Curr. Genet. 47: 381–388.

Grinyer, J., Kautto, L., Traini, M., Willows, R.D., Te'o, J., Bergquist, P. and Nevalainen, H. 2007. Proteome mapping of the *Trichoderma reesei* 20S proteasome. Curr. Genet. 51: 79–88.

Horie, K., Rakwal, R., Hirano, M., Shibato, J., Nam, H.W., Kim, Y.S., Kouzuma, K., Agrawal, G.K., Masuo, Y. and Yonekura, M. 2008. Proteomics of two cultivated mushrooms *Sparassis crispa* and *Hericium erinaceum* provides insight into their numerous functional protein components and diversity. J. Proteome Res. 7(5): 1819–1835. Doi: 10.1021/pr070369o.

Huang, B., Lin, W., Cheung, P.C. and Wu, J. 2011. Differential proteomic analysis of temperature-induced autolysis in mycelium of *Pleurotus tuber-regium*. Curr. Microbiol., 62(4): 1160–7. Doi: 10.1007/s00284-010-9838-4.

Ichinohe, T., Ainai, A., Nakamura, T., Akiyama, Y., Maeyama, J., Odagiri, T., Tashiro, M., Takahashi, H., Sawa, H., Tamura, S., Chiba, J., Kurata, T., Sata, T. and Hasegawa, H. 2010. Induction of crossprotective immunity against influenza A virus H5N1 by intranasal vaccine with extracts of mushroom mycelia. J. Med. Virol., 82: 128–137.

Jia, D., Wang, B., Li, X., Peng, W., Zhou, J., Tan, H., Tang, J., Huang, Z., Tan, W., Gan, B., Yang, Z. and Zhao, J. 2017. Proteomic analysis revealed the fruiting-body protein profile of *Auricularia polytricha*. Curr. Microbiol. 74(8): 943–951. Doi: 10.1007/s00284-017-1268-0.

Kim, J.S., Kwon, Y.S., Bae, D.W., Kwak, Y.S. and Kwack, Y.B. 2017. Proteomic analysis of *Coprinopsis cinerea* under conditions of horizontal and perpendicular gravity. Mycobiology, 45(3): 226–231. Doi: 10.5941/MYCO.2017.45.3.226.

Lee, M.K., Li, X.J., Yap, A.C.S., Cheung, P.C.K., Tan, C.S., Ng, S.T., Roberts, R., Ting, K.N. and Fung, S.Y. 2018. Airway relaxation effects of water-soluble sclerotial extract from *Lignosus rhinocerotis*. Frontiers in Pharmacology, Section Ethnopharmacology 9: 461. Doi: 10.3389/fphar.2018.00461.

Lee, S.S., Tan, N.H., Fung, S.Y., Sim, S.M., Tan, C.S. and Ng, S.T. 2014. Anti-inflammatory effect of the sclerotium of *Lignosus rhinocerotis* (Cooke) Ryvarden, the tiger milk mushroom. BMC Complement Alternate Med., 14: 359. Doi: http://10.1186/1472-6882-14-359.

Lee, M.L., Tan, N.H., Fung, S.Y., Tan, C.S. and Ng, S.T. 2012. The antiproliferative activity of sclerotia of *Lignosus rhinocerus* (tiger milk mushroom). Evid Based Complement Alternat Med., 697603. Doi: http://10.1155/2012/697603.

Lieu, C.W., Lee, S.S. and Wang, S.Y. 1992. The effect of *Ganoderma lucidum* on induction of differentiation in leukemic U937 cells. Anticancer Res., 12: 1211–1216.

Lin, Y.L., Wen, T.N., Chang, S.T. and Chu, F.H. 2011. Proteomic analysis of differently cultured endemic medicinal mushroom *Antrodia cinnamomea* T.T. Chang et W.N. Chou from Taiwan. Int. J. Med. Mushrooms, 13(5): 473–81.

Liu, J.Y., Chang, M.C., Meng, J.L., Feng, C.P., Zhao, H. and Zhang, M.L. 2017. Comparative proteome reveals metabolic changes during the fruiting process in *Flammulina velutipes*. J. Agric. Food Chem., 21; 65(24): 5091–5100.

Liu, J.Y., Men, J.L., Chang, M.C., Feng, C.P. and Yuan, L.G. 2017. iTRAQ-based quantitative proteome revealed metabolic changes of *Flammulina velutipes* mycelia in response to cold stress. J. Proteomics, 156: 75–84.

Liu, J.Y., Chang, M.C., Meng, J.L., Feng, C.P. and Wang, Y. 2018. A comparative proteome approach reveals metabolic changes associated with *Flammulina velutipes* mycelia in response to cold and light stress. J. Agric. Food Chem., 11; 66(14): 3716–3725.

Liu, C., Chen, J., Chen, L., Huang, X. and Cheung, P.C.K. 2016. Immunomodulatory activity of polysaccharide—protein complex from the mushroom sclerotia of *Polyporus rhinocerus* in murine macrophages. J. Agri. Food Chem. 64: 3206–3214. Doi: http://10.1021/acs.jafc.6b00932.

Lu, Z., Kong, X., Lu, Z., Xiao, M., Chen, M., Zhu, L., Shen, Y., Hu, X. and Song, S. 2014. Para-aminobenzoic acid (PABA) synthase enhances thermotolerance of mushroom *Agaricus bisporus*. PLoS One, 10; 9(3): e91298. Doi: 10.1371/journal.pone.0091298.

Matis, M., Zakelj-Mavric, M. and Peter-Katalinić, J. 2005. Mass spectrometry and database search in the analysis of proteins from the fungus *Pleurotus ostreatus*. Proteomics, 5(1): 67–75.

Melin, P., Schnurer, J. and Wagner, E.G. 2002. Proteome analysis of *Aspergillus nidulans* reveals proteins associated with the response to the antibiotic concanamycin A, produced by Streptomyces species. Mol. Genet. Genomics, 267: 695–702.

Nandakumar, M.P., Shen, J., Raman, B. and Marten, M.R. 2003. Solubilization of trichloroacetic acid (TCA) precipitated microbial proteins via NaOH for two dimensional electrophoresis. J. Proteome Res., 2: 89–93.

O'Brien, M., Grogan, H. and Kavanagh, K. 2014. Proteomic response of *Trichoderma aggressivum* f. *europaeum* to *Agaricus bisporus* tissue and mushroom compost. Fungal Biol., 118(9-10): 785–91. Doi: 10.1016/j.funbio.2014.06.004.

Ooi, V.E. and Liu, F. 2000. Immunomodulation and anti-cancer activity of polysaccharide-protein complexes. Curr. Med. Chem. 7: 715–729.

Patterson, S.D. and Aebersold, R.H. 2003. Proteomics: the first decade and beyond. Nat. Genet., 33: 311–323.

Pim, L.M. 2004. Proteomics in multiplex. Nature, 429(6987): 101–7.

Rabilloud, T. 1996. Solubilization of proteins for electrophoretic analyses. Electrophoresis. 17: 813–829.

Rabilloud, T. 1998. Use of thiourea to increase the solubility of membrane proteins in two-dimensional electrophoresis. Electrophoresis, 19: 758–760.

Rahmad, N., Al-Obaidi, J.R., Nor Rashid, N.M., Zean, N.B., Mohd Yusoff, M.H., Shaharuddin, N.S., Mohd Jamil, N.A. and Mohd Saleh, N. 2014. Comparative proteomic analysis of different developmental stages of the edible mushroom *Termitomyces heimii*. Biol. Res. Jul., 3; 47: 30. Doi: 10.1186/0717-6287-47-30.

Ramkissoon, J.S., Mahomoodally, M.F., Ahmed, N. and Subratty, A.H. 2013. Antioxidant and anti-glycation activities correlate with phenolic composition of tropical medicinal herbs. Asian Pac. J. Trop. Med. 6(7): 561–569. Doi: http://10.1016/s1995-7645(13)60097-8.

Rowan, N.J., Smith, J.E. and Sullivan, R. 2003. Immunomodulatory activities of mushroom glucans and polysaccharide–protein complexes in animals and humans (a review). Int. J. Med. Mushrooms, 5: 95–110.

Sathesh-Prabu, C. and Lee, Y.K. 2016. Genetic variability and proteome profiling of a radiation induced cellulase mutant mushroom *Pleurotus florida*. Pol. J. Microbiol., 26; 65(3): 271–277. Doi: 10.5604/17331331.1215606.

Shimizu, M. and Wariishi, H. 2005. Development of a sample preparation method for fungal proteomics. FEMS Microbiol. Lett., 247: 17–22.

Singh, M.K., Kumar, M. and Thakur, I.S. 2017. Proteomic characterization and schizophyllan production by *Schizophyllum commune* ISTL04 cultured on *Leucaena leucocephala* wood under submerged fermentation. Bioresour. Technol., 236: 29–36.

Sugawara, R., Yamada, S., Tu, Z.H., Sugawara, A., Kousuke, S., Hoshiba, T., Eisaka, S. and Yamaguchi. A. 2016. Rapid and reliable species identification of wild mushrooms by matrix assisted laser desorption/ionization time of flight mass spectrometry (MALDI-TOF MS). Analytica Chimica Acta, 934: 163–169.

Sulc, M., Peslova, K., Zabka, M., Hajduch, M. and Havlicek, V. 2009. Biomarkers of *Aspergillus* spores: strain typing and protein identification. Int. J. Mass Spectrom. 280: 162–168.

Sullivan, R., Smith, J.E. and Rowan, N.J. 2006. Medicinal mushrooms and cancer therapy. Translating a traditional practice into Western medicine. Perspect Biol. Med., 49: 159–170.

Sun, L.L., Zhu, G.J., Yan, X.J. and Dovichi, N.J. 2013. High sensitivity capillary zone electrophoresis-electrospray ionization-tandem mass spectrometry for the rapid analysis of complex proteomes. Current Opinion in Chemical Biology, 17: 795–800.

Tang, L.H., Tan, Q., Bao, D.P., Zhang, X.H., Jian, H.H., Li, Y., Yang, R.H. and Wang, Y. 2016. Comparative proteomic analysis of light-induced mycelial brown film formation in *Lentinula edodes*. Biomed. Res. Int., 2016: 5837293.

The Food and Agriculture Organization of the United Nations. 2018. http://www.fao.org/faostat/en/#data/QC (Accessed 12th June 2018).

Ujor, V.C., Peiris, D.G., Monti, M., Kang, A.S., Clements, M.O. and Hedger, J.N. 2012. Quantitative proteomic analysis of the response of the wood-rot fungus, *Schizophyllum commune*, to the biocontrol fungus, *Trichoderma viride*. Lett. Appl. Microbiol., 54(4): 336–43. Doi: 10.1111/j.1472-765X.2012.03215.x.

Wang, Y.T., Huang, Z.J. and Chang, H.M. 2004. Proteomic analysis of human leukemic U937 cells incubated with conditioned medium of mononuclear cells stimulated by proteins from dietary mushroom of *Agrocybe aegerita*. J. Proteome Res., 3(4): 890–6.

Wang, X.M., Zhang, J., Wu, L.H., Zhao, Y.L., Li, T., Li, J.Q., Wang, Y.Z. and Liu, H.G. 2014. Review—A mini-review of chemical composition and nutritional value of edible wild-grown mushroom from China. Food Chemistry, 151: 279–285.

Westermeier, R. and Naven, T. 2002 . Introduction. *In*: Proteomics in Practice: A Laboratory Manual of Proteome Analysis, Wiley –VCH Verlag-GmBH Weinheim Germany.

Xie, C., Yan, L., Gong, W., Zhu, Z., Tan, S., Chen, D., Hu, Z. and Peng, Y. 2016. Effects of different substrates on lignocellulosic enzyme expression, enzyme activity, substrate utilization and biological efficiency of *Pleurotus eryngii*. Cell Physiol. Biochem., 39(4): 1479–94. Doi: 10.1159/000447851.

Xu, X., Yang, J., Ning, Z. and Zhang, X. 2016. Proteomic analysis of intestinal tissues from mice fed with *Lentinula edodes*-derived polysaccharides. Food Funct., (1): 250–61. Doi: 10.1039/c5fo00904a.

Yap, H.Y.Y., Fung, S.Y., Ng, S.T., Tan, C.S. and Tan, N.H. 2015. Genome-based proteomic analysis of *Lignosus rhinocerotis* (Cooke) Ryvarden Sclerotium. Int. J. Med. Sci., 12(1): 23–31. Doi: 10.7150/ijms.10019.

Yap, Y.H.Y., Tan, N.H., Fung, S.Y., Aziz, A.A., Tan, C.S. and Ng, S.T. 2013. Nutrient composition, antioxidant properties, and anti-proliferative activity of *Lignosus rhinocerus* Cooke *sclerotium*. J. Sci. Food Agric., 93(12): 2945–2952.

Yap, Y.H.Y., Tan, N.H., Ng, S.T., Tan, C.S. and Fung, S.Y. 2018a. Molecular attributes and apoptosis-inducing activities of a putative serine protease isolated from Tiger Milk mushroom (*Lignosus rhinocerus*) sclerotium against breast cancer cells *in vitro*. Peer J. 6: e4940. Doi: 10.7717/peerj.4940.

Yap, Y.H.Y., Tan, N.H., Ng, S.T., Tan, C.S. and Fung, S.Y. 2018b. Inhibition of protein glycation by tiger milk mushroom (*Lignosus rhinocerus* (Cooke) Ryvarden) and search for potential anti-diabetic activity related metabolic pathways by genomic and transcriptomic data mining. Frontier in Pharmacology: Section Ethnopharmacology, 9: 103. Doi: 10.3389/fphar.2018.00103.

Yuan, B., Zhao, L., Rakariyatham, K., Han, Y., Gao, Z., Muinde Kimatu, B., Hu, Q. and Xiao, H. 2017. Isolation of a novel bioactive protein from an edible mushroom *Pleurotus eryngii* and its anti-inflammatory potential. Food Funct., 21; 8(6): 2175–2183. Doi: 10.1039/c7fo00244k.

Zaidman, B.Z., Yassin, M., Mahajna, J., and Wasser, S.P. 2005. Appl. Microbiol. Biotechnol., 67: 453–468.

Zeng, X., Ling, H., Yang, J., Chen, J. and Guo, S. 2018. Proteome analysis provides insight into the regulation of bioactive metabolites in *Hericium erinaceus*. Gene. pii: S0378–1119(18)30497-9.

Zhang, B.B., Chen, L. and Cheung, P.C. 2012. Proteomic insights into the stimulatory effect of Tween 80 on mycelial growth and exopolysaccharide production of an edible mushroom *Pleurotus tuber-regium*. Biotechnol. Lett., 34(10): 1863–7. Doi: 10.1007/s10529-012-0975-7.

Zhang, M., Cui, S.W., Cheung, P.C.K. and Wang, Q. 2007. Antitumor polysaccharides from mushrooms: A review on their isolation, structural characteristics and antitumor activity. Trends Food Sci. Technol., 18: 4–19.

Zhang, P.L., Li, K.H., Yang, G., Xia, C.Q., Polston, J.E., Li, G.N., Li, S.W., Lin, Z., Yang, L.J., Bruner, S.D. and Ding, Y.S. 2017. Cytotoxic protein from the mushroom *Coprinus comatus* possesses a unique mode for glycan binding and specificity. Proc. Natl. Acad. Sci. USA, 114(34): 8980–8985. Doi: 10.1073/pnas.1706894114.

Zhu, C.S. and Ca, H. 2006. Protein Microarrays. BioTechniques 40.

Zorn, H., Peters, T., Nimtz, M. and Berger, R.G. 2005. The secretome of *Pleurotus sapidus*. Proteomics, 5(18): 4832–8.

Index

Printed and bound by CPI Group (UK) Ltd, Croydon, CR0 4YY

24/10/2024

01778304-0014